METHODS IN MOLECULAR BIOLOGY

Series Editor
John M. Walker
School of Life and Medical Sciences
University of Hertfordshire
Hatfield, Hertfordshire, AL10 9AB, UK

For further volumes:
http://www.springer.com/series/7651

Transgenic Plants

Methods and Protocols

Edited by

Sandeep Kumar, Pierluigi Barone, and Michelle Smith

Corteva Agriscience™, Agriculture Division of DowDuPont™, Johnston, IA, USA

Humana Press

Editors
Sandeep Kumar
Corteva Agriscience^{TM}
Agriculture Division of DowDuPont^{TM}
Johnston, IA, USA

Pierluigi Barone
Corteva Agriscience^{TM}
Agriculture Division of DowDuPont^{TM}
Johnston, IA, USA

Michelle Smith
Corteva Agriscience^{TM}
Agriculture Division of DowDuPont^{TM}
Johnston, IA, USA

ISSN 1064-3745 ISSN 1940-6029 (electronic)
Methods in Molecular Biology
ISBN 978-1-4939-8777-1 ISBN 978-1-4939-8778-8 (eBook)
https://doi.org/10.1007/978-1-4939-8778-8

Library of Congress Control Number: 2018958711

This Humana Press imprint is published by the registered company Springer Science+Business Media, LLC, part of Springer Nature.
The registered company address is: 233 Spring Street, New York, NY 10013, U.S.A.

Preface

Advances in plant transgenic technology in the twentieth century led to the first commercially grown transgenic crop in 1996. In the creation of transgenic plants, the major hurdle for transfer of genetic material between species was overcome. This not only enabled fundamental insights into plant biology but also revolutionized commercial agriculture. Adoption of transgenic plants in industrial agriculture has reduced pesticide use, improved crop yields, and increased or preserved farm profitability. Targeted on meeting needs for greater efficiency and wider application of transgenic technology in plants, continuous technological advances over more than 30 years have paved the way for high-frequency transformation, but still of only a very small number of plant species; the majority of plants or their genotypes remain recalcitrant to transformation and regeneration. In addition, producing the complex traits that provide broad spectrum insect control, herbicide tolerance, and other agronomic traits desired by modern agriculture requires precise genome engineering. Recent developments in genome editing tools provide unique opportunities to create precise on-demand transgenic plants to enhance crop productivity. Combining genome engineering with new tools that overcome some of the limitations in plant transformation, there is clearly significant opportunity yet to explore in the area of transgenic plants.

We intend this book to provide thorough coverage of the topic of transgenic plants with methods capturing the current status of plant transformation, including precise genome engineering technology, biotechnological application of transgenic plants, and future developments needed to enable genome editing technology in crops. Grouped into sections focused on transformation of model and crop plants, genome engineering, and transgenic event characterization, the chapters primarily comprise detailed protocols as is typical in this series. We also sought review-style chapters to broaden the utility for readers, provide additional references for further understanding of this fascinating area of research, and especially to present the technology's potential for solving some of our most urgent global challenges in food security.

Johnston, IA

Sandeep Kumar
Pierluigi Barone
Michelle Smith

Contents

Part IV Transgenic Event Characterization

Contributors

AJITH ANAND • *Corteva Agriscience™, Agriculture Division of DowDuPont™, Johnston, IA, USA*
MAREN ARLING • *Corteva Agriscience™, Agriculture Division of DowDuPont™, Johnston, IA, USA*
SCOTT A. BEVAN • *Corteva Agriscience™, Agriculture Division of DowDuPont™, Indianapolis, IN, USA*
CHRISTOPHER R. BOOMSMA • *American Society of Agronomy, Crop Science Society of America, and Soil Science Society of America, Madison, WI, USA*
MYRON A. BRUCE • *Montana State University, Bozeman, MT, USA; Kansas State University, Manhattan, KS, USA*
TOBIAS BRUEGMANN • *Thuenen Institute of Forest Genetics, Grosshansdorf, Germany*
JORDAN BRUNGARDT • *Department of Plant Pathology, Kansas State University, Manhattan, KS, USA*
M. BRUNGART-ROSENBERG • *Department of Plant Sciences, Sheridan Research and Extension Center, University of Wyoming, Sheridan, WY, USA*
SI NIAN CHAR • *Department of Genetics, Development and Cell Biology, Iowa State University, Ames, IA, USA*
QIUXIA CHEN • *Department of Horticulture, Plant Biotechnology Resource and Outreach Center, Michigan State University, East Lansing, MI, USA*
YUEYING CHEN • *Department of Plant Pathology, Kansas State University, Manhattan, KS, USA*
WEI CHEN • *Corteva Agriscience™, Agriculture Division of DowDuPont™, Johnston, IA, USA*
MARY-DELL M. CHILTON • *Syngenta Crop Protection, LLC, Research Triangle Park, NC, USA*
CORY A. CHRISTENSEN • *Corteva Agriscience™, Agriculture Division of DowDuPont™, Johnston, IA, USA*
C. CLAFLIN • *Sheridan College, Sheridan, WY, USA*
VLADIMIR A. DA COSTA • *Kemin Industries, Inc., Des Moines, IA, USA*
BÁRBARA ANDRADE DIAS BRITO DA CUNHA • *Genetics and Biotechnology Laboratory, Embrapa Agroenergy (CNPAE), Brazilian Agricultural Research Corporation (EMBRAPA), Parque Estação Biológica, Brasília, DF, Brazil*
JOHN P. DAVIES • *Corteva Agriscience™, Agriculture Division of DowDuPont™, Indianapolis, IN, USA*
ERIC A. DEAN • *Pairwise, Research Triangle Park, Wooster, NC, USA; Department of Horticulture and Crop Science, The Ohio State University, Ohio Agricultural Research and Development Center, Wooster, OH, USA*
KHIRA DEECKE • *Thuenen Institute of Forest Genetics, Grosshansdorf, Germany*
TARLOCHAN SINGH DHADIALLA • *Corteva Agriscience™, Agriculture Division of DowDuPont™, Indianapolis, IN, USA*
S. A. DHEKNEY • *Department of Plant Sciences, Sheridan Research and Extension Center, University of Wyoming, Sheridan, WY, USA*
SHUJIE DONG • *Syngenta Crop Protection, LLC, Research Triangle Park, NC, USA*

SIVAMANI ELUMALAI • *Syngenta Crop Protection, LLC, Research Triangle Park, NC, USA*
MARÍA PAULA FILIPPONE • *Instituto de Tecnología Agroindustrial del Noroeste Argentino (ITANOA), EEAOC-CONICET, Las Talitas, Tucumán, Argentina*
JOHN J. FINER • *Department of Horticulture and Crop Science, The Ohio State University, Ohio Agricultural Research and Development Center, Wooster, OH, USA*
MATTHIAS FLADUNG • *Thuenen Institute of Forest Genetics, Grosshansdorf, Germany*
CHUNXIANG FU • *Qingdao Institute of BioEnergy and Bioprocess Technology, Chinese Academy of Sciences, Qingdao, Shandong, China*
DANIEL J. GACHOTTE • *Corteva Agriscience™, Agriculture Division of DowDuPont™, Indianapolis, IN, USA*
SATYALINGA SRINIVAS GAMPALA • *Corteva Agriscience™, Agriculture Division of DowDuPont™, Indianapolis, IN, USA*
WILLIAM GORDON-KAMM • *Corteva Agriscience™, Agriculture Division of DowDuPont™, Johnston, IA, USA*
D. J. GRAY • *Mid-Florida Research and Education Center, University of Florida/IFAS, Apopka, FL, USA*
ANDIKA GUNADI • *Department of Horticulture and Crop Science, The Ohio State University, Ohio Agricultural Research and Development Center, Wooster, OH, USA*
MANJU GUPTA • *Corteva Agriscience™, Agriculture Division of DowDuPont™, Indianapolis, IN, USA*
ANTHONY J. HARVEY • *Corteva Agriscience™, Agriculture Division of DowDuPont™, Indianapolis, IN, USA*
MATTHEW J. HAYDEN • *Department of Economic Development, Jobs, Transport and Resources (DEDJTR), Agribio, Bundoora, VIC, Australia; School of Applied Systems Biology, La Trobe University, Agribio, Bundoora, VIC, Australia*
GEORGE HOERSTER • *Corteva Agriscience™, Agriculture Division of DowDuPont™, Johnston, IA, USA*
QINGZHEN JIANG • *Noble Research Institute, Ardmore, OK, USA*
TODD JONES • *Corteva Agriscience™, Agriculture Division of DowDuPont™, Johnston, IA, USA*
PATRICIA KEEN • *The Boyce Thompson Institute, Ithaca, NY, USA*
ADILSON KENJI KOBAYSHI • *Genetics and Biotechnology Laboratory, Embrapa Agroenergy (CNPAE), Brazilian Agricultural Research Corporation (EMBRAPA), Parque Estação Biológica, Brasília, DF, Brazil*
RACHELLE A. LEFFERT-SORENSON • *Corteva Agriscience™, Agriculture Division of DowDuPont™, Indianapolis, IN, USA*
BRIAN LENDERTS • *Corteva Agriscience™, Agriculture Division of DowDuPont™, Johnston, IA, USA*
Z. T. LI • *USDA-ARS, Kearneysville, WV, USA*
RIQING LI • *Department of Genetics, Development and Cell Biology, Iowa State University, Ames, IA, USA*
KEITH LOWE • *Corteva Agriscience™, Agriculture Division of DowDuPont™, Johnston, IA, USA*
TEJINDER MALL • *Corteva Agriscience™, Agriculture Division of DowDuPont™, Indianapolis, IN, USA*
DAVID G. J. MANN • *Corteva Agriscience™, Agriculture Division of DowDuPont™, Indianapolis, IN, USA*

POLYANA KELLY MARTINS • *Genetics and Biotechnology Laboratory, Embrapa Agroenergy (CNPAE), Brazilian Agricultural Research Corporation (EMBRAPA), Parque Estação Biológica, Brasília, DF, Brazil*

JOHN MASON • *Department of Economic Development, Jobs, Transport and Resources (DEDJTR), Agribio, Bundoora, VIC, Australia; School of Applied Systems Biology, La Trobe University, Agribio, Bundoora, VIC, Australia*

HUGO BRUNO CORREA MOLINARI • *Genetics and Biotechnology Laboratory, Embrapa Agroenergy (CNPAE), Brazilian Agricultural Research Corporation (EMBRAPA), Parque Estação Biológica, Brasília, DF, Brazil*

EWEN MULLINS • *Department of Crop Science, Teagasc, Oak Park, Carlow, Co. Carlow, Republic of Ireland*

MÓNICA NAVIA-URRUTIA • *Department of Plant Pathology, Kansas State University, Manhattan, KS, USA*

JULIA NIETSCH • *CGS Crop Genetic Systems, Hamburg, Germany*

STEPHEN NOVAK • *Corteva Agriscience™, Agriculture Division of DowDuPont™, Indianapolis, IN, USA*

MICHAEL PACHER • *Botanical Institute, Karlsruhe Institute of Technology, Karlsruhe, Germany*

OLAF POLAK • *Thuenen Institute of Forest Genetics, Grosshansdorf, Germany*

HOLGER PUCHTA • *Botanical Institute, Karlsruhe Institute of Technology, Karlsruhe, Germany*

QIUDENG QUE • *Syngenta Crop Protection, LLC, Research Triangle Park, NC, USA*

DHEERAJ SINGH RATHORE • *Department of Crop Science, Teagasc, Oak Park, Carlow, Co. Carlow, Republic of Ireland*

KIMBERLY A. RICHEY • *Corteva Agriscience™, Agriculture Division of DowDuPont™, Indianapolis, IN, USA*

SARRIA RODRIGO • *Corteva Agriscience™, Agriculture Division of DowDuPont™, Indianapolis, IN, USA*

SAREENA SAHAB • *Department of Economic Development, Jobs, Transport and Resources (DEDJTR), Agribio, Bundoora, VIC, Australia*

CARLA SCHMIDT • *Botanical Institute, Karlsruhe Institute of Technology, Karlsruhe, Germany*

LORENA NOELIA SENDIN • *Instituto de Tecnología Agroindustrial del Noroeste Argentino (ITANOA), EEAOC-CONICET, Las Talitas, Tucumán, Argentina*

S. K. SESSIONS • *Department of Plant Sciences, Sheridan Research and Extension Center, University of Wyoming, Sheridan, WY, USA*

LIANG SHI • *Syngenta Crop Protection, LLC, Research Triangle Park, NC, USA*

JESSICA L. SHOUP RUPP • *Montana State University, Bozeman, MT, USA; Kansas State University, Manhattan, KS, USA*

GUO-QING SONG • *Department of Horticulture, Plant Biotechnology Resource and Outreach Center, Michigan State University, East Lansing, MI, USA*

GERMAN SPANGENBERG • *Department of Economic Development, Jobs, Transport and Resources (DEDJTR), Agribio, Bundoora, VIC, Australia; School of Applied Systems Biology, La Trobe University, Agribio, Bundoora, VIC, Australia*

VIBHA SRIVASTAVA • *Department of Crop, Soil & Environmental Sciences, University of Arkansas, Fayetteville, AR, USA; Department of Horticulture, University of Arkansas, Fayetteville, AR, USA*

Yi Sui • *The National Key Facility for Crop Gene Resources and Genetic Improvement, Institute of Crop Sciences, Chinese Academy of Agricultural Sciences, Beijing, China*
Bin Tian • *Department of Plant Pathology, Kansas State University, Manhattan, KS, USA*
Michelle Tjahjadi • *The Boyce Thompson Institute, Ithaca, NY, USA*
Harold N. Trick • *Department of Plant Pathology, Kansas State University, Manhattan, KS, USA*
Joyce Van Eck • *The Boyce Thompson Institute, Ithaca, NY, USA; Plant Breeding and Genetics Section, School of Integrative Plant Science, Cornell University, Ithaca, NY, USA*
Terence A. Walsh • *Corteva Agriscience™, Agriculture Division of DowDuPont™, Indianapolis, IN, USA*
Ning Wang • *Corteva Agriscience™, Agriculture Division of DowDuPont™, Johnston, IA, USA*
Zeng-Yu Wang • *Noble Research Institute, Ardmore, OK, USA*
PoHao Wang • *Corteva Agriscience™, Agriculture Division of DowDuPont™, Johnston, IA, USA*
Laura L. Wayne • *Corteva Agriscience™, Agriculture Division of DowDuPont™, Johnston, IA, USA*
Emily Wu • *Corteva Agriscience™, Agriculture Division of DowDuPont™, Johnston, IA, USA*
Chuanyin Wu • *The National Key Facility for Crop Gene Resources and Genetic Improvement, Institute of Crop Sciences, Chinese Academy of Agricultural Sciences, Beijing, China*
Bryant Wulfkuhle • *Corteva Agriscience™, Agriculture Division of DowDuPont™, Indianapolis, IN, USA*
Bing Yang • *Department of Genetics, Development and Cell Biology, Iowa State University, Ames, IA, USA*
Heng Zhong • *Syngenta Crop Protection, LLC, Research Triangle Park, NC, USA*
Evelyn Zuniga-Soto • *Department of Crop Science, Teagasc, Oak Park, Carlow, Co. Carlow, Republic of Ireland*

Part I

Plant Transformation and Vector Construction

Chapter 1

Repurposing Macromolecule Delivery Tools for Plant Genetic Modification in the Era of Precision Genome Engineering

Qiudeng Que, Mary-Dell M. Chilton, Sivamani Elumalai, Heng Zhong, Shujie Dong, and Liang Shi

Abstract

Efficient delivery of macromolecules into plant cells and tissues is important for both basic research and biotechnology product applications. In transgenic research, the goal is to deliver DNA molecules into regenerable cells and stably integrate them into the genome. Over the past 40 years, many macromolecule delivery methods have been studied. To generate transgenic plants, particle bombardment and *Agrobacterium*-mediated transformation are the methods of choice for DNA delivery. The rapid advance of genome editing technologies has generated new requirements on large biomolecule delivery and at the same time reinvigorated the development of new transformation technologies. Many of the gene delivery options that have been studied before are now being repurposed for delivering genome editing machinery for various applications. This article reviews the major progress in the development of tools for large biomolecule delivery into plant cells in the new era of precision genome engineering.

Key words Direct delivery, Biolistic transformation, *Agrobacterium*-mediated transformation, Protoplast transformation, Macromolecule delivery, DNA-free delivery, Cell-penetrating peptides, Genome engineering

1 Introduction

It has been almost 35 years since the first reports of successful genetic transformation of recombinant DNA molecules into plant cells using *Agrobacterium*-mediated delivery [1–3]. The initial success in *Agrobacterium*-mediated transformation was followed by demonstration of direct DNA delivery tools for generation of transgenic plants [4, 5]. These include polyethylene glycol (PEG)-mediated protoplast transformation, liposome-mediated protoplast transformation, microinjection of protoplasts, electroporation of protoplasts and calli, particle bombardment, silicon carbide fiber- or whisker-mediated transformation, aerosol beam microinjection, and the "pollen tube pathway" method

Sandeep Kumar et al. (eds.), *Transgenic Plants: Methods and Protocols*, Methods in Molecular Biology, vol. 1864, https://doi.org/10.1007/978-1-4939-8778-8_1,

[6–10]. The two methods, particle bombardment (aka biolistics) and *Agrobacterium*-mediated delivery, are the most efficient and commonly used methods [6, 10, 11]. Over the years, these transformation technologies became essential research tools for many plant biologists to study plant genetics and functional genomics. In addition, their application in crops also gave birth to agricultural biotechnology and opened a new chapter in modern agricultural innovation. Nevertheless, despite the remarkable technological advancement, after more than 30 years of research, production of genetically modified plants through routine transformation and regeneration still remains an expensive and time-consuming task for many crops [12].

Today we find ourselves in another new era of crop genetic modification, stemming from the discovery of customizable sequence-specific nucleases, especially the highly flexible RNA-guided CRISPR-Cas9 nuclease system. In this era of precision genome engineering, some applications are placing ever more stringent demands on delivery tools due to somewhat different technical requirements compared to stable transgene integration. Further, there are intrinsic differences in the requirements for different types of genome editing applications. For example, targeted mutagenesis may only require the transient presence of nuclease to generate chromosomal double-strand cleavage; thus, it is not always necessary to generate stable transgenic lines for expression of editing machinery. The use of a DNA-free method such as direct delivery of protein, RNA, or ribonucleoprotein might be preferred so that there is no need to deal with the presence and fate of a transgene in subsequent studies after the editing step is done. In contrast, editing processes that depend on the use of a DNA donor and host homology-dependent repair process may require techniques very similar to those for generation of stable transformants. This article will briefly review major progress in the development of tools for the delivery of macromolecules into plant cells and discuss potential areas for improvement in repurposing them for delivering genome editing machinery when appropriate.

2 Direct Delivery

In order to generate stable and heritable genome modifications in any plant species, DNA or editing machinery must be delivered into cells that are actively dividing or can be subsequently induced to divide and eventually regenerated into plants so that the inserted genetic modifications can be inherited to the offspring. Several physical approaches to mediate direct delivery of macromolecules into plant cells have been studied including electroporation [13], microinjection [14], biolistic bombardment [15], microwounding with silicon carbide fibers [16], sonication [17], and aerosol beam injection [18].

Direct uptake of exogenous DNA into plant protoplasts was explored in the early days of genetic transformation [19], but no conclusive evidence could be obtained that stable transformation occurred [20, 21]. Delivery of macromolecules into protoplasts can be enhanced by using different chemical treatments of either DNA or RNA molecules or target cells including osmoticum such as mannitol or PEG (polyethylene glycol) or PVA (polyvinyl alcohol) [20–22], liposome [20, 23], and poly-L-ornithine [21]. Some of these chemical treatments might serve to both protect DNA molecules from nuclease digestion and at the same time enhance the delivery process itself [20, 21]. Through the use of improved delivery conditions including the use of PEG and high Ca^{2+} concentration, Krens et al. [22] demonstrated convincingly for the first time that *Agrobacterium* Ti plasmid DNA could be delivered into tobacco protoplasts and also stably integrated; the transformed tissue lines could also grow independently of exogenously supplied phytohormones; and some of the transformed lines were regenerated into shoots. Protoplasts were also important target materials for developing other gene delivery techniques. For example, Fraley et al. [24] successfully delivered neomycin phosphotransferase transgenes into *Petunia* protoplasts via *Agrobacterium* and generated transgenic callus lines using kanamycin selection. Gene delivery into protoplasts could also be improved by other physical treatments such as electroporation, heat shock treatment, and irradiation of recipient protoplasts [21]. Due to their relative ease of preparation and ability to introduce DNA through electroporation and/or PEG treatment, protoplasts have been and will remain a viable target for rapid transient evaluation of functionality of constructs [9].The technical challenges associated with regenerating protoplasts into fertile plants, however, still limit its broader use in most crops.

Electroporation was initially used in protoplasts to induce plant cell fusions [25]. A similar setup was adapted successfully by optimizing electric pulse amplitude and time to deliver genes into carrot, tobacco, and maize protoplasts [26]. Plasmid DNA delivery into protoplasts by electroporation, followed by protoplast culture and regeneration of transgenic calli, led to successful recovery of transgenic plants in the important cereal crops maize and rice [27, 28]. Electroporation can also be used to deliver genes through cell walls into cells and tissues such as microspores and embryos for the purpose of transient expression assay or stable transformation [29–32]. For example, electroporation was carried out using immature zygotic embryos and embryogenic calli pre-wounded either by brief pectinase (macerozyme) digestion or plasmolysis [31]. Fertile transgenic maize plants were regenerated from embryogenic calli derived from suspension culture cells or wounded embryos electroporated with DNA vectors encoding neomycin phosphotransferase using kanamycin selection [31] or with *bar* gene encoding phosphinothricin acetyltransferase using bialaphos selection [32].

Direct microinjection into plant cells such as protoplasts and embryonic tissues was also explored as a DNA delivery method in the 1980s. Morikawa and Yamada [14] demonstrated microinjection of DNA molecules through pressurized glass micropipettes into the intranuclear compartment of protoplasts isolated from cultured *Euphorbia milii* cells and from tobacco mesophyll cells. Crossway et al. [33] injected tobacco mesophyll protoplasts and also obtained callus lines from injected protoplasts using the hanging drop culture method. Southern blot analysis indicated that the injected plasmid DNA was integrated into the genome of some of the recovered tobacco callus lines. The study also showed that the frequency of DNA integration was higher with intranuclear injections compared to cytoplasmic injections. Neuhaus et al. [34] demonstrated recovery of transgenic rapeseed plants from microspore-derived embryoids microinjected with plasmid DNA. Griesbach [35] reported microinjection of *Petunia hybrida* protoplasts with isolated chromosomes from *P. alpicola* and obtained transformants with heritable changes in the relative activity of various enzymes in the flavonoid biosynthetic pathway. Besides intracellular injection, direct *in planta* injection or delivery of DNA was also reported in several studies. De la Peña et al. [36] obtained transgenic rye plants by injecting plasmid DNA containing the aminoglycoside phosphotransferase II (APH(3′)II) gene expression cassette into young floral tillers at 14 days before meiosis and then screening seeds for kanamycin resistance. Another direct DNA delivery method is the "pollen tube pathway" technique [37] and was exemplified in rice by cutting the stigma off the floret and applying a drop of DNA solution to the cut end of the style and then screening the roots of progeny seedlings for the presence of transgene [38]. However, it is possible that the hybridization signals present in the seedlings might be from the residual "nonintegrated" plasmid DNA used to treat florets in these "pollen tube pathway" studies [39].

Perhaps one of the most important advances in plant transformation following *Agrobacterium*-mediated transformation was the development of high-velocity particle gun technology (also called gene gun, microprojectile bombardment, or biolistic transformation) [15, 40–42]. Interestingly, tobacco mosaic virus RNA was the first large molecule used to demonstrate the effectiveness of particle bombardment for delivery of large biomolecules to plants. The bombarded RNA was shown to replicate to form viral particles in plant cells [15, 41]. Successful delivery of plasmid DNA followed immediately [15, 40]. A separate version of the gene gun apparatus was developed which relies on a high-voltage discharger to generate an explosive expansion of water vapor to accelerate gold microprojectiles into plant cells [43, 44]. Since the initial demonstration of microparticle bombardment for RNA and DNA delivery, it has been used to transform many different kinds of target tissues of many plant species, both transiently and stably [6, 9]. Stable fertile transformants of numerous important crop species including corn, wheat, rice, cotton, and soybean were successfully generated by microprojectile bombardment of embryogenic calli during the subsequent few years [45–51]. In theory, biolistic technology can introduce any macromolecule including DNA, RNA, and protein into virtually any tissue from any plant species. Successful generation of transgenic plants, however, depends upon the ability of the bombarded cells to integrate the transgene, proliferate, and give rise to a fertile plant. A potential advantage of biolistic delivery in comparison with *Agrobacterium*-mediated delivery is the lack of host plant tissue's hypersensitive defense response against bacterial cells, which may negatively impact recovery of stable transformants [6]. Also, the biolistic delivery method is so far the only successful way for carrying out organelle transformation. To achieve protein delivery with particle bombardment method, Martin-Ortigosa and Wang [52] developed a biolistic protocol called "proteolistics" for delivering proteins by employing a simple protein/projectile preparation step in which the protein was mixed with a gold particle microprojectile suspension, placed onto a gene gun cartridge, and then dehydrated using either lyophilization or room-temperature air-drying. Thanks to its flexibility and high efficiency, biolistic delivery is uniquely suited for genome editing applications in which DNA-free delivery is needed [53].

Similar to particle bombardment and microinjection, a direct DNA delivery technology called "aerosol beam injection (ABI)" was also developed to deliver DNA molecules into maize and soybean callus tissues for transgenic event generation [18]. An advantage of ABI is that injection to target tissue is from a continuous liquid microstream (aerosol beam); thus no reloading of DNA into micropipettes or carrier cartridge is required as for microinjection or bombardment. Further, large target tissues such as immature embryos can be placed directly on an agar surface for injection as for particle bombardment. For example, Held et al. [54] described that aerosol beam injection of immature embryos of Stine elite inbred corn line 963 achieved an average of around 3% transformation frequency, substantially higher than the 1% frequency achieved via particle gun. In addition to microfluid stream injections as exemplified by aerosol beam injection, it was shown that integration of microfluidics and other microinjection techniques including electroporation, magnetofection, sonoporation, mechanoporation, and optoinjection could lead to increased efficiency by simultaneously microinjecting many cells [55].

One of the drawbacks of particle bombardment is that it requires expensive equipment, costly supplies, and a lengthy process of particle and tissue preparation for gene delivery. A system based on silicon carbide fiber and a simple vortexing procedure was developed to rapidly and inexpensively deliver foreign DNA into intact Black Mexican Sweet (BMS) maize (*Zea mays*) and tobacco (*Nicotiana tabacum*) suspension culture cells [16, 56]. In this method, the mixture of suspension culture cells, DNA, and whiskers was shaken vigorously, during which the whiskers presumably poked tiny holes in the cells to allow DNA entry. Frame et al. [57] later described the development of a whisker-mediated maize transformation method which used embryogenic suspension cultures as target tissues and silicon carbide whiskers to deliver plasmid DNA. Successful stable transformation with whisker-mediated delivery has been reported with many plant species including maize, wheat, and rice [7]. For genome engineering, whisker-mediated transformation has been used successfully to generate targeted gene insertion events in maize at the *IPK1* locus mediated by zinc-finger nucleases (ZFNs) using embryogenic cell cultures derived from Hi-II [58].

Recently, nanoparticles have also been explored as a novel way of delivering DNA. Torney et al. [59] developed a mesoporous silica nanoparticle (MSN) system with 3 nm pores to deliver DNA and chemicals into plants. This study loaded the MSN with plasmid DNA and its chemical inducer and capped the ends with gold nanoparticles to keep the molecules from leaching out. Upon uncapping, the gold nanoparticles released the chemical inducer which triggered gene expression in the plant cells under controlled-release conditions. Later, Martin-Ortigosa et al. [60] reported the use of a gold nanoparticle-functionalized mesoporous silica nanoparticle (Au-MSN) with larger average pore diameters (10 nm) to deliver proteins (eGFP and fluorescently labeled bovine serum albumin) and plasmid DNA to plant cells using bombardment. Delivered fluorescent eGFP was later released into plant cells. Recently, Martin-Ortigosa et al. [52] further reported that Cre protein could be loaded into gold-plated MSNs and then bombarded into plant cells to mediate excision of sequences flanked by *lox*P sites in maize cells.

3 Facilitated Direct Delivery

In the 1990s, it was found that peptides isolated from certain proteins could penetrate eukaryotic cells' plasma membranes and translocate across the cell membrane in an energy-independent way [61, 62]. These peptides with protein transduction domain (PTD) were named cell-penetrating peptides (CPP) [63]. These peptides can be used to deliver cargo macromolecules including peptides, proteins, or even nucleic acids across cell membranes [63, 64]. CPP-mediated delivery has the advantage that it can be applied to different cell or tissue types for *in planta* applications. CPPs can be derived from naturally occurring proteins or can be artificially designed chimeric peptides composed of a hydrophilic and a hydrophobic domain. CPPs can be divided into several different categories: cationic peptides, hydrophobic sequences, amphipathic peptides, proline-rich and antimicrobial sequences, and chimeric or bipartite peptides [64]. Well-characterized protein-derived cell-penetrating peptides include these: (1) penetratin (RQIKI WFQNR RMKWK K) from *Drosophila* homeoprotein *Antennapedia* [61], (2) Tat CPP (CGRKK RRQRR RPPQC) derived from HIV-encoded Tat transcription-activating factor [62], and (3) pVC CPP derived from murine vascular endothelial cadherin (LLIIL RRRIR KQAHA HSK) [65]. There are no common amino acid sequence motifs among different CPPs. The common features that most CPPs seem to share are net positive charges and amphipathic property. Some artificially designed CPPs also have a nuclear localization signal (NLS) incorporated [64]. For CPP-mediated delivery, cargo molecules could be linked to these carrier peptides and proteins covalently or be associated with them noncovalently in a mixture before delivery. One potential issue to be overcome with CPP-mediated delivery for genome engineering applications is that the rate and efficiency of translocation of large molecules like whole protein and RNA are usually much lower compared to small-sized cargo such as short peptides [64].

Delivery of macromolecules into plants by CPP-type transduction peptides has been tested with protein, RNA, and nucleic acids. Unnamalai et al. [66] used cationic oligopeptide polyarginine-12mer to facilitate delivery of in vitro prepared double-stranded RNA to tobacco (*Nicotiana tabacum*) suspension cells to induce posttranscriptional gene silencing. Mäe et al. [67] demonstrated the uptake of several CPPs in tobacco protoplasts and also found that transportan had the highest uptake efficacy among the studied peptides. Chang et al. [68] demonstrated that both the Tat peptide and short synthetic nona-arginine peptide (R9) were capable of efficiently delivering fused fluorescent proteins into different tissues of tomato and onion plants in a temperature- and endocytosis-independent manner. The same group later demonstrated that this short arginine-rich peptide was capable of delivering non-covalently tethered plasmid DNA into living plant cells [69]. Chugh and Eudes [70] found that the cellular uptake of two tested CPP peptides *p*VEC and transportan was tissue-dependent in that they were internalized in triticale mesophyll protoplasts, onion epidermal cells, leaf bases, and root tips but showed negligible florescence in coleoptile and leaf tips. It was also shown that endocytic/macropinocytosis inhibitors did not reduce the cellular uptake of the *p*VEC and transportan peptides, suggesting that these CPPs directly penetrated plant cell membranes through a receptor-independent internalization mechanism. These authors also reported that permeabilization treatment of immature wheat embryos resulted in a remarkably higher uptake of cell-penetrating peptides, whereas embryos that were not permeabilized failed to show significant cell-penetrating peptide uptake [71]. It was also shown that Tat (RKKRR QRRR) and Tat2 (RKKRR QRRRR KKRRQ RRR) peptides efficiently delivered functionally active β-glucuronidase (GUS) enzyme and plasmid DNA carrying the GUS expression cassette into permeabilized immature embryos and freshly isolated triticale microspores in a noncovalent manner [71, 72]. Tat2 was able to deliver GUS gene into about 2% triticale microspores, and another CPP, Pep-1 (KETWW ETWWT EWSQP KKKRK V), a synthetic CPP, was able to translocate GUS enzyme in its active form into 31% of the microspores [72]. Delivery of plasmid DNA into permeabilized immature wheat embryos could be further enhanced by the addition of cationic transfecting agent Lipofectamine 2000 [71]. CPP can also be used to facilitate delivery of RNA molecules in intact whole plants. Numata et al. [73] described an ionic complex of synthetic dsRNA with a CPP fused with a positively charged polycation sequence as a gene carrier. Infiltration of the RNA-protein complex into *Arabidopsis thaliana* leaf cells induced the downregulation of transgenic reporter and endogenous genes such as yellow fluorescent protein and chalcone synthase. Interestingly, Ng et al. [74] recently demonstrated that CPP-fused proteins of different sizes could be delivered efficiently into organelles including nucleus and peroxisome by infiltrating leaves of intact *Arabidopsis* plants. It will be interesting to know if it is possible to directly deliver DNA into the meristematic tissues of an intact plant for generating transgenic plants with optimized CPP or delivery conditions.

4 Biological Delivery

Infection of plants with virulent strains of *Agrobacterium tumefaciens* and *Agrobacterium rhizogenes* results in the formation of crown gall tumors and "hairy roots," respectively. It has been 40 years since the discovery that the causative agent of crown gall disease, *Agrobacterium tumefaciens*, transfers a small segment of its virulence Ti plasmid DNA into plant cells, resulting in the formation of the tumor tissues [75]. The use of modified *Agrobacterium* strains and disarmed Ti plasmids for gene delivery in combination with in vitro plant tissue culture techniques led to the successful generation of fertile transgenic plants [1–3]. After these initial successes, *Agrobacterium*-mediated transformation has been applied to numerous plant species, and it has become an indispensable tool for both basic plant biology and applied agricultural biotechnology research [76]. During this same time, many studies into the mechanism of *Agrobacterium*-mediated transferred DNA (T-DNA) transfer into plant cells revealed the genes of this remarkable natural genetic engineering process [77]. The factors responsible for *Agrobacterium tumefaciens* T-DNA transfer are encoded by six transcriptional loci in Ti plasmid: *virA*, *virB*, *virC*, *virD*, *virE*, and *virG* [78, 79]. The *vir* genes can be supplied *in trans* from a separate plasmid in respect to T-DNA, thus forming the basis of "binary vector systems" [80, 81]. T-DNA transfer uses a mechanism similar to bacterial conjugation employing a VirB/VirD4 type IV secretion system (T4SS) which primarily functions to transfer DNA-protein complexes from donor to recipient cells during conjugation [82–84]. T-DNA transfer is initiated by the formation of a single-stranded DNA molecule, which is guided and protected by VirD2 and VirE2 proteins during and after the T-strand transfer into the plant cell nucleus [85, 86]. Understanding the T-DNA transfer process has greatly benefitted the development of efficient transformation technologies, from the use of virulence gene inducer acetosyringone [87], superbinary vectors with virulence genes from a hypervirulent *Agrobacterium* strain [88, 89], to co-expression of plant genes to improve transformation efficiency [90].

Besides delivering the T-DNA complex, the *Agrobacterium* VirB/VirD4 T4SS also functions as a protein secretion machinery to deliver virulence factors including VirE2, VirE3, and VirF independently of DNA into plant cells [91]. To take advantage of this machinery to deliver desirable cargo proteins into plant cells, Vergunst et al. [92] successfully demonstrated that Cre recombinase could be translocated into plant cells as fusion proteins with either VirE2 or VirF by *Agrobacterium* VirB/VirD4 T4SS independent of the DNA transfer. Another study also showed that a chimeric protein consisting of VirD2 fused to the homing endonuclease I-*Sce*I and the C-terminal translocation signal of VirF could be delivered to *Arabidopsis* cells through the type IV channel by *Agrobacterium* and induced target sequence mutations at low frequency [93]. It was also shown that disarmed Ti plasmid containing the virulence gene cluster could be transferred to nonpathogenic *Rhizobium* symbiotic species belonging to several phylogenetically distinct genera including *Rhizobium* sp. NGR234, *Mesorhizobium loti*, *Sinorhizobium meliloti*, and *Ensifer adhaerens* OV14, thus expanding the toolbox available for mediating the delivery of T-DNA into plant tissues for generating transgenic plants [94, 95].

In addition to the T4SS, bacteria employ other dedicated secretion systems for translocating proteins across cell membranes and into other cells [96, 97]. Many Gram-negative bacteria use the type III protein secretion system (T3SS) to secrete numerous proteins into plant and animal cells to modulate host cell functions [98]. Among these proteins are transcriptional activator-like effectors (TALEs) involved in modulating host defense responses [99]. DNA-binding domains of TALEs were used to create chimeric sequence-specific site-directed TALE nucleases (TALENs) for gene editing purpose [100]. Therefore, it is likely that TALEN or other site-directed nucleases can be directly delivered into plant cells using T3SS for gene editing purpose through the use of diverse bacterial species as shown for human cells [101]. It should be noted that the function of T3SS is not limited to bacterial pathogenesis. Many plant symbiotic bacteria also use the T3SS to interact with their host [102]. This should be useful for expanding the T3SS delivery tool into nonpathogenic bacteria for delivering genome editing machinery.

It should be noted that plant virus-based systems can provide a simple and efficient approach for biological molecule delivery, especially for the purpose of transient expression. Viral DNA or RNA molecules or vectors can be delivered into plant cells and intact plants by gentle wounding with carborundum, bombardment, or *Agrobacterium* infection. Many plant virus-based expression systems have been developed for both dicots and monocots and are readily available for use [103]. Replication of delivered viral vector within a few days of inoculation will rapidly produce large numbers of DNA or RNA molecules that lead to high levels of protein expression [104]. In fact, engineered DNA and RNA viral vectors have been successfully used for genome engineering studies in several plant species. For example, Marton et al. [105] demonstrated the use of an RNA virus, tobacco rattle virus (TRV)-based expression system, for the delivery of zinc-finger nuclease (ZFN) into a variety of tissues and cells of intact plants. TRV systemically infected its host plants, and the expression of ZFN resulted in targeted mutagenesis in infected tissues. Baltes et al. [106] and Čermák et al. [107] reported the use of geminivirus- based replicons for transient expression of sequence-specific nucleases and delivery of DNA repair templates for mediating gene targeting in tobacco and tomato plants and achieved greater than tenfold enhancements in GT frequencies. Gil-Humanes et al. [108] also demonstrated successful application of a wheat dwarf virus (WDV)-based viral replicon in wheat cells and achieved a 110-fold increase in expression of a reporter gene in comparison with non-replicating controls. WDV replicons carrying CRISPR/Cas9 nucleases and repair templates resulted in 12-fold increase in gene targeting frequency at the endogenous ubiquitin locus compared to non-viral delivery methods.

5 Perspectives

Recent advances in genome editing technology have made it possible to generate heritable changes in the plant genome efficiently without transgene integration either by transient expression of DNA-encoded site-directed nucleases or by direct delivery of "DNA-free" purified site-directed nucleases or preassembled CRISPR-Cas9 ribonucleoprotein complexes. With transient expression and DNA-free delivery methods, genomic DNA is only exposed to the nuclease cleavage for a limited time, thus potentially limiting the chance of off-target cleavage. DNA-free editing with CRISPR-Cas9 ribonucleoprotein complexes was initially shown to generate targeted mutations in animal cells and embryos when delivered by either electroporation or direct injection [109, 110]. In addition, simply treating human cell lines with cell-penetrating peptide (CPP)-conjugated Cas9 protein and CPP-complexed guide RNAs has been shown to disrupt endogenous gene targets efficiently with reduced off-target mutations compared to using plasmid transfections [111].

DNA-free (or transgene free) methods save time because they eliminate the need for one or more generations to segregate away any transgene inserts encoding the editing machinery. These methods are particularly useful for widely cultivated vegetatively propagated crops including potato, cassava, sugarcane, yam, and banana. For the development of gene-edited products in these crops, the use of DNA-free delivery is preferable because it does not lead to integration of a transgene which is hard to remove from clonally propagated tissues. Recent studies have demonstrated the feasibility of such approaches in other plants too. Luo et al. [112] demonstrated that purified meganuclease I-*Sce*I can be introduced into tobacco protoplasts using PEG-mediated delivery to digest the separately introduced episomal YFP reporter plasmid. The introduced active meganuclease I-*Sce*I protein was also able to generate targeted chromosomal double-strand breaks in the reporter transgenic locus containing I-*Sce*I cleavage target sequence, but only when combined with a plasmid encoding a 3′ repair exonuclease Trex2 to enlarge the chromosomal break by resecting the cleaved DNA. These authors further showed that a pair of TALEN proteins can be delivered into *N. benthamiana* protoplasts to generate targeted mutations in the ALS target sequences at a relatively low frequency of 1.4% in transient assays. Woo et al. [113] transfected protoplasts of *Arabidopsis thaliana*, tobacco, lettuce, and rice with preassembled CRISPR-Cas9 ribonucleoprotein complexes and achieved targeted mutagenesis in regenerated plants at frequencies of up to 46% through targeted deep sequencing analysis. However, protoplasts are difficult to handle. Most frequently used regeneration systems amenable to easy in vitro manipulation are based on rapidly dividing young plant tissues such as immature embryos, calli, young leaves, or meristematic tissues. Direct delivery of editing machinery into these tissues will save a lot of time and efforts in recovering edited variants. Recently, it was shown that efficient genome editing could be achieved through transient expression of CRISPR/Cas9 DNA or with in vitro synthesized RNAs encoding Cas9 along with single guide RNAs (sgRNAs) in wheat immature embryos [114]. Alternatively, direct bombardment of in vitro assembled CRISPR-Cas9 ribonucleoprotein complexes into immature embryos has been successfully applied to edit several gene targets in two economically important crops, corn and bread wheat [115, 116].

One of the major gaps in the genetic modification of plants is the lack of efficient technologies to modify mitochondrial genomes. Plastid transformation is also limited to a few plant species. It is important to be able to edit organelle genomes in important food or feed crops, e.g., to optimize the plant's photosynthesis apparatus for increased yields or to engineer novel cytoplasmic male sterility system for simpler hybrid seed production. It is widely accepted that many chloroplast and mitochondrial proteins are encoded by nuclear genes, synthesized in the cytosol in the corresponding precursor form containing a targeting signal called transit peptide (TP), and then imported into the organelle [117]. Interestingly, it was reported recently that certain noncoding RNA could mediate import of foreign mRNA into chloroplasts in plants [118]. The availability of these tools for protein and RNA delivery opens up the possibility of DNA-free editing of organelle genome by directly importing protein, mRNA, or CRISPR ribonucleoprotein complexes.

References

1. Barton KA, Binns AN, Matzke AJM, Chilton MD (1983) Regeneration of intact tobacco plants containing full length copies of genetically engineered T-DNA, and transmission of T-DNA to Rl progeny. Cell 32:1033–1043
2. Caplan A, Herrera-Estrella L, Inze D, Van Haute E, Van Montagu M, Schell J, Zambryski P (1983) Introduction of genetic material into plant cells. Science 222:815–821
3. Murai N, Kemp JD, Sutton DW, Murray MG, Slightom JL et al (1983) Phaseolin gene from bean is expressed after transfer to sunflower via tumor-inducing plasmid vectors. Science 222:476–482
4. Milfin BJ (1985) The potential use of novel techniques in plant breeding. Hereditas 103 (S3):97–107
5. Birch RG (1997) Plant transformation: problems and strategies for practical application. Annu Rev Plant Physiol Plant Mol Biol 48:297–326
6. Hansen G, Wright MS (1999) Recent advances in the transformation of plants. Trends Plant Sci 4:226–231
7. Rakoczy-Trojanowska M (2002) Alternative methods of plant transformation- a short review. Cell Mol Biol Lett 7:849–858
8. Barampuram S, Zhang ZJ (2011) Recent advances in plant transformation. Methods Mol Biol 701:1–35
9. Rivera AL, Gómez-Lim M, Fernández F, Loske AM (2012) Physical methods for genetic plant transformation. Phys Life Rev 9:308–345
10. Que Q, Elumalai S, Li X, Zhong H, Nalapalli S et al (2014) Maize transformation technology development for commercial event generation. Front Plant Sci 5:379. https://doi.org/10.3389/fpls.2014.00379
11. Ishida Y, Hiei Y, Komari T (2007) *Agrobacterium*- mediated transformation of maize. Nat Protoc 2:1614–1621
12. Altpeter F, Springer NM, Bartley LE, Blechl AE, Brutnell TP et al (2016) Advancing crop transformation in the era of genome editing. Plant Cell 28:1510–1520
13. Sugar IP, Newmann E (1984) Stochastic model for electric field-induced membrane pores electroporation. Biophys Chem 19:211–225
14. Morikawa H, Yamada Y (1985) Capillary microinjection into protoplasts and intranuclear localization of injected materials. Plant Cell Physiol 26:229–236
15. Sanford JC (1988) The biolistic process. Trends Biotechnol 6:299–302
16. Kaeppler H, Gu W, Somers DA, Rines HW, Cockburn AF (1990) Silicon carbide fiber-mediated DNA delivery into plant cells. Plant Cell Rep 9:415–418

17. Joersbo M, Brunstedt J (1990) Direct gene transfer to plant protoplasts by mild sonication. Plant Cell Rep 9:207–210
18. Mets LJ (1993) Aerosol beam microinjector. US Patent 5,240,842, 31 Aug 1993
19. Bottino PJ (1975) The potential of genetic manipulation in plant cell cultures for plant breeding. Radiat Bot 15:1–16
20. Lurquin PF (1979) Entrapment of plasmid DNA by liposomes and their interactions with plant protoplasts. Nucleic Acids Res 6:3773–3784
21. Davey MR, Cocking EC, Freeman J, Pearce N, Tudor I (1980) Transformation of *Petunia* protoplasts by isolated *Agrobacterium* plasmids. Plant Sci Lett 18:307–313
22. Krens FA, Molendijk L, Wullens GJ, Schilperoort RA (1982) *In vitro* transformation of plant protoplasts with Ti-plasmid DNA. Nature 296:72–74
23. Nagata T, Okada K, Takebe R, Matsui C (1981) Delivery of tobacco mosaic virus RNA into plant protoplasts mediated by reverse-phase evaporation vesicles (liposomes). Mol Gen Genet 184:161–165
24. Fraley RT, Rogers SG, Horsch RB, Sanders P, Flick J et al (1983) Expression of bacterial genes in plant cells. Proc Natl Acad Sci U S A 80:4803–4807
25. Zimmerman U (1982) Electric field-mediated fusion and related electrical phenomena. Biochim Biophys Acta 694:227–277
26. Fromm M, Taylor LP, Walbot V (1985) Expression of genes transferred into monocot and dicot plant cells by electroporation. Proc Natl Acad Sci U S A 82:5824–5582
27. Rhodes CA, Pierce DA, Metler IJ, Mascarenhas D, Detmer JJ (1988) Genetically transformed maize plants from protoplasts. Science 240:204–207
28. Toriyama KF, Arimoto Y, Uchilniya H, Hinata K (1988) Transgenic rice plants after direct gene transfer into protoplasts. Bio/-Technology 6:1072–1074
29. Dekeyser RA, Claes B, De Rycke RMU, Habets ME, Van Montagu MC, Caplan AB (1990) Transient gene expression in intact and organized rice tissues. Plant Cell 2:591–602
30. Fennell A, Hauptmann R (1992) Electroporation and PEG delivery of DNA into maize microspores. Plant Cell Rep 11:567–570
31. D'Halluin K, Bonne E, Bossut M, De Beuckeleer M, Leemans J (1992) Transgenic maize plants by tissue electroporation. Plant Cell 4:1495–1505
32. Laursen CM, Krzyzek RA, Flick CE, Anderson PC, Spencer TM (1994) Production of fertile transgenic maize by electroporation of suspension culture cells. Plant Mol Biol 24:51–61
33. Crossway A, Oakes JV, Irvine JM, Ward B, Knauf VC, Shewmaker CK (1986) Integration of foreign DNA following microinjection of tobacco mesophyll protoplasts. Mol Gen Genet 202:179–185
34. Neuhaus G, Spangenberg G, Mittelsten Scheid O, Schweiger H-G (1987) Transgenic rapeseed plants obtained by the microinjection of DNA into microspore-derived embryoids. Theor Appl Genet 75:30–36
35. Griesbach RJ (1987) Chromosome-mediated transformation via microinjection. Plant Sci 50:69–77
36. De la Peña A, Lörz H, Schell J (1987) Transgenic rye plants obtained by injecting DNA into young floral tillers. Nature 325:274–276
37. Gong Z, Shen W, Zhou G, Huang J, Qian S (1988) Introducing exogenous DNA into plants after pollination-DNA traverse pathway of pollen tube reach embryonic sac. Sci China B Chem Biol Agri Med Earth Sci 31:1080–1084
38. Luo Z-X, Wu R (1988) A simple method for the transformation of rice via the pollen-tube pathway. Plant Mol Biol Rep 6:165–174
39. Langridge P, Brettschneide R, Lazzeri P, Lorz H (1992) Transformation of cereals via *Agrobacterium* and the pollen pathway: a critical assessment. Plant J 2:631–638
40. Klein TM, Wolf ED, Wu R, Sanford JC (1987) High-velocity microprojectiles for delivering nucleic acids into living cells. Nature 327:70–73
41. Sanford JC, Klein TM, Wolf ED, Allen N (1987) Delivery of substances into cells and tissues using a particle bombardment process. J Particul Sci Technol 5:27–37
42. Klein TM, Gradziel T, Fromm ME, Stanford JC (1988) Factors influencing gene delivery into *Zea mays* cells by high-velocity microprojectiles. Bio/Technology 6:559–563
43. Christou P, McCabe DE, Swain WF (1988) Stable transformation of soybean callus by DNA-coated gold particles. Plant Physiol 87:671–674
44. McCabe DE, Martinell BJ, Christou P (1988) Stable transformation of soybean (*Glycine max*) by particle acceleration. Bio/Technology 6:923–926
45. Fromm M, Morrish F, Armstrong C, Williams R, Thomas J, Klein T (1990) Inheritance and expression of chimeric genes in the

progeny of transgenic maize plants. Bio/-Technology 8:833–839

46. Gordon-Kamm WJ, Spencer TM, Mangano ML, Adams TR, Daines RJ et al (1990) Transformation of maize cells and regeneration of fertile transgenic plants. Plant Cell 2:603–618
47. Finer J, McMullen M (1990) Transformation of cotton (*Gossypium hirsutum L.*) via particle bombardment. Plant Cell Rep 8:586–589
48. Christou P, Ford TL, Kofron M (1991) Production of transgenic rice (*Oryza sativa* L.) plants from agronomically important indica and japonica varieties via electric discharge particle acceleration of exogenous DNA into immature zygotic embryos. Nat Biotechnol 9:957–962
49. Finer JJ, McMullen MD (1991) Transformation of soybean via particle bombardment of embryogenic suspension culture tissue. In Vitro Cell Dev Biol Plant 27:175–182
50. Vasil V, Castillo AM, Fromm ME, Vasil IK (1992) Herbicide resistant fertile transgenic wheat plants obtained by microprojectile bombardment of regenerable embryogenic callus. Bio/Technol 10:667–674
51. McCabe DE, Martinell BJ (1993) Transformation of elite cotton cultivars via particle bombardment of meristems. Bio/Technology 11:596–598
52. Martin-Ortigosa S, Wang K (2014) Proteolistics: a biolistic method for intracellular delivery of proteins. Transgenic Res 23:743–756
53. Kanchiswamy CN, Malnoy M, Velasco R, Jin-Soo Kim J-S, Viola R (2015) Non-GMO genetically edited crop plants. Trends Biotechnol 33:489–491
54. Held BM, Wilson HM, Hou L, Lewnau CJ, Eby JC (2004) Methods and compositions for the introduction of molecules into cells. US Patent 7,314,969, 1 Jan 2008
55. Meacham JM, Durvasula K, Degertekin FL, Fedorov AG (2014) Physical methods for intracellular delivery: practical aspects from laboratory use to industrial-scale processing. J Lab Autom 19:1–18
56. Kaeppler H, Somers DA, Rines HW, Cockburn AF (1992) Silicon carbide fiber-mediated stable transformation of plant cells. Theor Appl Genet 84:560–566
57. Frame BR, Drayton PR, Bagnall SV, Lewnau CJ, Bullock WP et al (1994) Production of fertile transgenic maize plants by silicon carbide whisker-mediated transformation. Plant J 6:941–948
58. Shukla VK, Doyon Y, Miller JC, DeKelver RC, Moehle EA et al (2009) Precise genome modification in the crop species *Zea mays* using zinc-finger nucleases. Nature 459:437–441
59. Torney F, Trewyn BG, Lin VS-Y, Wang K (2007) Mesoporous silica nanoparticles deliver DNA and chemicals into plants. Nat Nanotechnol 2:295–300
60. Martin-Ortigosa S, Valenstein JS, Lin VS-Y, Trewyn BG, Wang K (2012) Gold functionalized mesoporous silica nanoparticle mediated protein and DNA codelivery to plant cells via the biolistic method. Adv Funct Mater 22:3576–3582
61. Derossi D, Joliott MH, Chassaingl G, Prochiantztn A (1994) The third helix of the *Antennapedia* homeodomain translocates through biological membranes. J Biol Chem 269:10444–10450
62. Vives E, Brodin P, Lebleu B (1997) A truncated HIV-1 Tat protein basic domain rapidly translocates through the plasma membrane and accumulates in the cell nucleus. J Biol Chem 272:16010–16017
63. Lindgren M, Hällbrink M, Prochiantz A, Langel Ü (2000) Cell-penetrating peptides. Trends Pharmacol Sci 21:99–103
64. Zorko M, Langel U (2005) Cell-penetrating peptides: mechanism and kinetics of cargo delivery. Adv Drug Deliv Rev 57:529–545
65. Elmquist A, Lindgren M, Bartfai T, Langel Ü (2001) VE-cadherin derived cell-penetrating peptide, pVEC, with carrier functions. Exp Cell Res 269:237–244
66. Unnamalai N, Kang BG, Lee WS (2004) Cationic oligopeptide-mediated delivery of dsRNA for post-transcriptional gene silencing in plant cells. FEBS Lett 566:307–310
67. Mäe M, Myrberg H, Jiang Y, Paves H, Valkna A, Langel Ü (2005) Internalisation of cell-penetrating peptides into tobacco protoplasts. Biochim Biophys Acta 1669:101–107
68. Chang M, Chou JC, Lee HJ (2005) Cellular internalization of fluorescent proteins via arginine-rich intracellular delivery peptide in plant cells. Plant Cell Physiol 46:482–488
69. Chen C-P, Chou J-C, Liu BR, Chang M, Lee H-J (2007) Transfection and expression of plasmid DNA in plant cells by an arginine-rich intracellular delivery peptide without protoplast preparation. FEBS Lett 581:1891–1897
70. Chugh A, Eudes F (2008a) Cellular uptake of cell-penetrating peptides pVEC and transportan in plants. J Pept Sci 14:477–481
71. Chugh A, Eudes F (2008b) Studies of uptake of cell-penetrating peptides and their cargoes in wheat immature embryos. FEBS J 275:2403–2414

72. Chugh A, Amundsen E, Eudes F (2009) Translocation of cell-penetrating peptides and delivery of their cargoes in triticale microspores. Plant Cell Rep 28:801–810
73. Numata K, Ohtani M, Yoshizumi T, Demura T, Kodama Y (2014) Local gene silencing in plants via synthetic dsRNA and carrier peptide. Plant Biotechnol J 12:1027–1034
74. Ng KK, Motoda Y, Watanabe S, Sofiman Othman A, Kigawa T, Kodama Y et al (2016) Intracellular delivery of proteins via fusion peptides in intact plants. PLoS One 11:e0154081
75. Chilton M-D, Drummond MH, Merio DJ, Sciaky D, Montoya AL, Gordon MP, Nester EW (1977) Stable incorporation of plasmid DNA into higher plant cells: the molecular basis of crown gall tumorigenesis. Cell 11:263–271
76. Binns AN (2002) T-DNA of *Agrobacterium tumefaciens*: 25 years and counting. Trends Plant Sci 7:231–233
77. Binns AN, Thomashow MF (1988) Cell biology of *Agrobacterium* infection and transformation. Annu Rev Microbiol 42:575–606
78. Stachel SE, Nester EW (1986) The genetic and transcriptional organization of the *vir* region of the A6 Ti plasmid of *Agrobacterium tumefaciens*. EMBO J 5:1445–1454
79. Rogowsky PM, Close TJ, Chimera JA, Shaw JJ, Kado CI (1987) Regulation of *vir* genes of *Agrobacterium tumefaciens* plasmid pTiC58. J Bacteriol 169:5101–5112
80. DeFramond AJ, Barton KA, Chilton MD (1983) Mini-Ti: a new vector strategy for plant genetic engineering. Bio/Technology 1:262–269
81. Hoekema A, Hirsch PR, Hooykaas PJJ, Schilperoort RA (1983) A binary vector strategy based on separation of *vir* and T-region of the *Agrobacterium tumefaciens* Ti plasmid. Nature 303:179–180
82. Christie PJ, Vogel JP (2000) Bacterial type IV secretion: conjugation systems adapted to deliver effector molecules to host cells. Trends Microbiol 8:354–360
83. Zupan JR, Doyle W, Zambryski P (1998) Assembly of the VirB transport complex for DNA transfer from *Agrobacterium tumefaciens* to plant cells. Curr Opin Microbiol 1:649–655
84. Juhas M, Crook DW, Hood DW (2008) Type IV secretion systems: tools of bacterial horizontal gene transfer and virulence. Cell Microbiol 10:2377–2386
85. Stachel SE, Timmerman B, Zambryski P (1986) Generation of single stranded T-DNA molecules during the initial stages of T-DNA transfer from *Agrobacterium tumefaciens* to plant cells. Nature 322:706–712
86. Tzfira T, Citovsky V (2006) *Agrobacterium*-mediated genetic transformation of plants: biology and biotechnology. Curr Opin Biotechnol 17:147–154
87. Sheikholeslam SN, Weeks DP (1987) Acetosyringone promotes high efficiency transformation of *Arabidopsis thaliana* explants by *Agrobacterium tumefaciens*. Plant Mol Biol 8:291–298
88. Hiei Y, Ohta S, Komari T, Kumashiro T (1994) Efficient transformation of rice (*Oryza sativa* L.) mediated by *Agrobacterium* and sequence analysis of the boundaries of the T-DNA. Plant J 6:271–282
89. Komori T, Imayama T, Kato N, Ishida Y, Ueki J, Komari T (2007) Current status of binary vectors and superbinary vectors. Plant Physiol 145:1155–1160
90. Gelvin SB (2003) *Agrobacterium*-mediated plant transformation: the biology behind the "gene-jockeying" tool. Microbiol Mol Rev 67:16–37
91. Cascales E, Christie PJ (2003) The versatile bacterial type IV secretion systems. Nat Rev Microbiol 1:137–149
92. Vergunst AC, Schrammeijer B, den Dulk-Ras A, de Vlaam CMT, Regensburg-Tuïnk TJG, Hooykaas PJJ (2000) VirB/D4-Dependent protein translocation from *Agrobacterium* into plant cells. Science 290:979–982
93. Van Kregten M (2011) VirD2 of *Agrobacterium tumefaciens*: functional domains and biotechnological applications. Doctoral thesis, Leiden University
94. Broothaerts W, Mitchell HJ, Weir B, Kaines S, Smith LMA et al (2005) Gene transfer to plants by diverse species of bacteria. Nature 433:629–633
95. Wendt T, Doohan F, Mullins E (2012) Production of *Phytophthora infestans*- resistant potato (Solanum tuberosum) utilising *Ensifer adhaerens* OV14. Transgenic Res 21:567–578
96. Salmond GPC (1994) Secretion of extracellular virulence factors by plant pathogenic bacteria. Annu Rev Phytopathol 32:181–200
97. Green ER, Mecsas J (2016) Bacterial secretion systems – an overview. Microbiol Spectr 4:1. https://doi.org/10.1128/microbiolspec.VMBF-0012-2015

98. He SY, Nomura K, Whittam TS (2004) Type III protein secretion mechanism in mammalian and plant pathogens. Biochim Biophys Acta 1694:181–206
99. Sugio A, Yang B, Zhu T, White FF (2007) Two type III effector genes of *Xanthomonas oryzae pv. oryzae* control the induction of the host genes OsTFIIA_1 and OsTFX1 during bacterial blight of rice. Proc Natl Acad Sci U S A 104:10720–10725
100. Li T, Liu B, Spalding MH, Weeks DP, Yang B (2012) High-efficiency TALEN- based gene editing produces disease-resistant rice. Nat Biotechnol 30:390–392
101. Jia J, Jin Y, Bian T, Wu D, Yang L et al (2014) Bacterial delivery of TALEN proteins for human genome editing. PLoS One 9: e91547. https://doi.org/10.1371/journal.pone.0091547
102. Marie C, Broughton WJ, Deakin WJ (2001) *Rhizobium* type III secretion systems: legume charmers or alarmers? Curr Opin Plant Biol 4:336–342
103. Gleba Y, Klimyuk V, Sylvestre Marillonnet S (2007) Viral vectors for the expression of proteins in plants. Curr Opin Biotechnol 18:134–141
104. Marillonnet S, Thoeringer C, Kandzia R, Klimyuk V, Gleba Y (2005) Systemic *Agrobacterium tumefaciens*-mediated transfection of viral replicons for efficient transient expression in plants. Nat Biotechnol 23:718–723
105. Marton I, Zuker A, Shklarman E, Zeevi V, Tovkach A, Roffe S, Ovadis M, Tzfira T, Vainstein A (2010) Nontransgenic genome modification in plant cells. Plant Physiol 154:1079–1087
106. Baltes NJ, Gil-Humanes J, Čermák T, Atkins PA, Voytas DF (2014) DNA replicons for plant genome engineering. Plant Cell 26:151–163
107. Čermák T, Baltes NJ, Čegan R, Zhang Y, Voytas DF (2015) High-frequency, precise modification of the tomato genome. Genome Biol 16:232. https://doi.org/10.1186/s13059-015-0796-9
108. Gil-Humanes J, Wang Y, Liang Z, Shan Q, Ozuna CV et al (2017) High-efficiency gene targeting in hexaploid wheat using DNA replicons and CRISPR/Cas9. Plant J 89:1251–1262
109. Kim S, Kim D, Cho SW, Kim J, Kim JS (2014) Highly efficient RNA-guided genome editing in human cells via delivery of purified Cas9 ribonucleoproteins. Genome Res 24:1012–1019
110. Sung YH, Kim JM, Kim HT, Lee J, Jeon J et al (2014) Highly efficient gene knockout in mice and zebrafish with RNA-guided endonucleases. Genome Res 24:125–131
111. Ramakrishna S, Dad A-BK, Beloor J, Gopalappa R, Lee S-K, Kim H (2014) Gene disruption by cell-penetrating peptide-mediated delivery of Cas9 protein and guide RNA. Genome Res 24:1020–1027
112. Luo S, Li J, Stoddard TJ, Baltes NJ, Demorest ZL et al (2015) Non-transgenic plant genome editing using purified sequence-specific nucleases. Mol Plant 8:1425–1427
113. Woo JW, Kim J, Kwon SI, Corvalán C, Cho SW, Kim H, Kim S-G, Kim S-T, Choe S, Kim J-S (2015) DNA-free genome editing in plants with preassembled CRISPR-Cas9 ribonucleoproteins. Nat Biotechnol 33:1162–1164
114. Zhang Y, Liang Z, Zong Y, Wang Y, Liu J, Chen K, Qiu J-L, Gao C (2016) Efficient and transgene-free genome editing in wheat through transient expression of CRISPR/Cas9 DNA or RNA. Nat Commun 7:12617
115. Svitashev S, Schwartz C, Lenders B, Young JK, Cigan AM (2016) Genome editing in maize directed by CRISPR-Cas9 ribonucleoprotein complexes. Nat Commun 7:13275
116. Liang Z, Chen K, Li T, Zhang Y, Wang Y, Zhao Q, Liu J, Zhang H, Liu C, Ran Y, Gao C (2017) Efficient DNA-free genome editing of bread wheat using CRISPR/Cas9 ribonucleoprotein complexes. Nat Commun 8:14261
117. Bruce BD (2000) Chloroplast transit peptides: structure, function and evolution. Trends Cell Biol 10:440–447
118. Gómez G, Pallás V (2010) Noncoding RNA mediated traffic of foreign mRNA into chloroplasts reveals a novel signaling mechanism in plants. PLoS One 5:e12269. https://doi.org/10.1371/journal.pone.0012269

Chapter 2

The Use of an Automated Platform to Assemble Multigenic Constructs for Plant Transformation

David G. J. Mann, Scott A. Bevan, Anthony J. Harvey, and Rachelle A. Leffert-Sorenson

Abstract

Compared to traditional means, modern DNA assembly methods allow cloning of large, multigenic vectors for plant transformation in rapid fashion. These methods are often robust and efficient and can assemble multiple DNA fragments into a single vector in one reaction. Here we describe the use of an automated DNA assembly platform for the generation of complex, multigenic T-DNA binary vectors using a hierarchical Golden Gate cloning strategy. These DNA constructs contained diverse DNA elements for the expression of multiple genes for trait stacking in the crop of interest. This platform streamlines the DNA assembly and validation process through high-efficiency cloning methods, integrated automation equipment, and increased throughput. The implementation of this platform removes bottlenecks for routine molecular biology and opens new possibilities for downstream experimental idea testing.

Key words DNA assembly, Golden Gate cloning, Plasmid, T-DNA binary vectors, Multigenic construct, Plant biotechnology, Synthetic biology, Automation, Liquid handler

1 Introduction

The stable expression of transgenes in plants is a common method for understanding the biological function of heterologous genes and producing traits in crops of interest. Expressing these transgenes requires cloning DNA fragments into a plasmid backbone and introducing the DNA into the plant using techniques such as *Agrobacterium*-mediated transformation and biolistic bombardment. The process of ligation and cloning of recombinant DNA fragments is a well-established molecular biology method, and the technologies available in this field have rapidly advanced as the number, size, and complexity of the DNA fragments have continued to increase [1–7]. In addition, a number of proposed procedures and rulesets have been implemented for the standardization of DNA library collections in microbes and plants [8–15], allowing for reusability and flexibility of DNA fragments across

Sandeep Kumar et al. (eds.), *Transgenic Plants: Methods and Protocols*, Methods in Molecular Biology, vol. 1864,
https://doi.org/10.1007/978-1-4939-8778-8_2,

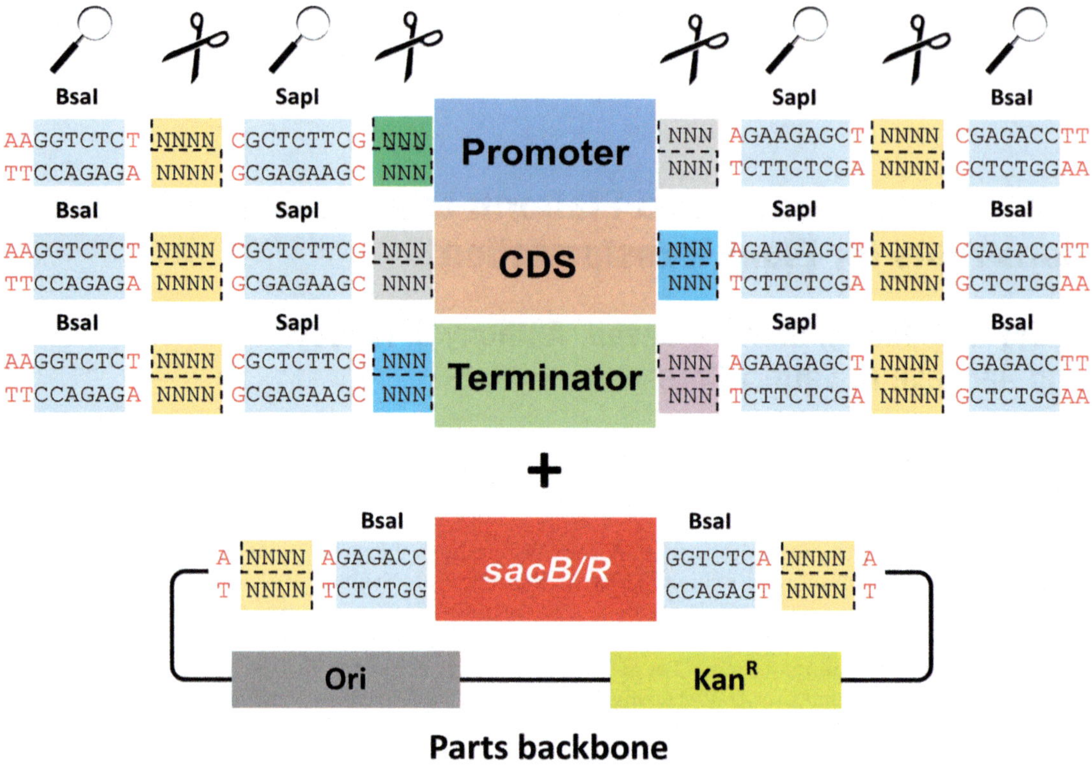

Fig. 1 Diagram depicting design and assembly of Level 1 Components into Level 2 Parts. Type IIS restriction endonuclease (BsaI, SapI) recognition sites are highlighted in light blue, while the cleavage sites are displayed as "N's" by their corresponding color. Upon endonuclease cleavage, the nucleotide sequences represented by "N's" with matching colors contain compatible overhangs which can be ligated together during the DNA assembly reaction. The *sacB/R* cassette acts as a negative selection marker, and the Kan^R cassette acts as a positive selection marker

projects and laboratories. As an example, the Golden Gate methodology [6, 7] has contributed significantly to the simplification and standardization of DNA cloning. This method utilizes Type IIS restriction enzymes which have the unique property of cutting adjacent to their nucleotide recognition sequences (Fig. 1), allowing for the engineering of distinct overhangs for every DNA fragment as well as simultaneous ligation of multiple DNA fragments within the same reaction tube. Additionally, these methods have the benefit of removing the restriction enzyme recognition sites from the final DNA molecule, resulting in "scarless" final products which contain no undesirable nucleotides between the ligated DNA fragments. In order for these methods to work at optimal efficiency, internal Type IIS restriction endonuclease recognition sequences are often removed from the target DNA fragment sequences (a process referred to as "domestication"), so that these internal sites are not targeted for enzyme cutting during the cloning process. Subsequently, overhangs are designed and added to the

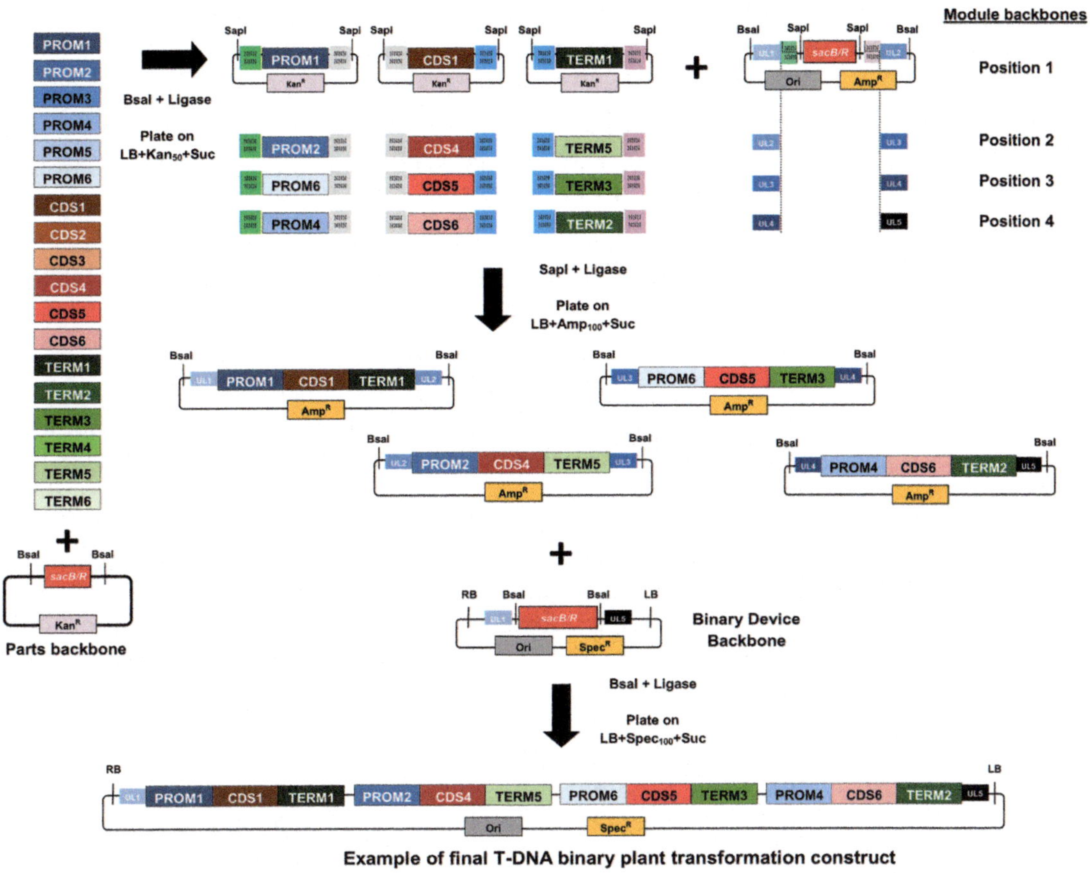

Fig. 2 Diagram of the strategy for hierarchical assembly of Level 1 Components into final Level 4 Binary Devices

nucleotide sequence for the DNA fragments to be cloned together. Tool kits based on Golden Gate cloning, such as GoldenBraid [14–16] and MoClo [17], have standardized the method so that once DNA fragments have been subcloned into a plasmid backbone using a Type IIS restriction endonuclease, they are compatible with other DNA fragments and can be added to a DNA library for reuse in the future. These tool kits can also be used in iterative fashion through alternating different Type IIS restriction endonuclease recognition sites and unique overhangs in the plasmid backbones, resulting in the hierarchical assembly of large, complex multigenic DNA constructs (Fig. 2). Since the DNA library is flexible and can be rapidly reused, it is particularly attractive for cloning projects where DNA fragments are difficult to synthesize or amplify (e.g., repetitive elements, GC- or AT-rich, etc.) or are frequently used. For our purposes, a reusable DNA library was generated utilizing the Golden Gate cloning method for hierarchical assembly as inspired by Engler et al. [11] and defined with the following levels:

1. Level 1 (linear fragment): Component—e.g., promoter or CDS (coding sequence) + adapter/linker sequences.
2. Level 2 (Kan^R): Parts—e.g., promoter or CDS subcloned in replicating plasmid backbone.
3. Level 3 (Amp^R): Modules—combination of Parts (promoter/CDS/terminator) into a transcriptional unit (TU).
4. Level 4 (Kan^R): Binary Devices—combination of many TUs into a multigenic construct within a T-DNA binary backbone for plant transformation.

Levels 1, 2, and 3 include specific Type IIS restriction endonuclease recognition and cleavage sites which come from either the generated linear fragment (Level 1) or are contributed through the plasmid backbone (Levels 2 and 3). In Fig. 1, the 4-nucleotide linker sequences at the 5′ and 3′ ends (NNNN) are unique from each other but identical for Components A, B, and C. These linkers allow for directional cloning into the Part backbone using BsaI enzyme. The subsequent 3-nucleotide linker sequences cleaved by SapI enzyme allow for directional and positional cloning of DNA fragments out of the Part backbone and into the subsequent Module backbone.

This paradigm assembles three Parts at a time (A + B + C) into a Module backbone in their respective order (i.e., ABC). Finally, all the Part, Module, and Device backbones contain the *sacB-sacR* negative selection marker between the SapI and BsaI recognition sites in order to select against any uncut replicating backbones upon transformation into *Escherichia coli*. At each level (i.e., Part, Module, Device), an antibiotic resistance marker that is different than the previous or subsequent level has been utilized, in order to provide positive selection for desirable *E. coli* clones (these are shown above in the numbered list of "levels"). Once Parts A + B + C are subcloned into a Module backbone, each Module backbone contains a pair of BsaI linkers which predetermine what position the Module will have in the subsequent Binary Device. Binary Device backbones are available to accommodate assembly of up to eight modules into a single backbone. While this hierarchical assembly paradigm requires an upfront investment of a defined ruleset, it results in the benefit of a simple, streamlined method for DNA library building and "agnostic" plug-and-play capability ("agnostic" meaning all Parts (A, B, or C); Modules (position 1, 2, 3, 4, 5, 6, 7, or 8) and Devices (containing 1, 2, 3, 4, 5, 6, 7, or 8 modules) can be treated equally and interchanged from one assembly project to another).

In addition to standardizing the DNA assembly methods, numerous labs have validated and implemented automated methods for the DNA assembly process. This includes the development and utilization of software for enabling high-throughput DNA design [18–21], lab information management systems (LIMS) for sample tracking, and implementation of equipment for streamlining and automating the physical assembly of DNA fragments and validation of the resulting DNA plasmids [4, 22–24]. Since various commercial and customized pieces of equipment can be utilized for automating the DNA assembly process, the resulting integrated platforms can be quite diverse in their final form. However, common elements typically implemented for enabling automation include the standardization of (1) sample format (i.e., plate or tube type) and (2) method protocols, which become increasingly more feasible once the DNA assembly method has first been standardized (as described above). For our purposes, we have implemented a variety of automated liquid handlers for different aspects of the process, as well as niche-specific automation equipment such as the Molecular Devices QPix 460 colony picker or the Thermo Fisher Scientific KingFisher™ Flex automated 96-well DNA extraction system (automation examples are shown in Table 1). The method below describes the materials and methods used for automated DNA assembly and validation of a combinatorial set of multigenic binary constructs for plant transformation. The automated platform results in reduced hands-on work required for routine molecular biology techniques such as DNA assembly and cloning and allows researchers to redeploy their time and effort toward idea generation and downstream testing.

2 Materials

All solutions should be properly stored at room temperature unless otherwise noted. Biological and chemical waste should be properly disposed of following specific regulations. Proper personal protective equipment (PPE) should be worn at all times under laboratory conditions.

2.1 DNA Vectors and Cloning Reagents

1. DNA samples in TE buffer (extracted and eluted using standard commercially available DNA extraction kit.
2. Bovine serum albumin (BSA) and molecular biology grade.
3. BsaI restriction endonuclease.
4. T4 DNA ligase and 10× ligase buffer.
5. Golden Gate Reaction Master Mix. The recipe for a single reaction is as follows:

   0.2 μL of each DNA fragment (Module plasmid, backbone vector).

   0.15 μL of 10× BSA.

   0.15 μL of 10× ligation buffer.

Table 1
List of methods and automation equipment used for each task

Method	Task	Automation tool	References
DNA assembly	Transfer DNA fragments and reagents	Labcyte Echo Liquid Handler	Subheading 3.1
DNA transformation and culture preparation	Transform DNA into *E. coli*	Agilent Bravo Liquid Handler	Subheading 3.2
	Plate cells	Molecular Devices QPix Colony Picker	Subheading 3.2
	Pick colonies, and inoculate cultures	Molecular Devices QPix Colony Picker	Subheading 3.2
DNA plasmid isolation	Lyse clone cultures	Tecan Freedom EVO Liquid Handler	Subheading 3.3
	Prepare reagents	BioTek uFill Reagent Dispenser	Subheading 3.3
	Purify DNA plasmids	KingFisher DNA Extraction System	Subheading 3.3
DNA plasmid validation	Normalize DNA plasmid concentration	Tecan Freedom EVO Liquid Handler	Subheading 3.4
	Digest DNA plasmids	Agilent Bravo Liquid Handler	Subheading 3.4
	Gel electrophoresis and imaging	Hamilton NIMBUS Select and Coastal Genomics' Workstation	Subheading 3.4

The reference refers to Subheading 2

0.05 μL of T4 ligase enzyme.
0.05 μL of BsaI enzyme.
X μL of water.

2.2 *E. coli* Cell Transformation and Media

1. Chemically competent *E. coli* cells in 96-well format.
2. 2-YT broth (Teknova, Hollister, CA).
3. LB agar with spectinomycin + sucrose solid media: 25 g of LB broth powder, 50 g of D-sucrose, and 15 g of agar. Bring the solution to 1 L with sterile molecular biology-grade water. Sterilize all components in autoclave at 118 ° C for 4 min. Spectinomycin stock is prepared by adding 1 g of spectinomycin dihydrochloride pentahydrate to 10 mL of sterile molecular biology-grade water. Once dissolved in solution, the spectinomycin stock is passed through a 0.22 μm filter and added to molten LB agar at a final concentration of 100 μg/mL.
4. LB + CHY broth: Mix 80% LB broth and 20% Cinnabar high-yield (CHY) production media broth (Teknova, Hollister, CA).

5. 1% bleach solution: Add 10 mL of Clorox bleach to 990 mL of sterile water, and mix.
6. 70% ethanol: Add 300 mL of sterile water to 700 mL of 100% ethanol.

2.3 DNA Plasmid Isolation and Validation Supplies

1. Sterile 25% glycerol solution: 25% glycerol and 75% sterile water.
2. MagJET 96-well Plasmid DNA Kit (Thermo Fisher Scientific, Waltham, MA).
3. Dual-Dye Loading Buffer (Coastal Genomics, Burnaby, British Columbia, Canada).
4. 12- or 24-well Agarose Cassette (Coastal Genomics, Burnaby, British Columbia, Canada).

2.4 Lab Supplies and Consumables

1. 384-well echo qualified low dead volume microplate (384LDV, Labcyte, Sunnyvale, CA).
2. 96- and 384-well PCR plates.
3. Plate-sealing films.
4. GeneMate aluminum cooler block for 96- or 384-well microplates (BioExpress-VWR, Radnor, PA).
5. 96-well flat-bottom UV-transparent plates.
6. 96-well deep-well square round-bottom plates.
7. Vented QTrays with covers and 48-well dividers (Molecular Devices, Sunnyvale, CA).
8. Microplate roller for plate sealing.
9. Standard and wide-bore sterile filtered Tecan and Bravo compatible tips (Thermo Fisher Scientific, Waltham, MA).
10. Rack of 0.5 mL Matrix V-bottom 2D screw tubes (Thermo Fisher Scientific, Waltham, MA).
11. 300 mL flat-bottom robotic reservoir (Thermo Fisher Scientific, Waltham, MA).
12. 96-well KingFisher DNA elution plate (Thermo Fisher Scientific, Waltham, MA).
13. KingFisher 96 tip comb for DW magnets (Thermo Fisher Scientific, Waltham, MA).

2.5 Lab Equipment

1. Barcode printer.
2. Echo® 550 Liquid Handler (Labcyte, Sunnyvale, CA).
3. Benchtop centrifuge.
4. Thermal cycler.
5. Bravo Automated Liquid Handling Platform (Agilent Technologies, Inc., Santa Clara, CA).

6. QPix 460 Microbial Colony Picker (Molecular Devices, Sunnyvale, CA).
7. Multitron Pro 96-well incubation shaker (INFORS HT, Switzerland).
8. Tecan Freedom EVO 150 Automated Liquid Handler (Tecan Trading AG, Switzerland).
9. LabElite I.D. Capper Automated Screw Cap Decapper (Hamilton, Reno, NV).
10. SampleScan Mini 2D Barcode Reader (BioMicroLab, Inc., Concord, CA).
11. KingFisher™ Flex automated 96-well DNA extraction system (Thermo Fisher Scientific, Waltham, MA).
12. Synergy HT Microplate Reader (BioTek Instruments, Inc., Winooski, VT).
13. NIMBUS Select featuring Coastal Genomics' Ranger Technology Automated Gel Electrophoresis Workstation (Hamilton, Reno, NV).
14. 96-well uFill Reagent Dispenser (BioTek Instruments, Inc., Winooski, VT).

3 Methods

All procedures and methods should be carried out at room temperature unless otherwise specified. Biological and chemical waste should be properly disposed of following specific regulations. Proper personal protective equipment (PPE) should be worn at all times under laboratory conditions. Consult your designated institutional safety department for any additional safety precautions.

3.1 DNA Assembly Using Golden Gate Methods

Since the overall method described here involves the hierarchical assembly of DNA fragments (Elements → Components → Parts → Modules → Binary Devices) and the details of each round of cloning are redundant, the main focus of this chapter is on the details of the final round of DNA assembly (Modules → Binary Devices). Although the DNA assembly details are not given, all elements (promoters, coding sequences, 3′ UTRs) were domesticated (internal Type IIS restriction endonuclease recognition sites were removed), synthesized or PCR amplified with linker sequences (Fig. 1), and subcloned into the Parts backbone. Validated Parts were used to generate Modules (Promoter/CDS/3′UTR), and validated Modules were used to generate Binary Devices as described below:

1. Organize all Module plasmids and backbone vectors, and dilute to 25 ng/μL in sterile, biology-grade water (TE buffer is an additional option).
2. Array 12 μL of each module in a Labcyte 384LDV source plate (*see* **Note 1**).
3. Create a worklist for the Labcyte Echo 550 liquid handler (*see* **Note 2**).
4. Ensure proper placement of source and destination plates, and initiate software for Labcyte Echo 550.
5. Use the Labcyte Echo 550 liquid handler to transfer 0.2 μL of each module in the source plate to the appropriate well of a destination plate (384-well Armadillo PCR plate).

   While the DNA samples are transferring from the source plate to destination plate, prepare a Golden Gate Reaction Master Mix. The volume of water added depends on the number (volume) of Modules used in the reaction. Water is added to the reaction mix so that the final reaction volume (including Modules) is 2 μL. Once the water volume is calculated, multiply the volume for each component (e.g., BSA, Ligation Buffer, etc.) of the single reaction mix by the number of final Binary Devices to be assembled.
6. Transfer up to 60 μL of total reaction master mix (maximum volume) to a single well of a Labcyte 384PP source plate (*see* **Note 3**). If 40 μL is not enough volume for the Binary Device assembly reactions, make additional master mix, and aliquot it into additional wells in the Labcyte 384PP source plate.
7. Create a worklist for the Labcyte Echo 550 liquid handler to transfer the master mix to each well of the destination plate (same 384-well Armadillo PCR plate containing the Module plasmids and vector backbones). As mentioned above, the volume to transfer for each reaction depends on the number (volume) of Module plasmids being used in the DNA assembly. The master mix is added to bring the final reaction volume to 2 μL.
8. Use the Labcyte Echo 550 to transfer the master mix volume to each well of the destination plate that contains the assembled DNA from **step 4**.
9. Once the DNA and enzyme reaction mix transfer to the destination plate is complete, seal the plate with Titer-Tops® plate-sealing film. Proceed immediately to the following step.
10. Place the destination plate on the Tetrad® 2 Peltier Thermal Cycler, and program the reaction protocol for 3 min at 37 °C followed by 4 min at 16 °C. Repeat these alternating temperature incubations for a total of 25 cycles. Finish with a single cycle of 5 min at 50 °C followed by 5 min at 80 °C to finish the reactions and inactivate the enzymes.

11. After thermal cycling is complete, proceed immediately to the *E. coli* transformation step. Alternatively, the reactions can be stored overnight at 4 °C on the thermal cycler or in the refrigerator. For long-term storage, place the reactions at −20 °C.

3.2 DNA Transformation and Culture Preparation

1. Remove the chemically competent *E. coli* cells (in 96-well format) from −80 °C storage, and incubate on ice until the cells are thawed.
2. Place plate of competent cells and 384-well destination plate (containing the DNA assembly reactions) on the Agilent Bravo liquid handler deck (*see* **Note 4**).
3. Use the Agilent Bravo liquid handler to transfer 20 μL of the competent cells from the 96-well plate, and overlay onto the 2 μL DNA assembly reaction in the corresponding quadrant of the 384-well destination plate (*see* **Note 5**).
4. Seal plate, and incubate at 4 °C for 30 min (*see* **Note 6**).
5. Incubate plate at 42 °C for 30 s and then immediately at 4 °C for 2 min (*see* **Note 7**).
6. Transfer the reaction from 384-well plate to 96-well deep-well square round-bottom plate(s) containing 250 μL of 2-YT broth in each well (*see* **Note 8**).
7. Seal deep-well plates with breathable seals (such as AeraSeal™ multiwell plate film), and incubate at 37 °C for 1 h while shaking at 225 rpm.
8. Following incubation, transfer 150–250 μL of each individual well of transformed *E. coli* culture to a 96-well flat-bottom Greiner plate using the Agilent Bravo liquid handler.
9. Insert the 96-well Greiner plates into the shallow plate stacker of the molecular Devices Q-Pix 460 Microbial Colony Picker.
10. Sterilize the Q-Pix 460 deck and place pre-labeled 48-region Q-Trays with appropriate antibiotic (spectinomycin for Binary Device plasmids) on the deck of the Q-Pix 460 (*see* **Note 9**).
11. Using the Q-Pix 460, transfer 100 μL of culture into each region of the Q-Tray, and spread out using the Q-Pix 460 plating head (*see* **Note 10**).
12. Invert Q-Trays, and incubate overnight at 37 °C (*see* **Note 11**).
13. Fill each well of a 96-well deep-well plates with 1.5 mL of LB + CHY broth with appropriate antibiotic (spectinomycin, 100 μg/mL) (*see* **Note 12**).
14. Create, and initiate a Q-Pix 460 protocol for regional picking (*see* **Note 13**).

15. Following colony picking, remove the 96-well deep-well plates from the Q-pix 460 deep-well plate stacker, and seal plates with breathable seals.
16. Place the 96-well deep-well clone culture plates in the Multitron Pro incubator, and shake for 9–12 h at 750 rpm.

3.3 DNA Plasmid Isolation

1. Remove the 96-well deep-well clone culture plates, and visually confirm for culture growth (*see* **Note 14**).
2. Seal the clone culture plates, and centrifuge in the Sorvall Legend XTR Benchtop Centrifuge at 3000 × *g* for 5 min (*see* **Note 15**).
3. Following centrifugation, pour off the supernatant into an appropriate waste container, and place the clone culture plates (containing pellets) onto the deck of the Tecan Freedom EVO 150 Automated Liquid Handler.
4. Create an automated liquid handling method for plasmid DNA isolation in 96-well plate format on Tecan Freedom EVO 150 Automated Liquid Handler. In this case, utilize the magnetic bead-based ThermoFisher MagJET 96-well Plasmid DNA Kit.
5. Prepare the solutions from MagJET 96-well Plasmid DNA Kit according to the manufacturer's provided procedure. Fill 300 mL reservoirs with resuspension buffer, lysis buffer, neutralization buffer, and isopropanol, and place on the Tecan Freedom EVO 150 Automated Liquid Handler deck.
6. Ensure proper placement of all lab wares (such as pipette tips), and clone culture plates according to the deck layout prior to initiating the method.
7. Run the automated method from cell pellet resuspension to neutralization using the Tecan Freedom EVO 150 Automated Liquid Handler. Add 50 μL of isopropyl to the neutralized lysate.
8. Remove the clone culture plates from the Tecan Freedom EVO 150 Automated Liquid Handler deck, and seal plates with Thermo Scientific™ 96-well Sealing Mats.
9. Invert the clone culture plates to mix and centrifuge at 4000 × *g* for 15 min.
10. Return the clone culture plates to the Tecan Freedom EVO 150 Automated Liquid Handler deck, and transfer 500 μL of supernatant from the clone culture plates to new 96-well deep-well square round-bottom plates. Add 250 μL of isopropanol and 25 μL of MagJET magnetic beads to each well using the Tecan Freedom EVO 150 Automated Liquid Handler (*see* **Note 16**).

11. While the Tecan Freedom EVO 150 Automated Liquid Handler method is running, prepare buffer solution plates from MagJET 96-well Plasmid DNA Kit using the BioTek uFill 96-well dispenser to transfer 750 μL of Wash Buffer 1 and Wash Buffer 2 solutions (×2) to each well of a 96-well deep-well square round-bottom plate. Transfer 100 μL of elution buffer to KingFisher DNA elution plate.
12. Remove the sample plates from Tecan Freedom EVO 150 Automated Liquid Handler deck, and place on Thermo-Fisher Kingfisher deck along with Wash Buffer 1, Wash Buffer 2, DNA elution plate, and KingFisher 96 tip comb for DW magnets according to manufacturer's instructions.
13. Run the ThermoFisher KingFisher protocol according to MagJET 96-well Plasmid DNA Kit instructions.
14. Once the protocol is complete, remove the 96-well KingFisher DNA elution plate containing 100 μL of isolated DNA.
15. Seal the plate, and store purified DNA at 4 °C for short-term storage or −20 °C for long-term storage.

3.4 DNA Plasmid Validation

1. Remove the DNA elution plate from storage, and transfer 6 μL of each plasmid DNA sample to a 96-well Low Profile PCR Plate using the Agilent Bravo Automated Liquid Handler (*see* **Note 17**).
2. Prepare an enzyme master mix using appropriate NEB enzyme (s) (according to manufacturer's instructions) for a final reaction volume of 10 μL for each sample well.
3. Transfer 6 μL of enzyme master mix (from reservoir) and 4 μL of each DNA sample well to a new 96-well Low Profile PCR Plate using the Tecan Freedom EVO 150 Automated Liquid Handler.
4. Seal the 96-well plates, and incubate at the appropriate temperature (depending on enzyme) using a thermal cycler.
5. Following incubation, select the appropriate Coastal Genomics' Dual-Dye Loading Buffer and Coastal Genomics' Agarose Cassette for each set of digested DNA samples (*see* **Note 18**).
6. Add 3 μL of Coastal Genomics' Dual-Dye Loading Buffer to each sample well in the 96-well plate using the Agilent Bravo liquid handler.
7. Vortex the 96-well DNA digest plates, and centrifuge briefly using the Sorvall Legend XTR Benchtop Centrifuge.
8. Ensure proper placement of all lab ware on the Hamilton NIMBUS Select featuring Coastal Genomics' Ranger Technology Automated Gel Electrophoresis Workstation according to

the deck layout prior to initiating the method. This includes filtered pipette tips, agarose cassettes, electrodes, and the 96-well DNA digest plate. Initiate method for DNA digest analytics (*see* **Note 19**).

9. Download and review the analytical results using the Coastal Genomics Ranger software outputs.
10. Select positive clone(s) for each DNA plasmid assembled and proceed to plant transformation (*see* **Note 20**).

4 Notes

1. To remove human error and increase efficiency and throughput of DNA assembly, the LabCyte Echo 550 instrument is used to automate liquid handling and miniaturize the reactions. In our experience, the best results were achieved when a maximum volume of 12 μL was added to each well of the Labcyte 384LDV mutiwell plate. At this volume, each DNA plasmid can be used approximately 35 times. Additionally, the dead volume is approximately 3 μL for each well.
2. The worklist is a .csv file and describes the location of each Module plasmid and backbone vector in the source plate, as well as the volume of Module plasmid and backbone vector to transfer to the destination plate and the well location for transfer to the destination plate of each resulting Binary Device. Since the 384-well plate is made up of four 96-well quadrants, ensure that the worklist transfers the DNA volumes into the same quadrant (e.g., quadrant 1) of the 384-well destination plate. This will make any downstream steps involving 96-well plate format substantially easier.
3. Due to the dead volume of the Labcyte 384PP plate, only approximately 40 μL of the reaction master mix will be usable from each well.
4. Ensure proper placement of all lab ware (such as pipette tips) and sample plates according to the deck layout prior to initiating the method. In order to recover the highest *E. coli* cell transformation efficiency, it is critical to keep the cells at 4 °C while on the liquid handler deck. In order to accomplish this, we use cooler blocks such as the aluminum versions made by GeneMate (for 96- or 384-well microplates).
5. The 384-well plate is made up of four quadrants of 96 wells. The number of quadrants is based on the number of samples being assembled and the number of 96-well plates of *E. coli* cells being used.
6. Can be accomplished via an ice bucket or thermal cycler.

7. Can be accomplished via a 42 °C water bath and ice bucket or thermal cycler.
8. 2-YT broth can be added to the deep-well plate using the Agilent Bravo liquid handler or BioTek uFill reagent dispenser. The number of deep-well plates needed will depend on the original number of DNA assembly reactions. Each set of 384 DNA assembly reactions (1 destination plate) can be transformed and recovered in *E. coli* using four 96-well deep-well plates.
9. The recipe for preparation of 48-region Q-Trays is as follows: In a sterile, laminar flow hood, pour 300 mL of molten (55 °C) LB agar with appropriate antibiotic (spectinomycin, 100 μg/mL) into the Q-Tray. Add 48-region divider to Q-tray and allow to cool to room temperature. Once Q-tray has solidified, immediately use or place at 4 °C for short-term storage. Prior to each use of the Q-Pix 460 Colony Picker, the UV-sterilization method and camera alignment protocols are performed per manufacturer's instructions. On-deck reservoirs are filled with 1% bleach, 70% ethanol and sterile water for cleaning of the plating and picking heads during machine use.
10. The exact volume of culture that is plated for each Q-Tray region can vary based on the transformation efficiency which depends on the type (low copy, high copy), size (3–60 kilobases) and complexity (2–10 fragments) of DNA plasmid being assembled. With these variables in mind, we plate a volume ranging from 25 to 125 μL.
11. If the volume plated on each Q-Tray region is greater than 100 μL, we will often allow air drying in a laminar flow hood. However, the amount of time to air dry should be minimized because the agar medium of Q-Tray will dry out rapidly and make colony picking more difficult (i.e., uneven surface or colony growth across the Q-Tray). Placing the Q-tray in a sealed plastic bag for overnight incubation at 37 °C can help prevent drying of agar medium as well.
12. CHY Broth is a mild irritant. Read Material Safety Data Sheet (MSDS) and handle with caution according to manufacturer's instructions.
13. Consult the Q-Pix 460 Colony Picker manufacturer's manual. Different protocol settings (e.g., depth of contact for colony picking, depth of inoculation in deep well) or imaging settings (e.g., colony diameter, colony axis ratio) can be optimized for the regional picking protocol based on the microbial strain and plasmid type. For Binary Devices containing 4–5 TUs, we typically pick 2–4 colonies per region. Avoid overgrowing the colonies on the Q-tray, as we have observed that the Q-Pix

460 imaging software detection is more accurate when colonies are smaller in size.

14. Visual observation of culture turbidity is a simple means of growth confirmation. In addition, automated plate assays can be performed to quantitate optical density (OD_{600}) by using a liquid handler (e.g., Agilent Bravo) for aliquoting culture samples from the 96-well deep-well plate into a 96-well clear flat-bottom plate (such as the Nunc 96-well UV Transparent plate) and running a protocol on the BioTek Synergy HT multiwell plate reader per manufacturer's instructions.
15. Alternatively, glycerol stocks of the fresh clone cultures can be prepared prior to the DNA extraction process using an automated liquid handler platform such as the Tecan Freedom EVO 150 integrated with a Hamilton LabElite I.D. Capper-Decapper. To achieve this, we use 0.5 mL Matrix V-bottom 2D Screw-tubes for glycerol stock sample preservation. Methods can be developed for seamless barcode scanning and tube decapping/capping on the LabElite I.D. Capper-Decapper, as well as liquid transfer and mixing of glycerol solutions and clone culture into the 2D Screw-tubes using the Tecan Freedom EVO 150 Automated Liquid Handler. Once glycerol stock samples of clone cultures have been prepared, they can be stored at −80 °C in an automated sample management storage system (such as the Hamilton SAM) for long-term storage.
16. For best results, vigorously vortex the MagJET magnetic beads prior to transfer of magnetic beads to each well of the sample plates.
17. Following DNA isolation using the automated method of the ThermoFisher MagJET 96-well DNA Plasmid Kit, the concentration of purified DNA is fairly consistent across each 96-well DNA elution plate. However additional methods can be used to ensure consistent DNA concentrations for downstream analyses. With this in mind, we have integrated automated methods for 96-well quantification of DNA concentrations and 96-well normalization of DNA concentrations as follows: Place the 96-well KingFisher DNA elution plate containing 100 μL of purified DNA on the ThermoFisher MagPlate for 10 min prior to transferring DNA. This will reduce the carry-over of magnetic beads, which can interfere with downstream analyses of the DNA samples. Using the Agilent Bravo Automated Liquid Handler, Transfer 100 μL of purified DNA from the DNA elution plates to Nunc 96-well UV Transparent plates. Place new 96-well plates in the Synergy HT Microplate Reader and quantify absorbance at A230, A260, and A280. We have created a method to use these values for automatically

determining the DNA concentration and generating a worklist of transfer volumes to normalize all DNA concentrations to 25 ng/μL. Following the worklist generation, place 96-well UV Transparent plates on the Tecan Freedom EVO 150 Automated Liquid Handler deck and run method to normalize DNA samples to 25 ng/μL. Transfer samples into a new 96-well Low Profile PCR Plate for downstream analyses.

18. The Coastal Genomics Dual-Dye Loading Buffer size range (measured in DNA base pairs) and Coastal Genomics Agarose Cassette (number of wells, % agarose) is determined by the digest pattern and size of DNA bands being analyzed.
19. Currently, only one 96-well plate of samples can be run at a time on the Hamilton NIMBUS Select featuring Coastal Genomics' Ranger Technology Automated Gel Electrophoresis Workstation.
20. In our workflow, DNA plasmid sequence validation is also performed following the enzyme digest clone screen and prior to plant transformation.

Acknowledgments

The authors would like to thank Bill Moskal for his intellectual input on the integration of an automated platform, Siobhan Davis for her collaboration on sample management and archiving, Ryan Blue and Patrick Westfall for their intellectual input on DNA assembly methods, and Ann Owens Merlo and Tom Meade for their leadership and support.

References

1. Gibson DG, Young L, Chuang RY, Venter JC, Hutchison CA 3rd, Smith HO (2009) Enzymatic assembly of DNA molecules up to several hundred kilobases. Nat Methods 6 (5):343–345. https://doi.org/10.1038/nmeth.1318
2. Juhas M, Ajioka J (2017) High molecular weight DNA assembly in vivo for synthetic biology applications. Crit Rev Biotechnol 37 (3):277–286. https://doi.org/10.3109/07388551.2016.1141394
3. Ellis T, Adie T, Baldwin GS (2011) DNA assembly for synthetic biology: from parts to pathways and beyond. Integr Biol (Camb) 3 (2):109–118. https://doi.org/10.1039/c0ib00070a
4. Chao R, Liang J, Tasan I, Si T, Ju L, Zhao H (2017) Fully automated one-step synthesis of single-transcript TALEN pairs using a biological foundry. ACS Synth Biol 6 (4):678–685. https://doi.org/10.1021/acssynbio.6b00293
5. Kosuri S, Eroshenko N, LeProust EM, Super M, Way J, Li JB, Church GM (2010) Scalable gene synthesis by selective amplification of DNA pools from high-fidelity microchips. Nat Biotechnol 28(12):1295–1299. https://doi.org/10.1038/nbt.1716 http://www.nature.com/nbt/journal/v28/n12/abs/nbt.1716.html#supplementary-information
6. Engler C, Gruetzner R, Kandzia R, Marillonnet S (2009) Golden gate shuffling: a one-pot DNA shuffling method based on type IIs restriction enzymes. PLoS One 4(5):e5553. https://doi.org/10.1371/journal.pone.0005553

7. Engler C, Kandzia R, Marillonnet S (2008) A one pot, one step, precision cloning method with high throughput capability. PLoS One 3 (11):e3647. https://doi.org/10.1371/journal.pone.0003647
8. Binder A, Lambert J, Morbitzer R, Popp C, Ott T, Lahaye T, Parniske M (2014) A modular plasmid assembly kit for multigene expression, gene silencing and silencing rescue in plants. PLoS One 9(2):e88218. https://doi.org/10.1371/journal.pone.0088218
9. Boyle PM, Burrill DR, Inniss MC, Agapakis CM, Deardon A, Dewerd JG, Gedeon MA, Quinn JY, Paull ML, Raman AM, Theilmann MR, Wang L, Winn JC, Medvedik O, Schellenberg K, Haynes KA, Viel A, Brenner TJ, Church GM, Shah JV, Silver PA (2012) A BioBrick compatible strategy for genetic modification of plants. J Biol Eng 6(1):8. https://doi.org/10.1186/1754-1611-6-8
10. Casini A, Storch M, Baldwin GS, Ellis T (2015) Bricks and blueprints: methods and standards for DNA assembly. Nat Rev Mol Cell Biol 16 (9):568–576. https://doi.org/10.1038/nrm4014
11. Engler C, Youles M, Gruetzner R, Ehnert T-M, Werner S, Jones JDG, Patron NJ, Marillonnet S (2014) A Golden Gate modular cloning toolbox for plants. ACS Synth Biol 3 (11):839–843. https://doi.org/10.1021/sb4001504
12. Guo Y, Dong J, Zhou T, Auxillos J, Li T, Zhang W, Wang L, Shen Y, Luo Y, Zheng Y, Lin J, Chen GQ, Wu Q, Cai Y, Dai J (2015) YeastFab: the design and construction of standard biological parts for metabolic engineering in Saccharomyces cerevisiae. Nucleic Acids Res 43(13):e88. https://doi.org/10.1093/nar/gkv464
13. Patron NJ, Orzaez D, Marillonnet S, Warzecha H, Matthewman C, Youles M, Raitskin O, Leveau A, Farré G, Rogers C, Smith A, Hibberd J, Webb AAR, Locke J, Schornack S, Ajioka J, Baulcombe DC, Zipfel C, Kamoun S, Jones JDG, Kuhn H, Robatzek S, Van Esse HP, Sanders D, Oldroyd G, Martin C, Field R, O'Connor S, Fox S, Wulff B, Miller B, Breakspear A, Radhakrishnan G, Delaux P-M, Loqué D, Granell A, Tissier A, Shih P, Brutnell TP, Quick WP, Rischer H, Fraser PD, Aharoni A, Raines C, South PF, Ané J-M, Hamberger BR, Langdale J, Stougaard J, Bouwmeester H, Udvardi M, Murray JAH, Ntoukakis V, Schäfer P, Denby K, Edwards KJ, Osbourn A, Haseloff J (2015) Standards for plant synthetic biology: a common syntax for exchange of DNA parts. New Phytol 208(1):13–19. https://doi.org/10.1111/nph.13532
14. Sarrion-Perdigones A, Vazquez-Vilar M, Palaci J, Castelijns B, Forment J, Ziarsolo P, Blanca J, Granell A, Orzaez D (2013) GoldenBraid 2.0: a comprehensive DNA assembly framework for plant synthetic biology. Plant Physiol 162(3):1618–1631. https://doi.org/10.1104/pp.113.217661
15. Sarrion-Perdigones A, Palaci J, Granell A, Orzaez D (2014) Design and construction of multigenic constructs for plant biotechnology using the GoldenBraid cloning strategy. Methods Mol Biol 1116:133–151. https://doi.org/10.1007/978-1-62703-764-8_10
16. Sarrion-Perdigones A, Falconi EE, Zandalinas SI, Juarez P, Fernandez-del-Carmen A, Granell A, Orzaez D (2011) GoldenBraid: an iterative cloning system for standardized assembly of reusable genetic modules. PLoS One 6:e21622
17. Weber E, Engler C, Gruetzner R, Werner S, Marillonnet S (2011) A modular cloning system for standardized assembly of multigene constructs. PLoS One 6:e16765
18. Chen J, Densmore D, Ham TS, Keasling JD, Hillson NJ (2012) DeviceEditor visual biological CAD canvas. J Biol Eng 6:1. https://doi.org/10.1186/1754-1611-6-1
19. Galdzicki M, Clancy KP, Oberortner E, Pocock M, Quinn JY, Rodriguez CA (2014) The Synthetic Biology Open Language (SBOL) provides a community standard for communicating designs in synthetic biology. Nat Biotechnol 32:545. https://doi.org/10.1038/nbt.2891
20. Gansner ER, North SC (2000) An open graph visualization system and its applications to software engineering. Softw Pract Exper 30 (11):1203–1233
21. Hillson NJ, Rosengarten RD, Keasling JD (2012) j5 DNA assembly design automation software. ACS Synth Biol 1(1):14–21. https://doi.org/10.1021/sb2000116
22. Johnson James R, D'Amore R, Thain Simon C, Craig T, McCue Hannah V, Hertz-Fowler C, Hall N, Hall Anthony JW (2016) GeneMill: a 21st century platform for innovation. Biochem Soc Trans 44(3):681–683. https://doi.org/10.1042/bst20160012
23. Linshiz G, Jensen E, Stawski N, Bi C, Elsbree N, Jiao H, Kim J, Mathies R, Keasling JD, Hillson NJ (2016) End-to-end automated microfluidic platform for synthetic biology: from design to functional analysis. J Biol Eng 10(1):3. https://doi.org/10.1186/s13036-016-0024-5
24. Linshiz G, Stawski N, Goyal G, Bi C, Poust S, Sharma M (2014) PR-PR: cross-platform laboratory automation system. ACS Synth Biol 3:515. https://doi.org/10.1021/sb4001728

Chapter 3

Ensifer-Mediated Transformation (EMT) of Rice (Monocot) and Oilseed Rape (Dicot)

Dheeraj Singh Rathore, Evelyn Zuniga-Soto, and Ewen Mullins

Abstract

Ensifer adhaerens OV14 underpins the successful crop transformation protocol, termed *Ensifer*-mediated transformation (EMT). The adaptability and efficiency of EMT technology to successfully transform both monocot and dicots have been previously reported. To facilitate community users' transition to EMT, the modified rice and oilseed rape plants generated in this work were developed using EMT protocols that were grounded in standard *Agrobacterium*-mediated transformation (AMT) processes. Therefore, this chapter describes simple yet crucial steps involved in transferring the use of EMT of rice and oilseed rape for generation of fertile and independent transgenic lines.

Key words *Ensifer adhaerens*, Transformation, Monocot, Dicot, Plant

1 Introduction

Historically, *Agrobacterium* species were considered unique in their ability to transfer and integrate foreign DNA into plant genomes [1, 2]. However, the increasing number of non-*Agrobacterium* species capable of genetically engineering host cells [3–7] suggests that this ability might be broadly spread among the bacterial community [8]. More recently, *Rhizobium etli* has been added to the group that encodes its own virulence protein machinery to promote DNA transfer and stable integration into *Nicotiana tabacum* [9]. It is noteworthy, however, that the transformation frequencies were low for the non-*Agrobacterium* species in these reports (reviewed in [10]). In contrast, Wendt, Doohan, and Mullins [11] reported the ability of the nonpathogenic plant-associated bacteria *Ensifer adhaerens* OV14 when is equipped with *Agrobacterium*-derived virulence machinery, to generate transgenic events (via *Ensifer*-mediated transformation, EMT) in *Arabidopsis*, tobacco, and potato at frequencies similar to those produced by the classic genetic engineer *A. tumefaciens*. Prior to this study, scientific understanding of the bacterium *E. adhaerens* was limited

Sandeep Kumar et al. (eds.), *Transgenic Plants: Methods and Protocols*, Methods in Molecular Biology, vol. 1864,
https://doi.org/10.1007/978-1-4939-8778-8_3,

to its propensity to act as a predator of other microbes [12, 13] or its ability to exhibit plant beneficial properties [14–16].

Interestingly, Rudder, Doohan, Creevey, Wendt, and Mullins [17] showed that the whole-genome (7.7 Mb) sequence of *E. adhaerens* OV14 is composed of two circular chromosomes (3.96 and 2.01 Mb) and two plasmids (1.61 Mb and 125 Kb) with phylogenetic analysis revealing that *E. adhaerens* OV14 forms a separate clade within the group *Ensifer/Sinorhizobium*, with *E. adhaerens* OV14 genetically distinct from *A. tumefaciens* and its related strains within the *Agrobacterium* group [17]. Furthermore, *E. adhaerens* OV14 possesses a series of genes homologous to chromosomal-based genes cited as essential to AMT. More interestingly, the genes that positively influence the virulence and the ability to transfer the T-DNA into a host's genome, while present in *E. adhaerens* OV14, are not essential but were found to be absent in *S. meliloti* 1021 [17]. More recently, rice has been added to the host range of EMT, where authors have successfully transformed three rice cultivars, viz., Nipponbare, Curinga, and IR64, with transformation efficiencies of 16, 7, and ~1%, respectively [18]. Random T-DNA integration patterns were observed across the rice genome and were comparable to that of AMT in regard to partial or complete deletion of right or left borders or adjacent T-DNA. This was the first report of successful monocot transformation via EMT, demonstrating the broader applicability of EMT across the plant kingdom. In addition to expanding the host range of EMT, an important goal is to investigate the biology supporting EMT and identify optimal growth conditions for *E. adhaerens* OV14, which includes the modification of growth media to address the pleomorphic trait of *E. adhaerens* OV14 [19]. Cultivation conditions have been optimized to deliver efficient rates of electroporation with plant transformation plasmids of up to 53 Kb in size. While *E. adhaerens* OV14 is resistant to ampicillin, paromomycin, streptomycin, spectinomycin, ticarcillin-clavulanate, and kanamycin, phenotype screens confirm the strain's susceptibility to gentamicin (≥10 mg/L), tetracycline (≥10 mg/L), chloramphenicol (≥100 mg/L), and cefotaxime (≥500 mg/L). As part of our efforts to assist present and future users of EMT, we are expanding the toolkit to include alternative vectors and growth protocols while investigating the potential of EMT to achieve non-genotype-dependent gene transfer in target crop species. We continuously provide updates of EMT research on our website www.emt4crops.com.

The method described in this chapter details the *Ensifer*-mediated transformation of monocot (rice) and dicot (oilseed rape) species from target explants to bacterial inoculum preparation, co-cultivation, selection of putative transformants, and regeneration of transgenic plants.

2 Materials

2.1 Plant Material

1. Mature seeds of rice and oilseed rape (OSR) to obtain target tissue (*see* **Note 1**).
2. Target tissue for monocot, viz., rice transformation: embryogenic calli derived from scutellum of mature embryos (ME) [18].
3. Target tissue for dicot, viz., oilseed rape (OSR): 5-day-old cotyledonary petioles as explants [20].

2.2 Ensifer *Strains and Vectors*

1. *Ensifer adhaerens* strain OV14 electro-transformed [19] with the unitary vector pCambia5105 (here onward E5105) [21] which contains a suite of virulence operons plus a T-DNA comprised of the *hptII* gene (conferring hygromycin resistance under 2× 35 s promoter) adjacent to the left border (LB) and the *gus* gene (encoding β-glucuronidase under control of a 35 s promoter) adjacent to the right border (RB) (Fig. 1a).
2. E5105 equipped with the binary vector pCambia2201 (here onward E5105_pC2201) [22] which harbors a T-DNA composed of the *nptII* gene under control of an enhanced promoter (2× 35 s) and the *gus* gene driven by a single 35 s promoter (Fig. 1b).

2.3 Bacterial Culture Medium

1. *E. adhaerens* OV14 Teagasc-tryptone yeast extract (TTY) [19]: dissolve 10 g/L tryptone and 5 g/L yeast extract in 4× Elga distilled water. Autoclave, and add 20 mL of 1 M sterile $CaCl_2$.
2. Yeast extract peptone (YEP) medium: add 10 g/L yeast extract, 10 g/L Bacto-peptone, and 5 g/L NaCl to 4× distilled water. Autoclave to sterilize (*see* **Note 2**).
3. TTY or YEP agar: add 15 g/L agar to TTY/YEP prior to autoclaving.

2.4 Plant Tissue Culture Media

2.4.1 EMT of Rice (Monocot) Stocks

All stocks were filter-sterilized and stored at −20 °C (*see* **Note 3**).

1. *100 mg/mL myo-inositol*: 1 g in 10 mL in sterile distilled water.
2. *2.5 mg/mL 2,4-D*: dissolve the powder in few drops of 1 N NaOH, and bring up to final volume with sterile distilled water.
3. *500 mg/mL L-proline*: 5 g in 10 mL 4× distilled water.
4. *500 mg/mL L-glutamine*: 5 g in 10 mL 4× distilled water.
5. *50 mg/mL hygromycin*: stock provided by supplier (Duchefa).
6. *250 mg/mL cefotaxime*: 2.5 g in 10 mL 4× distilled water. Use freshly prepared stock only.
7. *100 mM acetosyringone*: dissolve 196 mg of 3,5-dimethoxy-4-hydroxy-acetophenone in 10 mL DMSO.

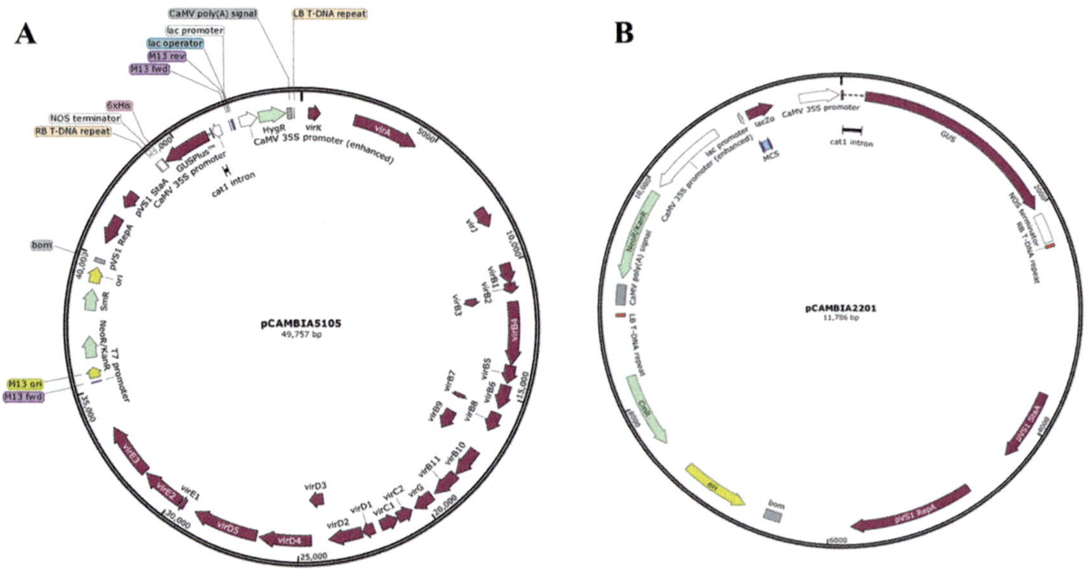

Fig. 1 Schematic representation of pCambia plasmids. (**a**) pCambia5105 unitary vector containing the virulence genes required for transformation along with T-DNA comprised of hygromycin gene (*hptII*) near the left border and β-glucuronidase (*gus*) gene with an intron at the right border and (**b**) pCambia2201 binary vector comprised of neomycin phosphotransferase (*nptII*) gene at the left border and an intron-containing β-glucuronidase (*gus*) gene at the right border and lacks virulence genes

2.4.2 EMT of Rice (Monocot) Media

Unless otherwise stated, prepare all media and components/stocks using 4× Elga distilled water.

1. *Chu-induction and proliferation media*: 3.9 g/L Chu (N6) basal salt mixture [23], 1 mL of 1000× stock solution of Chu (N6) vitamin mixture, 100 mg/L myo-inositol, 2.5 mg/L 2,4-D, 300 mg/L casein hydrolysate, 30 g/L maltose, and 3 g/L gelrite. Adjust the pH to 5.8 and autoclave. Add supplements as 500 mg/L L-proline and 500 mg/L L-glutamine.
2. *Chu-infection media*: 3.9 g/L Chu (N6) basal salt mixture, 1 mL of 1000× stock solution of Chu (N6) vitamin mixture [23], 100 mg/L myo-inositol, 2.5 mg/L of 2,4-D, 1 g/L casamino acids, 15 g/L of maltose, and 15 g/L glucose. Adjust pH to 5.2. Media was filter-sterilized and 100 μM sterilized acetosyringone was added.
3. *Chu-co-cultivation medium*: 3.9 g/L Chu (N6) basal salt mixture [23], 1 mL of 1000× stock solution of Chu (N6) vitamin mixture, 100 mg/L myo-inositol, 2.5 mg/L 2,4-D, 300 mg/L casein hydrolysate, 30 g/L maltose, and 3 g/L gelrite. Adjust pH to 5.8 and autoclave to sterilize. Add supplements 500 mg/L L-proline, 500 mg/L L-glutamine, and 100 μM acetosyringone.

4. *Chu-selection medium*: 3.9 g/L Chu (N6) basal salt mixture [23], 1 mL of 1000× stock solution of Chu (N6) vitamin mixture, 100 mg/L myo-inositol, 2.5 mg/L 2,4-D, 300 mg/L casein hydrolysate, 30 g/L maltose, and 3 g/L gelrite. Autoclave to sterilize, and afterward 500 mg/L L-proline, 500 mg/L L-glutamine, 20 mg/L hygromycin (for cv. Curinga), and 250 mg/L cefotaxime were added.
5. *MS-shoot induction/regeneration medium*: 4.3 g MS basal salt mixture [24], 1 mL of 1000× stock solution of MS vitamin mixture, 100 mg/L myo-inositol, 1 mg/L of NAA, 30 g/L sucrose, and 3 g/L gelrite. Autoclave post-adjusting pH to 5.8, and supplement with 4 mg/L kinetin, 20 mg/L hygromycin (for Curinga), and 250 mg/L cefotaxime.
6. *MS-rooting medium*: MS salts and vitamins [24], 30 g/L sucrose, and 3 g/L gelrite. Adjust pH to 5.8 prior to autoclaving. Add 20 mg/L hygromycin (for Curinga) and 250 mg/L cefotaxime.

2.4.3 EMT of Oilseed Rape (Dicot) Media Stocks

Unless otherwise stated, prepare all components/stocks and media using 4× Elga distilled water (*see* **Note 3**).

1. *Vitamin stock*: 1 g/L nicotinic acid, 10 g/L thiamine hydrochloride, 1 g/L pyridoxine, and 100 g/L myo-inositol. This stock should be filter sterilized, ready for use, and stored frozen at −20 °C.
2. *$AgNO_3$: dissolve* 5 mg/mL in sterile distilled water. Filter sterilize, and store frozen.
3. *10 mg/mL BAP*: *dissolve* 0.1 g of BAP in 1 mL of 1 N NaOH. Add 9 mL sterile distilled water, filter sterilize, and store frozen.
4. *10 mg/mL NAA*: *dissolve* 0.1 g of NAA in 1 mL of 1 N NaOH. Add 9 mL sterile distilled water, filter sterilize, and store frozen.
5. *2 mg/mL GA3*: *dissolve* 2 mg of GA3 in 1 mL of 50% ethanol.
6. *10 mg/mL IBA*: *dissolve* 0.1 g of IBA in 1 mL of 1 N NaOH. Add 9 mL sterile distilled water, filter sterilize, and store frozen.
7. *500 mg/mL cefotaxime*: *dissolve* 5 g in 10 mL of water. Filter sterilize, and use fresh.
8. 50 mg/L *kanamycin*: dissolve 0.5 g in 10 mL water. Filter sterilize to store frozen ready for use.

2.4.4 EMT of Oilseed Rape (Dicot) Media

1. *Seed germination media* (*SGM*): 2.2 g/L MS basal salts, 10 g/L sucrose, and 4 g/L phytagel (pH 5.8). Autoclave to sterilize.
2. *Transfection and bacterial suspension media*: 4.4 g/L MS basal salts and 20 g/L sucrose. Adjust the PH to 5.8 prior to autoclaving. Add 1 mL/L vitamin and 200 μM acetosyringone.

3. *Co-cultivation media* (*CCM*): 4.4 g/L MS basal salts, 20 g/L sucrose, and 4 g/L phytagel. Autoclave after adjusting pH 5.8. Add 1 mL/L vitamins, 5 mg/L $AgNO_3$, 0.75 mg/L BAP, 0.2 mg/L NAA, 0.01 mg/L GA3, and 200 μM acetosyringone.
4. *Callus induction media* (*CIM*): 4.4 g/L MS basal salts, 20 g/L sucrose, and 4 g/L phytagel (pH 5.8). Post-autoclaving, add 1 ml/L vitamins, 5 mg/L $AgNO_3$, 0.75 mg/L of BAP, 0.2 mg/L NAA, 0.01 mg/L GA_3, and 500 mg/L cefotaxime.
5. *Shoot initiation media* (*SIM*): 4.4 g/L MS basal salts, 20 g/L sucrose, and 4 g/L phytagel (pH 5.8). Post-autoclaving, add 1 ml/L vitamins, 5 mg/L $AgNO_3$, 0.75 mg/L BAP, 10 mg/L NAA, 0.01 mg/L GA_3, 500 mg/L cefotaxime, and 25 mg/L kanamycin.
6. *Shoot outgrowth media* (*SOM*): 4.4 g/L MS basal salts, 20 g/L sucrose, and 4 g/L phytagel (pH 5.8). After autoclaving, add 1 mL/L vitamins, 40 mg/L adenine hemisulfate, 500 mg/L PVP-40000, 0.0125 mg/L BAP, 500 mg/L cefotaxime, and 25 mg/L kanamycin.
7. *Transformant selection media* (*TSM*): 4.4 g/L MS basal salts, 20 g/L sucrose, and 4 g/L phytagel (pH 5.8). Then, after autoclaving add 1 mL/L vitamins, 40 mg/L adenine hemisulfate, 500 mg/L PVP-40000, 0.0125 mg/L BAP, 500 mg/L cefotaxime, and 50 mg/L kanamycin.
8. *Root initiation media* (*RIM*): 2.2 g/L MS basal salts, 10 g/L sucrose, and 4 g/L phytagel (pH 5.8). Post-autoclaving, add 1.0 mg/L IBA, 500 mg/L cefotaxime, and 25 mg/L kanamycin.

3 Methods

3.1 EMT of Rice

1. Manually dehusk healthy seeds of rice (cv. Curinga), and sterilize them using 50% Clorox solution containing a drop of Tween-20 for 20 min with continuous shaking.
2. Wash the seeds in sterile distilled water six to seven times. Line up the seeds in rows of five on a wet Whatman® filter paper, and remove the seed coat using fine forceps and sterile scalpel blade under a stereomicroscope in order to isolate the mature embryos (MEs).
3. For callus formation place MEs with scutellum side up on Chu-induction and proliferation media for 20–25 days (Fig. 2a) with subculturing onto fresh media every 8–10 days, and incubate at 24 °C in dark.

Fig. 2 Rice cv. Curinga transformation via EMT process. (**a**) Mature embryos (ME) on Chu-induction and proliferation media, (**b**) callus derived from MEs after 25 days selected for inoculation with E5105, (**c**) embryogenic calli post-3-day co-cultivation with E5105. (**d**) selection of transformed callus (white embryogenic growth) and non-transformed callus (necrotic/browning), (**e**) shoot regeneration from transformed embryogenic callus under hygromycin selection, (**f**) root formation from the healthy shoots under hygromycin selection, and (**g**) healthy plantlets transferred to glasshouse to grow to seed maturity. Stages from **d** to **f** contain hygromycin 25 mg/L and cefotaxime 250 mg/L

4. Pre-culture proliferated embryogenic calli for 3 days on Chu-co-cultivation medium containing 100 μM acetosyringone (Fig. 2b) before co-cultivation.
5. Pick a single colony of E5105 to inoculate YEP medium containing 50 mg/L kanamycin and 200 mg/L spectinomycin/streptomycin and grow at 28 °C and 220 rpm.
6. Once the OD_{600nm} reaches 0.5, centrifuge the culture at 3225 RCF for 10 min. Discard the supernatant, and resuspend the pellet in 20 mL of Chu-infection medium with 100 μM acetosyringone (*see* **Note 4**).
7. Inoculate pre-cultured calli from **Step 4** using a single drop (~5 μL) of E5105.

8. Co-cultivate the plates for 3 days in the dark at 21 °C (*see* Fig. 2c, **Note 5**).
9. Wash calli in sterile distilled water containing 500 mg/L cefotaxime to eradicate the bacteria. Dry them on sterile filter paper, and transfer to Chu-selection medium for 40–50 days with a subculturing at 15–20 days, and incubate in the dark at 24–26 °C (*see* Fig. 2d, **Notes 6–8**).
10. Transfer embryogenic selection resistant calli to MS-shoot induction/regeneration medium containing 20 mg/L hygromycin and 250 mg/L cefotaxime, and incubate them at 24–26 °C in 16 h photoperiod initially for 20–30 days or until the green shoot primordia appeared (Fig. 2e).
11. Continue to grow shoots on MS-shoot induction/regeneration medium under selection of 20 mg/L hygromycin and 250 mg/L cefotaxime for further 15–20 days.
12. Transfer individual healthy shoots of 2–3 cm length to MS-rooting medium in sterile culture tubes to allow further growth and root formation on selective media (Fig. 2f).
13. Transfer to glass jars the healthy rooted ~10 cm long plantlets, and cover with transparent bags to maintain high humidity and moisture. Later, introduce the plantlets into a hydroponic solution.
14. Transfer plants to pot in moist compost and allow seeds to set (Fig. 2g).

3.2 EMT of Oilseed Rape

1. Sterilize *Brassica napus* (cv. Delight) seeds in 12.5% sodium hypochlorite for 20 min with vigorous shaking. Rinse the seeds with sterile distilled water five times under aseptic conditions.
2. Plate seeds onto *SGM* and incubate at 24 °C in the dark for 5 days.
3. On day 4, pick a single colony of E5105_pC2201 to inoculate TTY media containing 100 mg/L kanamycin and 200 mg/L streptomycin + spectinomycin for pCambia5105 and 50 mg/L chloramphenicol in order to select for pCambia2201. Incubate at 28 °C, 220 rpm overnight.
4. Pellet E5105_pCambia2201 culture at 3225 RCF for 15 min to prepare *Ensifer* inoculum. Discard the supernatant, and resuspend the bacterial pellet in MS-based transfection and bacterial suspension media containing 200 μM acetosyringone to obtain a final OD of 0.8, and incubate at 28 °C for 2 h with shaking (*see* **Note 4**).
5. Excise cotyledons with ~2 mm petioles using sterile scalpel blade (Fig. 3a).

Fig. 3 Oilseed rape cv. Delight transformation via EMT process. (**a**) Five-day-old plantlets as source of target explants, viz., cotyledonary petioles. (**b**) Inoculation of cotyledonary petioles by immersing cut ends in *E. adhaerens* OV14_pCambia5105 (E5105)_pCambia2201 culture, (**c**) cotyledonary explant after 5 days of co-cultivation with E5105_pCambia2201, (**d**) cotyledonary explants on CIM supplemented with 500 mg/L cefotaxime post-co-cultivation step, (**e**) shoot initiation from the swollen cotyledonary petioles, (**f**) dissected shoot transferred on shoot outgrowth media, (**g**) healthy shoots transferred to rooting media for root development, and (**h**) healthy plant grown to maturity in glasshouse conditions

6. Immerse cut end only (*see* Fig. 3b, **Note 9**) in E5105_pCambia2201 (OD_{600nm} 0.8) culture for 30 s, and transfer the explants to *CCM*.
7. Co-cultivate plates in the dark at 24 °C for 5 days (*see* Fig. 3c, **Note 5**).
8. After 5 days of co-cultivation, transfer explants to *CIM* supplemented with 500 mg/L cefotaxime to eradicate bacteria. Allow incubation of Petri dishes under dim light (~650 lux) at 24 °C for 2 weeks (*see* Fig. 3d, **Note 6**).

9. Transfer explants to *SIM* containing suitable antibiotics, and incubate plates under 16 h ~4000 lux photoperiod at 24 °C for 4 weeks with subculturing every 2 weeks onto fresh media containing freshly prepared antibiotics (*see* Fig. 3e, **Notes 7** and **8**).
10. Transfer explants with emerging shoots to *SOM* containing 500 mg/L cefotaxime and 25 mg/L kanamycin for 4–8 weeks with subculturing every 2 weeks onto fresh *SOM* (Fig. 3f).
11. Excise shoots of 2–4 cm length, and transfer to Steri Vent containers on TSM supplemented with 500 mg/L cefotaxime, 50 mg/L kanamycin, and low sucrose (10 mg/L). Selection should continue for a minimum of 3 weeks with no subculturing under same conditions as above.
12. Transfer the surviving shoots to RIM with 500 mg/L cefotaxime and 25 mg/L kanamycin. Incubate in light at 24 °C for up to 4 weeks (Fig. 3g).
13. Remove the green shoots with well-developed roots from the Steri Vent containers, and wash with water to remove any phytagel.
14. Transfer healthy green shoots to pots containing water-soaked potting mix, and place in a growth chamber for hardening before transferring to contained glasshouse for flowering and seed formation (Fig. 3h).

3.3 Examination of GUS Expression and Molecular Confirmation

Once the plants are established in glasshouse (15–20 days after transplanting into pots), leaf samples can be taken to perform histochemical testing for GUS activity and subsequent DNA extractions for molecular analysis to confirm T-DNA copy number and/or insertion site sequencing per transgenic line.

4 Notes

1. The quality of seeds is a critical factor in ensuring healthy explants (calli/cotyledonary petioles) as starting tissue for EMT. Seed viability should be tested, and fresh seed stock is preferably maintained at 4–10 °C for a maximum of 12 months. Moreover, the seeds of rice or OSR should not be harvested from fungicide-treated plants; the seed-coated plants with fungicides are not recommended to be used as donor plant material as this can significantly reduce the transformation frequencies.
2. TTY medium can be used as an alternate to YEP to grow E5105.

3. All stocks should be freshly prepared. Any stocks more than 3 months old are not recommended for use in EMT.
4. Maintaining the optical density of bacterial cultures (0.5–0.8) for inoculation of explants is key, as lower OD will lead to lower transformation frequencies, while higher OD values will lead to bacterial overgrowth post-co-cultivation.
5. Three days co-cultivation is optimum for rice, while 5 days is optimal for OSR, but it is important to note that this may vary with the variety of choice. If testing EMT for the first time on a crop/variety of choice, then it is therefore recommended that a preliminary evaluation (e.g., transient GUS assay on treated explants) is conducted to ensure the optimal period of co-cultivation is attained.
6. If an explant is showing signs of strong bacterial overgrowth after the co-cultivation stage, it is advised that such tissue is discarded to avoid bacterial contamination through the subsequent stages of the protocol.
7. Always use freshly prepared cefotaxime each time a media is prepared or explant washings are performed. Also, strict subculturing to fresh medium (initially every 7–10 days in case of OSR) is recommended to mitigate the risk of bacterial overgrowth.
8. A hygromycin/kanamycin kill curve should be generated for each type of tissue explant/variety selected, prior to commencing the EMT protocol. This will identify the optimum concentration of antibiotic required with which to identify putatively transformed cell and hence minimize the occurrence of false positives and false negatives.
9. Do not allow OSR cotyledonary petiole explants to be fully immersed in the bacterial solution as this will encourage bacterial overgrowth and necrosis of explants. In the case of rice, embryogenic calli are fully submerged in bacterial cultures, and to eradicate bacteria post-co-cultivation, the calli are washed with water containing 500 mg/L cefotaxime.

References

1. Gelvin SB (2003) *Agrobacterium*-mediated plant transformation: the biology behind the “gene-jockeying” tool. Microbiol Mol Biol Rev 67(1):16–37
2. Nester E (2008) Agrobacterium: the natural genetic engineer 100 years later. in APSnet Features. University of Washington, Seattle, WA
3. Hooykaas P, Klapwijk P, Nuti M, Schilperoort R, Rorsch A (1977) Transfer of the *Agrobacterium tumefaciens* Ti plasmid to avirulent *Agrobacteria* and to *Rhizobium* ex planta. J Gen Microbiol 98(477–474):484
4. Van Veen R, den Dulk-Ras H, Bisseling T, Schilperoort R, Hooykaas P (1988) Crown gall tumor and root nodule formation by the bacterium Phyllobacterium myrsinacearum after the introduction of an Agrobacterium Ti plasmid or a Rhizobium Sym plasmid. Mol Plant Microbe Interact 1:231–234

5. Broothaerts W et al (2005) Gene transfer to plants by diverse species of bacteria. Nature 433(7026):629–633
6. Rahmawati S, Jefferson O, Sopandie D, Suharsono S, Loedin I (2010) Comparative analysis of rice transformation using *Agrobacterium tumefaciens* and *Rhyzobium leguminosarum*. Indian J Biotechnol 15(1):37–45
7. Wendt T, Doohan F, Winckelmann D, Mullins E (2011) Gene transfer into *Solanum tuberosum* via *Rhizobium* spp. Transgenic Res 20 (2):377–386
8. Lacroix B, Citovsky V (2016) Transfer of DNA from bacteria to eukaryotes. mBio 7(4): e00863-16
9. Lacroix B, Citovsky V (2016) A functional bacterium-to-plant DNA transfer machinery of rhizobium etli. PLoS Pathog 12(3): e1005502
10. Altpeter F et al (2016) Advancing crop transformation in the era of genome editing. Plant Cell 28(7):1510–1520
11. Wendt T, Doohan F, Mullins E (2012) Production of *Phytophthora infestans*-resistant potato (*Solanum tuberosum*) utilising *Ensifer adhaerens* OV14. Transgenic Res 21 (3):567–578
12. Casida JLE (1982) *Ensifer adhaerens* gen. Nov., sp. nov.: a bacterial predator of bacteria in soil. Int J Syst Bacteriol 32(1):339–345
13. Germida JJ, Casida LE (1983) *Ensifer adhaerens* predatory activity against other bacteria in soil, as monitored by indirect phage analysis. Appl Environ Microbiol 45(4):1380–1388
14. Rogel MA, Hernandez-Lucas I, Kuykendall LD, Balkwill DL, Martinez-Romero E (2001) Nitrogen-fixing nodules with *Ensifer adhaerens* harboring *Rhizobium tropici* symbiotic plasmids. Appl Environ Microbiol 67 (7):3264–3268
15. Zhou G et al (2013) Biodegradation of the neonicotinoid insecticide thiamethoxam by the nitrogen-fixing and plant-growth-promoting rhizobacterium *Ensifer adhaerens* strain TMX-23. Appl Microbiol Biotechnol 97 (9):4065–4074
16. Zhou GC et al (2014) The metabolism of neonicotinoid insecticide thiamethoxam by soil enrichment cultures, and the bacterial diversity and plant growth-promoting properties of the cultured isolates. J Environ Sci Health B 49 (6):381–390
17. Rudder S, Doohan F, Creevey CJ, Wendt T, Mullins E (2014) Genome sequence of *Ensifer adhaerens* OV14 provides insights into its ability as a novel vector for the genetic transformation of plant genomes. BMC Genomics 15:268–285
18. Zuniga-Soto E, Mullins E, Dedicova B (2015) Ensifer-mediated transformation: an efficient non-agrobacterium protocol for the genetic modification of rice. Springerplus 4(1):1–10
19. Rathore DS et al (2015) Profiling antibiotic resistance and electrotransformation potential of Ensifer adhaerens OV14; a non-agrobacterium species capable of efficient rates of plant transformation. FEMS Microbiol Lett 362(17):fnv126
20. Rathore DS, Doohan F, Mullins E (2016) Capability of the plant-associated bacterium, *Ensifer adhaerens* strain OV14, to genetically transform its original host *Brassica napus*. Plant Cell Tiss Organ Cult 127(1):85–94
21. http://www.snapgene.com/resources/plasmid_files/plant_vectors/pCAMBIA5105/ (Plasmid pCambia5105 sequence)
22. http://www.snapgene.com/resources/plasmid_files/plant_vectors/pCAMBIA2201/ (Plasmid pCambia2201 sequence)
23. Chu C (1978) The N6 medium and its applications to anther culture of cereal crops. In: Proceedings of symposium on plant tissue culture. Science Press, Peking, pp 45–50
24. Murashige T, Skoog F (1962) A revised medium for rapid growth and bio assays with tobacco tissue cultures. Physiol Plant 15:473–497

Chapter 4

Setaria viridis as a Model Plant for Functional Genomic Studies in C4 Crops

Polyana Kelly Martins, Bárbara Andrade Dias Brito da Cunha, Adilson Kenji Kobayshi, and Hugo Bruno Correa Molinari

Abstract

Setaria viridis is an emerging model for C4 species, and it is an important model to validate some genes for further C4 crop transformation, such as sugarcane, maize, and wheat. Here, we describe two protocols for stable transformation of *S. viridis* mediated by *Agrobacterium tumefaciens* with three different reporter genes and two selectable markers. Routine transformation efficiency reaching 29% was achieved using embryogenic callus in *S. viridis* (accession A10.1). Alternatively, we developed a transformation method by floral dip with 0.6% efficiency. The developed protocols could be useful for genetic and genomics studies of important food-feed-fiber-fuel C4 crops.

Key words Embryogenic callus, Floral dip, Transformation, *Agrobacterium*, Grasses

1 Introduction

Model plants for genetic transformation like *Arabidopsis thaliana* are used in proof of concept studies for many traits in various important crops. However, there is still a need for additional model plants to decode and translate traits that are absent in some species [1]. *Setaria viridis* is an emerging monocot plant model for molecular and genetic studies [2]. It is a short, fast-growing C4 plant with fully sequenced genome available [3] making it a reliable model for genetic studies. In this chapter, we provide detailed protocols for *Agrobacterium*-mediated transformation based on embryogenic callus cultures and spikes as explants. Both methods were first published in 2015 [4, 5] and have been used in a number of gene/promoter validation studies [6–8]. In the first method, mature seeds were used for embryogenic callus induction, and *A. tumefaciens* (EHA105 strain)-carrying binary vectors containing expression cassette with bialaphos resistance gene (*bar* gene from *Streptomyces hygroscopicus*) and hygromycin resistance gene (*hpt*

Sandeep Kumar et al. (eds.), *Transgenic Plants: Methods and Protocols*, Methods in Molecular Biology, vol. 1864,
https://doi.org/10.1007/978-1-4939-8778-8_4,

gene from *Streptomyces hygroscopicus*) as selectable marker genes (responsible for conferring resistance to glufosinate-ammonium herbicide and hygromycin B, respectively) were used in the transformation assays of embryogenic callus. This transformation method, with up to 29% transformation efficiency and an average of approximately 15%, can provide a valuable tool for genomics and physiological studies [4]. Such transformation efficiency is three times higher than previously reported [9]. The second method utilizes *S. viridis* spikes at boot stage for floral dip (*Agrobacterium*-mediated transformation with vacuum infiltration). The appropriate stage for dipping is from mid- to late-uninucleate microspore stage when the spike has not emerged from the flag sheath (boot stage) [10]. This stage coincides with the highest frequency of uninucleate microspore in the anther [11]. In addition, in rice, the anther/microspores were the preferential targets of the floral dip transformation [12]. *A. tumefaciens* (AGL1 strain) carrying the binary vector containing the reporter gene *rfp* (red fluorescent protein) driven by *Zm*Ubi1 promoter and hygromycin resistance gene were used for floral dip reaching 0.6% of transformation efficiency. The developed protocols could be useful for genetic and genomics studies of important food-feed-fiber-fuel C4 crops.

2 Materials

2.1 Plant Materials

1. *S. viridis* (accession A10.1) plants cultivated in a greenhouse or growth chamber.
2. Mature seeds of *S. viridis*.
3. Dehulled mature seed culture (*see* **Note 1**).
4. Embryogenic callus induced from mature seeds.

2.2 Binary Vector Constructs and Agrobacterium Strains

1. Binary vectors containing:
   (a) The plant selectable marker gene (*bar* gene from *Streptomyces hygroscopicus*) controlled by constitutive CaMV35S and 3′ T35S transcription termination signal; the reporter gene *gus* (β-glucuronidase) controlled by the constitutive *Zea mays* ubiquitin 1 promoter (*Zm*Ubi1) containing a intron [13] and nopaline synthase transcription termination signal (NOS T), ColE1, and pVS1 origins of replication; and the resistance gene to spectinomycin/streptomycin (*Sp/Sm*) antibiotic (Fig. 1a).
   (b) (b) The plant selectable marker gene (*hpt* gene from *Escherichia coli*) controlled by the constitutive double CaMV35S (2× CaMV35S) promoter and 3′ rbcsE9 transcription termination signal. The reporter gene *gfp* (green

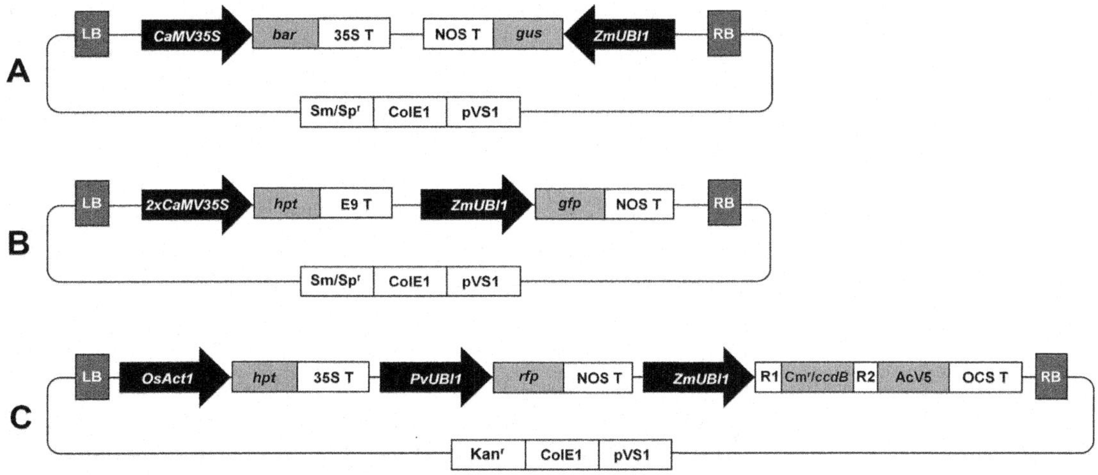

Fig. 1 Basic structure of the binary vectors. (**a**) Vector with *gus* reporter gene controlled by constitutive *Zea mays* ubiquitin 1 promoter (*Zm*Ubi1) and *bar* as selection marker gene controlled by constitutive CaMV35S promoter. (**b**) Vector with *gfp* reporter gene controlled by constitutive *Zm*Ubi1 promoter and *hpt* as selection marker gene controlled by constitutive double CaMV35S promoter. (**c**) Gateway-compatible pANIC 6A with *rfp* as reporter gene controlled by constitutive *Panicum virgatum* L. ubiquitin 1 promoter (*Pv*Ubi1) and *hpt* as selection marker gene controlled by constitutive *Oryza sativa* actin-1 promoter (*Os*Act1) promoter. Sm/Sp^r and Kan^r (streptomycin/spectinomycin and kanamycin bacterial resistance); NOS T (nopaline synthase terminator), E9 T (rbcsE9 transcription terminator), OCS T (octopine synthase terminator), AcV5 (epitope tag), ColE1 (origins of replication in *E. coli*), PVS1 (origin of replication in *A. tumefaciens*), Cm^r (chloramphenicol resistance gene), ccdB (negative selection marker), LB (left border), RB (right border), R1 and R2 (attR1 and attR2 recombination sites)

fluorescent protein) controlled by *Zm*Ubi1 containing an intron [13] and NOS T, ColE1, and pVS1 origins of replication and *Sp/Sm* antibiotic resistance genes (Fig. 1b).

(c) (c) Gateway vector pANIC 6A [15] with selectable marker gene (*hpt* gene from *Escherichia coli*) controlled by constitutive *Oryza sativa* actin-1 promoter (*Os*Act1) and 3′ T35S transcription termination signal; the reporter gene *rfp* (red fluorescent protein) controlled by the constitutive *Panicum virgatum* L. ubiquitin-1 promoter (*Pv*Ubi1) containing a intron and nopaline synthase transcription termination signal (NOS T), ColE1, and pVS1 origins of replication; and resistance gene to kanamycin (Kan) antibiotic (Fig. 1c).

2. *A. tumefaciens* EHA105 and AGL1.

2.3 Culture Media and Buffers

The media components listed are on a per liter basis of deionized H_2O.

1. Callus Induction Medium (CIM): MS basal salt powder mixture [14], 1 mg/L d-biotin stock solution (1 mg/mL), 1 mL/

L *Setaria* vitamin stock solution (1000×), 100 mg/L myo-inositol, 1 mL/L cupric sulfate stock solution (1000×), 3% sucrose (30 g/L), 2 mL/L 2,4-dichlorophenoxyacetic acid stock solution (1 mg/mL), and 0.5 mL/L kinetin stock solution (1 mg/mL). Adjust pH to 5.8 with 1 M NaOH and then add 4 g/L Phytagel™. Autoclave at 121 °C for 20 min. Pour 20 mL per 90 mm × 15 mm polystyrene Petri dishes. The medium can be stored at 4 °C in the dark for up to 2 weeks.

2. Yeast Extract Broth (YEB) *Agrobacterium* Growth Medium Liquid: 5 g/L meat extract, 1 g/L yeast extract, 5 g/L peptone, 5 g/L sucrose, and 240 mg/L magnesium sulfate ($MgSO_4$). Adjust pH to 6.8 with 1 M NaOH and autoclave at 121 °C for 20 min. The medium can be stored for up to 1 month at room temperature. Add the rifampicin antibiotic (1 μL stock solution/mL culture medium) required for the *Agrobacterium* EHA105 and AGL1 strains and the binary vector (spectinomycin and/or streptomycin) only at the time of use. For YEB solid medium, add 15 g/L agar after adjusting pH to 7.0, autoclave, and then after the medium cools to 60 °C, add the appropriate antibiotics, and store at room temperature or at 4 °C for up to 1 month.
3. Liquid Callus Induction Medium (LCIM): MS basal salt powder mixture, 1 mg/L d-biotin stock solution (1 mg/mL), 1 mL/L *Setaria* vitamin stock solution (1000×), 100 mg/L myo-inositol, 1 mL/L cupric sulfate stock solution (1000×), 3% sucrose (30 g/L), 2 mL/L 2,4-dichlorophenoxyacetic acid stock solution (1 mg/mL), and 0.5 mL/L kinetin stock solution (1 mg/mL). Adjust pH to 5.8 with 1 M NaOH and autoclave at 121 °C for 20 min. The medium can be stored at 4 °C in the dark for up to 1 month.
4. Co-Cultivation Medium (CCM): MS basal salt powder mixture, 1 mg/L d-biotin stock solution (1 mg/mL), 1 mL/L *Setaria* vitamin stock solution (1000×), 100 mg/L myo-inositol, 1 mL/L cupric sulfate stock solution (1000×), 3% sucrose (30 g/L), 2 mL/L 2,4-dichlorophenoxyacetic acid stock solution (1 mg/mL), and 0.5 mL/L kinetin stock solution (1 mg/mL). Adjust pH to 5.8 with 1 M NaOH and then add 4 g/L Phytagel™. Autoclave at 121 °C for 20 min. Cool to 60 °C, and then add 2 mL/L acetosyringone stock solution (100 mM). Pour 20 mL per 90 mm × 15 mm polystyrene Petri dishes. The medium can be stored at 4 °C in the dark for up to 2 weeks.
5. Callus Induction Medium with Timentin (CIMT): MS basal salt powder mixture, 1 mg/L d-biotin stock solution (1 mg/mL), 1 mL/L *Setaria* vitamin stock solution (1000×), 100 mg/L myo-inositol, 1 mL/L cupric sulfate stock solution

(1000×), 3% sucrose (30 g/L), 2 mL/L 2,4-dichlorophenoxyacetic acid stock solution (1 mg/mL), and 0.5 mL/L kinetin stock solution (1 mg/mL). Adjust pH to 5.8 with 1 M NaOH and then add 4 g/L Phytagel™. Autoclave at 121 °C for 20 min. Cool to 60 °C, and then add 0.5 mL/L of Timentin stock solution (300 mg/L). Pour 20 mL per 90 mm × 15 mm polystyrene Petri dishes. The medium can be stored at 4 °C in the dark for up to 2 weeks.

6. Selective Callus Induction Medium (SCIM): MS basal salt powder mixture, 1 mg/L d-biotin stock solution (1 mg/mL), 1 mL/L *Setaria* vitamin stock solution (1000×), 100 mg/L myo-inositol, 1 mL/L cupric sulfate stock solution (1000×), 3% sucrose (30 g/L), 2 mL/L 2,4-dichlorophenoxyacetic acid stock solution (1 mg/mL), and 0.5 mL/L kinetin stock solution (1 mg/mL). Adjust pH to 5.8 with 1 M NaOH and then add 4 g/L Phytagel™. Autoclave at 121 °C for 20 min. Cool to 60 °C, and then add 0.5 mL/L of Timentin stock solution (300 mg/L), 0.6 mL/L hygromycin B solution (50 mg/mL), or 3 mL/L glufosinate-ammonium stock solution (1 mg/mL). Pour 20 mL per 90 mm × 15 mm polystyrene Petri dishes. The medium can be stored at 4 °C in the dark for up to 1 month.
7. Selective Callus Regeneration Medium (SCRM): MS basal salt powder mixture, 1 mg/L d-biotin stock solution (1 mg/mL), 1 mL/L *Setaria* vitamin stock solution (1000×), 100 mg/L myo-inositol, 2% sucrose (20 g/L), and 2 mL/L kinetin stock solution (1 mg/mL). Adjust pH to 5.8 with 1 M NaOH and then add 2 g/L Phytagel™. Autoclave at 121 °C for 20 min. Cool to 60 °C, and then add 0.5 mL/L of Timentin stock solution (300 mg/L), 0.6 mL/L hygromycin B solution (50 mg/mL), or 3 mL/L glufosinate-ammonium stock solution (1 mg/mL). Pour 25 mL per 100 mm × 20 mm polystyrene Petri dishes. The medium can be stored at 4 °C in the dark for up to 1 month.
8. Selective Development Medium (SDM): MS basal salt powder mixture, MS vitamin stock solution (1000×), 100 mg/L myo-inositol, and 1.5% sucrose (15 g/L). Adjust pH to 5.8 with 1 M NaOH and then add 2 g/L Phytagel™. Autoclave at 121 °C for 20 min. Cool to 60 °C, and then add 0.5 mL/L of Timentin stock solution (300 mg/L), 0.6 mL/L hygromycin B solution (50 mg/mL), or 3 mL/L glufosinate-ammonium stock solution (1 mg/mL). Pour 40 mL per Magenta™ vessel GA-7. The medium can be stored at 4 °C in the dark for up to 1 month.
9. Solid Half-Strength Murashige and Skoog Basal Medium: 1/2 MS basal salt powder mixture, 1 mL/L MS vitamin stock

solution (1000×), and 100 mg/L myo-inositol. Adjust pH to 5.8 with 1 M NaOH and then add 2 g/L Phytagel™. Autoclave at 121 °C for 20 min. Cool to 60 °C, and then add 1.2 mL/L hygromycin B solution (50 mg/mL) or 6 mL/L glufosinate-ammonium stock solution (1 mg/mL). Pour 60 mL per 150 mm × 20 mm polystyrene Petri dishes. The medium can be stored at 4 °C in the dark for up to 1 month.

10. Luria-Bertani (LB) *Agrobacterium* Growth Medium: 10 g/L tryptone, 10 g/L NaCl, and 5 g/L yeast extract. Adjust pH to 7.0 with 1 M NaOH, autoclave at 121 °C for 20 min, and store at room temperature.
11. MM Buffer: 10 mL/L 10 mM MES and 10 mL/L 10 mM $MgCl_2$. Adjust pH to 5.6 with 1 M NaOH and store at room temperature.
12. Tobacco Leaf Extract: In 1 L of MM buffer, cut into 2 cm^2 pieces 40 g of fresh leaves cultured in vitro and stirred gently for 1 h. Sift the solution and collect the liquid phase (*see* Fig. 4c–e).
13. X-Gluc Buffer: 1.38 g of $NaH_2PO_4{\cdot}H_2O$, 21 mg of $K_4Fe(CN)_6{\cdot}3H_2O$, 372 mg of $Na_2EDTA{\cdot}2H_2O$, 100 μL of Triton® X-100, and 1 mL of X-Gluc salt (50 mg/mL); bring volume to 100 mL with sterile water. Adjust pH to 7.0 with 1 M NaOH, filter through 0.22 μm Millex-GP syringe filter, aliquot in 10 mL conical tubes, and store at −20 °C. The tissues are collected and incubated in the buffer at 37 °C overnight and in the dark. After staining, rinse the tissues in 70% ethanol until the chlorophyll is removed.

2.4 Stock Solutions

1. 500× d-biotin: dissolve 500 mg/L d-biotin in sterile water, aliquot in 1.5 mL tubes, and store at −20 °C for up to 5 months.
2. 1000× *Setaria* vitamins: dissolve 100 mg/L thiamine, 500 mg/L pyridoxine, and 500 mg/L nicotinic acid in sterile water, aliquot in 1.5 mL tubes, and store up to 5 months at −20 °C.
3. 1000× cupric sulfate pentahydrate: dissolve 600 mg/L of $CuSO_4{\cdot}5H_2O$ in sterile water, and store at 4 °C for up to 3 months.
4. 1 mg/mL 2,4-dichlorophenoxyacetic acid: dissolve 30 mg of 2,4-dichlorophenoxyacetic salt in 3 mL of 1 M NaOH solution while heating in a small glass beaker. Bring to volume (30 mL) with sterile water, and store for up to 3 months at 4 °C.
5. 1 mg/mL kinetin: dissolve 10 mg of kinetin in 1 mL of 1 M NaOH, bring to volume (10 mL) with sterile water, aliquot in 1.5 mL tubes, and store at −20 °C for up to 5 months.

6. 1 M NaOH: dissolve 3.99 g of NaOH in 100 mL of sterile distilled water, and store up to 6 months at room temperature.
7. 50 mg/mL rifampicin: dissolve rifampicin salt in dimethyl sulfoxide (DMSO), aliquot in 1.5 mL tubes, and store at −20 °C for up to 5 months.
8. 50 mg/mL spectinomycin: dissolve spectinomycin salt in sterile water, filter through 0.22 μm Millex-GP syringe filter, aliquot in 1.5 mL tubes, and store at −20 °C for up to 5 months.
9. 300 mg/mL streptomycin: dissolve streptomycin salt in sterile water, filter through 0.22 μm Millex-GP syringe filter, aliquot in 1.5 mL tubes, and store at −20 °C for up to 5 months.
10. 100 mM acetosyringone: dissolve 19.6 mg of acetosyringone (3′,5′-dimethoxy-4′-hydroxyacetophenone) and 1 mL of DMSO or absolute ethanol, aliquot in 1.5 mL tubes, and store at −20 °C for up to 5 months. Acetosyringone should be added to the medium only at the time of use of this.
11. 300 mg/mL Timentin: dissolve Timentin salt (Tioxin-Novafarma) in sterile water, filter through 0.22 μm Millex-GP syringe filter, aliquot in 1.5 mL tubes, and store at −20 °C for up to 4 months.
12. We purchase hygromycin B (50 mg/mL) stock solution (Product #PN10687-010) from Thermo Fisher Scientific (https://www.thermofisher.com). Store at 4 °C.
13. 15.5 mL/L glufosinate-ammonium: dissolve Liberty® 200 g/L, in sterile water, filter through 0.22 μm Millex-GP syringe filter, aliquot in 1.5 mL tubes, and store at −20 °C for up to 3 months.
14. 1000× MS vitamin stock solution: dissolve 1 g/L thiamine, 500 mg/L pyridoxine, 500 mg/L nicotinic, 2 g/L glycine, 50 g/L L-arginine, and 150 mg/L citric acid in sterile water, aliquot in 1.5 mL tubes, and store up to 5 months at −20 °C.
15. 50 mg/mL X-Gluc solution: dissolve 50 mg of X-Gluc sodium salt in 1 mL of DMSO or dimethyl formamide or methanol, and add to buffer.

2.5 Other Solutions, Reagents, and Supplies

1. Sodium hypochlorite 10%.
2. Concentrate sulfuric acid (H_2SO_4).
3. Sodium hypochlorite 2% (v/v).
4. Tween® 20.
5. Eppendorf (2 mL) for disinfection of mature seeds.
6. Laminar airflow cabinet.
7. Sterile distilled water.
8. Sterile Whatman filter paper.

9. Sterile stainless steel scalpel blade n.11.
10. Tweezers.
11. Stereomicroscope.
12. Centrifuge.
13. Orbital shaker with controlled temperature.
14. Spectrophotometer to measure bacterial culture optical density.
15. Poloxamer 188 solution 10%.
16. Growth chamber set at 25 °C ± 2 °C without light.
17. Growth chamber set at 25 to 27 °C and with 16/8 h light/dark photoperiod (luminous intensity of at least 150 μmol m^{-2} s^{-1}).
18. Soil, substrate, and vermiculite mixture (*see* **Note 2**).
19. 0.22 μm Millex-GP syringe filter.
20. Silwet-L77.
21. Leica M205 FA stereomicroscope equipped with a long-pass filter set with a 488 nm excitation filter and a 460–500 nm emission filter and Leica DFC7000T camera.
22. Leica MZ16F stereomicroscope equipped with an XCite EXFO fluorescence illumination source, a filter combination for fluorescence in the red region of the spectrum (excitation, 558 nm; emission, 580 nm), and a digital camera (Retiga 2000 Fast Cooled 12-bit).
23. Secador® Techni-dome® 360 vacuum desiccator.
24. Plastic bags.
25. Magenta™ vessel GA-7.
26. 150 mm × 20 mm polystyrene Petri dishes for seed segregation.
27. Growth chamber set at 25 to 27 °C, with 16/8 h light/dark photoperiod (luminous intensity of at least 400-–600 μmol m^{-2} s^{-1}), and with controlled relative humidity 65%.
28. Plastic pots for plant growth.
29. Greenhouse.
30. Supersoil® (Scotts Miracle-Gro Company).

3 Methods

3.1 Agrobacterium-Mediated Transformation of Setaria viridis Embryogenic Callus

3.1.1 Establishment of Embryogenic Callus Cultures from Mature Seeds

1. Collect mature seeds from 3- to 4-month-old *S. viridis* plants cultivated in a growth chamber or greenhouse (*see* **Note 3**) (Fig. 2a).
2. Remove the lemmas and paleas of seeds (*see* **Note 4**) (Fig. 2b).
3. Transfer the dehulled seeds to an Eppendorf (2 mL), close, and transfer to the laminar airflow cabinet (*see* **Note 5**).
4. Add 1 mL of 10% sodium hypochlorite and 1 drop of Tween® 20, close the Eppendorf, and shake vigorously by hand for 3 min.
5. Discard the hypochlorite and rinse five times with sterile distilled water.
6. Remove the water, and transfer the seeds to a sterile Petri dish containing five sheets of sterile Whatman filter paper to dry them (*see* **Note 6**).
7. Transfer 25 seeds to Petri dishes CIM with the embryo side facing upward (*see* **Note** 7) (Fig. 2c).
8. Incubate the plates in the dark at 25 ± 2 °C for 4 to 5 weeks (*see* **Note 8**).
9. After 30–35 days of culture (Fig. 2d), isolate the embryogenic callus (Fig. 2e), and divide into approximately 3 mm in size pieces. Transfer them to Petri dishes containing CIM medium without any antibiotic (Fig. 2f). Incubate at 25 ± 2 °C in the dark (*see* **Note 9**) until the formation of a translucent embryogenic callus approximately 7 days (Fig. 2g).

3.1.2 Agrobacterium Preparation and Callus Infection

1. Streak the inoculum from the glycerol stock. Transfer a single colony of *A. tumefaciens* EHA105 strain transformed with the desired binary vector to a sterile 15 mL conical tube containing 3 mL of YEB medium supplemented with 50 mg/L rifampicin and binary vector selective antibiotic at its appropriate concentration. Incubate at 28 °C on an orbital shaker at 180 rpm (*see* **Note 10**).
2. Transfer 2 mL of the pre-inoculum to a sterile 125 mL Erlenmeyer flask containing 18 mL of YEB medium supplemented with selective antibiotics and 200 μM acetosyringone. Incubate again with shaking at 28 °C in the dark for ~3 h or until OD_{600} 0.6 (maximum) is reached (*see* **Note 11**).
3. Transfer the bacterial culture to sterile 50 mL conical tubes, centrifuge at 2000 × *g* for 15 min at 20 °C, discard the supernatant, and resuspend the pellet in LCIM adjusting the volume to reach OD_{600} 0.6 (*see* **Note 12**).

Fig. 2 Generation of stable transgenic *S. viridis* by *Agrobacterium tumefaciens*-mediated transformation of embryogenic callus. (**a**) Spikes from 3- to 4-month-old *S. viridis* plants. (**b**) Dehulling of seeds using a double layer sandpaper device. (**c**) Seeds in CIM, detail of the seed in the bottom right. (**d**) Callus induction after 4 weeks in CIM. (**e**) Isolated embryogenic callus (bar = 2 mm). (**f**) Fragmented embryogenic callus in CIM. (**g**) Translucent embryogenic callus most suitable for transformation is indicated by arrowheads (bar = 0.5 mm). (**h**) Embryogenic callus agro-inoculated and maintained in CIM for 7 days with 30 mg/L hygromycin B. (**i**) Shoot regeneration in SCRM (bar = 5 mm). (**j**) Regenerated putative transgenic plantlets. (**k**) Plant ready for acclimatization. (**l**) Whole hygromycin-resistant plant maintained in greenhouse

4. Transfer 50–100 embryogenic calli to a sterile 50 mL conical tube, and then add the bacterial culture.
5. Add 10–15 mL of *Agrobacterium* culture with optical density adjusted to 10 μL of Poloxamer 188 solution 10% and 200 μM of acetosyringone per 1 mL of bacterial culture.
6. Mix gently and incubate for 5 min without shaking.
7. Discard the *Agrobacterium* culture, blot dry the callus on sterile Whatman filter paper for 5 min, and transfer the agro-inoculated callus to Petri dishes containing CCM (*see* **Note 13**).
8. Incubate the Petri dishes at 22 °C in the dark for 3 days.
9. Transfer the agro-inoculated callus to CIMT but without selective agent, and incubate at 25 °C for 7 days in the dark (*see* **Note 14**).

3.1.3 Callus Regeneration, Selection, and Plant Recovery

1. After 7 days in the dark, transfer the agro-inoculated callus to SCIM supplemented with 150 mg/L Timentin and selective agent (30 mg/L hygromycin B or 3 mg/L glufosinate-ammonium). Incubate at 25 °C for another 7 days in the dark (Fig. 2h) (*see* **Note 15**).
2. After 7 days in the dark, transfer agro-inoculated callus to SCRM, and incubate at 26 ± 1 °C and with 16/8 h light/dark photoperiod at 100 μmol m^{-2} s^{-1} in a growth chamber (*see* **Note 16**).
3. Subculture after 15 days in fresh SCRM, and keep the Petri dishes under the same conditions (*see* **Note 17**) (Fig. 2i).
4. Transfer the regenerated putative transgenic plants to Magenta™ vessel GA-7 containing selective development medium (SDM) (*see* **Note 18**).
5. Keep the Magenta™ vessel in a growth chamber at 26 ± 1 °C and with 16/8 h light/dark photoperiod at 150 m^{-2} s^{-1} until rooting for acclimatization (Fig. 2j).
6. Acclimate the regenerated plants (Fig. 2k) in soil/Plantmax® substrate/vermiculite mixture (3:1:0.5 ratio), and keep in the growth chamber at 26 ± 2 °C, with 16/8 h light/dark photoperiod at light intensity of at least 500 μmol m^{-2} s^{-1}, and with 65% relative humidity (*see* **Note 19**).
7. Transfer the regenerated plants to greenhouse, proceed with molecular analysis (e.g., PCR, real-time qPCR, and Southern blot), and harvest the seeds after 30–40 days (Fig. 2l).
8. Tissues of positive transgenic plants can be analyzed by histochemical GUS assays or GFP visualization (Fig. 3).

3.1.4 Segregation Analysis

1. Harvest the mature seeds of T_1 progeny from each T_0 generation.
2. Transfer 100 seeds to a 10 mL glass Becker, and add 1 mL of concentrate sulfuric acid (H_2SO_4). Incubate for 15 min at room temperature stirring every 5 min (*see* **Note 20**).
3. Remove the excess of H_2SO_4, and transfer the seeds to a 100 mL glass Becker, and rinse thoroughly with tap water.
4. Transfer the seeds to 2 mL Eppendorf, and proceed to sterilization (*see* **Note 21**).
5. Add 1 mL of 2% sodium hypochlorite and 1 drop of Tween® 20, close the Eppendorf tube, and shake vigorously by hand for 5 min.
6. Discard the hypochlorite and rinse five times with sterile distilled water.

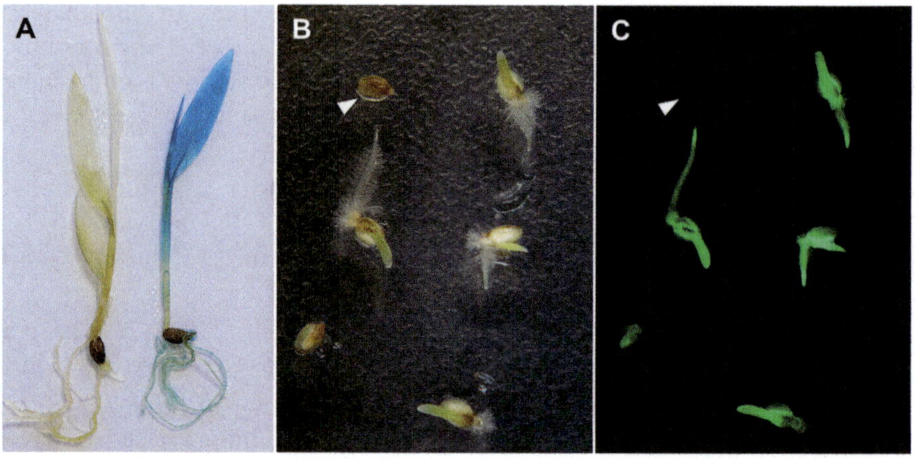

Fig. 3 Histochemical GUS assay and GFP visualization in transgenic *S. viridis*. (**a**) Transgenic plant expressing GUS (right) and non-transgenic (left). (**b** and **c**) Transgenic T_1 seeds observed under visible and UV light, respectively. Arrowheads indicate non-transgenic seed

7. Remove the water, and transfer the seeds to a sterile Petri dish containing five sheets of sterile Whatman filter paper to dry the seeds.
8. Transfer the seeds to 150 mm × 20 mm polystyrene Petri dishes containing solid half-strength Murashige and Skoog basal medium (1/2SMS) (*see* **Note 22**).
9. After 7–10 days, segregation analysis was conducted based on the chi-squared (χ^2) test ($p < 0.05$) in which observed values were compared to theoretical values corresponding to the integration of one or more copies of the transgene.
10. Ten seedlings of each T_1 event with 3:1 segregation are transferred to soil/Plantmax® substrate/vermiculite mixture (3:1:0.5 proportions) and kept in the growth chamber at 26 ± 2 °C, with 16/8 h light/dark photoperiod at light intensity of at least 500 μmol m^{-2} s^{-1}, and with 65% relative humidity until the maturity of the seeds.
11. Seeds of each T_2 progeny are harvested, and the same procedure is conducted to identify the homozygous lines.

3.2 Agrobacterium-Mediated Transformation of Setaria viridis by Floral Dip

1. Sow the seeds of *S. viridis* (accession A10.1) in plastic pots (230 mL) with Supersoil®, and grow in a greenhouse with a 26 °C/22 °C day/night cycle with a 14/10 h photoperiod.
2. *S. viridis* spikes at the boot stage are used for floral dipping (*see* **Note 23**) (Fig. 4a).
3. Pre-inoculum preparation: Streak the inoculum from the glycerol stock. Transfer a single colony of *A. tumefaciens* AGL1 strain transformed with the desired binary vector to a sterile

Fig. 4 *Setaria viridis* floral dip transformation. (**a**) Inflorescence developmental stage (boot stage) selected for transformation. (**b**) *Agrobacterium* culture. (**c**) In vitro tobacco cultures. (**d**) Tobacco leaves cut into approximately 2 cm^2 pieces. (**e**) Tobacco leaf extract preparation. (**f**) Vacuum desiccator system. (**g**) Aerial part of plants at boot stage immersed in *Agrobacterium* suspension. (**h**) Spikes dipped inside a desiccator for vacuum-assisted *Agrobacterium* infiltration. (**i**) Spikes covered with plastic bags after infiltration. (**j**) Co-cultivation for 3 days under greenhouse conditions without irrigation. (**k**) Spike showing RFP-expressing immature seed. (**l**) RFP-expressing mature seeds (*NT* non-transgenic, *T* transgenic)

50 mL conical tube containing 5 mL of LB medium supplemented with 50 mg/L of rifampicin and selective antibiotic at its appropriate concentration (*see* **Note 24**). Incubate at 28 °C on an orbital shaker at 180 rpm.

4. Transfer 5 mL of the pre-inoculum to a sterile 1000 mL Erlenmeyer flask containing 500 mL of LB medium supplemented with selective antibiotics, and incubate again with shaking at 28 °C in the dark until OD_{600} 1.0 (maximum) is reached (Fig. 4b).
5. Transfer the bacterial culture to sterile 50 mL conical tubes, centrifuge at 5000 × *g* for 15 min at 20 °C, discard the supernatant, wash the pellet with 10 mL of MM buffer, and resuspend in 50 mL of fresh MM buffer without tobacco extract.
6. Add the 50 mL of resuspended bacterial culture in 950 mL of MM buffer plus tobacco extract supplemented with 1 mL of 50 mM acetosyringone, 200 μL of Silwet-L77, and 50 g of sucrose (*see* **Note 25**).
7. Inside of a Secador® Techni-dome® 360 vacuum desiccator, immerse the aerial part of plants at boot stage in *Agrobacterium* suspension, and subject to vacuum infiltration (80 kPa) for 10 min (*see* **Note 26**) (Fig. 4f–h).
8. Remove the spikes of desiccator, and cover them with plastic bags for 24 h (Fig. 4i). The co-cultivation step was performed for 3 days under greenhouse conditions without irrigation (Fig. 4j).
9. After 1.5–2 months, screen the seeds for red fluorescence using a Leica MZ16F stereomicroscope equipped with an XCite EXFO fluorescence illumination source and a filter combination for fluorescence in the red region of the spectrum (excitation, 558 nm; emission, 580 nm) (Fig. 4k, l).
10. Harvest the mature seeds, and germinate in solid half-strength Murashige and Skoog basal medium (1/2MS) containing 30 mg/L hygromycin B for selection of transformed seedlings.
11. Proceed to molecular analysis of surviving plants on hygromycin-containing medium.

4 Notes

1. The 3 M™ Sandpaper (180 grit) is cut and glued to the petri dish (60 mm × 15 mm) to peel the seeds.
2. The soil and Plantmax® substrate are firstly sterilized by autoclaving at 121 °C for 20 min. After cooling, they are mixed to vermiculate at 3:1:0.5 ratio.
3. From this stage, the procedures are performed in the laboratory. *S. viridis* mature seeds are responsive to callus induction. However, we recommend initiating tissue culture with seeds at least 3 months old. Young seeds do not produce embryogenic calli.

4. The seeds are gently peeled with sandpaper avoiding injuring the embryo. The integrity of the embryo is necessary for the induction of embryogenic calli.
5. From this stage all the procedures are performed in the laminar airflow cabinet.
6. Use sterile Tweezers by autoclaving or flaming.
7. Do not use high density of seeds per plate. It is very important that the side of the embryo is facing upward for induction of embryogenic callus. Cupric sulfate used in SCIM shows beneficial effects with respect to embryogenic callus induction.
8. At this moment, no subculture is required. The seeds are not removed for fresh medium until 4–5 weeks.
9. The addition of antibiotics may affect the compatible interaction of *Agrobacterium* with embryogenic callus, decreasing gene delivery and agrotransformation efficiency. Confirm that the culture is at the appropriate phase (7–10 days) to transform observing under stereomicroscope.
10. Glycerol stocks of *Agrobacterium*-harboring constructs should be stored at −80 °C. *A. tumefaciens* EHA105 and AGL1 are hypervirulent strains commonly used in monocot transformation. Use a fresh *Agrobacterium* culture obtained from streaked colonies grown on YEB agar medium.
11. The volume ratio of the *A. tumefaciens* pre-inoculum to the volume of liquid YEB medium should be 1:10.
12. Resuspend the pellet gently. Do not use cupric sulfate in the liquid medium because it can interfere with the *Agrobacterium* activities.
13. Blot dry the excess of bacterial culture since this contributes to the overgrowth of *Agrobacterium* in the subsequent culture medium.
14. Do not use high density of callus per plate. Timentin (150 mg/L) has been used successfully immediately after co-cultivation and maintained in all subsequent media to control the growth of *A. tumefaciens*.
15. Hygromycin acts by inhibiting polypeptide synthesis. It stabilizes the tRNA-ribosomal acceptor site, thereby inhibiting translation. Glufosinate-ammonium (phosphinothricin-alanyl-alanine) is an inhibitor of the enzyme glutamine synthase (GS) which plays an essential role in the metabolism of nitrogen by catalyzing the condensation of glutamate and ammonia to form glutamine. Glufosinate-treated plants die due to an accumulation of ammonium ions in the plant tissues under development.

16. Do not use cupric sulfate in SCRM because it can disrupt plant regeneration.
17. Subcultivate after 15 days to increase the efficiency of selection by the selective agent. Discard the necrotic callus that remains in the regenerated calli. At the beginning of callus regeneration, a few escape shoots might appear, but they tend to die with subculture.
18. Transfer four to six plants/Magenta™. Do not use high density of shoots per box.
19. The plantlets should be covered with plastic bags, for 1 week, to maintain moisture and allow a gradual acclimatization.
20. For germination of *S. viridis* mature seeds, removing paleae and lemma with sandpaper is not necessary.
21. The procedure is performed in laminar airflow cabinet.
22. The appropriated selective agent is added in the 1/2 MS medium. In this analysis, non-transformed seedlings became yellow, and roots ceased to elongate because of their sensitiveness, while transformed seedlings remained green and developed elongated roots.
23. At this stage, the efficiency of transformation is higher than other developmental stages.
24. The binary vector used for floral dip is pANIC6A (*see* Fig. 1c). This vector has kanamycin resistance gene as selection marker. The concentration used for selection of transformants is 50 mg/L.
25. Tobacco (*Nicotiana tabacum* cv. SR1) extract is known to improve the gene transfer by *Agrobacterium* in monocots [16].
26. At the end of 10 min, the vacuum should be released very slowly.

References

1. Diao X, Schnable J, Bennetzen JL (2014) Initiation of *Setaria* as a model plant. Front Agr Sci Eng 1:16–20. https://doi.org/10.15302/j-fase-2014011
2. Muthamilarasan M, Prasad M (2015) Advances in *Setaria* genomics for genetic improvement of cereals and bioenergy grasses. Theor Appl Genet 128:1–14. https://doi.org/10.1007/s00122-014-2399-3
3. Bennetzen JL, Schmutz J, Wang H, Percifield R, Hawkins J, Pontaroli AC et al (2012) Reference genome sequence of the model plant *Setaria*. Nat Biotechnol 30:555–561. https://doi.org/10.1038/nbt.2196
4. Martins PK, Ribeiro AP, da Cunha BADB, Kobayashi AK, Molinari HBC (2015) A simple and highly efficient *Agrobacterium*-mediated transformation protocol for *Setaria viridis*. Biotechnol Rep 6:41–44. https://doi.org/10.1016/j.btre.2015.02.002
5. Martins PK, Nakayama TJ, Ribeiro AP, da Cunha BADB, Nepomuceno AL, Harmon FG, Kobayashi AK, Molinari HBC (2015) *Setaria viridis* floral-dip: a simple and rapid *Agrobacterium*-mediated transformation method. Biotechnol Rep 6:61–63. https://doi.org/10.1016/j.btre.2015.02.006
6. Ribeiro AP, de Souza WR, Martins PK, Vinecky F, Duarte KE, Basso MF et al (2017)

Overexpression of *BdMATE* gene improves aluminum tolerance in *Setaria viridis*. Front Plant Sci 8:865. https://doi.org/10.3389/fpls.2017.00865

7. Vasconcelos VDB (2014) Avaliação funcional do promotor do gene *anatrC* de *Aspergillus nidulans* na planta modelo *Setaria viridis*. Dissertation, Federal University of Lavras
8. Duarte KE (2014) Análise funcional do gene órfão *CcUNK8* de *Coffea canephora* via transformação genética de *Setaria viridis*. Dissertation, Federal University of Lavras
9. Van Eck J, Swartwood K (2015) Setaria viridis. In: Wang K (ed) Methods in molecular biology, *Agrobacterium* protocols, vol 1. Springer Science + Business Media, New York, pp 57–67
10. Zale JM, Agarwal S, Loar S, Steber CM (2009) Evidence for stable transformation of wheat by floral dip in *Agrobacterium tumefaciens*. Plant Cell Rep 28:903–913. https://doi.org/10.1007/s00299-009-0696-0
11. Liu W, Zheng MY, Polle EA, Konzak CF (2002) Highly efficient doubled-haploid production in wheat (*Triticum aestivum* L.) via induced microspore embryogenesis. Crop Sci 42(3):686–692
12. Rod-in W, Sujipuli K, Ratanasut K (2014) The floral-dip method for rice (*Oryza sativa*) transformation. J Agric Technol 10(2):467–474
13. Christensen AH, Quail PH (1996) Ubiquitin promoter-based vectors for high-level expression of selectable and/or screenable marker genes in monocotyledonous plants. Transgenic Res 5:213–218. https://doi.org/10.1007/BF01969712
14. Murashige T, Skoog F (1962) A revised medium for rapid growth and bio assays with tobacco tissue cultures. Physiol Plantarum 15:473–497
15. Mann DGJ, LaFayette PR, Abercrombie LL, King ZR, Mazarei M, Halter MC et al (2012) Gateway-compatible vectors for high-throughput gene functional analysis in switchgrass (*Panicum virgatum* L.) and other monocot species. Plant Biotechnol J 10:226–236. https://doi.org/10.1111/j.1467-7652.2011.00658.x
16. Fursova O, Pogorelko G, Zabotina OA (2012) An efficient method for transient gene expression in monocots applied to modify the *Brachypodium distachyon* cell wall. Ann Bot 110:47–56. https://doi.org/10.1093/aob/mcs103

Chapter 5

Transient Transformation Using Particle Bombardment for Gene Expression Analysis

Andika Gunadi, Eric A. Dean, and John J. Finer

Abstract

Transient transformation or transient expression results in rapid and fleeting gene expression. This approach is often used as a first-tier screening tool for evaluation of components that affect gene expression. Here, we describe the use of particle bombardment of lima bean cotyledons with constructs containing the *green fluorescent protein* (*gfp*) coding region for evaluation of promoter components that influence gene expression. Although this approach is conceptually quite simple, this lima bean transient expression system may not work well, if our methods and notes are not carefully read and followed. Our laboratory has successfully optimized this method over the past 10 years, resulting in a transient expression system, which works like no other that we have seen.

Key words Promoter analysis, Transient expression, Gene introduction, Green fluorescent protein, Lima bean cotyledon

1 Introduction

Although the ultimate goal of plant transformation is the recovery of a transgenic plant from a single transformed plant cell, the process of plant formation from that single cell requires a minimum of 4–6 weeks for direct regeneration from even the most responsive cells and tissues [1, 2]. Unfortunately, 3–6 months or longer may be necessary for recovery of some transgenic plants [3]. To deal with this often lengthy time requirement, alternate approaches have been developed for many applications using transient transformation or transient expression. Transient transformation, by definition, results in temporary alterations in gene expression, and this approach is often useful, as rapid analysis (hours or days) of transient gene expression can provide valuable information on the functionalities of the coding region or gene components that affect the timing or intensity of expression. Transient transformation approaches include agroinfiltration [4], VIGS [5], as well as direct DNA introduction using protoplast electroporation [6] and

Sandeep Kumar et al. (eds.), *Transgenic Plants: Methods and Protocols*, Methods in Molecular Biology, vol. 1864, https://doi.org/10.1007/978-1-4939-8778-8_5,

particle bombardment [7]. For analysis of gene regulatory components like promoters, introns, or terminators, transient transformation using particle bombardment of the appropriate target tissue has provided a wealth of extremely rapid and reliable information on specific sequences that impact the intensity of gene expression [8]. Although this approach does have limitations, analysis takes days, instead of months or years. Transient transformation has therefore often been used to rapidly analyze large numbers of DNA constructs as a first-tier screen [9], prior to selecting constructs or genes for stable introduction and plant recovery.

In this chapter, we describe in detail the method that we have developed for introduction of large numbers of promoter constructs into lima bean (*Phaseolus lunatus*) cotyledons using particle bombardment. Although this approach is conceptually quite simple, the target tissue and bombardment conditions have been thoroughly optimized, yielding results that are entirely consistent and amazingly efficient. Lima bean cotyledonary tissue was selected from among hundreds of different potential targets because it provides a large, flat, homogeneous target for evaluation of reporter gene expression. The cv 'Henderson Bush' was selected due to the large seed size and consistency of expression. Large numbers of seed can easily be obtained from commercial suppliers or from plants grown and selfed in the greenhouse or growth chamber. Plants will produce flushes of fresh seed, if maintained under low pathogen stress and adequate moisture/fertilizer conditions. Finally, even though this target has been specifically selected for this approach, seeds must be precisely imbibed, visually selected, and partially dried to provide the best target. Although any good particle bombardment device is suitable for DNA introduction, we will present on the use of a simple, easy to assemble Particle Inflow Gun [10], including the use of baffles for highly efficient introduction of DNAs into lima bean cotyledons.

2 Materials

Prepare cotyledon tissue and materials prior to bombardment. Sterilize autoclavable materials for 15–20 min at approximately 121 °C and 15 PSI. Dry microcentrifuge tubes prior to use. Use a laminar flow hood to prevent contamination of samples.

2.1 Bombardment Apparatus

1. Particle Inflow Gun (PIG) [10].
2. Vacuum pump: Connect the pump to the vacuum chamber of the PIG using vacuum tubing, able to withstand more than 30 in. Hg vacuum (*see* **Note 1**).

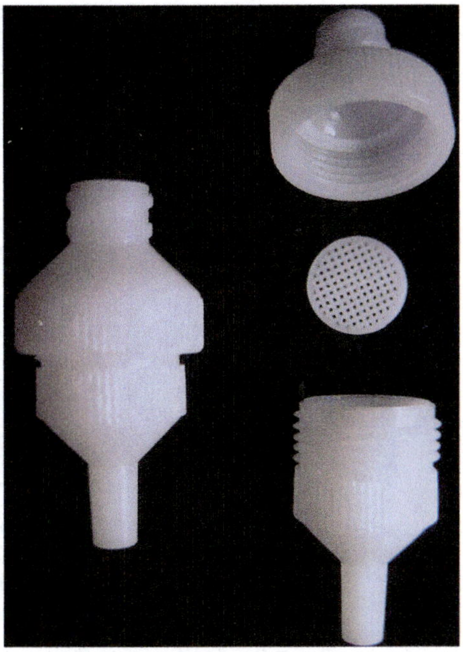

Fig. 1 Assembled syringe filter unit (left) and disassembled components (right) showing the screen where the particles are placed

3. Compressed helium: Connect pressurized helium from a helium tank to the solenoid on the PIG using flexible copper tubing, capable of tolerating 150 PSI.
4. Syringe filters (Fig. 1, *see* **Note 2**).
5. Bombardment baffles (Fig. 2, *see* **Note 3**).
6. Dedicated pipettor and sterile pipette tips.

2.2 Cotyledon Tissue Preparation

1. Lima bean seeds (*Phaseolus lunatus* cv. 'Henderson Bush'): Store seeds at −20 °C prior to use (*see* **Note 4**).
2. Bleach (8.25% sodium hypochlorite).
3. 50 mL plastic centrifuge tubes.
4. Magenta™ GA-7 culture containers containing folded papers towels, moistened with 15–20 mL of water and autoclaved for 1 h (*see* **Note 5**).
5. Sterilized forceps and scalpel with sharp blade.
6. Empty sterile 100 mm × 25 mm Petri dishes.
7. MS (Murashige and Skoog) medium [11] with 0.2% Gelrite™ in 100 mm × 25 mm Petri dishes (35 mL per dish).

2.3 Bombardment Mixture

1. Purified pFLEV (Finer Laboratory Expression Vector) regulating *gfp* or another suitable scorable marker gene at ~1 μg/μL [8] (*see* **Note 6**).

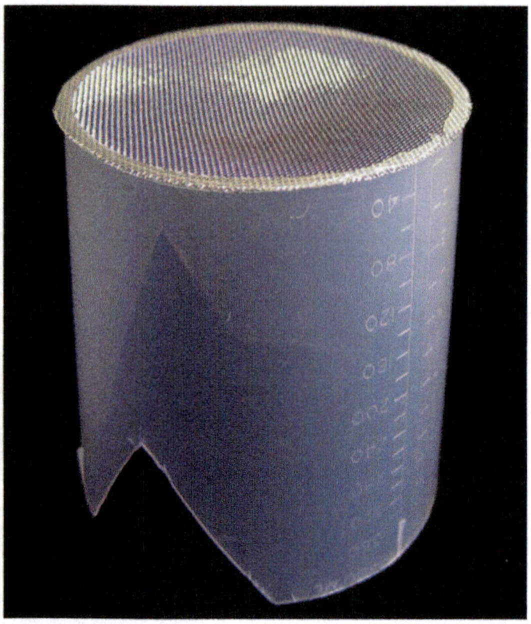

Fig. 2 Bombardment baffle consisting of a stainless steel mesh melted to the bottom of a polypropylene beaker, with inverted "V" cuts for venting

2. 1 M spermidine solution: Thaw spermidine (≥99.0% purity, stored in desiccator at −20 °C) to liquid form by brief immersion in 60 °C water bath. Dilute 156 μL solution into 844 μL of sterile water, mix thoroughly, and filter through sterile 0.2 μm cellulose acetate syringe filter to yield 1 M sterile spermidine solution. Aliquot 10 μL of the filtered solution into autoclave-sterilized 0.6 mL microcentrifuge tubes, and store at −20 °C.
3. 2.5 M calcium chloride solution: Dissolve 18.44 g of calcium chloride dihydrate (≥99.0% purity, stored at room temperature) in water for a total volume of 50 mL, and autoclave for 15 min. Aliquot 125 μL into autoclave-sterilized 0.6 mL microcentrifuge tubes, and store at room temperature.
4. Sterilized M10 tungsten particles (*see* **Notes** 7 and **8**): Aliquot 50 mg of M10 tungsten particles per autoclaved 1.5 mL microcentrifuge tubes. For each tube, add 1 mL of 95% ethanol, cap the tube, vigorously vortex the mixture, and let stand for 20 min. Centrifuge the mixture for 15 min at 3000 rpm in a microcentrifuge (~850 × *g*), and remove the supernatant. Place the ethanol-sterilized pelleted tungsten in opened microcentrifuge tubes in the PIG chamber that is connected to a vacuum pump, and bring the chamber to a vacuum of greater than 30 in. Hg vacuum, or vacuum dry using another method for 15 min. Cap the microcentrifuge tubes, and store the tungsten particles at room temperature.

3 Methods

3.1 Pre-bombardment Tissue Preparation and Culture: Four Days Prior to Gene Delivery

1. Select high-quality dry lima bean seeds for germination (Fig. 3, *see* **Note 4**).
2. Immerse seeds in 5% bleach solution with agitation for 20 min inside a 50 mL centrifuge tube.
3. Rinse seeds 10 times with sterile water, until the odor of the bleach is gone.
4. Place five sterilized lima bean seeds in between layers of moistened paper towel in each sterilized GA-7 container (Fig. 4). Damaged or excessively wrinkled seeds should not be used.
5. After 4 days, excise the cotyledons, and discard the hypocotyl and roots using scalpel and forceps. Remove and discard seed coats, being careful not to damage the cotyledons.
6. Remove the embryo axis and first true leaf by cutting through the base of the cotyledons, above the cotyledonary node (Fig. 5).
7. Place cotyledons, adaxial (or flat, interior) side facing upward, in an empty Petri dish being cautious to return the lid to the dish after placement of each cotyledon.
8. After all cotyledons have been prepared, remove lid so that cotyledons may dry for approximately 15 min, or until the adaxial surfaces have dried thoroughly (*see* **Note 9**).
9. For subsequent particle bombardment, select cotyledons that are uniform in size and flatness and light green in color at the adaxial side (Fig. 6, *see* **Note 10**). Cover the Petri dish containing the excised cotyledons while preparing the bombardment mixture.

Fig. 3 Sorted lima bean seeds. Uniform seeds of acceptable quality to be used for germination and bombardment (left) and unacceptable damaged or misshapen seeds that should be discarded (right)

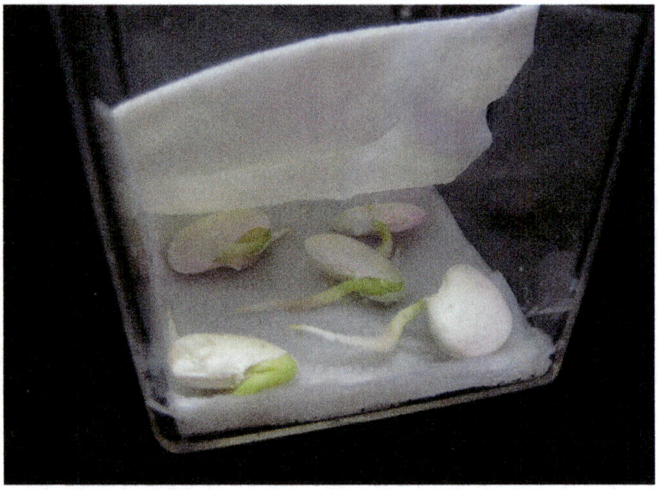

Fig. 4 Germinating lima bean seeds 3 days after placement into GA-7 culture container with moistened paper towels

Fig. 5 Excision of embryo axis from cotyledons

3.2 Pre-bombardment Materials and DNA Preparation

1. Autoclave syringe filters and baffles for 1 h, and ensure these materials are dry and ready for use.
2. Adjust the pFLEV plasmid concentration to ~1 μg/μL (*see* **Note 11**).
3. Place aliquots of sterile water, the pFLEV plasmid, 2.5 M calcium chloride, 1 M spermidine, and sterilized M10 tungsten particles on ice.
4. Add 90 μL of sterile water into microcentrifuge tube containing 10 μL of spermidine (to make a 100 μM solution), and mix well.
5. Add 500 μL of sterile water to the microcentrifuge tube containing the M10 tungsten particles, and thoroughly mix by

Fig. 6 High-quality lima bean cotyledons prior to bombardment

vortexing. While the tungsten particles are still suspended, aliquot 25 μL of the tungsten mixture into 0.6 mL microcentrifuge tubes (one for every bombardment mixture).

6. For each bombardment mixture, add 5 μL of plasmid DNA into the tungsten mixture, followed by addition of 25 μL of 2.5 M calcium chloride and 10 μL of 100 μM spermidine into the mixture. Work quickly and mix after each addition. After all additions, mix thoroughly by vortexing, and place on ice for 5 min for DNA to precipitate onto tungsten particles.
7. Tungsten particles should accumulate at the sides and bottom of the microcentrifuge tube. Tap the bottom of the tube lightly on the hood surface to move the particles to the bottom of the tube—do not centrifuge. Remove 50 μL of the supernatant using a pipette, taking care not to disturb the precipitated tungsten particles. The bombardment mixture is now ready for particle bombardment (*see* **Notes 12** and **13**).
8. In an experiment requiring more than one bombardment mixture, repeat **steps 6** and 7 to generate more mixtures, once the previous mixture has been used during the bombardment procedure.

3.3 Bombardment Procedure

1. Cotyledons may remain in the empty dish with the lid on until bombardment.
2. Turn on vacuum pump, making sure the valve connecting the pump to the vacuum chamber is closed, so that the pump will not be continuously removing air.
3. Adjust the incoming helium pressure to the PIG to 50 PSI.
4. Mix the concentrated DNA-coated tungsten particles thoroughly by vortexing either with a mechanical vortexer or

Fig. 7 Droplet of 2 μL mixture containing DNA-coated tungsten particles on the syringe filter screen

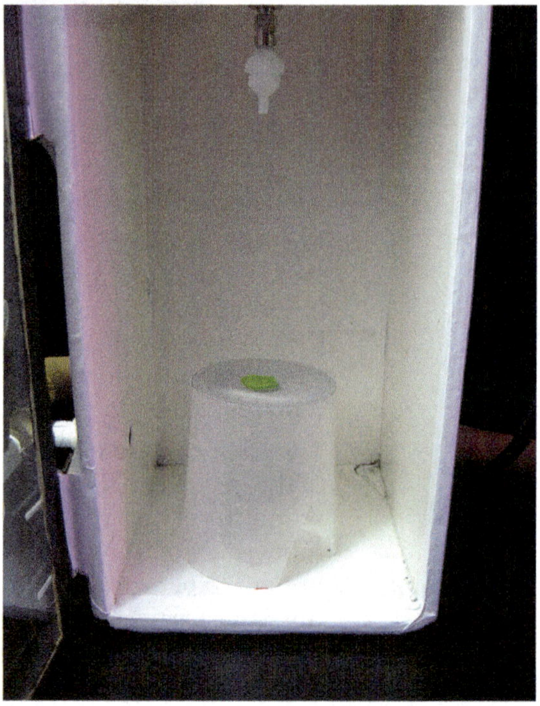

Fig. 8 Cotyledon on the screen of the baffle, inside the vacuum chamber prior to bombardment

strong finger vortex or washboarding the tube along a microfuge tube rack. Quickly tap the bottom of the tube a few times on the hood surface, and immediately place 2 μL of the mixture on the center of the screen in the opened syringe filter (Fig. 7). Reassemble the filter and fasten to the syringe adaptor in the PIG (*see* **Note 14**).

5. Place the cotyledon on the top of the bombardment baffle, adaxial side facing upward, approximately 14 cm below the syringe filter (Fig. 8).

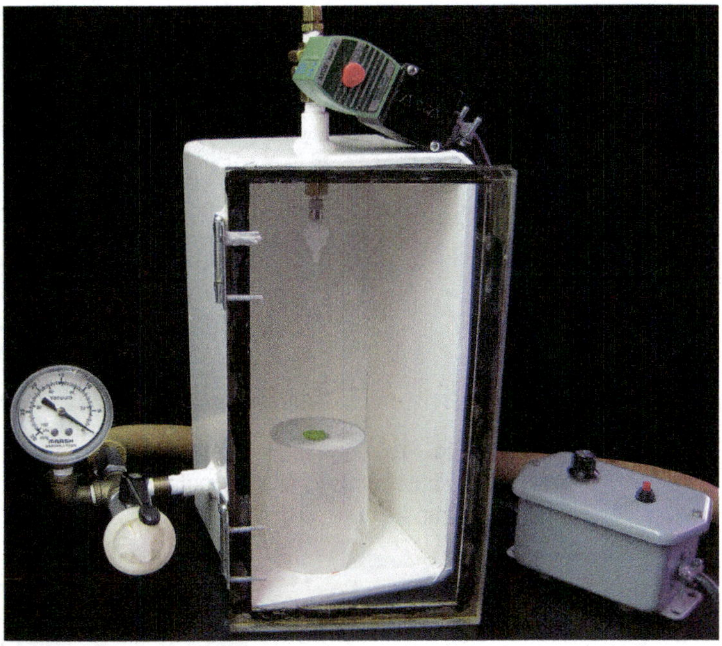

Fig. 9 PIG containing lima bean cotyledon on the screen of the baffle prior to bombardment, showing the vacuum gauge (left of the chamber), solenoid connecting the helium line and the chamber (top of the chamber), and the timer relay switch (right of the chamber)

6. Seal the vacuum chamber (Fig. 9). Once the pressure gauge reaches 30 in. Hg, briefly tap the switch for the timer relay to release the helium through the solenoid, carrying the particles to the target tissue (*see* **Note 15**).
7. Release the vacuum inside the bombardment chamber by closing the vacuum valve and slowly opening the ventilation valve.

3.4 Post-bombardment

1. Place bombarded cotyledons with the adaxial side up on MS medium in 100 mm × 25 mm Petri dishes immediately following bombardment (*see* **Note 16**).
2. Once all cotyledons have been bombarded, remove the Petri dish lid and dry cotyledons for an additional 15 min in the laminar flow hood until adaxial surfaces are visibly dry (*see* **Note 17**).
3. Cover the Petri dish with the lid, and seal the cotyledons inside the Petri dish with Parafilm® or plastic wrap (Fig. 10, *see* **Note 18**).
4. Transient *gfp* expression should be observable within 3 h and peaks at approximately 24 h (Fig. 11, *see* **Note 19**).

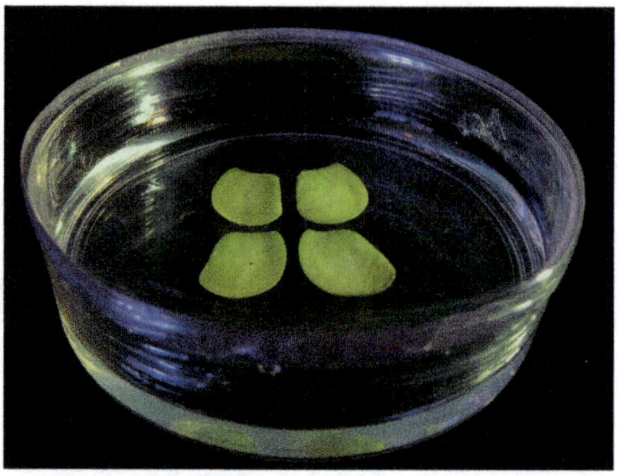

Fig. 10 Cotyledons post-bombardment on MS medium. The condensation reducing thick polycarbonate lid is attached to the dish bottom using plastic wrap

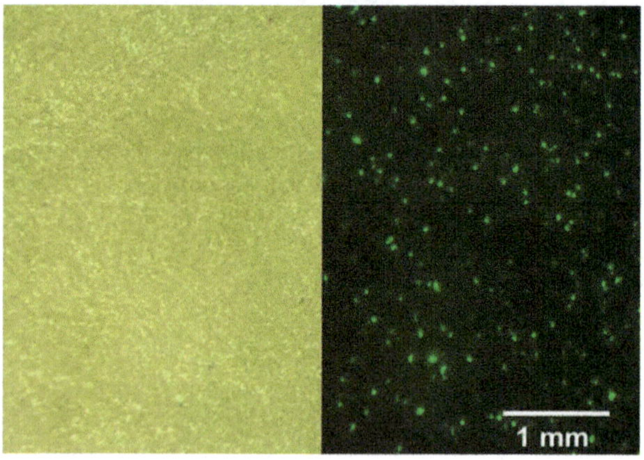

Fig. 11 The cotyledon adaxial surface under white light (left) and blue light (right) showing large numbers of GFP foci, 24 h after bombardment with pFLEV containing *gfp* regulated by a strong promoter

4 Notes

1. We use a 1/3 horsepower vacuum pump to evacuate the PIG chamber, prior to acceleration of the particles. A building vacuum line or a 1/10 horsepower vacuum pump will not provide adequate vacuum.
2. The syringe filter assembly (Fig. 1) that we use (13 mm plastic Swinney filter holder from PALL® #4317) provides a consistent spread of particles. The 2 μL droplet of DNA-coated tungsten particles is placed on the filter holder screen (Fig. 7). The syringe filter holders are autoclavable and reusable but do have a limited life span.

3. The bombardment baffles (Fig. 2) are made from 400 mL polypropylene Tri-Pour® beakers with the top and bottom cutoff. Stainless steel mesh (~500 μm opening) is melted to the bottom of the beaker by heating on a hot plate, covered with aluminum foil. Inverted "V"s are cut in the beaker (Fig. 2) to allow helium to escape outward instead of upward toward the sample during bombardment.
4. Perhaps the greatest consistency in results comes from the use of appropriate target tissue. Lima bean seeds should be uniform in size and color, and the seed coat should be intact and of uniform surface texture (Fig. 3). Selection of high-quality dry seeds will lead to a consistent and high seed germination. Lima bean seeds are screened prior to (as described here) ***and*** again after germination (Fig. 6, *see* **Note 10**) to ensure the highest quality target tissue.
5. Paper towels should be damp, but not too wet. Ideally, minimal water should run out of the GA-7 culture container if inverted. For our purposes, 15 mL of water per culture container is optimal, but this depends on the absorbency of the paper towels.
6. pFLEV stands for Finer Laboratory Expression Vector (Genbank accession no. KX156843.1), but this can be substituted by any suitable vector.
7. M10 tungsten particles (0.6–0.9 μm) have given extremely consistent results with our transient expression analysis, but DNA-coated particles need to be used quickly as the tungsten may interact with the DNA. Gold particles of similar size range (0.5–0.8 μm) may be used in place of tungsten and may be less reactive with the DNA. This protocol has been optimized for use with tungsten particles, and gold particles do not work as efficiently as tungsten particles when using the methods described here. Economical gold particles may be purchased from Alfa Aesar (Tewksbury, MA, USA). Tungsten particles may be purchased from Bio-Rad (Hercules, CA, USA).
8. Whether using tungsten or gold, particle size and helium pressure affect the depth of particle penetration into the tissue. Higher pressures and larger particles are more likely to penetrate into the subepidermal parenchyma cell layers of the cotyledon tissue.
9. Partial drying of target tissue prior to bombardment is absolutely critical for optimum transient expression by minimizing cell rupture and loss of cytoplasm upon particle impact [7]. If the tissue is not adequately dried, efficiency of gene expression declines considerably. If the tissue is overdried, that will lead to damage or death of the target cells.

10. Flatness of the cotyledon surface is very important for efficient gene introduction and gene expression documentation. Cotyledons may curl slightly after desiccation; however, some degree of flatness is required to provide an even surface of target cells. Cotyledons should be homogenous light green in color, rather than dark green or yellow (Fig. 6). Cotyledons that are mottled in color following germination should be avoided.
11. The pFLEV expression vector is approximately 3.7 kb, and the promoters inserted into pFLEV are approximately 1.5–2.5 kb; the total size of the plasmid we use for bombardment is approximately 5.2–6.2 kb.
12. Conduct **steps 6** and 7 as quickly as possible, as tungsten particles may clump over time and be difficult to resuspend. This is problematic if the plasmid DNA preparation is contaminated with large chromosomal DNA. In addition, the DNA may interact negatively with the tungsten.
13. We use a single bombardment mixture to bombard up to three cotyledons (replicates), producing relatively consistent *gfp* expression between the replicates.
14. The droplet can be very accurately placed on the center of the screen of the syringe filter unit when the top is removed (Fig. 7). It is faster (but more difficult) to leave the top on, and pipette a 2 μL aliquot on the screen directly through the opening in the top of the syringe filter unit.
15. The timer relay is set to the lowest duration of 50 milliseconds, allowing the release of ~400 mL of helium at ~50 PSI.
16. Excess moisture in the Petri dishes containing MS medium can lead to condensation on the Petri dish lid, which can adversely affect observation of GFP. Prior to bombardment, ensure that Petri dishes do not contain condensation or excess liquid on the medium surface. Overnight "drying" of covered plates in a laminar flow hood is sufficient for resolving this issue, but care must be taken to avoid overdrying.
17. Cotyledons retaining moisture on the adaxial surface will have lower levels of transient expression.
18. Replacement of the standard Petri dish lid with thick polycarbonate disk [12] is useful for automated image collection and image analysis. For observation of GFP over time using automated image collection, Petri dishes can be covered using 6-mm-thick disks of Makrolon® polycarbonate, wrapped with five layers of Parafilm® or plastic wrap (Fig. 10). We prefer Glad® wrap, which can be easily cut into 1 in. rolls using a ratcheting PVC pipe cutter.

19. This procedure was developed for promoter analysis and tracking *gfp* expression over time in the same tissue using automated image collection. Lima bean cotyledons constitute an idea target tissue because the adaxial surface is flat, providing a large surface area where all of the *gfp*-expressing cells are in the same focal plane.

Acknowledgments

Salaries and research support were provided by the United Soybean Board and by state and federal funds appropriated to the Ohio State University/Ohio Agricultural Research and Development Center. Mention of trademark or proprietary products does not constitute a guarantee or warranty of the product and also does not imply approval to the exclusion of other products that may also be suitable.

References

1. Takebe I, Labib G, Melchers G (1971) Regeneration of whole plants from isolated mesophyll protoplasts of tobacco. Naturwissenschaften 58:318–320
2. Horsch RB, Fry JE, Hoffmann NL, Eichholtz D, Rogers SG, Fraley RT (1985) A simple and general method for transferring genes into plants. Science 227:1229–1231
3. Finer JJ (2016) Generation of transgenic soybean via particle bombardment of embryogenic cultures. Curr Protoc Plant Biol 1:592–603
4. Yang Y, Li R, Qi M (2000) *In vivo* analysis of plant promoters and transcription factors by agroinfiltration of tobacco leaves. Plant J 22:543–551
5. Ruiz MT, Voinnet O, Baulcombe DC (1998) Initiation and maintenance of virus-induced gene silencing. Plant Cell 10:937–946
6. Fromm M, Taylor LP, Walbot V (1985) Expression of genes transferred into monocot and dicot plant cells by electroporation. Proc Natl Acad Sci U S A 82:5824–5828
7. Vain P, McMullen MD, Finer JJ (1993) Osmotic treatment enhances particle bombardment-mediated transient and stable transformation of maize. Plant Cell Rep 12:84–88
8. Hernandez-Garcia CM, Bouchard RA, Rushton PJ, Jones ML, Chen X, Timko MP, Finer JJ (2010) High level transgenic expression of soybean (*Glycine max*) GmERF and Gmubi gene promoters isolated by a novel promoter analysis pipeline. BMC Plant Biol 10:237
9. Gunadi A, Rushton PJ, McHale LK, Gutek AH, Finer JJ (2016) Characterization of 40 soybean (*Glycine max*) promoters, isolated from across 5 thematic gene groups. Plant Cell Tissue Organ Cult 127:145–160
10. Finer JJ, Vain P, Jones MW, McMullen MD (1992) Development of the particle inflow gun for DNA delivery into plant cells. Plant Cell Rep 11:323–328
11. Murashige T, Skoog F (1962) A revised medium for rapid growth and bio assays with tobacco tissue cultures. Physiol Plant 15:473–497
12. Finer JE, Finer JJ (2007) A simple method for reducing moisture condensation on Petri dish lids. Plant Cell Tissue Organ Cult 91:299–304

Chapter 6

Maize Transformation Using the Morphogenic Genes *Baby Boom* and *Wuschel2*

Todd Jones, Keith Lowe, George Hoerster, Ajith Anand, Emily Wu, Ning Wang, Maren Arling, Brian Lenderts, and William Gordon-Kamm

Abstract

Despite the fact that maize transformation has been available for over 25 years, the technology has remained too specialized, labor-intensive, and inefficient to be useful for the majority of academic labs. Compounding this problem, future demands in maize genome engineering will likely require a step change beyond what researchers view as "traditional" maize transformation methods. Recently, we published on our use of constitutively expressed morphogenic transcription factors Baby Boom (*Bbm*) and Wuschel2 (*Wus2*) to improve maize transformation, which requires CRE-mediated excision before regeneration of healthy, fertile T0 plants. Moving beyond this first-generation system, we have developed a new expression system for *Bbm* and *Wus2*, using a non-constitutive maize phospholipid transferase protein promoter ($Pltp_{pro}$) driving *Bbm* expression and a maize auxin-inducible promoter ($Axig1_{pro}$) for WUS2 expression. Using this combination of expression cassettes, abundant somatic embryos rapidly form on the scutella of *Agrobacterium*-transformed zygotic immature embryos. These somatic embryos are uniformly transformed and can be directly germinated into plants without a callus phase. Transformed plants are sent to the greenhouse in as little as 1 month, and these T0 plants match the seed-derived phenotype for the inbred and are fertile. T1 seeds germinate normally and have a uniformly wild-type inbred phenotype. This new system represents a rapid, user-friendly transformation process that can potentially facilitate high-throughput production of transgenic T0 plants in B73, Mo17, and the recently developed Fast-Flowering Mini-Maize.

Key words Maize, Transformation, Bbm, Wus2, $PLTP_{pro}$, $Axig1_{pro}$, Morphogenic genes

1 Introduction

Since the first report in 1996 that maize immature embryos were a suitable target for direct transformation, followed by production of transgenic callus and ultimately regeneration of fertile plants [1], little has changed in this basic system. Typically, efficient transformation methods still require explants such as immature embryos obtained from greenhouse-grown material and after transformation at least 2 months of callus selection before producing plants, making maize transformation impractical for most academic labs. Additionally, transformation, callus growth, and plant regeneration

Sandeep Kumar et al. (eds.), *Transgenic Plants: Methods and Protocols*, Methods in Molecular Biology, vol. 1864, https://doi.org/10.1007/978-1-4939-8778-8_6,

frequencies vary greatly among different inbreds and genotypes, often limiting the scope of experiments to less commercially relevant transformation-competent genotypes.

We have recently described a transformation method using slices of the embryo axis from mature maize seed as a starting explant for *Agrobacterium*-mediated transformation. This is facilitated by the maize transcription factors *Bbm* and *Wus2* (morphogenic genes) which act synergistically to stimulate embryogenic growth [2]. In this case, cells are directly induced to form calli by the transgenes rather than transforming cells being induced to divide by conventional hormonal manipulation. In these studies, it became apparent that regenerable embryogenic callus is readily recovered after transformation of immature embryos, seedling-derived leaf segments, and from leaf primordia in mature seed-derived embryos, and that expression of *Bbm* and *Wus2* is beneficial to maintaining an embryogenic callus growth pattern. However, because constitutive promoters were used to express the morphogenic genes in this first-generation system, CRE-mediated excision of *Bbm* and *Wus2* was required in order to produce healthy, fertile plants. Based on these observations, we began searching for promoters that expressed only in desired tissues (young leaves, embryos, and callus) but were not expressed in tissues where we observed detrimental effects of ectopic morphogenic gene expression (roots, meristems, and reproductive tissues). As a result of this search, identified new promoter that met our criteria (from a maize phospholipid transferase gene, $Pltp_{pro}$) to drive expression of Bbm [3]. Used in combination with a maize auxin-inducible promoter $Axig1_{pro}$ [4] to drive *Wus2* expression, transformation of immature embryos using the combination of $Pltp_{pro}$::*Bbm* and $Axig1_{pro}$::*Wus2* now results in the direct and rapid formation of single cell-derived somatic embryos from the scutellar epithelial layer. These embryos can be germinated directly into fertile plants without excision of the morphogenic genes, eliminating the conventional 2- to 3-month-long callus proliferation stage of the tissue culture process. This process is very rapid and transition stage somatic embryos are visible just 5 days after transformation growing at approximately the same rate as their zygotic counterparts. This callus-free transformation process appeared to be genotype independent (working in all pioneer and public inbreds tested to date including B73, FFMM and Mo17), efficient, and fast, producing T0 plants in roughly half the time of conventional methods.

In addition to transformation of immature embryos from greenhouse-derived immature ears, we have continued to extend *Bbm* and *Wus2 Agrobacterium*-mediated transformation into new target explants. These include mature seed-derived embryo sections and leaf segments from seed-derived seedlings. These new explants can rapidly be used to generate a brief callus phase followed by rapid regeneration of healthy, fertile plants. These new results are finally producing the first tangible evidence that immature embryos can be replaced as the starting material for maize transformation and that genotype barriers will continue to be pushed back.

2 Materials

2.1 Plant Material

1. Seed of maize inbreds B73, Mo17, and the A and B parent lines of FFMM (Fast-Flowering Mini-Maize, see [5]) is obtained through the USDA-GRIN website.

2.2 Sterilization of Seeds and Ears

1. 20% commercial bleach (1.6% sodium hypochlorite) with 0.1% Tween 20.
2. Double-distilled sterile water for rinsing.
3. Straight, sturdy forceps (or screwdriver).
4. Bucket, pitcher, or beaker.
5. Sterile scalpels, tweezers, and spatulas.
6. Vacuum desiccator and variable speed shaker platform.

2.3 Agrobacterium Strain and Vector

1. *Agrobacterium* strain LBA4404 THY- (a thymidine-auxotrophic mutant) containing virulence plasmid pVIR9 (*see* **Note 1**) plus the T-DNA-containing plasmid PHP79066 (in which the T-DNA contains $Axig1_{pro}$::*Wus2* + $Pltp_{pro}$::*Bbm* + *Sb-Als* $_{pro}$::*Zm-Hra* + *Hv-Ltp2* $_{pro}$::ZS-YELLOW1 N1, *see* **Note 2**).
2. Solid YP culture medium (5g/L sodium chloride, 5g/L yeast extract, 10g/L peptone, 50mg/L spectinomycin, 50mg/L thymidine and 50mg/L gentamicin), pH 6.8.

2.4 Pollination

1. Shoot bag—white paper bag 6¾″ × 2½″ (Canvasback#S26 usually made of parchment), also comes in other sizes, use appropriate size.
2. Tassel bag—plain brown bag (Canvasback #T514) or green striped bags for transgenic plants (Canvasback Cat. #T514G), approximate size 5¾″ × 14″ × 4″.
3. KleenGrow 5%.
4. Paper clips or stapler.
5. Scissors.

2.5 Media

For the media components, refer to Table 1. Media components listed are on a per liter basis of deionized H_2O.

Table 1
Media formulations used for maize transformation and tissue culture

Medium name	Use	Medium base	Ingredients
700A	Infection	MS	68.5 g/L sucrose, 36 g/L glucose, 1.5 mg/L 2,4-D, 50 mg/L thymidine, pH 5.8
700F	Leaf infection		10 mM magnesium sulfate, 5 g/L maltose, pH 5.6
710I	Co-cultivation	MS	20 g/L sucrose, 10 g/L glucose, 2.0 mg/L 2,4-D, 100 um acetosyringone, 10 mg/L ascorbic acid, 50 mg/L thymidine, 8 g/L agar, 0.5 g/L MES buffer, pH 5.8
562V	Co-cultivation	N6	3% sucrose, 2 mg/L 2,4-D, 1 mg/L silver nitrate, 100 uM acetosyringone, 50 mg/L thymidine, 8 g/L agar, pH 5.8
605T	Somatic embryo development	MS + N6	MS salts, 0.6X N6 macro salts, 1.68g/L potassium nitrate, 0.6X B5 minor salts, 0.4X Eriksson's vitamins, 0.6X S&H vitamins, 0.2mg/L thiamine HCl, 0.3g/L casein hydrolysate, 20g/L sucrose, 2g/L proline, 0.6g/L glucose, 0.8mg/L 2,4-D, 1.2mg/L dicamba, 100 mg/L Cefotaxime, 150 mg/L Timentin, 8 g/L agar, pH 5.8
13329A	Shoot formation	MS	60 g/L sucrose, 0.5 mg/L zeatin, 0.1 mg/L thidiazuron, 1 mg/L BAP, 100 mg/L carbenicillin, 0.05 to 0.1 mg/L imazapyr, 8 g/L agar, pH 5.8
13158	Seed germination	MS	40 g/L sucrose, hormone free, with 100 mg/L benomyl, pH 5.8

Media are made by adding salt/mineral mixes (MS or N6), sugars, and agar, autoclaving and cooling to 45 °C, and additional ingredients are then added as filter-sterilized stock solutions. Approximately 35 mL of medium is dispensed into sterile 100 × 15 mm petri dishes

3 Methods

All steps from explant sterilization through the completion of root formation in T0 plantlets should be conducted in a sterile laminar flow hood.

3.1 Growing Donor Plants and Harvesting Immature Ears

1. Grow the publicly available maize inbreds B73, Mo17, and FFMM in the greenhouse in pots containing a soilless substrate composed of (by vol.) 77% Canadian Sphagnum peat, 16% perlite, and 7% vermiculite, adjusted with line to a starting pH of 6.0 greenhouse day/night average temperatures which are approximately 25.5 °C day and 20 °C night with a 16-h day photoperiod.
2. Cover emerging ears with shoot bags to prevent unwanted contaminating pollinations. Cover emerging tassels with a tassel bag, affixing the base around the stalk using a stapler or paper clip.

3. After silks have emerged from the ear, cut the silks with clean scissors approximately ½ in. above the top of the husk leaves the day before pollination. Scissors should be cleaned with 5% KleenGrow in between each use.
4. When anthers emerge from the tassel, collect pollen by gently bending the plant over so the tassel (still in the bag) is roughly parallel with the floor, tapping the bag to release pollen into the paper tassel bag. Fold the bag to prevent pollen from spilling, and carry the bag to the plant being pollinated.
5. Use pollen from the same plant (selfing) or from a wild-type plant of the same inbred (cross-pollination) to pollinate the silks. Remove the parchment ear bag, and pour pollen from the tassel bag onto the silks.
6. Immediately cover the ear with the tassel bag, and staple closed at the base until ready to harvest the immature ear.
7. Immature ears are harvested 9–12 days after pollination, producing immature embryos with an average length of 1.5–2.0 mm (*see* **Note 3**).

3.2 Ear Sterilization and Immature Embryo Isolation

1. Remove husk leaves from the immature ear, and insert some straight, sturdy forceps (or screwdriver) into the cob in order to provide a handle during later embryo isolation.
2. Surface sterilize the immature ears by immersion in 20% commercial bleach (1.6% sodium hypochlorite) with 0.1% Tween 20 for 20 min.
3. Rinse immature ears three times in sterile distilled water (5 min per rinse).
4. With the immature kernels still attached to the cob, slice off using a sharp sterile scalpel the top 1–2 mm of the kernel crowns.
5. Insert a small (i.e., 2–3 mm in diameter) spatula between the developing endosperm and the pericarp farthest away from the embryo (on the side of the kernel closest to the base of the ear), and gently twist the spatula to dislodge the endosperm and the embryo.
6. Use the spatula to transfer the extracted immature embryos directly into the liquid infection medium (medium 700A above).

3.3 Agrobacterium-Mediated Transformation of Immature Embryos

1. Grow *Agrobacterium tumefaciens* strain LBA4404 THY- containing a helper plasmid harboring the virulence components, a gentamycin resistance gene, and the pVS1 ORI (PHP71539, *see* Anand et al. [6]) and the T-DNA-containing plasmid (PHP79066) with the following expression cassettes from right-to-left border: (a) the maize Axig1 promoter, the maize

Wus2 gene, and the maize In2 terminator; (b) the maize Plpt promoter, the maize Bbm gene, and a rice T28 terminator; (c) the sorghum Als promoter with the maize HRA and a sorghum PepC terminator; and (d) the barley Ltp2 promoter with Zs-YELLOW N1 and the potato PinII terminator in addition to the RepABC ORI and a spectinomycin resistance gene outside the T-DNA overnight days on YP medium +50 mg/L thymidine +50 mg/L spectinomycin + 50 mg/L gentamycin.

2. Collect individual *Agrobacterium* colonies and suspend in 700A liquid, and adjust the volume until an OD of 0.4 at 550 nm is achieved.
3. Add immature embryos to the *Agrobacterium* suspension in 700A liquid medium for 5 min, and then remove them from the liquid and place them scutellum side up on 710I (or 562 V) solid medium overnight at 21 °C in the dark.
4. Culture embryos on resting medium 605T for 1 week (4–10 days) at 26 °C in the dark. During this period, somatic embryos can be observed developing on the surface of the zygotic scutellum (Fig. 1a, b showing low magnification under the stereomicroscope and Fig. 1c, d showing higher magnifications under stereomicroscope and transmitted light microscope).
5. Move embryos onto maturation medium with selective agent (13329A) for 2 weeks, and then onto rooting medium (13158 plus 0.05 mg/L imazapyr) for 1–2 weeks.
6. Once vigorous roots have formed, transfer plantlets to pots containing a soilless substrate composed (by vol.) of 38% Canadian sphagnum peat, 51% composted bark, 8% perlite, and 3% vermiculite, adjusted with line to a pH of 6.9.

3.4 Preparation of Mature Seed-Derived Explants for Transformation

1. Surface sterilize dry seeds of maize inbreds B73, Mo17, or FFMM with 80% ethanol for 3 min, followed by exposure to 50% commercial bleach (2.6% sodium hypochlorite) with 0.1% Tween 20 for 20 min, and then wash three times with sterile water.
2. Soak overnight (18–24 h) the surface-sterilized seeds in sterile water at room temperature with stirring.
3. Dissect out mature embryos from the softened seeds by manually cutting out a rectangle that contains the mature embryo axis.
4. Prepare longitudinal slices of approximately 300–400 μm in thickness by holding the rectangular segment with a sterile forceps and using sterile, sharp forceps for hand sectioning the embryo. Each slice contains exposed leaf primordia, the

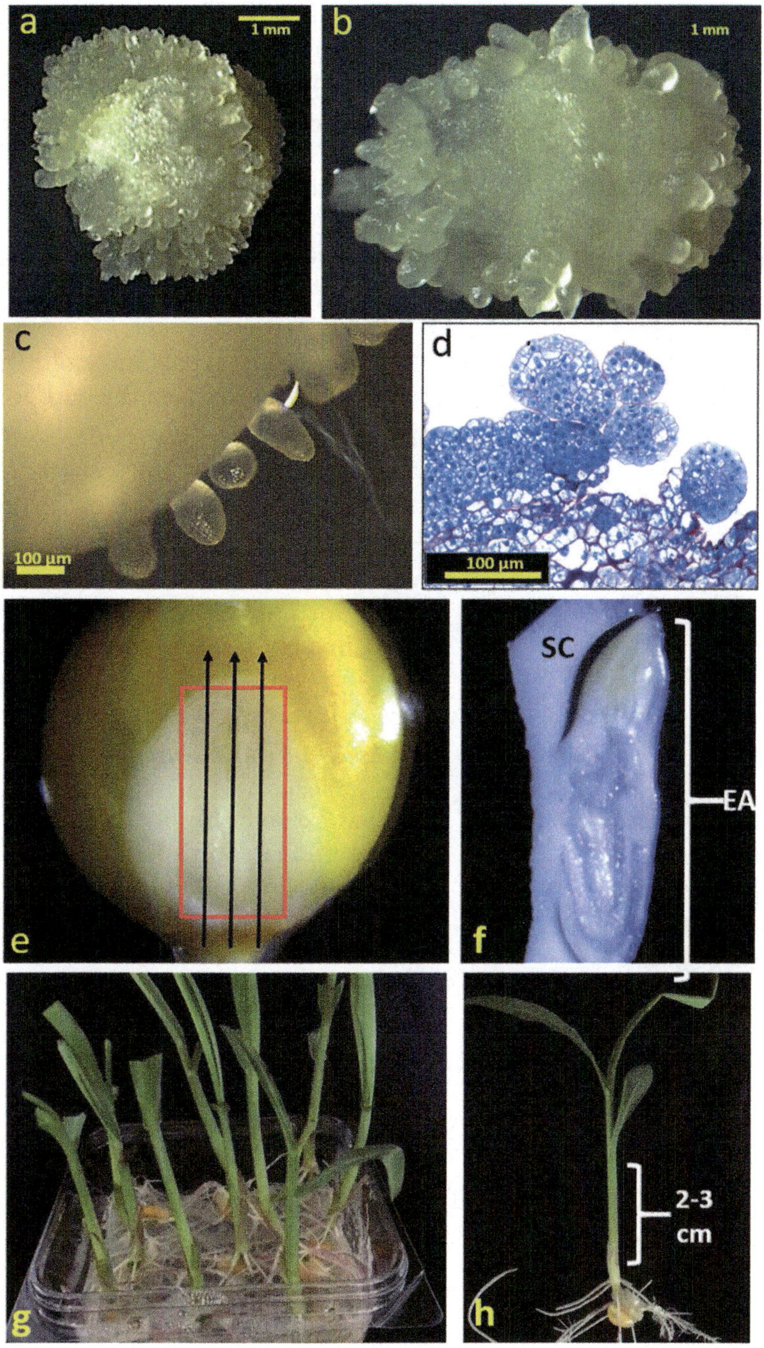

Fig. 1 (**a**, **b**) Immature embryos of maize inbred FFMM, showing the prolific growth of somatic embryos from the scutellar surface. At higher magnifications (**c**, **d**), the somatic embryos are observed to be comprised of small, rapidly dividing cells. In (**e**, **f**) thin longitudinal cross sections are prepared from mature maize seed exposing sections through the embryo axis for treatment with *Agrobacterium*. In (**g**, **h**) sterile seedlings are used as a source of leaf cross sections dicing the first 2–3 cm of the leaf whorl into 1 mm segments, which are then treated with *Agrobacterium*

scutellar node, and root primordia regions (*see* Fig. 1e, f and **Note 4**).

5. Transfer immediately fresh slices into liquid infection medium (700A) before *Agrobacterium* infection.

3.5 Preparation of Leaf Segments for Transformation

1. Surface sterilize maize inbred seed of B73, Mo17, or FFMM with 80% ethanol for 3 min, followed by exposure to 50% commercial bleach (2.6% sodium hypochlorite) with 0.1% Tween 20 for 20 min, and then wash three times with sterile water.
2. Place sterilized seeds onto solid medium 13158 for direct germination at 26 °C in the dark for 12–18 days (Fig. 1g).
3. Manually cut from the sterile seedling using a sharp, sterile scalpel A segment of the leaf whorl approximately 2–3 cm in length (directly above the mesocotyl and the first leaf node) (Fig. 1h).
4. Further dissect the 2–3 cm-long whorl segments by splitting the segment lengthwise and then cutting cross sections at approximately 1–2 mm intervals.
5. Tease apart the leaf segments within each whorl cross section using sterile forceps, and transfer immediately into the prepared *Agrobacterium* suspension for infection (15–20 min).

3.6 Agrobacterium-Mediated Transformation of Mature Embryo or Leaf Segments with Agrobacterium

1. For mature embryo slice infection, grow *Agrobacterium* strain LBA4404 THY- on solid YP culture medium with 50 mg/L spectinomycin + 50 mg/L gentamycin + 50 mg/L thymidine at 28 °C in the dark for 1 day. Adjust the *Agrobacterium* suspension to an OD 0.7 at 550 nm using liquid medium (700A) containing 200 μM acetosyringone and 0.04% Silwet L-77.
2. For leaf segment infection, grow *Agrobacterium* strain AGL1 on solid medium and incubate at 28 °C in the dark for 1 day. Adjust the *Agrobacterium* suspension to an OD of 0.4 at 550 nm using 700F liquid medium plus 200 uM acetosyringone and 0.02% Silwet L-77.
3. Decant the liquid medium bathing either to the embryo sections or leaf segments and replace with the freshly prepared *Agrobacterium* suspension.
4. Transfer the culture plates containing mature embryo slices into a vacuum desiccator, connect to house vacuum (24 in. Hg) and maintain on a shaker platform with a speed of 100 rpm for 15–30 min.
5. After infection, decant the *Agrobacterium* suspension and blot dry with sterile filter paper the embryo slices or leaf pieces

before transferring them onto solid medium 710I, and incubate at 21 °C in the dark for 3 days.

6. After 3 days co-cultivation on medium 710I, transfer embryo slices or leaf pieces onto culture medium 605T to kill the *Agrobacterium*.
7. After 1 week, subculture tissue onto fresh 605T medium plus 0.05 to 0.1 mg/L imazapyr.
8. Subculture tissue onto fresh 605T plus 0.05–0.1 mg/L imazapyr every 2 weeks.
9. After 4–6 weeks of culture, transfer fast-growing embryogenic calli to maturation media 13329A for 2–3 weeks in the dark to induce shoot formation.
10. When shoots are 1–2 cm in length, transfer them to hormone-free medium 13158 plus 0.05 mg/L imazapyr for further development of shoots and roots under low lights (10–30 mE m^{-2} s^{-1}).
11. Sample plantlets with well-developed shoots and roots for comprehensive molecular analysis before being transferred to the greenhouse for seed production.
12. Perform detailed PCR analyses to determine copy number of transgenes and to confirm the absence of plasmid backbone (all plasmid sequence outside the T-DNA borders).

3.7 Molecular Analysis

1. Use qPCR [7] (a) to estimate the copy number of the transgenes, (b) to determine if T-DNA integrations were intact or truncated, (c) to determine if recombinase-mediated excision had occurred, and (d) to screen for the presence of *Agrobacterium* binary vector backbone integration.
2. Extract genomic DNA from a single piece (approximately 200–300 ng) of fresh leaf tissue from each plant [8].
3. Use non-transgenic maize inbred lines for inbreds B73, Mo17, or FFMM as the negative controls.
4. Quantification is based on the detection of amplification of gene sequences using gene-specific forward and reverse primers along with the corresponding gene-specific FAM/Vic-based MGB fluorogenic probes (Applied Biosystems, Foster City, CA).
5. To estimate the copy number, use the $2 - \Delta\Delta CT$ method ([9]; ABI's user bulletin #2, www.appliedbiosystems.com/cms/groups/mcb_support/documents/generaldocuments/cms_040980.pdf).
6. Identify plants that contain intact, single-copy T-DNA integrations with no *Agrobacterium* backbone. These plants represent quality events that will be grown to maturity for production of T1 progeny.

7. Discard plants with multiple-copy scores for any of the structural genes within the T-DNA as having multiple-copy integrations, which could either be intact or truncated.
8. For plants derived from PHP79066, primer sets are used that amplified segments of either the BBM:Terminator junction, the WUS:terminator junction, the HRA:terminator junction, or sequences within the ZS-YELLOW1 NI coding sequence.
9. For all maize transgenic events, the detection of *Agrobacterium* vector backbone integration is based on screening for sequences from five regions outside of the T-DNA (5′ of the RB, virG, SPC, and 3′ of the LB).
10. Plants with negative qPCR signals for regions outside the T-DNA are considered to be backbone-negative. Otherwise, they were classified as backbone-positive. Plants with intact single-copy T-DNA integrations without vector backbone were defined as quality events (QEs).

3.8 Transplanting to the Greenhouse and Production of T1 Seed

1. Remove T0 plantlets from the agar plates and wash the roots to remove the agar. Transplant the plantlets in approximately 3 in. square containers containing a standard soilless mixture (e.g., a mix of peat, perlite, and vermiculite) and grown in the greenhouse for 2 weeks. For details on care of transgenic plants in the greenhouse, *see* **Note 5**.
2. Transplant the entire soilless plug into a larger pot in the greenhouse. Water plants as necessary when soil is dry to the touch.
3. Use controlled-release fertilizer such as Osmocote Plus 15-9-12 either incorporated into the substrate mix or added to the surface of the mixture after planting.
4. When ear shoots first emerge from the corn plant, use ear bags to cover the developing ear shoots until controlled pollinations are performed, with a semi-transparent bag (Canvasback Cat. No. #S26, usually made of parchment) being used so that the emerging silks can be observed without removing the bag (reducing the risk of contamination).
5. Shortly after silk emergence (1–2 days), trim the silks to a uniform length of approximately 1/2 in. above the husk leaves, using scissors that have been decontaminated using 70% ethanol or 5% KleenGrow. Trimming the silks in this manner provides a uniform tuft of silk hairs for pollination the next day.
6. Ideally, for most maize genotypes, a 2–3 days after tassel or silk emergence is optimal for pollination.
7. Collect pollen from the same plant (selfing) or from wild-type plants of the same inbred (outcrossing or incrossing, *see* **Note 6**) in paper bags and apply to the silks of the T0 plant. If pollen

from a wild-type (non-transgenic) inbred plant is collected, use a plain brown paper tassel bag, and if pollen from a transgenic plant is collected, use a paper bag with green stripes to clearly demarcate this collected pollen as transgenic.

8. Collect pollen by gently bending the top of the plant over and tapping the tassel bag to aid in pollen shed.
9. Remove the ear bag and slip the tassel bag (with shed pollen inside) over the top of the ear and staple in place near the base of the bag.
10. The source of the pollen and date of pollination are recorded on the ear bag.
11. Approximately 2 weeks after pollination, remove the ear bags from the ears to aid in dry-down.
12. To enhance dry-down further, watering of the plant can be discontinued at approximately 3 weeks after pollination, and the husk leaves can be stripped back to expose the seed.
13. At approximately 45 days after pollination, seed can be harvested and put into cold storage (4–12 °C).

4 Notes

1. An auxotrophic (THY-) version *Agrobacterium tumefaciens* strain of LBA4404 is used for all transformations. This *Agrobacterium* strain contains pVir9 (PHP71539, a helper plasmid containing Bo542 virulence genes, see WO 2017/078836 A1) and PHP79066.
2. Both the *Bbm* and *Wus2* genes were cloned from the Pioneer/DuPont EST maize EST library. Identity was confirmed relative to other AP2-domain proteins by alignment of the encoded *Bbm* protein to the published BBM protein sequences of *Brassica napus* and Arabidopsis [10]. The cloned maize *Bbm* nucleotide sequence can be found as GenBank Accession Number CS155772 and the protein as GenBank Number CAJ29869 [11]. Likewise, the identity of the *Wus2* gene was confirmed by alignment with Arabidopsis WUS family members [12, 13] and has been archived in GenBank as Accession Number EA275154 and the protein as GenBank Number ABW43772 [14].
3. The growth rate of the ear and the development of embryos within ears vary between genotypes and with the season of the year (e.g., due to seasonal differences in light quality). Embryo size is a better indicator as to the suitability for transformation.
4. The regions of the embryo axis exposed on the surface of each longitudinal slice are the primary target for T-DNA delivery

during *Agrobacterium*-mediated transformation and contained cells that were culture responsive. Surprisingly, the cross section of the mature scutellum was virtually totally recalcitrant to *Agrobacterium*-mediated T-DNA delivery (as assessed by transient ZS-GREEN expression).

5. For more detail on care and handling of transgenic plants in the greenhouse, please refer to "Greenhouse care for transgenic maize plants" in the link maintained by Dr. Kan Wang's Transformation Facility at the Department of Agronomy, Iowa State University; http://ptf.agron.iastate.edu/protocol/Greenhouse%20Protocol.pdf.
6. If homozygous T1 seed is desired and the tassel and silk have emerged from the T0 plant, pollen from the same T0 plant is used for pollination (selfing). If the tassel and silks do not mature simultaneously, pollen from a wild-type inbred is carried onto the T0 ear (incrossing) producing hemizygous T1 seed. Likewise, pollen from the T0 can be carried out to wild-type plants (outcrossing).

References

1. Songstad DD, Armstrong CL, Peterson WL, Hairston B, Hinchee MAW (1996) Production of transgenic maize plants and progeny by bombardment of Hi-II immature embryos. In Vitro Cell Dev Biol Plant 32:179–183
2. Lowe K, Wu E, Wang N, Hoerster G, Hastings C, Cho M-J, Scelonge C, Lenderts B, Chamberlin M, Cushatt J, Wang L, Ryan L, Khan T, Chow-Yiu J, Hua W, Yu M, Bahn J, Bao Z, Brink K, Igo E, Rudrappa B, Shamseer PM, Bruce W, Newman L, Shen B, Zheng P, Bidney D, Falco C, Register J, Zhao Z-Y, Xu D, Jones T, Gordon-Kamm W (2016) Morphogenic regulators Baby boom and Wuschel improve monocot transformation. Plant Cell 28:1–19
3. Lowe K, La Rota M, Hoerster G, Hastings C, Wang N, Chamberlin M, Wu E, Jones T, Gordon-Kamm W (2018) Rapid genotype independent maize transformation via direct somatic embryogenesis. In Vitro Cell Dev Biol Plant 54(3):240–252
4. Garnaat C, Lowe K, Roth B (2002) *Zm-AXIG1*-specific polynucleotides and methods of use. WO2002006499
5. McCaw ME, Wallace JG, Albert PS, Buckler ES, Birchler JA (2016) Fast-flowering mini-maize: seed to seed in 60 days. Genetics 204:35. https://doi.org/10.1534/genetics.116.191726
6. Anand A, Bass SH, Cho H-J, Klein TM, Lassner M, McBride KE (2017) Methods and compositions of improved plant transformation. PCT/US2016/049132
7. Yang H, Schmuke JJ, Flagg LM, Roberts JK, Allen EM, Ivashuta S, Gilbertson LA, Armstrong TA, Christian AT (2009) A novel real-time polymerase chain reaction method for high throughput quantification of small regulatory RNAs. Plant Biotechnol J 7:621–630
8. Truett GE, Heeger P, Mynatt RL, Truett AA, Walker JA, Warmman ML (2000) Preparation of PCR-quality mouse genomic DNA with hot sodium hydroxide and tris (HotSHOT). BioTechniques 29:52–54
9. Livak KJ, Schmittgen TD (2001) Analysis of relative gene expression data using real-time quantitative PCR and the 2(−Delta Delta C(T)) method. Methods 25:402–408
10. Boutilier K, Offringa R, Sharma VK, Kieft H, Ouellet T, Zhang L, Hattori J, Liu CM, van Lammeren AA, Miki BL, Custers JB, van Lookeren Campagne MM (2002) Ectopic expression of BABY BOOM triggers a conversion from vegetative to embryonic growth. Plant Cell 14:1737–1749

11. Gordon-Kamm WJ, Helentjaris TG, Lowe KS, Shen B, Tarczynski MC Zheng P (2005) Ap2 domain transcription factor Odp2 (ovule development protein 2) and methods of use. Patent no. WO2005/075655-A2
12. Laux T, Mayer KF, Berger J, Jürgens G (1996) The WUSCHEL gene is required for shoot and floral meristem integrity in Arabidopsis. Development 122:87–96
13. Gallois JL, Woodward C, Reddy GV, Sablowski R (2002) Combined SHOOT MERISTEMLESS and WUSCHEL trigger ectopic organogenesis in Arabidopsis. Development 129:3207–3217
14. Lowe KS, Cahoon RE, Scelong CJ, Tao Y, Gordon-Kamm WJ, Bruce WB, Newman LJ (2007) Wuschel (WUS) Gene Homologs Patent No US7256322

Chapter 7

Efficient and Fast Production of Transgenic Rice Plants by *Agrobacterium*-Mediated Transformation

Chuanyin Wu and Yi Sui

Abstract

Genetic transformation plays a key role in deciphering regulation of agronomic traits at molecular level in rice, a model monocot cereal crop. Here we describe an efficient and fast protocol for producing transgenic japonica rice plants using the *Agrobacterium*-mediated transformation method. The protocol simplifies medium compositions and transformation steps and can be easily followed by a lab technician with little tissue culture experience. Using this protocol, we have transformed thousands of gene constructs in the past 10 years and edited hundreds of genes with the CRISPR-*Cas9* system recently.

Key words Rice, *Agrobacterium*-mediated transformation, Callus induction, Selection of transformants, Plant regeneration

1 Introduction

The first fertile transgenic rice plants produced using the protoplast electroporation method were reported almost 30 years ago [1]. The disadvantage of the protoplast method is the requirement of skills and extensive labor in establishing fast-growing suspension cultures, protoplast preparation and culture, and plant regeneration from protoplast-derived colonies. In addition, protoplast method produces plants frequently associated with somatic variation. Soon after, the particle bombardment method was successfully employed, allowing the transformation of immature embryos and production of transgenic plants [2]. However the biolistic method requires the use of a special device and gold particles to deliver the DNA into the plant cells making the process expensive, and the resulting transgenic plants often have a high copy number of the transgene [3]. This method is less popular and sometimes used for special purposes such as transfer of multiple genes at once [4–6] and DNA-free genome editing due to the flexibility to coat particles with RNA or RNA–protein complex [7–9]. Following success in dicot transformation, *Agrobacterium*-mediated transformation was

Sandeep Kumar et al. (eds.), *Transgenic Plants: Methods and Protocols*, Methods in Molecular Biology, vol. 1864, https://doi.org/10.1007/978-1-4939-8778-8_7,

explored in monocot plants, with early reports on transgenic calli or plants [10, 11]. A reliable and efficient transformation system was not established, however, until 1994 [12]. Since then, the *Agrobacterium* method has become the most popular technique to transform rice due to easy implementation and resulting fewer copies of transgene. Although various modifications have been made based on the protocol of Hiei et al. [13], efficient transformation can be achieved regardless of the strategy employed [14, 15]. This is because japonica rice is highly transformable compared to indica rice and other cereal crops. For this reason, we have focused on simplification of medium compositions and the transformation procedure. Using the simplified protocol, one full-time employee can transform over 250 gene constructs annually, with efficiencies varying from 50 to 90% depending on cultivars used and constructs transformed. Our transformation platform, with efficiency, has greatly facilitated our research on gene function analysis, genome editing, and gene replacement [16–20].

2 Materials

We routinely use the japonica varieties Kita-ake, Nipponbare, and Zhonghua 11. Various other japonica varieties are also used, as requested by customers. Milli-Q water is used for preparation of all media and solutions.

2.1 Stock Solutions

1. Clorox (20%): Add 200 mL of Clorox bleach (available in grocery stores) to 800 mL of water. Store at room temperature.
2. Acetosyringone (100 mM): Add 196.2 mg of acetosyringone to 10 mL of dimethyl sulfoxide. Divide it into ten aliquots. Store at −20 °C.
3. 2,4-dichlorophenoxyacetic acid (2 mg/mL 2,4-D): Add 200 mg of 2,4-D to 20 mL of 0.1 N NaOH. Dissolve 2,4-D completely with a magnetic stirrer. Add 80 mL of water to a final volume of 100 mL. Store at 4 °C.
4. 6-benzylaminopurine (2 mg/mL 6-BAP): Add 200 mg of 6-BAP to 2 mL of 1 N NaOH. Dissolve 6-BAP completely with a magnetic stirrer. Add 98 mL of water to a final volume of 100 mL. Store at 4 °C.
5. Naphthalene acetic acid (0.2 mg/mL NAA): Add 20 mg of NAA to 1 mL of 1 N NaOH. Dissolve NAA completely. Add 98 mL of water to a final volume of 100 mL. Store at 4 °C.
6. Vitamins and glycine (1000×): Dissolve 500 mg of thiamine HCI, 100 mg of pyridoxine HCI, 100 mg of nicotinic acid, and 200 mg of glycine in 100 mL of water. Store at 4 °C.

7. Cu/Mo/Co (1000×): Dissolve 25 mg of $CuSO_4{\cdot}5H_2O$, 250 mg of $NaMoO_4{\cdot}2H_2O$, and 25 mg of $CoCl_2{\cdot}6H_2O$ in 1000 mL of water. Store at 4 °C.
8. KI (0.8 mg/mL): Dissolve 80 mg of potassium iodide in 100 mL of water. Store at 4 °C.
9. Carbenicillin (250 mg/mL): Dissolve 25 g of carbenicillin in 100 mL of water. Filter-sterilize and divide it into 100 aliquots. Store at −20 °C.
10. Hygromycin (75 mg/mL): Dissolve 7.5 g of hygromycin in 100 mL of water. Filter-sterilize and divide it into 100 aliquots. Store at −20 °C.
11. Paromomycin (100 mg/mL): Dissolve 10 g of paromomycin in 100 mL of water. Filter-sterilize and divide it into 100 aliquots. Store at −20 °C.
12. Kanamycin (50 mg/mL): Dissolve 500 mg of kanamycin in 10 mL of water. Filter-sterilize and divide it into 10 aliquots. Store at −20 °C.
13. Rifampicin (50 mg/mL): Dissolve 500 mg of rifampicin in 10 mL of water. Filter-sterilize and divide it into 10 aliquots. Store at −20 °C.
14. Spectinomycin (75 mg/mL): Dissolve 750 mg of spectinomycin in 10 mL of water. Filter-sterilize and divide it into 10 aliquots. Store at −20 °C.
15. Bialaphos (1000×): Dissolve 500 mg of bialaphos in 100 mL of water. Filter-sterilize and divide it into 100 aliquots. Store at −20 °C.

2.2 Media

1. YEB: 5 g/L peptone, 1 g/L yeast extract, 5 g/L beef extract, 0.443 g/L $MgSO_4{\cdot}7H_2O$ and 5 g/L sucrose. Adjust pH to 7.0. Autoclave it and store at room temperature.
2. N6-C: Add 4 g of N6 salts (Phytotechnology), 1 mL of each of the Cu/Mo/Co, vitamins and glycine, KI, and 2,4-D stock solutions, 100 mg of myoinositol, 300 mg of casamino acid, 2.8 g of proline, and 30 g of sucrose to 800 mL of water. Add water to a final volume of 1 L. Adjust pH to 5.6. Add 4.0 g/L Gelrite for the solid N6-C medium. Autoclave at 121 °C for 15 min.
3. N6-AS: Same as N6-C but without proline and with casamino acid concentration increased to 1 g/L. For solid N6-AS, add 4.0 g/L Gelrite. After autoclaving and cooling down of the medium to about 65 °C, add the acetosyringone solution to a final concentration of 200 μM (*see* **Note 1**).
4. N6-S: Same as solid N6-C, but add 1 mL of the hygromycin, paromomycin, or bialaphos stock solution, depending on the

selectable marker used (*see* **Note 2**), and 1 mL of the carbenicillin solution after autoclaving and cooling down of the medium to about 65 °C.

5. MS-R: Add 4.33 g of MS salts (Phytotechnology), 1 mL of the vitamins and glycine stock, 1 mL of the 6-BAP stock, 250 μL of the NAA stock, 100 mg of myoinositol, 500 mg of casamino acid, 20 g of sorbitol, and 30 g of sucrose to 800 mL of water. Add water to a final volume of 1 L. Adjust pH to 5.6 and then add 2.5 g of Gelrite. Autoclave at 121 °C for 15 min.
6. MS-RG: Add 2.165 g of MS salts, 1 mL of the vitamins and glycine solution, 100 mg of myoinositol, and 15 g of sucrose to 800 mL of water. Add water to a final volume of 1 L. Adjust pH to 5.6. Add 2.5 g/L Gelrite. Autoclave at 121 °C for 15 min.

3 Methods

Carry out all procedures at room temperature unless otherwise specified. After autoclaving and adding supplements, pour all media into Petri dishes, 30 mL each unless otherwise specified, or culture containers (such as Magenta boxes for rooting), 40 mL each, at around 60 °C. Media can be stored at room temperature for up to 2 months, if not used immediately after autoclaving. Upon use, melt stored solid media in a microwave.

3.1 Seed Sterilization

1. Dehusk mature seeds and discard the spotted ones. Seeds that have been stored at room temperature for over 2 years may not be used (*see* **Note 3**).
2. Transfer up to 300 dehusked seeds to a 50 mL conical tube (such as CORNING 430290) (*see* **Note 4**).
3. Rinse seeds in 70% ethanol for 1 min. Pour out ethanol, and rinse seeds with autoclaved Milli-Q water.
4. Add 35 mL 20% Clorox solution. Shake the tube by hand occasionally or on a rotary shaker for 30 min.
5. Pour out the Clorox solution, and rinse seeds with autoclaved water, and repeat four times.
6. Dry the seeds on autoclaved Kimwipes for a few minutes.

3.2 Callus Induction

1. Transfer seeds to the solid N6-C callus induction medium, 10–15 seeds each Petri dish (100 × 20 mm, FALCON or similar). Seal dishes with surgical tapes to allow air exchange.
2. Culture the seeds in dark at 28 °C for 4 weeks. Callus initiation starts from the scutella in a few days (Fig. 1a), and embryogenic calli with a granular structure and smooth surface can be seen in

Fig. 1 Different stages of rice transformation with *Agrobacterium*-mediated method using mature seeds. (**a**) Callus induction from mature seeds, 10 days after inoculation. (**b**) A single seed-derived callus pieces with granular structure and smooth surface, 25 days after inoculation. (**c**) Embryogenic calli maintained on callus induction medium, 7 days after subculture. (**d**) *Agrobacterium*-infected calli at the end of 3-day co-cultivation. (**e**) Over-infected calli. Arrow indicates visible *Agrobacterium* growth that can result in low transformation efficiency. (**f**) Callus cells transiently expressing GFP during co-cultivation. (**g**) A callus 10 days on the first selection, showing effective killing of non-transformed cells and active growth of stably transformed sectors. (**h**) Image of (**g**) under UV light. (**i**) A callus 15 days on the first selection, showing clustered transformed events of independent origin. (**j**) Image of (**i**) under UV light. (**k**) Calli 20 days on the first selection. The new pieces of calli, as indicated by arrows, can be easily distinguished from browning original calli and are ready to transfer to the second selection. (**l**) Shoot formation 12 days on regeneration medium. (**m**) Hardening of transgenic plants 4 days after the covers were removed. (**n**) Transgenic plants in soil

3–4 weeks (Fig. 1b). The calli can be used directly for *Agrobacterium* infection or subcultured on the same N6-C medium for later use (Fig. 1c) (*see* **Note 5**).

3.3 Infection and Co-cultivation of Calli with Agrobacterium Cells

1. Store *Agrobacterium* cells containing constructs of interest in YEB medium with 15% glycerol (*see* **Note 6**). Transfer about 50 μL of *Agrobacterium* stock to a sterile 15 mL tube containing 5 mL of YEB medium containing 5 μL of rifampicin and 5 μL of kanamycin or spectinomycin stock (depending on vectors used). Incubate the tube on a shaker at 140 rpm overnight until an OD_{600} reading between 0.8 and 1.2.
2. Transfer 1 mL of liquid culture to a 1.5 mL Eppendorf tube, and centrifuge it at 4000 × *g* for 5 min.
3. Discard the supernatant and resuspend *Agrobacterium* cells in N6-AS liquid medium.
4. Adjust cell density with the liquid N6-AS medium to the reading of 0.02 at OD_{600}.
5. Transfer up to 15 g embryogenic calli to a sterile 50 mL conical tube, and add 25 mL of the diluted *Agrobacterium* culture. Shake the tube by hand gently for 3 min.
6. Pour out the liquid culture. Transfer calli onto autoclaved Kimwipes paper with a spatula to remove extra liquid.
7. Transfer calli from the Kimwipes paper to an autoclaved filter paper placed on the solid N6-AS medium in a Petri dish. Each dish contains approximately 20 mL of the N6-AS medium. Seal the dish with surgical tape.
8. Culture calli at 22 °C in dark for 3–4 days. Calli should remain healthy during co-cultivation (Fig. 1d). The infection must be stopped as long as over-infected calli are seen (Fig. 1e). Infection efficiency can be monitored using a GFP vector (Fig. 1f) (*see* **Note 7**).

3.4 Selection of Transformed Calli

1. Transfer the infected calli from co-cultivation to a sterile 50 mL tube containing 35 mL of autoclaved water. Shake the tube for a few seconds and pour out the water. Repeat washing for four times. For the final wash, add the carbenicillin stock to water at a final concentration of 250 mg/L.
2. Transfer calli to autoclaved Kimwipes papers to remove extra water.
3. Transfer calli to the N6-S medium, with 20 pieces in each Petri dish. Seal the dishes with surgical tape, and culture the calli at 28 °C in the dark for 20 days. The effectiveness of a selection agent can be monitored with a GFP vector during selection (Fig. 1g–j). At the end of the first selection, actively growing sectors should be seen against the dying original calli (Fig. 1k).

4. Subculture the newly formed calli on the same N6-S medium under the same condition for 5 days. The resistant calli should survive well during the second selection (*see* **Note 8**).

3.5 Shoot Regeneration from Resistant Calli and Hardening of Transgenic Plants

1. Transfer resistant calli from the second selection to the N6-R plant regeneration medium, with ten pieces in each Petri dish.
2. Culture calli at 28 °C under cold fluorescent light for 15–20 days. Shoots are seen in 7–12 days (Fig. 1l), depending on varieties used.
3. Transfer plantlets to 125 mL flasks or Magenta boxes containing 40 mL of the MS-RG medium for root growth at 28 °C under cold fluorescent light. Flasks, if used, can be covered with aluminum foil. Plants grow vigorously on this medium and are ready for transplanting in 10–14 days.
4. Open flasks or Magenta boxes and add 25 mL of autoclaved water. Keep them at 28 °C under cold fluorescent light for hardening. Plants are ready to transplant to pots or rice field in 4–5 days, and they usually survive well without delay in growth (Fig. 1m, n).

4 Notes

1. N6-AS medium can be stored at room temperature before addition of acetosyringone and be melted with a microwave at each use. Once added with acetosyringone, however, it is recommended to be used immediately.
2. We have used three selectable markers, hygromycin phosphotransferase (HPT), neomycin phosphotransferase II (NPT II), and phosphinothricin *N*-acetyltransferase (PAT), for our routine transformation. When paromomycin is used for selection, Gelrite for medium solidification must be replaced with agar (7 g/L). Paromomycin precipitation is always seen with use of Gelrite.
3. Healthy mature seeds are important for high-frequency and contamination-free induction of quality calli. High air temperature and humidity during seed maturation often result in spotted seeds. Fungus contamination occurs on the spotted seeds in 5–7 days on callus induction medium. This kind of contamination cannot be avoided even if the sterilization time is extended to 2 h. Harvesting light green, not fully matured seeds is one way to effectively reduce fungus contamination.
4. For large-scale transformation, a bigger container such as a 500 mL centrifuge bottle is a choice to sterilize thousands of seeds at once.

5. We recommend to use embryogenic calli in vitro cultured for less than 2 months. Long-term in vitro culture reduces regeneration ability of transgenic calli and leads to formation of albinos and somatic mutation.
6. We have tried the *Agrobacterium* strains EHA101, EHA105, LBA4404, GV3101, and C5 and didn't see significant difference between them in transformation efficiency. EHA105 is the most used strain in our lab.
7. Observe calli during co-cultivation with a dissection microscope. An over-infected callus means its slightly brown color and visible *Agrobacterium* growth surrounding it. When the GFP gene is used for testing, green spots are seen in 2 days, and maximum GFP spots are seen in 4 days. Extended co-cultivation time does not necessarily increase infection efficiency, as monitored by transient GFP expression.
8. Due to high transformation efficiency in japonica rice, multiple resistant sectors are often formed on a single original callus (Fig. 1i, j). In this case, we transfer only one piece of new callus to the second selection to ensure event independence.

Acknowledgment

This work was supported by the Innovation Program of Chinese Academy of Agricultural Sciences.

References

1. Shimamoto K, Terada R, Izawa T, Fujimoto H (1989) Fertile transgenic rice plants regenerated from transformed protoplasts. Nature 338:274–276
2. Christou P, Ford T, Kofron M (1991) Production of transgenic rice (*Oryza Sativa* L.) plants from agronomically important indica and japonica varieties via electric discharge particle acceleration of exogenous DNA into immature zygotic embryos. Nat Biotechnol 9:957–962
3. Dai S, Zheng P, Marmey P, Zhang S, Tian W, Chen S, Beachy RN, Fauquet C (2001) Comparative analysis of transgenic rice plants obtained by Agrobacterium-mediated transformation and particle bombardment. Mol Breed 7:25–33
4. Chen L, Marmey P, Taylor NJ, Brizard JP, Espinoza C, D'Cruz P, Huet H, Zhang S, Kochko A, Beachy RN, Fauquet CM (1998) Expression and inheritance of multiple transgenes in rice plants. Nat Biotechnol 16:1060–1064
5. Zhu C, Naqvi S, Breitenbach G, Sandmann J, Christou P, Capell T (2008) Combinatorial genetic transformation generates a library of metabolic phenotypes for the carotenoid pathway in maize. Proc Natl Acad Sci U S A 105:18232–18237
6. Naqvi S, Zhu C, Farre G, Bassie L, Ramessar K, Breitenbach J, Perez-Conesa D, Ros-Berruezo G, Sandmann G, Capell T, Christou P (2009) Transgenic multivitamin corn through biofortification of endosperm with three vitamins representing three distinct metabolic pathways. Proc Natl Acad Sci U S A 106:7762–7767
7. Zhang Y, Liang Z, Zong Y, Wang Y, Liu J, Chen K, Qiu JL, Gao C (2016) Efficient and transgene-free genome editing in wheat through transient expression of CRISPR/Cas9 DNA or RNA. Nat Commun 7:12617. https://doi.org/10.1038/ncomms12617
8. Svitashev S, Schwartz C, Lenderts B, Young JK, Cigan AK (2016) Genome editing in maize directed by CRISPR–Cas9

ribonucleoprotein complexes. Nat Commun 7:13274. https://doi.org/10.1038/ncomms13274

9. Liang Z, Chen K, Li T, Zhang Y, Wang Y, Zhao Q, Liu J, Zhang H, Liu C, Ran Y, Gao C (2017) Efficient DNA-free genome editing of bread wheat using CRISPR/Cas9 ribonucleoprotein complexes. Nat Commun 8:14261. https://doi.org/10.1038/ncomms14261
10. Rained DM, Bottino P, Gordon MP, Nester EW (1990) Agrobecterium-mediated transformation of rice (Oryza sativa L.). Nat Biotechnol 8:33–38
11. Chan MT, Chang HH, Ho SL, Tong WF, Yu SM (1993) *Agrobacterium*mediated production of transgenic rice plants expressing a chimeric α-amylase promoter/β glucuronidase gene. Plant Mol Biol 22:491–506
12. Hiei Y, Ohta S, Komari T, Kumashiro T (1994) Efficient transformation of rice (*Oryza sativa* L.) mediated by *Agrobacterium* and sequence analysis of the boundaries of the T-DNA. Plant J 6:271–282
13. Hiei Y, Komari T (2008) *Agrobacterium*-mediated transformation of rice using immature embryos or calli induced from mature seed. Nat Protoc 3:824–834
14. Roy M, Jain RK, Rohila JS, Wu R (2000) Production of agronomically superior transgenic rice plants using Agrobacterium transformation methods: present status and future perspectives. Curr Sci 79:954–960
15. Toki S, Hara N, Ono K, Onodera H, Tagiri A, Oka S, Tanaka H (2006) Early infection of scutellum tissue with Agrobacterium allows high-speed transformation of rice. Plant J 47:969–976
16. Zhou F, Lin Q, Zhu L et al (2013) D14–SCFD3-dependent degradation of D53 regulates strigolactone signaling. Nature 504:406–410
17. Gao H, Jin M, Zheng X et al (2014) Days to heading 7, a major quantitative locus determining photoperiod sensitivity and regional adaptation in rice. Proc Natl Acad Sci U S A 46:16337–16342
18. Liu Y, Wu H, Chen H et al (2015) A gene cluster encoding lectin receptor kinases confers broad-spectrum and durable insect resistance in rice. Nat Biotechnol 33:301–305
19. Wu S, Xie Y, Zhang J et al (2015) VLN2 regulates plant architecture by affecting microfilament dynamics and polar auxin transport in rice. Plant Cell 27:2829–2845
20. Sun Y, Zhang X, Wu C, He Y, Ma Y, Hou H, Guo X, Du W, Zhao Y, Xia L (2016) Engineering herbicide-resistant rice plants through CRISPR/Cas9-mediated homologous recombination of acetolactate synthase. Mol Plant 9:628–631

Chapter 8

Protocol for *Agrobacterium*-Mediated Transformation and Transgenic Plant Production of Switchgrass

QiuXia Chen and Guo-Qing Song

Abstract

Switchgrass (*Panicum virgatum* L.) is one of the most important bioenergy crops for lignocellulose ethanol production. Molecular breeding provides a powerful tool to supplement conventional switchgrass breeding by introducing or editing genes of interest. In this chapter, we describe *Agrobacterium tumefaciens*-mediated transformation protocols for lowland tetraploid switchgrass cultivar Alamo.

Key words *Agrobacterium tumefaciens*, Genetic transformation, Plant regeneration, Somatic embryogenesis

1 Introduction

Switchgrass (*Panicum virgatum* L.) is a C4 perennial grass native to North America. High biomass productivity has made switchgrass an attractive alternative to corn grain as a bioenergy crop for production of cellulosic biofuels [1–3]. Accordingly, extensive studies on switchgrass have been conducted to produce lignocellulose-derived biofuels through metabolic engineering [4].

Cultivated switchgrass varieties include two taxonomically distinct ecotypes: lowland and upland varieties [1, 3]. A major lowland, tetraploid cultivar Alamo ($2n = 4x = 36$) has been reported to be amenable for genetic transformation using either biolistic- or *Agrobacterium*-mediated gene delivery [5–12]. In this chapter, we describe our protocol for routine *Agrobacterium*-mediated stable transformation of 'Alamo' at frequencies of 1–5% for seedling explants and 10–25% of callus explants.

2 Materials

2.1 Plant Material

Starting plant materials: mechanically dehulled (preferable) or intact mature caryopses of switchgrass cultivar Alamo (*see* **Note 1**).

Sandeep Kumar et al. (eds.), *Transgenic Plants: Methods and Protocols*, Methods in Molecular Biology, vol. 1864, https://doi.org/10.1007/978-1-4939-8778-8_8, © Springer Science+Business Media, LLC, part of Springer Nature 2019

2.2 Solutions

1. Sterile water: autoclaved double-distilled water (ddH_2O).
2. 70% ethanol.
3. Sterilizing solution: 50% Clorox® (v/v) containing 3.08% of sodium hypochlorite and 0.1% Tween 20.
4. Dehulling solution: 60% sulfuric acid (H_2SO_4) (v/v). Never add water to acid.
5. Stock solutions of plant growth regulator (PGR) (1 mg/mL): dissolve 6-benzylaminopurine (BAP), 2, 4-dichlorophenoxyacetic acid (2, 4-D), and gibberellic acid (GA) in solvent (3 mL of 1 *N* NaOH for 100 mg of PGR), dilute to 1 mg/mL with ddH_2O as needed, and store at 4 °C up to 6 months.
6. Stock solution of antibiotics: dissolve 50 mg/mL kanamycin (Km), 50 mg/mL hygromycin (Hyg), and 250 mg/mL cefotaxime (Cef) in sterile ddH_2O. Dissolve 30 mg/mL rifampicin (Rif) in DMSO. Filter-sterilize stock solutions of Km, Hyg, and Cef through 0.22 μm MILLEX®-GV filters (Millipore Corporation). Filter-sterilize Rif through a 0.22 μm MILLEX®-LG filter (Millipore Corporation). Store antibiotics in aliquots at −20 °C for Km, Rif, and Cef and 4 °C for Hyg.
7. Stock solution of herbicide: dissolve 1 mg/mL glufosinate ammonium (GS) in sterile ddH_2O, filter-sterilize through 0.22 μm MILLEX®-GV filters (Millipore Corporation), and store in aliquots at 4 °C.
8. Stock solution of other chemicals: dissolve 100 mM acetosyringone and 20 mM calcium chloride ($CaCl_2$) in DMSO and ddH_2O, respectively. Filter-sterilize acetosyringone and $CaCl_2$ through 0.22 μm MILLEX®-LG and 0.22 μm MILLEX®-GV filters (Millipore Corporation), respectively, and store in aliquots at 4 °C.
9. 10 μM of each primer for PCR amplification of target gene as described in Table 1.

2.3 Culture Medium

1. Adjust media pH with 1 *N* NaOH or 1 *N* HCl before adding the agar and autoclaving. Use double-distilled water unless otherwise mentioned. Autoclave all media at 121 °C for 20 min at 105 kPa. Keep media at room temperature, and use within a week; otherwise, store at 4 °C for up to 2 weeks. Add sterile stock solutions of Km, Cef, GS, Rif, and acetosyringone to agar medium cooled to 50–60 °C or to liquid medium at room temperature after autoclaving. Pour 30 mL per plate for solid agar plates unless otherwise stated.
2. LB *Escherichia coli* culture medium: Luria-Bertani medium (10 g/L tryptone, 10 g/L NaCl, and 5 g/L yeast extract) and 18 g/L agar for solid medium, pH 7.0.

Table 1
List of primers

Target gene	Primer name	Sequence	Expected size
Hygromycin phosphotransferase (*hpt*)	*Hpt* forward primer	5′-CGC ATA ACA GCG GTC ATT GAC TGG AGC-3′	375 bp
Hygromycin phosphotransferase (*hpt*)	*Hpt* reverse primer	5′-GCC TGA ACT CAC CGC GAC-3′	375 bp
Bialaphos resistance (*bar*)	*Bar* forward primer	5′-ACC ATC GTC AAC CAC TAC ATC-3′	438 bp
Bialaphos resistance (*bar*)	*Bar* reverse primer	5′-GAA GTC CAG CCA GAA AC-3′	438 bp
Switchgrass *actin* gene	*Act* forward primer	5′-CAC TGG AAT GGT CAA GGC TG-3′	201 bp
Actin gene of switchgrass	*Act* reverse primer	5′-CTC CAT GTC ATC CCA GTT G-3′	201 bp

3. YEP *Agrobacterium* culture medium: 5 g/L NaCl, 10 g/L Bacto™ Peptone (Becton, Dickinson and Company, Sparks, MD, USA), 10 g/L yeast extract, and 18 g/L Bacto™ agar for solid medium, pH 7.0.
4. MS basal medium: MS salts and vitamins [13], 30 g/L sucrose, and 8 g/L agar (Sigma-Aldrich, St. Louis, MO, USA), pH 5.7.
5. Seed germination medium: 1/2 MS salts and vitamins [13], 15 g/L sucrose, and 8 g/L agar, pH 5.7.
6. Callus induction medium (CIM): MS basal medium containing 5 mg/L 2, 4-D; and 1.1 mg/L BAP, pH 5.7.
7. Co-cultivation medium: CIM containing 100 μM acetosyringone.
8. Selection CIM: CIM, 250 mg/L Cef (*see* **Note 2**), and 2 mg/L GS or 50 mg/L Hyg when the *bar* or *hpt* gene is the selectable marker, respectively.
9. Selection regeneration medium (RM): MS salts and vitamins, 30 g/L maltose, 8 g/L agar, 250 mg/L Cef, and 2 mg/L GS or 50 mg/L Hyg when the *bar* or *hpt* gene is the selectable marker, respectively.

2.4 Bacterial Strains and Binary Vector

1. Use two binary vectors, pDHB321.1 and pCAMBIA1391, for successful transformation of switchgrass as previously reported [11] (*see* **Note 3**). The pDHB321.1 vector contains the *bar* encoding resistance to the herbicide glufosinate ammonium. The pCAMBIA1391 vector contains a hygromycin phosphotransferase (*hpt*) encoding resistance to Hyg in plants and a

promoterless *gus*A. Both vectors contain a kanamycin-resistant marker for bacteria.

2. Transform pDHB321.1 and pCAMBIA1391 into competent cells of *Agrobacterium tumefaciens* strain EHA105 using the freezing-thaw method [14]. Select transformed colonies (EHA105:pDHB321.1 and EHA105:pCAMBIA1391) on YEP agar plate containing 100 mg/L Km and 30 mg/L Rif. Store single colony cultures of EHA105:pDHB321.1 and EHA105:pCAMBIA1391 in liquid YEP containing 100 mg/L Km and 30 mg/L Rif in 30% sterile glycerol (v/v) at −80 °C (*see* **Note 4**).

2.5 Other Materials

1. Sterile filter paper: cut sterile filter paper to 9 cm diameter, and autoclave in glass Petri-dish (100 mm × 20 mm) at 121 °C for 1 h at 105 kPa.
2. Container: use sterile Petri plates (100 mm × 15 mm) unless otherwise noted.
3. Planting medium: Suremix Perlite planting medium (Michigan Grower Products Inc., Galesburg, MI, USA) in plastic flats (72 cell insert) and 4 in. pots.
4. DNeasy Plant Mini Kit (Qiagen, Valencia, CA, USA).
5. RNeasy Plant Mini Kit (Qiagen, Valencia, CA, USA).

3 Method

3.1 Establishment of Sterile Cultures

1. Soak 3–5 g mature caryopses in 100 mL of dehulling solution in a 100 mL glass bottle, while on a rotary shaker at 200 rpm at room temperature for 45 min, decant the solution into a waste container, and rinse the seeds four times in sterile ddH_2O (*see* **Note 5**).
2. Mechanically dehulled mature caryopses are preferable (*see* **Note 6**).
3. Soak the seeds in 20 mL of 70% ethanol for 1 min in a sterile 50 mL tube and decant the ethanol.
4. Add sterilizing solution to the tube until it is full.
5. Soak the seeds in sterilizing solution for 30 min and decant the solution.
6. Rinse the seeds by soaking with 50 mL sterile ddH_2O for 2 min and decanting the solution. Repeat four times for a total of five washes.
7. Transfer sterilized seeds to a Petri dish.

3.2 Preparation of Explants

1. Place 30 surface-sterilized seeds onto CIM or seed germination medium in each Petri dish.
2. Induce callus or germinate seeds in the dark at 25 °C.
3. Subculture seed-derived calluses on fresh CIM every 4 weeks (Fig. 1a) (*see* **Note 7**).
4. Use 10-day-old sterile seedlings on seed germination medium to get seedling explants (Fig. 2a).
5. Obtain 0.5–0.8 cm seedling explants from the white node on the seedling stem (Fig. 2b).

3.3 Preparation of Agrobacterium Cells

1. Grow the *Agrobacterium* culture in a 15 mL sterile tube containing 3 mL of LB medium, 30 mg/L Rif, and 60 μL of EHA105 stock frozen in 30% glycerol. Place the tube on an orbital shaker at 200 rpm at 28 °C for 48 h.
2. Add 1 mL of culture to 99 mL of fresh LB medium containing 30 mg/L Rif in a 250 mL flask capped with foil. Place the flask

Fig. 1 *Agrobacterium*-mediated transformation of switchgrass calluses. (**a**) 10-week-old embryogenic calluses induced from dehulled seeds of 'Alamo' on CIM. (**b**) Inoculation of embryogenic calluses with *Agrobacterium* cells EHA105/pDHB321.1. (**c**) Embryogenic calluses after 4-day co-cultivation in the dark. (**d**) Glufosinate ammonium-resistant embryogenic calluses induced after 8-week selection on selection CIM. (**e**) Plant regeneration from glufosinate ammonium-resistant embryogenic calluses after 3-week culture on selection RM. (**f**) Rooting of transformants on selection RM

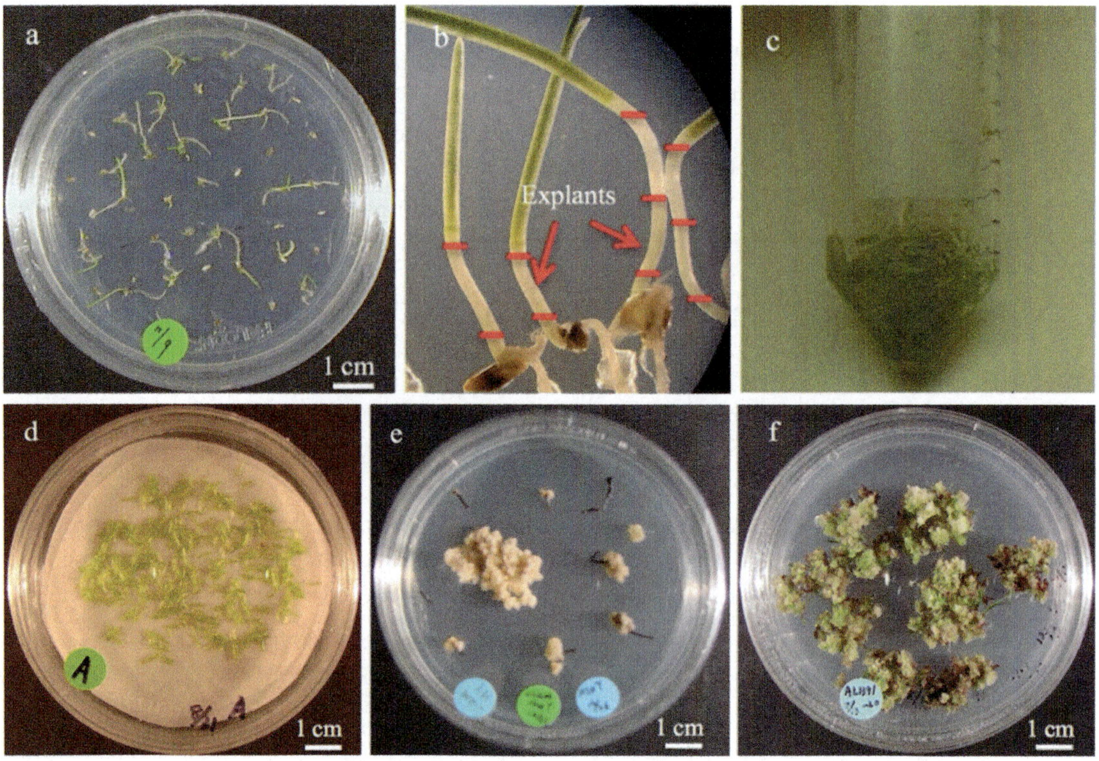

Fig. 2 *Agrobacterium*-mediated transformation of seedling explants of switchgrass. (**a**) 10-day seedlings from dehulled seeds of 'Alamo' on germination medium. (**b**) Seedling explants from the base part of each seedling. (**c**) Inoculation of seedling explants with *Agrobacterium* cells EHA105/pCAMBIA1391. (**d**) Embryogenic calluses after 4-day co-cultivation in the dark. (**e**) Hygromycin-resistant embryogenic calluses induced after 12-week selection on selection CIM. (**f**) Plant regeneration from hygromycin-resistant embryogenic calluses after 4-week culture on selection RM

on an orbital shaker at 200 rpm at 28 °C for 14–16 h. Grow the culture to OD_{600} (optical density) of 0.5–0.6.

3. Chill the culture on ice for 30 min.
4. Pour 50 mL of culture solution into two pre-chilled 50 mL tubes. Pellet the cells in a centrifuge at 2500 × *g* at 4 °C for 10 min.
5. Remove as much of the supernatant as possible.
6. Resuspend the cells with 2 mL of pre-chilled (0 °C) 20 mM sterile $CaCl_2$ in each 50 mL tube.
7. Aliquot 100 μL of the competent cell solution in pre-chilled sterile 1.5 mL tubes.
8. Place the tubes on ice if they are going to be used for transformation immediately. Otherwise, drop the tubes into liquid nitrogen, store the tubes at −80 °C, and thaw the cells on ice for 30 min prior to transformation.

9. Add 2–5 μg (<10 μL) plasmid DNA to competent *Agrobacterium* cells and mix by flicking the tube gently (*see* **Note 8**).
10. Incubate the mixture on ice for 20 min, and then freeze the tube in liquid nitrogen for 5 min.
11. Transfer the tube to a 37 °C water bath and incubate the mixture for 5 min.
12. Remove the tube from the water bath, and add 900 μL of LB (*see* **Note 9**), and place on an orbital shaker at 200 rpm at 28 °C for 4 h.
13. Centrifuge the tubes at 5510 × *g* at room temperature for 2 min to pellet the cells, discard 800 μL of supernatant, and resuspend the cells. Spread 100 μL on one LB plate containing 100 mg/L Km and 30 mg/L Rif.
14. Place the plate in an incubator at 28 °C for 2–3 days to form colonies.
15. The transformed cells on the plate can be stored in a refrigerator for up to 6 months.
16. Culture a single colony obtained from the plate of transformed EHA105 (containing either the pDHB321.1 or pCAMBIA1391) in 10 mL of liquid YEP containing 100 mg/L Km at 28 °C for 48 h in the dark.
17. Centrifuge at 2500 × *g* at room temperature for 5 min, and decant the liquid medium to collect *Agrobacterium* cells. Resuspend the culture to OD_{600} of 0.5–0.6 using 10–20 mL of liquid co-cultivation medium for infection.

3.4 Inoculation, Co-cultivation, and Plant Regeneration

1. Transfer explants (10–20 g calli or 200–400 freshly prepared seedling explants) into a 50 mL tube.
2. Add 10–20 mL of *Agrobacterium* suspension culture, and inoculate the explants for 10 min (Figs. 1b and 2c).
3. Decant the *Agrobacterium* suspension culture, and transfer the explants onto two layers of sterile filter paper in a Petri dish.
4. Transfer the calluses or seedling explants onto one piece of sterile filter paper overlaid on solidified co-cultivation medium in each Petri dish (Figs. 1c and 2d).
5. Co-cultivate the explants with *Agrobacterium* cells at 25 °C for 4 days in the dark (*see* **Note 10**).
6. Transfer the co-cultivated explants to a 50 mL tube.
7. Wash the explants (with gentle inversion by hand) 3–5 times in 50 mL of liquid CIM and a final wash in 50 mL of liquid CIM containing 500 mg/L Cef (*see* **Note 11**).
8. Transfer the explants onto two layers of sterile filter paper in a Petri dish to lightly blot dry.

9. Place the explants, 20 callus clusters or seedling explants per Petri dish, on selection CIM in each Petri dish. Culture all explants on selection CIM at 25 °C for 4 weeks in the dark.
10. Subculture the explants to fresh selection CIM at 4-week intervals (Figs. 1d and 2e) (*see* **Note 12**).
11. After 2–3 months of selection, transfer resistant calli onto selection RM in Petri dishes (Figs. 1e and 2f). Culture all calli on selection RM at 25 °C for 4–6 weeks under a 16 h photoperiod of 30 μmol m^{-2} s^{-1} from cool white fluorescent tubes.
12. Transfer the regenerated plants onto 50 mL of fresh selection RM in each GA7 Magenta box for rooting (Fig. 1f).

3.5 Transplanting and Greenhouse Care

1. Select rooted regenerants, 5–10 for each regenerated callus cluster (*see* **Note 13**), and wash them using tap water to remove all agar attached (Fig. 3a).
2. Insert individual regenerant directly in planting medium in six-pack flats (Fig. 3b).
3. Cover the flats with transparent plastic covers, and grow at 25 °C for 5 weeks under a 16 h photoperiod of 50 μmol m^{-2} s^{-1}.
4. Progressively open and remove the plastic covers in the first 7 d, and water the plants at 3-d intervals.
5. Transplant the plants from the cell packs into 4 in. pots.
6. Transfer the plants to a secured greenhouse [21–32 °C/10–21 °C (day/night), 25–65 μmol/m^2/s on 10–14 h/d] (Fig. 3c).
7. Water the plants as needed, and fertilize weekly using a nutrient solution of 0.2 g/L 20-20-20 fertilizer.
8. Grow the plants for molecular analysis and to harvest seeds (Fig. 3d) (*see* **Note 14**).

3.6 Molecular Analysis of Transgenic Plants

1. Isolate genomic DNA from young leaves using the DNeasy Plant Mini Kit or the CTAB method [15].
2. Use 150–200 ng of DNA for each 20 μL of PCR reaction (10 μL of 2× PCR master mix, 1 μL of forward primer, 1 μL of reverse primer, 1 μL of DNA, and 7 μL of ddH_2O).
3. Isolate total RNA from young leaves using the RNeasy Plant Mini Kit.
4. Purify the RNA using DNase.
5. Synthesize cDNA using 1 μg of purified RNA. Use 2 μL of diluted cDNA (1:5 in ddH_2O) product for each 20 μL RT-PCR reaction (10 μL of 2× PCR master mix, 1 μL of forward primer, 1 μL of reverse primer, 2 μL of cDNA, and 6 μL of ddH_2O).

Fig. 3 Rooting and greenhouse care of transgenic switchgrass plants cv. Alamo.(**a**) Rooted plants. (**b**) Transplanted transgenic plants. (**c**) Transgenic plants growing in the greenhouse. (**d**) Transgenic plants flowering

6. The reaction conditions for PCR and RT-PCR: 94 °C for 2 min, 35 cycles of 94 °C for 45 s, 58 °C for 60 s and 72 °C for 90 s, with a final extension at 72 °C for 10 min.
7. Run on a 1.0% agarose gel containing ethidium bromide to separate and visualize amplified PCR and RT-PCR products.

4 Notes

1. To date, octoploid switchgrass cultivars are still recalcitrant for transformation due to the lack of efficient regeneration systems [16].
2. Alternatively, 250 mg/L Timentin also works effectively. For both Cef and Timentin, 500 mg/L can be used for eliminating

Agrobacterium during selection stages after co-cultivation, since they do not have significant inhibition on embryogenic callus induction and plant regeneration for switchgrass.

3. The selectable marker *hpt* and *bar* are preferable to the neomycin phosphotransferase II gene (*npt*II), because non-transgenic calluses of switchgrass shows high tolerance to Km (e.g., 100 mg/L) [11].
4. The glycerol stocks of *Agrobacterium* cultures can be stored at −80 °C for up to 2 years. To initiate single colonies, streak the stocks on LB plate containing appropriate antibiotics, and culture the cells at 28 °C for 2–3 days. All three commonly used *A. tumefaciens* strains can be used for gene delivery to switchgrass calluses. In previous transient transformation studies, EHA105 appeared to be more efficient compared to other strains [11].
5. It is important to use glassware, protective eyewear glasses, and gloves to handle sulfuric acid. The waste solution of sulfuric acid needs to be neutralized before disposal.
6. We prefer to mechanically dehull seeds although it is not easy to perform. Direct sterilization of mature caryopses without dehulling caused fungi contamination in over 95% of seeds.
7. Regeneration ability of these embryogenic calluses can be maintained for 2 years, as 2-year-old calluses did not show significant decrease in regeneration frequency compared to 10-week-old calluses.
8. A smaller amount of plasmid (e.g., 100 ng) is needed for an electroporator.
9. Alternatively, add 900 μL LB to the tube, and transfer the culture to a 15 mL tube.
10. Use filter paper to prevent *Agrobacterium* overgrowth.
11. The last wash using CIM containing 500 mg/L Cef is important for successful elimination of *Agrobacterium* cells after co-cultivation.
12. During the subculture, treat each callus cluster as a separate explant, remove dead callus cells, and keep the rest of the callus cluster intact.
13. Plants from separate callus clusters were considered plants from independent transgenic events. Plants from the same callus cluster may or may not belong to the same transgenic event.
14. Only a very low percentage of mature seeds ranging from 1.7 to 8.7% for open pollinated plants were obtained.

References

1. Vogel KP (2004) Switchgrass. In: Moser LE, Burson BL, Sollenberger LE (eds) Warm-season (C4) grasses, Agronomy monograph, vol 45. ASA, CSSA, and SSSA, Madison
2. McLaughlin SB, Kszos LA (2005) Development of switchgrass (Panicum virgatum) as a bioenergy feedstock in the United States. Biomass Bioenerg 28:515–535
3. Bouton J (2008) Improvements of switchgrass as a bioenergy crop. In: Vermerris W (ed) Genetic improvement of bioenergy crops. Springer, Berlin, Heidelberg
4. Merrick P, Fei SZ (2015) Plant regeneration and genetic transformation in switchgrass – a review. J Integr Agric 14(3):483–493. https://doi.org/10.1016/S2095-3119(14)60921-7
5. Somleva MN, Tomaszewski Z, Conger BV (2002) Agrobacterium-mediated genetic transformation of switchgrass. Crop Sci 42:2080–2087
6. Somleva MN (2006) Switchgrass (Panicum virgatum L.). In: Wang K (ed) Methods in molecular biology, Agrobacterium protocols, vol 344. Humana Press Inc, Totowa
7. Somleva M, Snell K, Beaulieu J, Peoples O, Garrison B, Patterson N (2008) Production of polyhydroxybutyrate in switchgrass, a value-added co-product in an important lignocellulosic biomass crop. Plant Biotechnol J 6:663–678
8. Burris JN, Mann DGJ, Joyce BL, Stewart CN (2009) An improved tissue culture system for embryogenic callus production and plant regeneration in switchgrass. BioEnergy Res 2:267–274
9. Xi Y, Fu C, Ge Y, Nandakumar R, Hisano H, Bouton J, Wang Z-Y (2009) Agrobacterium-mediated transformation of switchgrass and inheritance of the transgenes. BioEnerg Res 2:275–283
10. Richards HA, Rudas VA, Sun H, McDaniel JK, Tomaszewski Z, Conger BV (2001) Construction of a GFP-BAR plasmid and its use for switchgrass transformation. Plant Cell Rep 20:48
11. Song G, Walworth A, Hancock JF (2012) Factors influencing Agrobacterium-mediated transformation of switchgrass cultivars. Plant Cell Tiss Org Cult 108:445–453
12. Li RY, Qu RD (2011) High throughput agrobacterium-mediated switchgrass transformation. Biomass Bioenergy 35 (3):1046–1054. https://doi.org/10.1016/j.biombioe.2010.11.025
13. Murashige T, Skoog F (1962) A revised medium for rapid growth and bioassay with tobacco tissue cultures. Physiol Plant 15:473–497
14. Hood EE, Gelvin SB, Melchers LS, Hoekema A (1993) New agrobacterium helper plasmids for gene transfer to plants. Transgenic Res 2 (4):208–218. https://doi.org/10.1007/bf01977351
15. Doyle JJDJ (1987) A rapid DNA isolation procedure for small quantities of fresh leaf tissue. Phytochem Bull 19:11–15
16. Seo M, Takahara M, Takamizo T (2010) Optimization of culture conditions for plant regeneration of Panicum spp. through somatic embryogenesis. Grassl Sci 56(1):6–12

Chapter 9

Biolistic Transformation of Wheat

Bin Tian, Mónica Navia-Urrutia, Yueying Chen, Jordan Brungardt, and Harold N. Trick

Abstract

Biolistic transformation of wheat is one of the most commonly used methods for gene function study and trait discovery. It has been widely adapted as a fundamental platform to generate wheat plants with new traits and has become a powerful tool for facilitating the crop improvement. In this chapter, we present a complete and straightforward protocol for wheat transformation via biolistic bombardment system. Although wheat is still one of the hardest plant species to transform, this protocol offers an optimized and efficient system to produce transgenic plants. To demonstrate the application of this protocol, in this chapter we describe an example of obtaining transgenic wheat by the co-bombardment of two plasmids, containing a green fluorescent protein gene and a glufosinate herbicide selection gene, respectively. In addition, procedures for the screening and testing of putative transgenic plants are described. This protocol has been successfully applied to generate stable transgenic bread wheat (*Triticum aestivum*) in both spring and winter varieties.

Key words Wheat, Biolistics, Particle bombardment, Embryogenic callus, Genome editing, GFP

1 Introduction

Hexaploid wheat (*Triticum aestivum*, $2n = 6x = 42$; AABBDD) is one of the most important crops in the world. As a staple food worldwide, the production of wheat needs to be increased to meet the ever-growing population [1]. Like other crops, wheat faces myriad of biotic and abiotic factors that limit its production. Many of these factors may be overcome though the use of modern biotechnology, using plant transformation as the foundation of crop improvement. Plant transformation is a powerful tool for gene function study and trait discovery and is essential for applied and fundamental studies of wheat. Currently, two primary delivery methods have been established for wheat transformation: *Agrobacterium*-mediated system and a biolistics-based system. The first inheritable transgenic wheat was obtained through biolistic bombardment in 1992 [2] and in 1997 with *Agrobacterium*-mediated

Sandeep Kumar et al. (eds.), *Transgenic Plants: Methods and Protocols*, Methods in Molecular Biology, vol. 1864, https://doi.org/10.1007/978-1-4939-8778-8_9, © Springer Science+Business Media, LLC, part of Springer Nature 2019

transformation [3]. Although considerable debates on advantages and disadvantage of these two systems are still ongoing [4–6], the biolistic transformation of wheat has become a commonly used approach to generate transgenic wheat plants. Biolistic transformation is advantageous to integrate new technologies such as genome-editing tools (e.g., CRISPR/Cas9) and to provide a robust platform in the development and recovery of DNA- and marker-free genome-edited plants [7, 8]. The particle inflow gun-mediated [9] protocol described below has been successfully used to generate transgenic wheat in a number of spring and winter varieties [10–13] and is routinely used in our lab to produce transgenic wheat plants with an efficiency of 5–10% from embryogenic calli. Typically, the plasmid pAHC20 [14] can be co-bombarded with target gene vectors as the source of the selectable marker gene. The former vector carries the *bar* gene driven by the constitutive ubiquitin (Ubil) promoter from maize [14]. Although the independent integrations of these vectors may occur during the co-bombardment, more than 90% of the events contain both constructs during our tests. Nonetheless, the co-bombardment system has the potential to generate marker-free plants through segregation in later generations. This simple and comprehensive procedure produces young wheat plantlets within 10–14 weeks from callus induction, and transgenic seed can be recovered in a further 4–5 months post-transformation.

2 Materials

2.1 Plants Materials and Sterilization

1. Wheat plants (*Triticum aestivum* L. cv. 'Bobwhite') grown under controlled environment (e.g., growth chambers and greenhouses) (*see* **Notes 1** and **2**).
2. Tween 20.
3. Commercial bleach (sodium hypochlorite 5.25% by weight).
4. Sterile double-distilled water (ddH_2O).

2.2 Stock Solutions and Tissue Culture Medium

The stock solutions of basal culture media are all made as 100× stocks. Solutes are dissolved in ddH_2O as described in Table 1, and solutions are stored at 4 °C. The final tissue culture media are prepared from these stock solutions (*see* **Note 3**).

2.2.1 Induction Media

1. Wheat Initiation Media (CM4): add 10 mL each of 100× stock solutions of MS FeEDTA (add first), MS halides, MS sulfates, MS P, B, Mo, and B5 vitamins to approximately 800 mL of distilled water. While mixing with stir plate, add 1.65 g of NH_4NO_3, 1.9 g of KNO_3, 0.5 mL of 1 mg/mL 2,4-D, 2.2 mL of 1 mg/mL Picloram, and 40 g of maltose. Adjust pH to 5.7 and add water to 1 L volume. Divide medium into

Table 1
MS and B5 vitamin stock solutions (100× concentration)

Stock	Component	Amount (g)/1 L	Amount (g)/500 mL
MS sulfates			
	$MgSO_4{\cdot}7H_2O$	37.00	18.50
	$MnSO_4{\cdot}H_2O$	1.69	0.845
	$ZnSO_4{\cdot}7H_2O$	0.86	0.43
	$CuSO_4{\cdot}5H_2O$	0.0025	0.00125
MS halides			
	$CaCl_2{\cdot}2H_2O$	44.00	22.00
	KI	0.083	0.0415
	$CoCl_2{\cdot}6H_2O$	0.0025	0.00125
MS P, B, MO			
	KH_2PO_4	17.00	8.50
	H_3BO_3	0.62	0.31
	$Na_2MoO_4{\cdot}2H_2O$	0.025	0.0125
MS FeEDTA[a,b]			
	Na_2EDTA	3.724	1.862
	$FeSO_4{\cdot}7H_2O$	2.784	1.392
B5 vitamins[b]			
	Myoinositol	10.0	5.00
	Nicotinic acid	0.10	0.05
	Pyridoxine HCI	0.10	0.05
	Thiamine	1.00	0.50

[a]Completely dissolve EDTA first, slowly add $FeSO_4$ while stirring
[b]Store in foil-covered bottles

two 1 L media bottles, add 1 g of Gelrite into each bottle, and autoclave for 20–25 min.

2. CM4 + MS Media: add 36.44 g of mannitol and 36.44 g of sorbitol to CM4 medium above. Adjust pH to 5.7 and add water to 1 L volume. Divide medium into two 1 L media bottles, add 1 g of Gelrite into each bottle, and autoclave for 20–25 min.

2.2.2 Regeneration Media

1. Wheat Shoot Production Medium (MSP): add 10 mL 100× stock solutions of MS FeEDTA (add first), MS halides, MS sulfates, MS P, B, Mo, and B5 vitamins to approximately 800 mL of distilled water. While mixing with stir plate, add 1.65 g of NH_4NO_3, 1.9 g of KNO_3, 0.2 mL of 1 mg/mL 2,4-D, 1.9 g of MES, 40 g of maltose, and 100 mg of ascorbic acid. Adjust pH to 5.7, and add water to 1 L volume. Divide medium into two 1 L media bottles, add 1 g of Gelrite into each bottle, and autoclave for 20–25 min.

2. Wheat Root Production Medium (MSE): add 10 mL 100× stock solutions of MS FeEDTA (add first), MS halides, MS sulfates, MS P, B, Mo, and B5 vitamins to approximately

800 mL of distilled water. While mixing with stir plate, add 1.65 g of NH_4NO_3, 1.9 g of KNO_3, 1.95 g of MES, 40 g of maltose, and 100 mg of ascorbic acid. Adjust pH to 5.7 and add water to 1 L volume. Divide medium into two 1 L media bottles, add 1 g of Gelrite into each bottle, and autoclave for 20 to 25 min.

2.2.3 Selection Media

Allow autoclaved medium to cool down (about 50–55 °C) before adding any selection agents (e.g., herbicide). Pour the warm media into plates (100 mm × 15 mm) or pour into large borosilicate culture tubes (25 mm × 150 mm, autoclaved previously), about 13 mL per tube.

1. CM4 + 5G: add 250 μL of 10 mg/mL filter-sterilized glufosinate ammonium stock to 500 mL CM4 after autoclaving.
2. CM4 + 10G: add 500 μL of 10 mg/mL filter-sterilized Glufosinate ammonium stock to 500 mL of CM4 after autoclaving.
3. MSP + 10G: add 500 μL of 10 mg/mL filter-sterilized Glufosinate ammonium stock to 500 mL MSP after autoclaving.
4. MSE + 5G: add 250 μL of 10 mg/mL filter-sterilized Glufosinate ammonium stock to 500 mL MSE after autoclaving.

2.3 Bombardment Materials

1. The biolistics system (*see* Fig. 1) and as described by Finer et al. [7].
2. Tungsten particle (99.9% 1 micron M10 equiv., Atlantic Equipment Engineers, Bergenfield, NJ): weight out 50 mg of

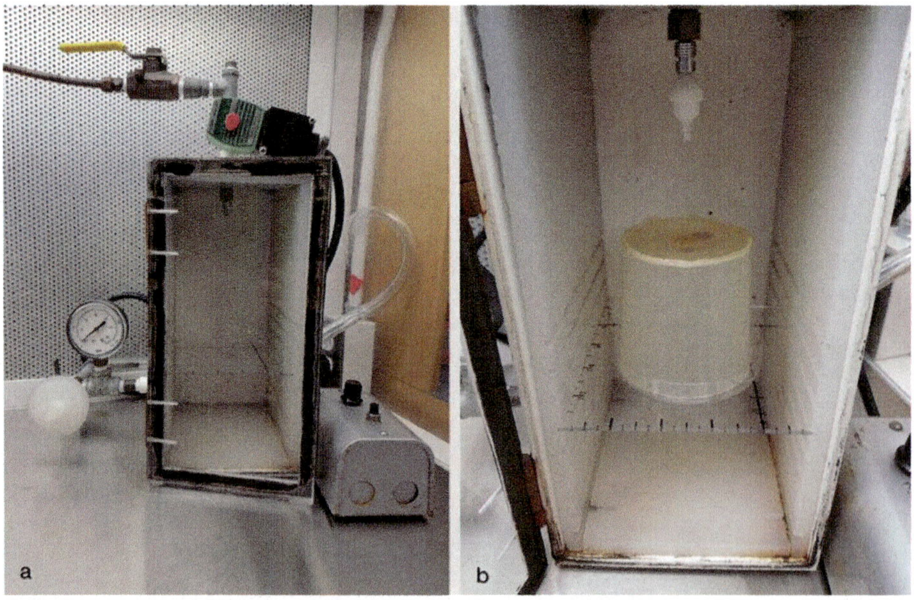

Fig. 1 Particle inflow gun used in biolistic transformation. (**a**) The general structure of particle gun. (**b**) Particle gun chamber with Swinney filter and baffle

particles into a sterile 1.5 mL microfuge tube, and sterilize the particles with 500 μL 95% (v/v) ethanol for 20 min. Centrifuge the tube at 10 K × *g* for 1 min, remove the ethanol, and dry the particles completely (*see* **Note 4**).

3. Baffles are constructed as described by Finer et al. [9].
4. High-purity plasmid DNA is used for the bombardment. The minipreps are extracted by any commercial plasmid DNA mini kit (e.g., QIAGEN or OMEGA), and adjust the concentration to 0.5–1 μg/μL in ddH_2O and storage at −20 °C.
5. 2.5 M calcium chloride: dissolve 3.68 g of $CaCl_2 \cdot 2H_2O$ in 10 mL sterile water, and filter-sterilize to aliquot by 50 μL in 500 μL tubes. Aliquots can be stored at room temperature for at least 6 months. Discard aliquot after use to prevent contamination.
6. 0.1 M spermidine is prepared from 1 M stock solution prior to each use.

2.4 Materials for Testing Gene Expression

1. Glufosinate ammonium-based herbicide solution (Liberty): make up a 0.2% Liberty solution from the Liberty stock stored at 4 °C (e.g., 200 μL Liberty into 100 mL H_2O).
2. pAHC20 containing *bar* gene for selection.
3. DNA extraction and plasmid extraction reagents and equipment using any commercial kits.
4. PCR and primers for detecting the presence of bar genes.

   Ubi-ABF: 5′- CCTGCCTTCATACGCTATTTATTTGC-3′.

   Bar-ABR: 5′- CTTCAGCAGGTGGGTGTAGAGCGTG-3′.

   Product size: 453 bp.
5. For GFP expression, the superfolder green fluorescent protein (sGFP) gene with full length 720 bp from HBT-sGFP-NOS construct (kind gift from Jen Sheen, Massachusetts General Hospital) was flanked with *Bam*HI, and blunt end ligated to the previously digested and filled *Bam*HI site of pAHC17 [15].

3 Methods

3.1 Wheat Seed Collection and Sterilization

1. Collect wheat spikes at 10–14 days post-anthesis (*see* **Note 5**). The immature embryos should be approximately 2–3 mm in length (Fig. 2a).
2. Isolate the immature seeds from the lemma and palea by hand, surface sterilize wheat seeds in 20% (v/v) bleach with 0.1% (v/v) Tween20 in a 50 mL sterilized tube, and fix the 50 mL tube in an orbital shaker, with continuous shaking at 100–150 rpm for 15–20 min.

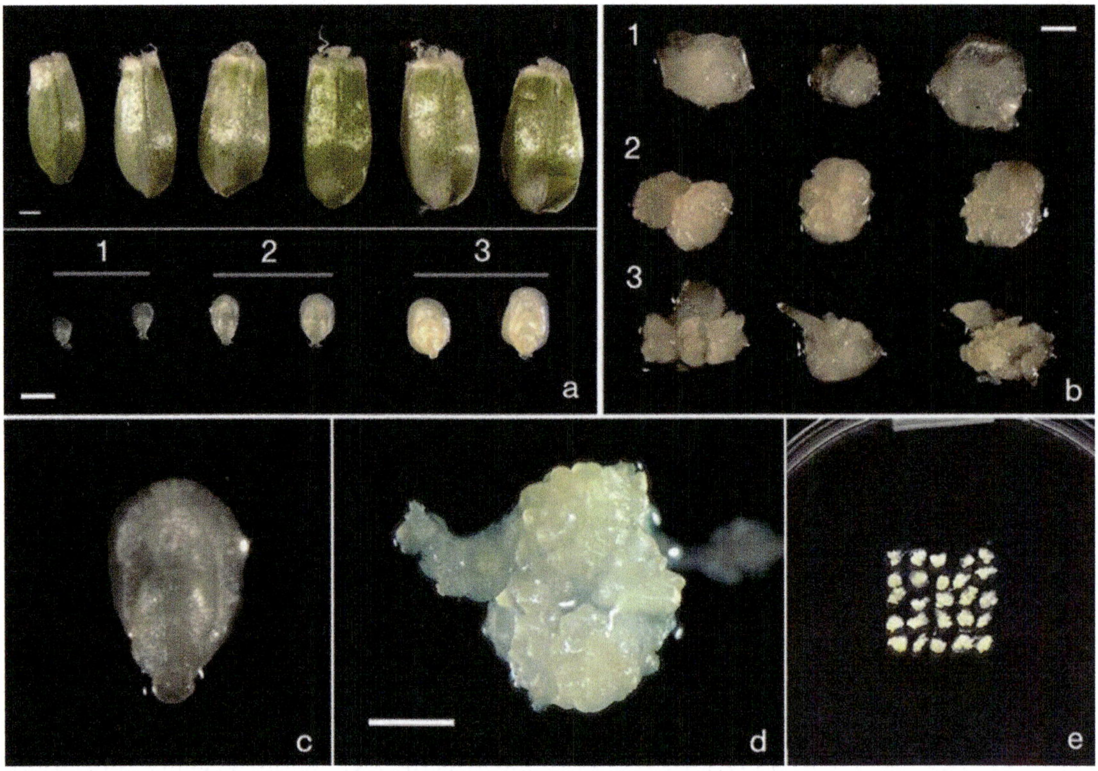

Fig. 2 Selection of immature seeds, isolation of immature embryos and embryogenic callus induction for bombardment. (**a**) Representation of different immature seed sizes (top) with respected isolated embryos (below). Embryos in group "2" represent most responsive to callus induction. (**b**) Embryogenic calli induced after 10 days. Calli in group "2" typically generate the best tissue for transformation. (**c**) Representation of embryos that respond the best for callus induction. Note, axis should be faced down in contact with media for best induction. (**d**) Ideal embryogenic callus tissue for bombardment. (**e**) Representation of selected calli was placed in a 5 × 5 grid in the center of petri dish prior to bombardment. Scale bar = 1 mm approximately

3. Remove the bleach from the sterilized seeds, and rinse copiously 3–5 times with sterile ddH_2O in a sterilizing hood. Remove excess water from seeds, but keep seeds in a moist condition (*see* **Note 6**).

3.2 Immature Embryos Selection and Pre-culture

1. In the laminar flow hood, excise the immature embryos from sterilized seeds under the microscope. Optimal embryo size is between 0.5 and 1.5 mm in length (*see* Fig. 1c). Place 100–150 embryos per CM4 plate (100× 15 mm Petri dish), with the embryo axis faced down in contact with the induction medium CM4 (*see* **Note 7**).
2. Seal the plates with polyethylene wrap. Initiate embryos on CM4 in the dark at room temperature. After 7–12 days of induction on CM4 medium, embryogenic calli should regenerate on the scutellum (*see* **Note 8**).

3. Select vigorously growing calli and separate them from the scutellum with tweezers. Place selected embryogenic calli in the center area of the plate (100 × 15 mm) with CM4 medium. Put 25 calli per plate, in a 5 × 5 pattern (*see* Fig. 2e). Seal the plates with polyethylene (plastic) wrap, and keep them in the dark at room temperature for 2–3 days (*see* **Note 9**). Seal plates with plastic wrap and keep in the dark.

3.3 Tissue Preparation

1. The CM4 plates with selected calli are maintained in the dark for 2–3 days at room temperature before bombardment.
2. Precondition the tissue before bombardment. Remove the cover of Petri dishes, and air dry calli in a laminar flow hood for 20 min to reduce any water on the calli surface (*see* **Note 10**). Air-dried calli are ready for the bombardment.

3.4 Preparing and Coating Particles with DNA for Bombardment

1. Wash the prepared tungsten particles 2–4 times with sterile ddH_2O before use by repeated centrifugation (at 9.6 K × *g* for 1 min) and resuspension by vortex.
2. Resuspend the particles in a final volume of 500 μL sterile water (use within 1 day). Aliquot 25 μL of resuspended particles to 5 μL of 1 μg/μL DNA (*see* **Note 11**) into a 1.5 mL sterilized microfuge tube, and mix by vortex. If additional samples need to be prepared, vortexing Tungsten particles between each aliquot is necessary to ensure complete homogenization of particles. Allow binding for 1 min at RT.
3. Precipitate the DNA by adding 25 μL of 2.5 M CaCl2 and then add 10 μL of 0.1 M spermidine. Place the tube on ice for 4 min and allow to settle (do not vortex).
4. Remove 50 μL of supernatant and discard. Keep coated particles on ice, and use mixture within 15 min. There should be enough volume for six shots per precipitated DNA sample, at 2 μL per shot.

3.5 Particle Bombardment, Recovery, and Induction of Embryogenic Calli

1. Clean the biolistic chamber and the microfuge tube rack with 70% (v/v) ethanol. Sterilize the Swinney filter and baffle by autoclave.
2. Turn on helium gas and adjust the helium line pressure to 60 PSI (*see* **Note 12**). Turn the vacuum pump on and close the vacuum and vent lines. Press the charge button a few times to make sure pressure remains steady.
3. Place a sterile Swinney filter in the sterilized microfuge tube rack. Washboard or vortex prepared coating particles from Subheading 3.4 immediately before (and during, if possible) removal of 2 μL of particles for bombardment (*see* **Note 13**).

4. Carefully aliquot 2 μL of coated particles onto the center of the Swinney filter by pipetting through the top of filter unit onto the screen.
5. Place the CM4 plate with wheat tissue that is prepared in Subheading 3.3 on the adjustable plexiglass shelves. Place a sterilized baffle on top of the medium covering the tissue. Place the Swinney filter unit containing the coated particles in the filter holding unit. Close the plexiglass door of the chamber (*see* Fig. 1).
6. Lightly push on the plexiglass door while opening the valve leading to the vacuum. The vent valve should be closed. After the vacuum reaches between 28 and 30 Hg, press the button to activate the solenoid and release the particles.
7. Close the vacuum line valve and slowly open the vent valve. It should take approximately 30 s for the chamber to vent and the door to open. Remove the CM4 plates from plexiglass shelves. Check the remaining tissues on the raffle before the move to the next round of bombardment.
8. After bombardment, seal the CM4 plates with polyethylene wrap (*see* **Note 14**) and recover at room temperature (RT, 22 ± 1 °C) in the dark for 2–4 days.

3.6 Regeneration, Rooting, and Selection of Transgenic Plants

1. Following recovery, transfer the embryogenic calli to plates with induction medium plus selection agent (CM4 + 5G), spreading them across the medium. Seal the plates with polyethylene wrap (*see* **Note 15**), and incubate at RT in the dark for 2 weeks.
2. After 2 weeks, transfer the embryogenic calli to plates with induction medium plus selection agent (CM4 + 10G), distributing them on the surface. Seal the plates and incubate at RT in the dark for 2 weeks.
3. Transfer the embryogenic calli to fresh CM4 + 10G plates, and incubate under the same conditions described before.
4. Transfer the callus clumps to shoot production medium plus selection agent (MSP + 10G). Seal the plates individually with polyethylene wrap, distribute the plates over a shelf (without stacking) in a regime of 16 h under light (~80 μmol m^{-2} s^{-1}) at 20 °C, and 8 h at 18 °C in darkness, and 40–50% relative humidity (RH) for 2 weeks. Shoots should begin to appear after 2 weeks and continue to appear for more than a month.
5. Transfer the callus clumps to shoot elongation selection medium plus selection agent, MSE+ 5G, distributing callus clumps in two or three plates, according to their size, to give them more space to grow and avoid nutrients depletion. Seal the plates and incubate under light with the same conditions

described in the previous step. The MSE+ 5G plates can be kept for up to 2 months.

6. Transfer shoot-regenerating calli to tubes with MSE + 5G when regenerated shoots are larger than 1.0 cm in length (*see* **Notes 15** and **16**). Seal the tube and incubate under light with the same conditions described in **step 4** until the development of the plantlets (*see* **Note 17**).

3.7 Transfer of Plants to Potting Mix Leaf Painting Bioassays

1. Transfer well-developed plantlets (more than 7 cm length and established root system) from tubes to soil in square Jiffy pots (2″ × 2″). Put the pots inside a clear plastic container, and cover it to maintain a high relative humidity. Place the container in a controlled environment (16 h in light at 20 °C, and 8 h in darkness at 15 °C), and keep the plantlets under high humidity for 5 days to promote acclimatization.
2. After 5 days, partially open the container, and keep the peat pots inside the container for 1 or 2 weeks until plants are well established and acclimated to ambient humidity levels.
3. Transfer the plants to big pots (0.5 gallon pots), and grow the plants under previously reported conditions (*see* **Note 2**).

3.8 Leaf Painting Bioassays

1. Test herbicide resistance in plants that reach a three-leaf or four-leaf stage using a leaf "painting" methodology. Select a leaf from each tiller, avoiding the flag leaf, and make a mark with a permanent marker at 2/3 of the leaf base.
2. Dip a cotton ball in the 0.2% Liberty® solution (*see* **Note 18**), remove excess liquid by squeezing the cotton, and apply the solution on the upper surface (adaxial) of the leaf, from the mark to the leaf tip. Keep the plants under previously reported conditions.
3. Evaluate the reaction to the herbicide after 7–10 days.
4. The treated leaves of the herbicide-resistant plants show no change in coloration/appearance (*see* Fig. 3). Consider these plants/tillers as positive (presence of the marker gene and the gene of interest, GOI), save them, and use them for the following analysis.
5. The treated leaves of herbicide-sensitive plants have a light green or brown coloration. Consider these plants/tillers as negative (absence of the marker gene and the GOI), and discard them. If positive and negative tillers are found in the same pot, cut the negative tillers and keep only the positive ones. Save some negative plants to use as an experimental control.
6. Consider plants with doubtful symptoms as positive, and confirm the diagnostic in the following analysis.

Fig. 3 Development and growth of transformed wheat. (**a**) Embryogenic calli after 2 weeks on MSP + 5G medium. (**b**) Plantlet development in culture tubes containing MSE + 5G media. (**c**) Liberty painting screening of leaves from putative transgenic plants showing Liberty-sensitive leaves (1) and Liberty-positive leaves (2). (**d**) Confirmed transgenic plant with heads bagged to prevent outcrossing

7. Sample a piece of leaf, avoiding the flag leaf and the leaf treated with the herbicide. The size of the sample depends on the protocol to extract DNA to be used. Confirm the presence of the marker gene *bar* and the GOI by PCR with proper primers (*see* **Note 19**).

4 Notes

1. Healthy plants are critical for successful transformation. Optimal and consistent conditions need to be established for plant

growth. 'Bobwhite' is the preferred wheat variety due to its good response to tissue culture in this protocol. This protocol is applicable to other spring varieties such as 'Fielder' and 'Forefront' and hard winter varieties, such as 'Jagger' and 'Everest.' Five to ten seeds are planted per 1 gallon plastic pots (PF310, Premier Nursery supplies, Hummert, USA) depending upon germination rates of seeds. Ideally three to five donor plants are grown in each pot.

2. Growth condition for donor plants should be set at a 16 h photoperiod, at a light intensity of approximately 600 μE m^{-2} s^{-1} at a day:night temperature regime of 20:18 °C. For winter wheat varieties, it is critical to vernalize at 4–10 °C for 4–12 weeks depending on the variety. Water daily if needed to avoid dry pots, and fertilize weekly with a 20-20-20 general purpose fertilizer at a rate of 1 teaspoon/gallon.
3. The stock solutions are stored at 4 °C should be used in 1–2 months.
4. Tungsten particles can be prepared in aliquots (washed and sterilized) beforehand. Particles in aliquoted 1.5 mL microcentrifuge tubes should be dried completely (e.g., air dry overnight) and stored at room temperature. Autoclaving or storage of particles in water (for more than 12 h) should be avoided as this will cause oxidation.
5. Plants infected with diseases need to be treated or discarded immediately. Powdery mildew can be a common disease, and the fungicide Baytan 30 (Bayer CropScience, NC, USA) should be applied as early as symptom first shown (Apply Baytan 30 at a rate of 2 mL fungicide per 1 L water). Alternatively, sulfur can be applied in sulfur burners (typically set for 4 h/night) as a preventative to powdery mildew. Neither of these methods of powdery mildew control appears to affect the regeneration of callus formation.
6. It is preferable to excise seeds the same day they are collected and surface sterilized but can be excised the following day if stored overnight at 4 °C. It is preferable to store the wheat spikes in 4 °C if they need to be collected in advance and surface sterilize seeds the day of excising.
7. Embryos size is critical for the callus generation. Optimum embryo size is 0.5–1.5 mm in length. Embryos length should not be larger than 4 mm as they will have poor callus induction.
8. Initiation time varies with cultivar and growth conditions of the parent plants.
9. The tungsten particles usually shot into an area of 2 cm^2 diameter with the gene gun. Twenty-five target tissues per plate are

typically arranged in a 5 × 5 grid within 2 cm^2 area to maximize efficiency of each bombardment.

10. Air drying of calli should not be more than 30 min. Overdried callus will lower the bombardment efficiency. Osmotic pre-treatment (0.2 M Mannitol + 0.2 M sorbitol in medium) for 4 h, followed by a 16 h posttreatment or recovery period, is another common method to precondition calli prior to bombardment.
11. Each sample preparation should contain 0.5–1 μg/μL DNA.
12. Helium pressure can range between 50 and 100 PSI. This protocol has been optimized at 60 PSI. The maximum rating for the solenoid is 140 PSI—do not exceed this setting.
13. The prepared coated particles need to be used as soon as possible. If multiple treatments or many samples are to be prepared, it is recommended to keep them on ice all the time prior to loading on the Swinney filter.
14. Culture plates requiring with 24 h photoperiod (e.g., cultures on CM4, CM4 + 5G, and CM4 + 10G media) can be stacked and wrapped together with polyethylene film to save space.
15. Before transferring shoots to MSE + 5G tubes (*see* Fig. 3b), an additional transfer to MSE + 5G medium in deep plates (100 mm × 25 mm) can be conducted to favor the growth of shoots.
16. Instead of transferring the tissue mass with shoots, shoots could be separated from the mass and transferred independently to MSE + 5G tubes. Only well-developed shoots should be transferred independently as this can decrease the viability of small shoots. In some instances, small roots can be observed in this tissue as well.
17. Check the condition of the plates frequently, if contamination is observed, transfer clean tissue to fresh medium when possible. Dispose of the obviously contaminated calli and those calli surrounding the contamination point even though they appear to be clean.
18. Liberty® 280 SL herbicide (Bayer CropScience, Raleigh, NC, USA) is a liquid formulation; the active ingredient glufosinate ammonium is 24.5%, equivalent to 280.4 gr/L (2.34 pounds per US gallon). The 0.2% Liberty® solution used for painting has 0.56 g of active ingredient per liter.
19. The primer pair Ubi-ABF and Bar-ABR (*see* Subheading 2.4, **item 4**) can be used for confirming the integrity of the *bar* gene on pAHC20. Figure 4 showed the *GFP* gene expression along time course after co-bombardment, and Fig. 5 exhibited the phenotype of stable transgenic wheat with GFP obtained from the experiment.

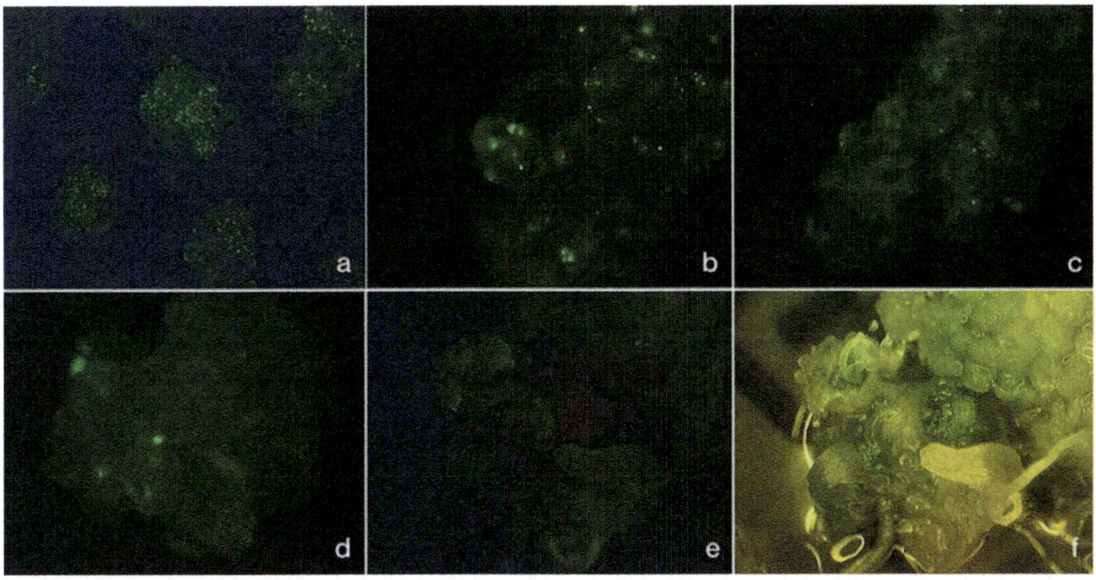

Fig. 4 Growth and development of transgenic wheat embryogenic calli expressing GFP. (**a**) Transient expression of GFP reporter in the calli 1 day after bombardment, (**b**) 2 days after bombardment, (**c**) 8 days after bombardment, (**d**) 30 days after bombardment, (**e**) shoots formed after 2 weeks in MSP + 5G plate under UV light, with the corresponding photo under the light (**f**)

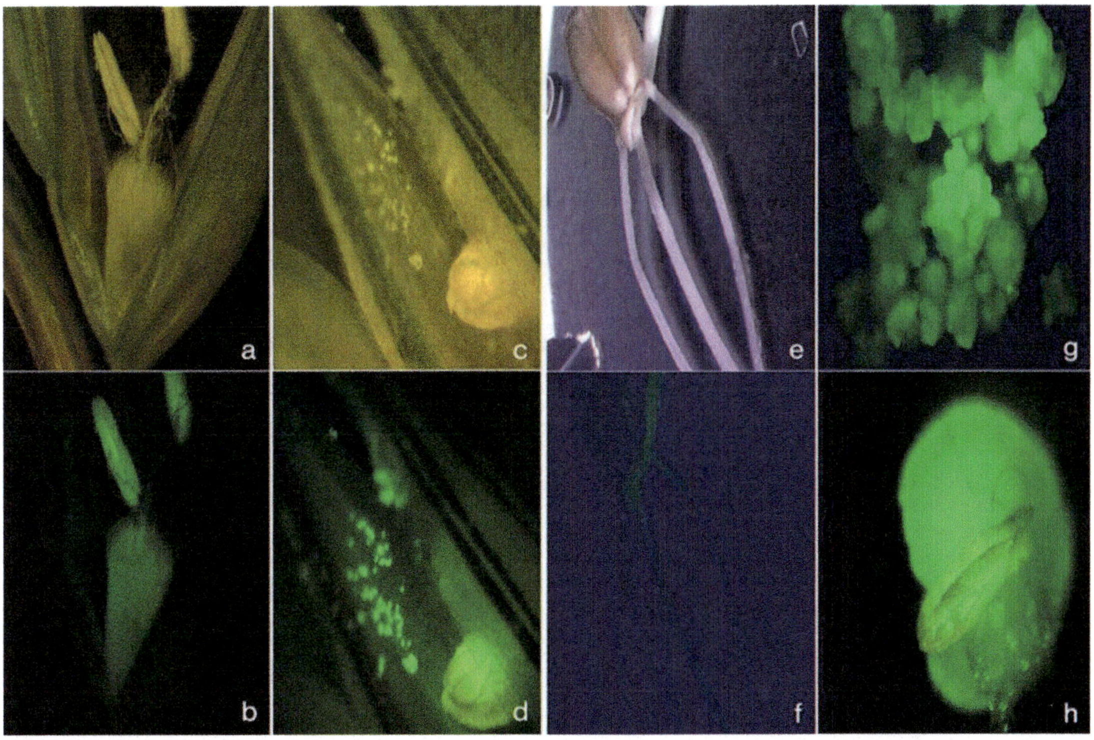

Fig. 5 Phenotype of different wheat organs expressing GFP. Floral tissue with GFP expressing anthers and carpel of the anther of GFP transgenic wheat under brightfield (**a**) and, corresponding photo under UV light (**b**). Pollen and sheath of GFP transgenic wheat under brightfield (**c**) and with the corresponding photo under UV light (**d**). Germinated transgenic seed viewed with brightfield microscopy (**e**) and with the corresponding photo under UV light (**f**). Calli expressing GFP (**g**) and transgenic embryos under UV light. Top view of transgenic anther under UV light (**h**)

References

1. Shewry PR (2009) Wheat. J Exp Bot 60 (6):1537–1553. https://doi.org/10.1093/jxb/erp058
2. Vasil V, Castillo AM, Fromm ME, Vasil IK (1992) Herbicide resistant fertile transgenic wheat plants obtained by microprojectile bombardment of regenerable Embryogenic callus. Nat Biotech 10(6):667–674
3. Cheng M, Fry JE, Pang S, Zhou H, Hironaka CM, Duncan DR, Conner TW, Wan Y (1997) Genetic transformation of wheat mediated by agrobacterium tumefaciens. Plant Physiol 115 (3):971–980. https://doi.org/10.1104/pp.115.3.971
4. Altpeter F, Baisakh N, Beachy R, Bock R, Capell T, Christou P, Daniell H, Datta K, Datta S, Dix PJ, Fauquet C, Huang N, Kohli A, Mooibroek H, Nicholson L, Nguyen TT, Nugent G, Raemakers K, Romano A, Somers DA, Stoger E, Taylor N, Visser R (2005) Particle bombardment and the genetic enhancement of crops: myths and realities. Mol Breed 15(3):305–327. https://doi.org/10.1007/s11032-004-8001-y
5. Harwood WA (2012) Advances and remaining challenges in the transformation of barley and wheat. J Exp Bot 63(5):1791–1798. https://doi.org/10.1093/jxb/err380
6. Ulker B, Li Y, Rosso MG, Logemann E, Somssich IE, Weisshaar B (2008) T-DNA-mediated transfer of Agrobacterium tumefaciens chromosomal DNA into plants. Nat Biotech 26 (9):1015–1017 http://www.nature.com/nbt/journal/v26/n9/suppinfo/nbt.1491_S1.html
7. Zhang Y, Liang Z, Zong Y, Wang Y, Liu J, Chen K, Qiu J-L, Gao C (2016) Efficient and transgene-free genome editing in wheat through transient expression of CRISPR/Cas9 DNA or RNA. Nat Commun 7:12617. https://doi.org/10.1038/ncomms12617 http://www.nature.com/articles/ncomms12617#supplementary-information
8. Liu W, Yuan JS, Stewart CN Jr (2013) Advanced genetic tools for plant biotechnology. Nat Rev Genet 14(11):781–793. https://doi.org/10.1038/nrg3583 http://www.nature.com/nrg/journal/v14/n11/abs/nrg3583.html#supplementary-information
9. Finer JJ, Vain P, Jones MW, McMullen MD (1992) Development of the particle inflow gun for DNA delivery to plant cells. Plant Cell Rep 11(7):323–328. https://doi.org/10.1007/bf00233358
10. Shi G, Zhang Z, Friesen TL, Raats D, Fahima T, Brueggeman RS, Lu S, Trick HN, Liu Z, Chao W, Frenkel Z, Xu SS, Rasmussen JB, Faris JD (2016) The hijacking of a receptor kinase–driven pathway by a wheat fungal pathogen leads to disease. Sci Adv 2(10):e1600822. https://doi.org/10.1126/sciadv.1600822
11. Makandar R, Nalam VJ, Chowdhury Z, Sarowar S, Klossner G, Lee H, Burdan D, Trick HN, Gobbato E, Parker JE, Shah J (2015) The Combined Action of ENHANCED DISEASE SUSCEPTIBILITY1, PHYTOALEXIN DEFICIENT4, and SENESCENCE-ASSOCIATED101 Promotes Salicylic Acid-Mediated Defenses to Limit Fusarium graminearum Infection in Arabidopsis thaliana. Mol Plant-Microbe Interact 28 (8):943–953. https://doi.org/10.1094/MPMI-04-15-0079-R
12. Saintenac C, Zhang W, Salcedo A, Rouse MN, Trick HN, Akhunov E, Dubcovsky J (2013) Identification of wheat gene *Sr35* that confers resistance to Ug99 stem rust race group. Science 341(6147):783–786. https://doi.org/10.1126/science.1239022
13. Lin Z, Li X, Shannon LM, Yeh C-T, Wang ML, Bai G, Peng Z, Li J, Trick HN, Clemente TE, Doebley J, Schnable PS, Tuinstra MR, Tesso TT, White F, Yu J (2012) Parallel domestication of the Shattering1 genes in cereals. Nat Genet 44(6):720–724 http://www.nature.com/ng/journal/v44/n6/abs/ng.2281.html#supplementary-information
14. Christensen AH, Sharrock RA, Quail PH (1992) Maize polyubiquitin genes: structure, thermal perturbation of expression and transcript splicing, and promoter activity following transfer to protoplasts by electroporation. Plant Mol Biol 18(4):675–689
15. Christensen AH, Quail PH (1996) Ubiquitin promoter-based vectors for high-level expression of selectable and/or screenable marker genes in monocotyledonous plants. Transgenic Res 5(3):213–218

Chapter 10

Mesophyll Protoplasts and PEG-Mediated Transfections: Transient Assays and Generation of Stable Transgenic Canola Plants

Sareena Sahab, Matthew J. Hayden, John Mason, and German Spangenberg

Abstract

Plant protoplasts are derived by controlled enzymatic digestion that removes the plant cell wall without damaging the cell membrane. Protoplasts represent a true single-cell system and are useful for various biochemical and physiological studies. Protoplasts from several agriculturally important crop species can be regenerated into a fertile whole plant, extending the utility of protoplasts from transient expression assays to the generation of stable transformation events. Here we describe procedures for transient and stable transformation of leaf mesophyll protoplasts obtained from axenic shoot cultures of canola (*Brassica napus*). Key steps including enzymatic digestion for protoplast release, density gradient-based protoplast purification, PEG-mediated transfection, bead-type culturing (sea-plaque agarose and sodium alginate), and the recovery of putative transgenic canola plants are described. This method has been used for double-stranded DNA break-mediated genome editing and for the routine generation of stable transgenic canola events at commercial scale.

Key words Mesophyll protoplasts, PEG-mediated transfection, Canola, Direct DNA delivery, Transient protoplast assays, FACS-based analysis, Reporter gene assays, Sodium alginate, Sea-plaque agarose, Stable transgenic events

1 Introduction

Protoplasts are plant cells from which the cell wall has been enzymatically removed. The procedure involves treating plant tissues such as leaves, stem, roots, anthers, pollen, flower petals with lytic enzymes, comprising a mixture of cellulases and pectinases, in the presence of an osmoticum. The released protoplasts are purified from cell wall debris by centrifugation or flotation. Isolation, purification, and culturing methods vary between plant species and the tissue source employed. Under optimal conditions, a large number of protoplasts (single-cell explants) can be obtained from a single gram of leaf tissue; approximately 8–10 $\times$ 10^6 protoplasts can be

Sandeep Kumar et al. (eds.), *Transgenic Plants: Methods and Protocols*, Methods in Molecular Biology, vol. 1864, https://doi.org/10.1007/978-1-4939-8778-8_10,

isolated from 1 g fresh weight of canola leaf tissue. Optimum conditions are established by testing key factors that contribute to protoplast yield and viability, such as explant/donor material quality (age, duration in culture, culturing conditions), tissue digestion (choice of lytic enzymes and their concentration, duration of digestion, type of osmoticum), and protoplast purification methods (centrifugation, density gradient). Plant protoplasts are useful for both basic research and plant genetic improvement studies. Protoplasts have been used as explant for transient assays involving gene expression [1], protein subcellular localization [2], protein-protein interaction [3, 4], signal transduction [5], and transcriptional regulatory studies [6, 7]. Protoplasts derived from some plant species are amenable to downstream culturing and the regeneration of whole plants, thus extending their utility to genetic engineering and recovery of stable transgenic events.

Canola (*Brassica napus*) is an important edible and industrial oilseed crop worldwide [8] and is grown as both a winter and spring annual crop. Owing to its lower saturated seed oil content, it is widely used as a cooking oil, in salad dressings, and as a butter substitute in spreads. Canola oil is also used for industrial and manufacturing applications including lubricants and hydraulic fluids and as a constituent in the manufacture of inks, biodegradable grease, pharmaceuticals, and cosmetics. Post seed oil extraction, the residual canola meal is used as a protein supplement in animal feed [8]. Over the last decade, conventional genetic engineering approaches have been used to improve canola germplasm for various agronomic and qualitative traits [8–20]. More recently, modern biotechnological tools such as zinc finger nucleases (ZFNs) and clustered regularly interspaced short palindromic repeats (CRISPRs) have been used to perform precision genome editing at specific gene targets [21, 22].

Cell biology techniques and gene delivery methods are well advanced in canola due to its amenability to tissue culture and higher totipotency response from a wide range of explant sources including hypocotyls [8, 9, 23, 24], cotyledons [25], stem internodes [26], stem segments [27], protoplasts [28, 29], and microspores [30–33]. Among the gene delivery techniques of biolistics, electroporation and *Agrobacterium tumefaciens*-mediated transformation, the latter has been widely used to generate transgenic canola events from a diverse range of explant sources. With the emergence of precision genome-editing techniques in crop plants, direct gene delivery methods using protoplasts are gaining momentum. Some of the key advantages of protoplast-based direct DNA delivery include a wider choice of gene delivery modes (DNA, RNA, protein), flexibility for vector DNA topology (linear or circular), ease for manipulating the molar ratio of individual vectors when multiple vectors are co-delivered, and an abundance of explant source (approx. 1×10^6 protoplasts per reaction) to maximize yield for downstream processes and analyses, the latter of which is especially useful for the detection of rare outcomes.

The protocol described here outlines the key steps for transient and stable gene delivery in canola using a PEG-mesophyll-derived protoplast system. Protoplasts required for this process are aseptically prepared from axenic shoot cultures. The use of a sucrose density gradient in the isolation method results in a debris-free source of protoplasts for downstream processing. This protocol is highly reproducible, robust, and suitable for large-scale transgenic pipeline experiments and only requires small quantities of starting biomaterials. The average regeneration and transformation efficiencies of the method range between 80–90% and 30–40%, respectively, with transgenic canola events being transferred to soil within 3 months. The transient transfection efficiency ranges from 10 to 20%, as measured using a fluorescent reporter gene and flow cytometry (Fig. 1).

2 Materials

2.1 Explant Preparation

1. Seed material (*see* **Note 1**).
2. Surface-sterilizing agents (70% (v/v) ethanol and 20% (v/v) commercial "Domestos" bleach solution (12.5 g/L active chlorine) with 0.1% (v/v) Tween 20).
3. Sterile double-distilled water.
4. Benchtop rotary shaker.
5. Sterile screw-capped tubes (50 mL).
6. Sterile spatula.
7. Sterile forceps.
8. Sterile Petri dish (15 × 100 mm).
9. Whatman™ filter paper Grade 1(sterilized).

2.2 Protoplast Isolation

1. Protoplasts digestion buffer (1% (w/v) cellulase "Onozuka" R10 (Source: Duchefa) and 0.1% (w/v) Macerozyme® R10 (Duchefa) dissolved in K4 media [K3 media (Table 1) with 0.4 M sucrose instead of 0.3 M]) (*see* **Note 2**).
2. Protoplast wash buffer (154 mM NaCl, 125 mM $CaCl_2$, 5 mM glucose) [34].
3. Round-bottom tubes (14 mL polypropylene tubes; 17 × 100 mm).
4. Falcon® sterile cell strainer (70 μm).
5. Sterile serological pipettes (10, 25 mL) and pipette aid.
6. Aerosol barrier pipette tips (1 mL) and pipette aid.

Task #	Task	Duration	Efficiency
Step:1	Tissue digestion	Week-1 Day-1	100%
Step:2	Protoplasts isolation	Week-1 Day-2	>90%
Step:3	Gene delivery (PEG)	Week-1 Day-2	20% loss
Step:4	Bead type cultures	Week-1 Day-2	20% loss
Step:5	Transient assays	Week-1 Day-4	10-20%
Step:6	Microcalli formation	Week-4	1-2ml SCV
Step:7	Callus picking	Week-5-7	0.1-0.3%
Step:8	Shoot Initiation	Week-6*	20-30%
Step:9	Shoot Elongation	Week-8*	>70%
Step:10	Root Initiation	Week-10*	>90%
Step:11	Post *in vitro* hardening	Week-12*	>90%
Step:12	Glasshouse	Week-42*	>90%

Fig. 1 Summary of key tasks, timelines, and efficiency cascade for generation of stable transgenic canola events using mesophyll protoplasts and PEG-mediated DNA delivery. Post transfection (**steps 1–3**), protoplasts are embedded in bead-type cultures (3–4 weeks) until visible microcalli are formed (**steps 4**, **6**). Microcalli are released from bead-type cultures and spread on callus proliferation media supplemented with a suitable selection agent. Callus picking begins at about week 5 (**step 7**) and continues for 2–3 weeks post release until all surviving calli are recovered to shoot regeneration media supplemented with a selection agent. Recovery of shoot meristems to shoot elongation media starts at week 6 (**step 8**) and continues until all surviving callus explants yield a normal looking shoot meristem. Rooted plants are recovered and transferred to soil at week 12. More than 90% of the putative transgenic events can be successfully grown to maturity and set seed in the glasshouse

Table 1
Composition of media used [35]

Media components	K3 media	H media	A media
Macroelements (mg/L final concentration)			
KNO_3	2500	1900	1010
NH_4NO_3	250	600	800
$CaCl_2·2H_2O$	900	600	440
$MgSO_4·7H_2O$	250	300	740
$(NH_4)_2SO_4$	250		
KH_2PO_4		170	136
$NaH_2PO_4·H_2O$	150		
(NH_4)Succinate			50
$CaHPO_4$	50		
Microelements (mg/L final concentration)			
Na_2EDTA	37.3	37.3	37.3
$FeSO_4·7H_2O$	27.8	27.8	27.8
H_3BO_4	3.0	3.0	3.0
KI	0.75	0.75	0.75
$MnSO_4·H_2O$	10	10	10
$ZnSO_4·7H_2O$	2.00	2.00	2.00
$CuSO_4·5H_2O$	0.025	0.025	0.025
$Na_2MoO_4·2H_2O$	0.25	0.25	0.25
$CoCl_2·6H_2O$	0.025	0.025	0.025
Carbohydrates (g/L final concentration)			
D (+) sucrose	102.69	0.125	30
D (+) glucose		68.4	
D-mannitol		0.125	50
D-sorbitol		0.125	
D-cellobiose		0.125	
D(−) fructose		0.125	
D (+) nannose		0.125	
D(+) rhamnose		0.125	
D(−) ribose		0.125	
D(+) xylose	0.25	0.125	

(continued)

Table 1
(continued)

Media components	K3 media	H media	A media
Myo-inositol	0.1	0.1	0.1
Hormones (mg/L final concentration)			
2, D-dichlorophenoxyacetic acid	0.1	0.1	
1-naphthylacetic acid	1.00	1.00	0.1
6-benzylaminopurine	0.2	0.2	1
Vitamins (mg/L final concentration)			
Pyridoxine HCl	1.0	1.0	1.0
Thiamine HCl	10.0	10.0	10.0
Nicotinamide		1.0	
Nicotinic acid	1.0		1.0
Folic acid		0.2	
D-Ca pantothenate		0.5	
D-aminobenzoic acid		0.01	
Choline chloride		0.5	
Riboflavin		0.1	
L(+) ascorbic acid		1.0	
Vitamin A		0.005	
Vitamin D_3		0.005	
Vitamin B_{12}		0.01	
D-biotin		0.003	
Organic acids (mg/L final concentration)			
Sodium pyruvate		5.0	
Citric acid		10.0	
Malic acid		10.0	
Other organics (mg/L final concentration)			
Casein hydrolysate		250	

7. Benchtop centrifuge with swing-out rotor (Eppendorf 5810R).
8. Sterile Petri dish (100 × 20 mm).
9. Parafilm®.
10. Sterile screw-capped tubes (50 mL).

11. Syringe filter (0.22 μm).
12. Syringe (50 mL).
13. Sterile inoculation loop.

2.3 Viability and Yield Assessments

1. Evans blue stain (400 mg/L dissolved in 0.5 M mannitol).
2. Hemocytometer.
3. Bright-field upright microscope with 10× objective lens.
4. Sterile Eppendorf tubes (1.5 mL).
5. Aerosol barrier pipette tips (20 μL) and pipette aid.

2.4 PEG-Mediated Transfection

1. Transformation buffer (15 mM $MgCl_2$, 0.1% (w/v) morpholinoethanesulfonic acid (MES), 0.5 M mannitol; pH 5.8) (*see* **Note 3**) [35].
2. Polyethylene glycol (40% (w/v) PEG 4000 (source: Sigma; Catalogue # 81240) in 0.4 M mannitol, 0.1 M Ca $(NO_3)_2$; pH 6–7) (*see* **Note 4**) [35].
3. Plasmid DNA (*see* **Note 5**).
4. Protoplast count required per transfection (1×10^6).
5. Protoplast wash buffer (154 mM NaCl, 125 mM $CaCl_2$, 5 mM glucose; pH 5.8) [34].
6. Aerosol barrier pipette tips (200 μL) and pipette aid.
7. Sterile serological pipettes (10 mL) and pipette aid.
8. Round-bottom tubes (14 mL polypropylene tubes; 17 × 100 mm).
9. Benchtop centrifuge with swing-out rotor (Eppendorf 5810R).
10. Water bath (Set at 40 °C).
11. Syringe filter (0.22 μm).
12. Syringe (50 mL).

2.5 Assessment of Transient Reporter Gene Expression

2.5.1 Epifluorescence Method

1. Glass slides.
2. Upright fluorescence microscope (equipped with 10× or 20× objective lens) with filters suitable for GFP (ex/em wavelength: 482/502).
3. Aerosol barrier pipette tips (20 μL) and pipette aid.

2.5.2 Flow Cytometry-Based Assays

1. BD Influx (use any benchtop flow cytometer equipped with lasers and filter assembly capable of detecting required fluorescence).
2. Laser (488 nm coherent sapphire solid-state laser), filter setting (517/18 nm band-pass filter for GFP).

3. Round-bottom tubes for FACS assays (5 mL polypropylene tubes; 12 × 75 mm).
4. Pipette tips (1 mL) with pipette aid.

2.6 Protoplasts Embedding/Bead-Type Cultures

2.6.1 Sea-Plaque Agarose Method

1. K3 media (Table 1).
2. A media (Table 1).
3. 1:1 mix of K3:H medium containing 0.6% SeaPlaque™ agarose (Table 1).
4. Petri dish (50 × 9 mm).
5. Parafilm®.
6. Micropore tape.
7. Water bath.
8. Microwave.
9. Aerosol barrier pipette tips (1 mL) and pipette aid.
10. Sterile serological pipettes (10 mL) and pipette aid.
11. Sterile screw-capped flat-bottom container (500 mL; 125 × 90 mm).
12. Incubator shaker (Infors Multitron Standard) (25 mm shaking throw).

2.6.2 Sodium Alginate Method

1. 0.5 M mannitol.
2. 1% sodium alginate.
3. Bead-forming solution (0.4 M mannitol, 50 mM $CaCl_2$; pH 5.8).
4. 1:2 mixtures of K3 + H:A media (Table 1).
5. Depolymerization buffer (0.3 M mannitol, 20 mM sodium citrate; pH 5.8).
6. 6-well plates (Falcon-Cat # 351146).
7. Parafilm®.
8. Aerosol barrier pipette tips (1.0 mL) and pipette aid.
9. Sterile serological pipettes (10 mL) and pipette aid.
10. Round-bottom tubes (14 mL polypropylene tubes; 17 × 100 mm).
11. Inverted microscope with 10× objective lens.
12. Benchtop centrifuge with swing-out rotor (Eppendorf 5810R).
13. Ice.
14. Incubator shaker (Infors Multitron Standard) (25 mm shaking throw).

2.7 Microcalli Recovery and Proliferation

1. Liquid A media (Table 1).
2. Sterile screw-capped tube (15, 50 mL).
3. Sterile disposable pipettes (10 mL) and pipette aid.
4. Sterile forceps.
5. Micropore tape.
6. Parafilm®.

2.8 Plant Culture Media and Components

1. Seed germination media (SGM) (½ strength MS/B5 vitamins, 1% sucrose, 0.8% agar; pH 5.8 with 50 mL of media poured into each 15 × 100 mm sterile Petri dish).
2. Shoot induction media (SIM) (MS/B5 vitamins, 3% sucrose, 500 mg/L MES, 13 μM BAP, 5 μM zeatin, 5 mg/L silver nitrate, 0.8% agar; pH 5.8 with 30 mL of media poured into each 100 × 20 mm sterile Petri dish).
3. Microcalli proliferation media (MPM) (MS/MS vitamins, 3.5% sucrose, 500 mg/L MES, 5 μM BAP, 5 μM NAA, 5 μm 2, 4-D, 0.7% agarose type I; pH 6.0 with 30 mL of media poured into each 100 × 20 mm sterile Petri dish). Supplement the MPM with an appropriate chemical selection agent if required.
4. Shoot regeneration media (SRM) (MS/MS vitamins, 3.0% sucrose, 500 mg/L MES, 500 mg/L PVP, 5 mg/L silver nitrate, 5 mg/L 2i P, 0.5 μM NAA, 0.3 μM GA-3, 0.7% agarose type I; pH 5.8 with 30 mL of media poured in each 100 × 20 mm sterile Petri dish). Supplement the MPM with an appropriate chemical selection agent if required.
5. Shoot elongation media (SEM) (MS/B5 vitamins, 2% sucrose, 500 mg/L MES, 2 μM BAP, 0.1 μM GA-3, 0.8% agar; pH 5.8, with 50 mL of media poured into each STERICON™ 8 sterilized culture vessel). Supplement the MPM with an appropriate chemical selection agent if required.
6. Root initiation media (RIM): (½ strength MS/B5 vitamins, 1% sucrose, 500 mg/L MES, 2.5 μM IBA, 0.6% agar; pH 5.8 with 100 mL of media poured in STERICON™ 13 sterilized culture vessel). Supplement the MPM with an appropriate chemical selection agent if required.
7. Shoot propagation media (SPM): (½ strength MS/B5 vitamins, B5 micronutrients, 1% sucrose, 500 mg/L MES, 2.5 μM IBA, 0.6% agar; pH 5.8 with 100 mL of media poured in each STERICON™ 13 sterilized culture vessel).
8. Sterile long forceps.
9. Sterile scalpel blade.

2.9 Post-In Vitro Propagation

1. Post-in vitro potting mix (commercial seedling mix: ingredients per 30 L bag—0.9 L of vermiculite, 0.6 L of perlite, 25 g of micro-Nutricote, 20 g of water holding granules, 2.5 g of trace elements).
2. General potting mix (AGX mix: ingredients per 30 L bag—0.5 L of vermiculite, 0.3 L of perlite, 33 g of slow-release fertilizer (Macracote™), 28 g of slow-release nitrogen source (Sirflor Plus®), 25 g of water-holding granules, 15 g of trace elements, 7 g of garden lime).
3. 85 mm pot.
4. 200 mm pots.
5. Wooden stakes.
6. Paper bags/cloth bags.
7. Conviron™ chamber for hardening.
8. Glasshouse (day/night temperature, 24/19 °C, 16–18 h photoperiod).

3 Methods

Carry out all steps under sterile conditions using laminar air flow bench at room temperature unless otherwise specified.

3.1 Establishment of In Vitro Donor Plants

3.1.1 Seed Germination

1. Place required quantity of seed in a 50 mL sterile Falcon tube (*see* **Note 1**).
2. Surface-sterilize seeds in 20 mL of 70% ethanol for 1 min by gentle shaking, followed by 3–4 rinses in sterile double-distilled water.
3. Add 20 mL of 20% (v/v) bleach and 10 μL of Tween 20 to the seeds, and place on a table top rotary shaker at 100 rpm for 15 min. Rinse 3–4 times in sterile double-distilled water.
4. Using a sterile spatula, transfer the seeds onto sterile filter paper to remove excess moisture.
5. Plate the sterilized seeds on seed germination medium (SGM) at approximately 50 seeds per plate. Incubate the plates upright at 22 °C in 16 h per day light (30 μmol m^{-2} s^{-1}) for 6 days (*see* **Note 6**) (Fig. 2).

3.1.2 Establishment of Donor Stocks from Hypocotyl Segments

1. Dissect hypocotyls from the 6-day-old seedlings into 0.5-cm-long segments, and transfer it on a shoot induction medium (SIM) (approximately 20 hypocotyl explants per plate).
2. Transfer the shoot meristems after 3–4 weeks onto shoot elongation medium (SEM). Maintain the cultures on SEM for 4 weeks with one round of subculturing at 2 weeks interval (*see* **Note 7**).

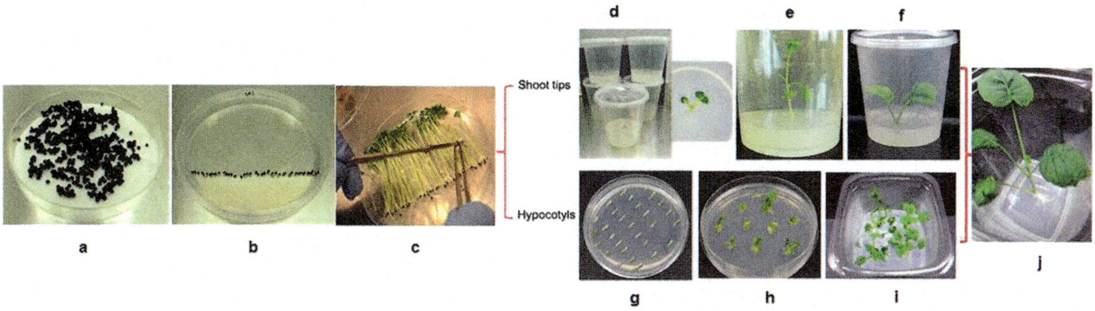

Fig. 2 Establishment of in vitro donor materials. (**a**) Canola seeds are surface sterilized; (**b**) seeds are arranged on germination media; (**c**) isolation of hypocotyl segments and shoot tips from seedlings; (**d–f**) shoot tip transfer onto rooting media and subsequent establishment of rooted plant (6 weeks); (**g–i**) establishment of rooted plants from hypocotyl segments: (**g**) generation of shoot primordia, (**h**) generation of elongated shoots, and (**i**) transfer onto rooting media; and (**j**) recovery of rooted plants, ready for transfer to soil

3. Transfer shoots of 2–3 cm height onto root initiation medium (RIM) for further root development.
4. Subculture rooted onto fresh shoot propagation medium (SPM) at 3–4 week intervals, as described for stem cuttings, and maintain for 2–3 rounds before being used for protoplast isolation (*see* **Note 8**). Maintain the cultures throughout at 22 °C in 16 h per day light (30 μmol m^{-2} s^{-1}) (*see* **Note 9**) (Fig. 2).

3.1.3 Tissue Digestion

Mesophyll-derived protoplasts are isolated from 3- to 4-week-old sterile shoot cultures of *Brassica napus* (DH10275) (*see* **Note 8**).

1. Take 2–3 fully expanded leaves from 3- to 4-week-old shoot cultures under sterile conditions, and place in a 100 × 20 mm plastic Petri dish. Wet the tissue using 1–2 mL of digestion buffer, and remove the leaf mid-ribs (*see* **Note 10**).
2. Cut the leaf halves into thin 0.5–1 mm strips with a sharp scalpel, and transfer the leaf strips to a fresh 100 × 20 mm Petri dish containing 30 mL of digestion buffer (place the leaf strips in such a manner that the lower epidermal side comes in contact with the digestion buffer).
3. Seal the Petri dish with Parafilm®, and incubate overnight (14–18 h) at 24 °C in the dark without agitation (Fig. 3).

3.2 Protoplast Isolation and Purification

1. Following overnight incubation of the leaf material for tissue digestion, gently agitate the digest mix by hand using a sterile inoculation loop, and let the digestion continue for 1 more hour. Gentle mixing at this stage enhances the release of protoplasts from the leaf debris.

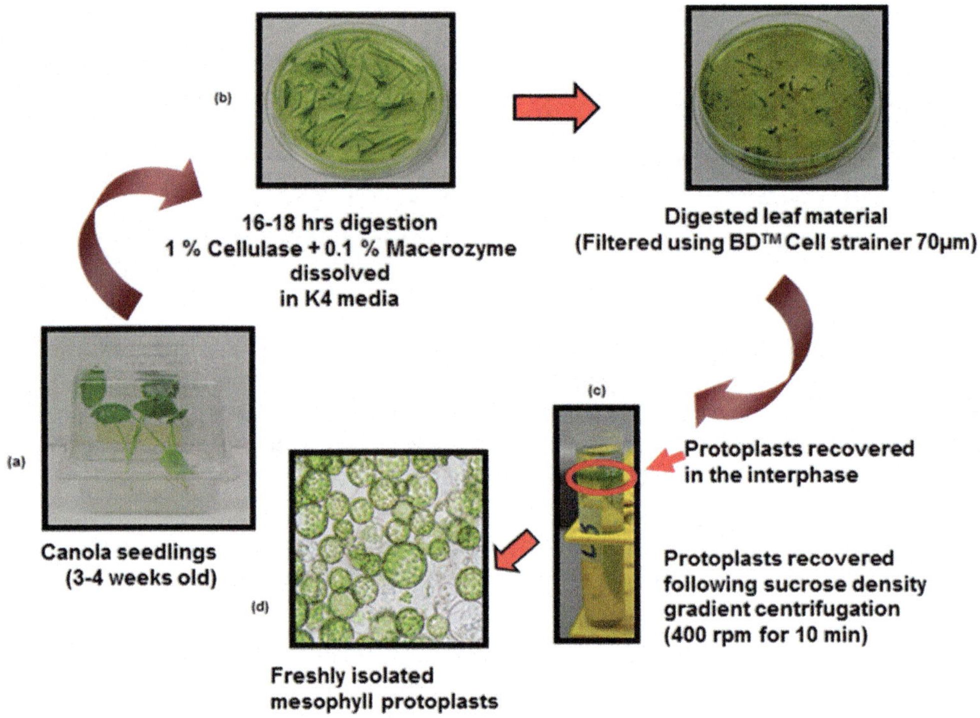

Fig. 3 Workflow for isolation of mesophyll protoplasts (**a**–**d**)

2. Aspirate the protoplast digests using a 25 mL sterile pipette, and dispense into a Falcon® cell strainer (mesh size 70 µm) placed in the mouth of a 50 mL sterile screw-capped tube.
3. Gently mix the protoplasts suspension that passes through the cell strainer and collects in the 50 mL sterile tube. Transfer 6–8 mL of the filtered digest mix into each 14 mL round-bottom tube. Carefully overlay the suspension with 1 mL of wash buffer. Centrifuge for 10 min at 70 × *g* using a benchtop centrifuge. Protoplasts collect at the interphase (Fig. 3).
4. Using a 1 mL pipette, remove protoplasts from the interphase, which are observed as a green-colored layer between the top and bottom colorless layers. Combine the protoplasts recovered from each round-bottom tube into one fresh 14 mL round-bottom tube (*see* **Note 11**).
5. Add 6–8 mL protoplast wash buffer to the protoplasts collected in the fresh 14 mL round-bottom tube. Cap the tube and resuspend the protoplasts by gently tilting the tube on its side. Pellet the protoplasts by centrifugation at 70 × *g* for 10 min using a swing-out rotor.
6. Carefully remove the supernatant and gently resuspend the protoplast pellet in 5 mL of fresh wash buffer. (Fig. 3).

3.3 Assessment of Protoplasts Yield and Viability

Protoplasts yield is assessed using a hemocytometer and the viability tested by staining using Evans blue.

1. Use a sterile pipette aliquot 50 μL of protoplast suspension into a clean 1.5 mL Eppendorf tube. Add an equal volume of Evans blue solution, and incubate at room temperature for 5 min.
2. Transfer 9 μL of treated protoplast onto a clean hemocytometer slide, and count the number of unstained (viable) and stained (non-viable) protoplasts (*see* **Note 12**) [36, 37].

3.4 Transient DNA Delivery and Gene Expression

1. Isolate the required plasmid DNA (*see* **Note 4**) for test constructs from *E.coli* cultures following the manufacturer's instructions.
2. For transfection, add 30 μg of plasmid DNA in a final volume of 60 μL to one million protoplasts (viability ≥95%) suspended in 100 μL of transformation buffer; to this suspension, add 150 μL of PEG solution, and gently tap the tube to mix all the ingredients uniformly (*see* **Notes 13** and **14**).
3. After 15 min of incubation at room temperature, add 5 mL of wash buffer in a drop wise manner. Gently mix the protoplasts, and then add another 5 mL of wash buffer in a slow stream to the protoplast suspension.
4. Gently mix the protoplast suspension, and then pellet the protoplasts by centrifugation using a swing-out rotor at 70 × *g* for 10 min. Remove the supernatant and resuspend the protoplast pellet in 1 mL of fresh wash buffer. Incubate the transfected protoplasts at 22 °C for 48 h. After 48 h, assess the protoplast viability (refer Subheading 3.3), and transient transfection efficiency using fluorescence microscopy and flow cytometry. In case of stable transfections, remove 200 μL of transfected sample for use in the above mentioned assays; use the remaining protoplasts for embedding following the method described in Subheading 3.5 on the same day of transfection.
5. Transfer 20 μL of the transfected protoplasts onto a clean glass slide, and observe for the fluorescent reporter gene expression using an upright microscope equipped with suitable fluorescence filters.
6. Analyze the protoplast suspensions using any benchtop flow cytometer equipped with suitable lasers and filters (*see* **Note 15**).

3.5 Culturing of Mesophyll-Derived Protoplasts to Recover Viable Microcalli

The transfected protoplasts are embedded following either sea plaque agarose or sodium alginate method.

3.5.1 Method 1 (S–P) = Sea-Plaque Agarose Method

1. Before embedding, centrifuge the transfected protoplasts (from Subheading 3.4) at 70 × *g* using a swing-out rotor for 10 min, and carefully remove the supernatant. Resuspend the protoplasts in 0.5 mL of K3 media [23] (Table 1).
2. Transfer the resuspended transfected protoplasts (ca. 5 × 10^5 protoplasts) into a 6 cm Petri dish, and add 4.5 mL of pre-warmed (melted in a microwave oven and incubated in a water bath at 40–45 °C) 1:1 mix of K3:H medium [23] (Table 1) containing 0.6% SeaPlaque™ agarose.
3. Gently mix the agarose and protoplast suspension, and allow to set for 20–30 min at room temperature (*see* **Note 16**). After solidification, seal the dishes with Parafilm®, and culture the protoplasts in the dark for 24 h at 24 °C, followed by 6 days in continuous dim light (5–10 μmol m^{-2} s^{-1}). After 6 days, assess for protoplast cell division under a bright-field inverted microscope.
4. Cut the agarose containing the dividing protoplasts into four quadrants, and transfer to a 500 mL sterile screw-capped culture vessel containing 100 mL of liquid A medium [23] (Table 1).
5. Incubate the cultures at 24 °C in continuous dim light on an incubator shaker (Infors Multitron Standard) at 70 rpm. After 1–2 weeks, a lawn of protoplast-derived colonies appears, and as the microcalli formed within the agarose bead culture become larger, they are released from the disintegrating agarose beads.
6. Carefully transfer the microcalli (diameter between 2 and 3 mm) onto microcalli proliferation medium (MPM) after gently breaking up the agarose beads. Remove any excess liquid A media from each plate before sealing using Parafilm or micropore tape, which enhances embryo maturation (Fig. 1).

3.5.2 Method 2 (S–A) = Sodium Alginate Method

1. Before embedding, centrifuge the transfected protoplasts (from Subheading 3.4) at 70 × *g* using a swing-out rotor for 10 min, and carefully remove the supernatant (*see* **Note 17**).
2. Resuspend the protoplasts in 5.0 mL of 0.5 M mannitol, and repeat the wash by centrifugation at 70 × *g* using a swing-out rotor (Eppendorf 5810 R) for 10 min. Remove the supernatant and resuspend the protoplast pellet in 1.0 mL of 0.5 M mannitol. Incubate the protoplast suspension on ice until needed (*see* **Note 18**).
3. Add an equal volume of 1.0% sodium alginate to the suspended protoplasts and mix gently. Place the protoplast suspension mix on ice until it is embedded.

4. Add 4 mL of bead-forming solution into a sterile six-well plate.
5. Aspirate the protoplast suspension using a 1 mL pipette, and add in a dropwise manner to the bead-forming solution at random positions within the plate.
6. Incubate the protoplast suspension for 20–30 min at room temperature until sodium alginate beads are formed.
7. Carefully remove the bead-forming solution, and replace with 4 mL of 1:2 mixtures of K3 + H:A media [23] (Table 1).
8. Incubate the protoplasts for 3–4 weeks in the dark at 22 °C in an incubator shaker (Infors Multitron Standard) (50 rpm). Observe for protoplast division and absence of any contaminations under an inverted microscope.
9. After 3–4 weeks, release the microcalli (0.5–1.0 mm-sized calli) by incubating the sodium alginate beads with 3–4 mL of depolymerization buffer for 45–60 min at room temperature.
10. Gently mix the beads using sterile forceps to enhance the release of microcalli.
11. Remove the gelling agent released into the depolymerization buffer using a sterile 1.0 mL pipette.
12. Resuspend the microcalli in 5 mL of liquid A media, and gently mix using a sterile 10 mL disposable pipette. Repeat this wash one more time if the alginate is not fully removed. Remove liquid A media using sterile 10 mL serological pipette, and resuspend the microcalli in fresh 5 mL of liquid A media.
13. Carefully transfer the released microcalli to a sterile 15 mL screw-capped tube, and allow to rest for 5 min in the laminar air flow bench. Note down the settled cell volume (SCV) using the graduations on the 15 mL screw-capped tube. Transfer the microcalli to a 50 mL tube, and resuspend it using suitable volume of liquid A media (30 mL of liquid A media is required per 1 mL of the settled cell volume (SCV) (*see* **Note 19**).
14. Gently mix the microcalli using a sterile 10 mL disposable pipette. Transfer 1 mL of the resuspended microcalli per microcalli proliferation medium plate (MPM) (*see* **Note 20**). Add an additional 2 mL of liquid A medium to the plate and gently swirl to uniformly spread the microcalli across the plate.
15. Remove excess liquid A media from each plate using a 1 mL pipette without disturbing the microcalli (*see* **Note 20**).
16. Repeat **steps 19** and **20** until all microcalli suspension is transferred onto MPM plates.
17. Seal the plates using Parafilm® or micropore tape. Maintain the cultures at 22 °C in 16 h/d light (30 μmol m^{-2} s^{-1}) (*see* **Note 21**) (Fig. 4).

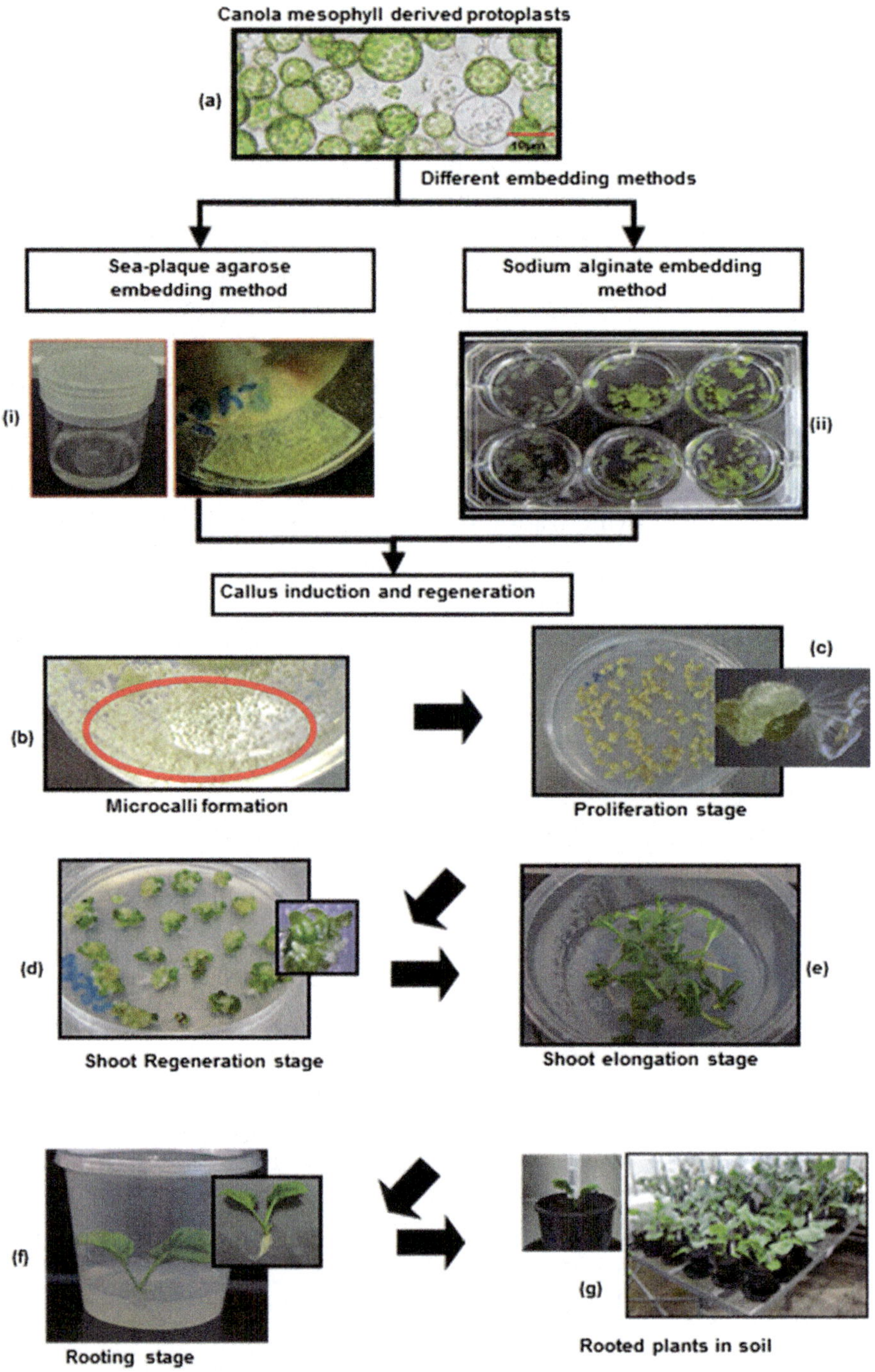

Fig. 4 Workflow for embedding mesophyll protoplasts in bead-type cultures and recovery of rooted plants in soil (**a**–**g**)

3.6 Proliferation/Regeneration of Shoots from Mesophyll-Derived Protoplasts (See Note 22)

1. Transfer green-colored microcalli derived from either the SA or SP method to shoot regeneration medium (SRM) after 1 week, at a density of 25–30 calli per plate. Seal the plates using Parafilm®, and incubate at 22 °C in 16 h/d light (30 μmol m^{-2} s^{-1}) (*see* **Note 23**).
2. Transfer green and healthy calli to fresh SRM plates at 2-week intervals until shoot primordia are produced, usually after 2–3 rounds of subculturing (Fig. 4) (*see* **Note 24**).
3. Transfer all green calli that has produced shoot primordia to shoot elongation medium (SEM).
4. Maintain the shoot cultures in SEM for 4 weeks with one round of subculturing at 2-week intervals (*see* **Note 25**).
5. Transfer the healthy shoots that survive and grow normally to root induction media (RIM). After successful root initiation is typically observed within 3–7 days (*see* **Note 26**) after which rooted plantlets are transferred to soil (Fig. 4).

3.7 Post-In Vitro Establishment

1. Transfer healthy plants to 85 mm pots filled with post-in vitro potting mix.
2. Transfer the potted plants to a Conviron™ chamber, and the plants are hardened for 7 days before transferring to glasshouse conditions.
3. Finally transfer well-established plants to 200 mm pots after 1–2 weeks.
4. Attach the plants to bamboo stakes for physical support in the later stages of growth.
5. Bag the inflorescence using a paper or cloth (10 μm pore size) bag to ensure self-pollination.
6. Harvest the fully matured seeds after 5–6 months post transfer to soil (Fig. 4).

4 Notes

1. This protocol is optimized for the transformation of canola genotype DH12075; the sterilization of seed material from other canola genotypes may require further optimization.
2. Wear a respiratory mask while preparing enzyme solutions, and use a fume hood to weigh and dispense the enzyme powders. Also, transfer and dissolve the enzymes into the required volume of K4 medium within the fume hood. Protoplast digestion buffer should be prepared and used fresh. Any excess buffer should be stored at 4 °C and used within 2 days of preparation.

3. Prepare fresh transformation buffer and PEG solution. Filter sterilize before use.
4. After preparing PEG solution, leave it in a water bath set at 40 °C until required for protoplast transfection.
5. Plasmid preparations can be done using any commercially available kit. Resuspend plasmid DNA in 60 μL of sterile double-distilled water at a final concentration of 0.5 μg/μL for all transfections. The physical structure of the DNA can be supercoiled or linear, depending on requirements.
6. After sterilization, seeds can be spread on moist sterile filter paper in a Petri dish and incubated in the dark at 24 °C for 8–10 h or overnight; transfer pre-germinated seeds to seed germination media.
7. Recovery of shoot primordia from hypocotyls (Subheading 2.1, **step 2**) and callus recovered from microcalli derived from protoplast bead-type culture (Subheading 3.7, **step 5**) is an ongoing process. Primary shoots and shoot primordia recovered in shoot elongation medium (SEM) are subcultured once every 2 weeks for up to 2 cycles. Well-established shoots can be transferred onto rooting medium earlier if required.
8. Rooted in vitro donor plants are multiplied in preparation for protoplast isolation by performing 2–3 rounds of subculturing every 2–3 weeks. In vitro multiplication ensure the generation of high-quality leaf tissue as a source for protoplast isolation. Following leaf sampling, donor plants can be recycled to produce more fresh leaf material via subculturing on fresh shoot propagation medium. Ensure the leaf sampled (third or fourth leaf) for protoplast isolation is not chlorotic (yellowing), since this can result in reduced transfection efficiency and poor division at later stages.
9. As an alternative to establishing donor plants from hypocotyl-derived shoots, shoot tips derived from seedling culture can be used. However, the shoot tips only yield one shoot per seed. Hypocotyl-based donor material production enables fast establishment of donor materials from a limited seed source within a short period of time. Approximately 25–30 shoot buds can be obtained from 5 to 6 hypocotyl segments derived from a single seedling/seed source within 1 month of culture initiation.
10. Perform all tissue sampling and digestion under aseptic conditions using a sterile scalpel blade and forceps. Use new blades for cutting leaf strips to avoid excessive tissue injury. Ensure the midrib is removed before initiating tissue digestion, as this helps reduce post-digestion debris.
11. When removing purified protoplasts from the interphase, take care not to disturb the bottom layer, as it can contain lysed protoplasts.

12. Protoplast yields and viability are calculated as below:

$$\text{Frequency of viable protoplast (\%)} = \frac{\text{No. of unstained protoplast}}{\text{Total number of protoplast counted}} \times 100$$

(Includes both stained and unstained samples)

Count at least 100 protoplasts including both stained and unstained protoplasts.

13. Take special care to avoid carryover of protoplast wash buffer into the transfection mix.
14. Minimize time between resuspending protoplasts in transfection buffer and addition of plasmid DNA and PEG solution.
15. Microscopy and FACS-based analysis can be used to qualitatively and quantitatively transiently assess the success of transfection, respectively. Assessment of transfection efficiency using a flow cytometry can be performed using any benchtop analyzer fitted with a suitable laser and filter source and an appropriate set of negative controls. Sampling for downstream assays such as FACS, microscopy, and viability should be done before embedding into bead-type cultures for stable transformation and for DNA, RNA, or protein extractions required for transient molecular assays. Typically 200 μL of protoplast suspension is retained from the 1 mL final volume of transfected sample before embedding; retained samples are further diluted by adding 300 μL of protoplast wash buffer and used for all three transient assays.
16. After mixing the protoplasts into the melted agarose-containing culture medium, do not move the culture dish until the medium has solidified.
17. Ensure the protoplasts and tube are well washed with 0.5 M mannitol before adding the alginate solution. Incomplete washing results in unwanted carryover of calcium that can cause instant alginate solidification, resulting in very poor microcalli formation.
18. Dilute the protoplasts using 0.5 M mannitol to a final concentration of one million protoplasts per 0.5 mL of 0.5 M mannitol.
19. Assess the microcalli yield by measuring the settled cell volume (SCV) after releasing the microcalli from the bead-type culture. This is achieved by transferring the released microcalli and liquid A solution into a graduated 50 mL sterile tube, allowing the microcalli to settle for 5 min and then measuring the settled cell volume. SCV should range between 1 and 2 mL for one million microcalli.

20. Care should be taken when spreading the microcalli to ensure they do not touch each other. Add liquid A medium, and use forceps to physically separate them from one another.
21. Ensure all liquid A medium is removed before sealing the plates. Use of breathable micropore tape further helps to remove moisture and aids faster callus growth.
22. The key steps described for the downstream recovery of regenerated shoots in Subheadings 3.6 and 3.7 are identical for both bead-type cultures.
23. Multiple rounds of callus picking are performed until all green and proliferated microcalli are transferred to shoot regeneration media (SRM). Usually 3–4 rounds of callus picking are required.
24. Shoot recovery is continued until all surviving callus with green shoot primordia are transferred to SEM.
25. Shoot primordia propagated in SEM typically produce multiple clones (5–10 clones per explant). Individual clones are separated from the main clump, and up to three clones are maintained in SEM for establishment into healthy shoots.
26. One healthy shoot is transferred to rooting medium, and the remaining clones are maintained on SEM. Following successful rooting and subsequent establishment of a plant in soil, backup clones can be discarded. Spindly shoots can be subcultured onto fresh RIM to enhance growth before transfer to soil.

References

1. Yoo S-D, Cho Y-H, Sheen J (2007) Arabidopsis mesophyll protoplasts: a versatile cell system for transient gene expression analysis. Nat Protoc 2:1565–1572
2. Chen J, Yi Q, Song Q, Gu Y, Zhang J, Hu Y, Hanmei L, Yinghon Y, Huang Y (2015) A highly efficient maize nucellus protoplast system for transient gene expression and studying programmed cell death-related processes. Plant Cell Rep 34:1239–1251
3. Fujikawa Y, Nakanishi T, Kawakami H, Yamasaki K, Sato MH, Tsuji H, Matsuoka M, Kato N (2014) Split luciferase complementation assay to detect regulated protein-protein interactions in rice protoplasts in a large-scale format. Rice 7:11
4. Li C, Lin H, Dubcovsky J (2015) Factorial combinations of protein interactions generate a multiplicity of florigen activation complexes in wheat and barley. Plant J 84:70–82
5. Cao J, Yao D, Lin F, Jiang M (2014) PEG-mediated transient gene expression and silencing system in maize mesophyll protoplasts: a valuable tool for signal transduction study in maize. Acta Physiol Plant 36:1271–1281
6. Nakashima K, Yamaguchi-Shinozaki K, Shinozaki K (2014) The transcriptional regulatory network in the drought response and its crosstalk in abiotic stress responses including drought, cold, and heat. Front Plant Sci 5:170
7. Pruneda-Paz JL, Breton G, Nagel DH, Kang SE, Bonaldi K, Doherty CJ, Ravelo S, Galli M, Ecker JR, Kay SA (2014) A genome-scale resource for the functional characterization of *Arabidopsis* transcription factors. Cell Rep 8:622–632
8. Cardoza V, Stewart CN (2004) *Agrobacterium*-mediated transformation of canola. In: Curtis IS (ed) Transgenic crops of the world – essential protocols. Kluwer, Dordrecht, pp 379–387
9. Facciotti MT, Bertain PB, Yuan L (1999) Improved stearate phenotype in transgenic

canola expressing a modified acyl–acyl carrier protein thioesterase. Nat Biotechnol 17:593–597

10. Knutzon DS, Hayes TR, Wyrick A, Xiong H, Davies HM, Voelker TA (1999) Lysophosphatidic acid acyltransferase from coconut endosperm mediates the insertion of laurate at the sn-2 position of triacylglycerols in lauric rapeseed oil and can increase total laurate levels. Plant Physiol 120:739–746
11. Stoutjesdijk PA, Hurlestone C, Singh SP, Green AG (2000) High-oleic acid Australian *Brassica napus* and *B. juncea* varieties produced by co-suppression of endogenous delta 12-desaturases. Biochem Soc Trans 28:938–940
12. Katavic V, Freisen W, Barton DL, Gosen KK, Giblin EM, Luciw T, An J, Zou JT, MacKenzie SL, Keller WA, Males D, Taylor DC (2001) Improving erucic acid content in rapeseed through biotechnology: what can the *Arabidopsis FAE1* and the yeast *SLC1-1* genes contribute? Crop Sci 41:739–747
13. Oelck MM, Phan CV, Eckes P, Donn G, Rakow G, Keller WA (1991) Field resistance of canola transformants (*Brassica napus* L.) to ignite (phosphinothricin). Proc GCIRC Int Rapeseed Congr 8:292–297
14. Falco SC et al (1995) Transgenic canola and soybean seeds with increased lysine. Nat Biotechnol 13(6):577–582
15. Zhang ZB, Kornegay ET, Radcliffe JS, Wilson JH, Veit HP (2000) Comparison of phytase from genetically engineered Aspergillus and canola in weanling pig diets. J Anim Sci 78:2868–2878
16. Ravanello MP, Ke D, Alvarez J, Huang B, Shewmaker CK (2003) Coordinate expression of multiple bacterial carotenoid genes in canola leading to altered carotenoid production. Metab Eng 5(4):255–263
17. Voelker TA, Hayes TR, Cranmer AM, Turner JC, Davies HM (1996) Genetic engineering of a quantitative trait: metabolic and genetic parameters influencing the accumulation of laurate in rapeseed. Plant J 9:229–241
18. Monsanto (2002) Safety assessment of Roundup Ready canola event GT73 http://www.monsanto.com/monsanto/content/our_pledge/roundupcanola_product.pdf
19. Stewart CN, Adang MJ, All JN, Raymer PL, Ramachandran S, Parrott WA (1996) Insect control and dosage effects in transgenic canola containing a synthetic *Bacillus thuringiensis cry1Ac* gene. Plant Physiol 112:115–120
20. Grison R, Grezesbesset B, Schneider M, Lucante N, Olsen L, Leguay JJ, Toppan A (1996) Field tolerance to fungal pathogens of *Brassica napus* constitutively expressing a chimeric chitinase gene. Nat Biotechnol 14:643–646
21. Gupta M, DeKelver RC, Palta A, Clifford C, Gopalan S, Miller JC, Novak S, Desloover D, Gachotte D, Connell J, Flook J, Patterson T, Robbins K, Rebar EJ, Gregory PD, Urnov FD, Petolino JF (2012) Transcriptional activation of *Brassica napus* β-ketoacyl-ACP synthase II with an engineered zinc finger protein transcription factor. Plant Biotechnol J 10 (7):783–791
22. Lawrenson T, Shorinola O, Stacey N, Li C, Østergaard L, Patron N, Uauy C, Harwood W (2015) Induction of targeted, heritable mutations in barley and *Brassica oleracea* using RNA-guided Cas9 nuclease. Genome Biol 16:258
23. Radke SE, Andrews BM, Moloney MM, Crouch ML, Krid JC, Knauf VC (1988) Transformation of *Brassica napus* L. using *Agrobacterium tumefaciens*: developmentally regulated expression of a reintroduced napin gene. Theor Appl Genet 75:685–694
24. De Block M, De Brower D, Tenning P (1989) Transformation of *Brassica napus* and *Brassica oleracea* using *Agrobacterium tumefaciens* and the expression of the *bar* and *neo* genes in the transgenic plants. Plant Physiol 91:694–701
25. Moloney MM, Walker JM, Sharma KK (1989) High efficiency transformation of *Brassica napus* using *Agrobacterium* vectors. Plant Cell Rep 8:238–242
26. Fry J, Barnason A, Horsch RB (1987) Transformation of *Brassica napus* with *Agrobacterium* based vectors. Plant Cell Rep 6:321–325
27. Pua EC, Mehra Palta A, Nagy F, Chua NH (1987) Transgenic plants of *Brassica napus* L. Biotechnology 5:815–817
28. Hervé C, Rouan D, Guerche P, Montané MH, Yot P (1993) Molecular analysis of transgenic rapeseed plants obtained by direct transfer of 2 separate plasmids containing, respectively, the cauliflower mosaic virus coat protein gene and a selectable marker gene. Plant Sci 91:181–193
29. Spangenberg G, Koop H-U, Lichter R, Schweiger HG (1986) Microculture of single protoplasts of *Brassica napus*. Physiol Plant 66:1–8
30. Souvre A, Jardinaud MF, Alibert G (1996) Transformation of rape (*Brassica napus* L.) through the haploid embryogenesis pathway. Acta Soc Bot Polon 65:194–195
31. Jones-Villeneuve E, Huang B, Prudhomme I, Bird S, Kemble R, Hattori J, Miki B (1995)

Assessment of microinjection for introducing DNA into uninuclear microspores of rapeseed. Plant Cell Tissue Organ Cult 40:97–100

32. Fukuoka H, Ogawa T, Matsuoka M, Ohkawa Y, Yano H (1998) Direct gene delivery into isolated microspores of rapeseed (*Brassica napus* L.) and the production of fertile transgenic plants. Plant Cell Rep 17:323–328
33. Nehlin L, Möllers C, Bergman P, Glimelius K (2000) Transient *β-gus* and *gfp* expression and viability analysis of microprojectile bombarded microspores of *Brassica napus* L. J Plant Physiol 156:175–183
34. Menczel L, Nagy F, Kiss Z, Maliga P (1981) Streptomycin-resistant and sensitive somatic hybrids of *N. tabacum* + *N. knightiana*: correlation of resistance to *N. tabacum* plastids. Theor Appl Genet 59:191–195
35. Spangenberg G, Potrykus I (1996) In: Potrykus I, Spangenberg G (eds) Polyethylene glycol-mediated direct gene transfer to tobacco protoplasts and regeneration of transgenic plants: gene transfer to plants. Springer-Verlag, Berlin, Heidelberg, New York, pp 59–65
36. Sambrook J, Russell DW (2006) Estimation of cell number by hemocytometry counting. Cold Spring Harb Protoc. https://doi.org/10.1101/pdb.prot4454
37. Huang CN, Corenjo MJ, Bush DS, Jones RL (1996) Estimating viability of plant protoplasts using double and single staining. Protoplasma 135:80–87

Chapter 11

A Unified *Agrobacterium*-Mediated Transformation Protocol for Alfalfa (*Medicago sativa* L.) and *Medicago truncatula*

Qingzhen Jiang, Chunxiang Fu, and Zeng-Yu Wang

Abstract

Simplification of transformation procedures greatly improves work efficiency. In this chapter, we introduce a unified *Agrobacterium*-mediated transformation protocol that is used for both alfalfa (*Medicago sativa* L., Regen SY4D) and *Medicago truncatula* (ecotype R108). Whole trifoliates or leaflets are used as explants. Sonication is applied to enhance *Agrobacterium* infection and cytokinins are added to the medium to facilitate shoot regeneration. More than 90% transformation efficiency is achieved for alfalfa, while the relatively lower efficiency of up to 60% is obtained for *M. truncatula*, which depends on different selectable markers in the gene constructs. Transgenic plants are produced within 4–8 months with average timeline of 6 months. Using this unified protocol, the same types of media are used for both species which results in significant saving in time and resources.

Key words Alfalfa, *Medicago sativa*, *Medicago truncatula*, *Agrobacterium tumefaciens*, Genetic transformation, Regeneration, Transgenic plant, Somatic embryogenesis

1 Introduction

Because of its high quality as forage, alfalfa (*Medicago sativa* L.) is a widely grown legume crop around the world. This also brings strong research interest in improving its agronomic traits. However, traditional breeding efforts have resulted in little or no gain in alfalfa forage yield in the Northern Midwestern USA [1]. Genetic engineering technology has been used in the breeding program to target improvement of interesting agronomic traits such as herbicide tolerance and forage quality. Roundup ready alfalfa [2] and low-lignin alfalfa have been deregulated and commercialized in the USA. Various methods have been tried for genetic transformation of alfalfa [3–7]. After identifying a highly regenerable genotype Regen SY4D [8, 9], *Agrobacterium*-mediated transformation of alfalfa became easier and more efficient. To date, the transformation method that uses infection of leaf as explant by *Agrobacterium tumefaciens* followed by the production of somatic embryos for

Sandeep Kumar et al. (eds.), *Transgenic Plants: Methods and Protocols*, Methods in Molecular Biology, vol. 1864, https://doi.org/10.1007/978-1-4939-8778-8_11,

transgenic plant regeneration has been widely used in many labs [7].

Medicago truncatula is a diploid and annual legume species with a short life cycle and small genome. Its available genomic (e.g., sequences available) and genetic resources (e.g., tnt1 mutants) make it a model to study other legume species including alfalfa. In many cases, gene functions are investigated or confirmed in *M. truncatula* before they are studied in alfalfa. Several transformation protocols have been developed using either leaf [10–13], cotyledon [14, 15], root [16], or flower [17] as explants for infection and producing transgenic plants through either somatic embryogenesis or shoot organogenesis.

As a Transformation Core Facility working with several species, we always aim to improve work efficiency by optimizing the procedures for transformation. Here we describe a simple protocol that has been routinely used for both alfalfa (Regen SY4D) and *M. truncatula* (R108) in our facility. Occasionally, alfalfa line R2336 and *M. truncatula* ecotype A17 are used according to our customer's requests. Sonication is used to enhance the infection procedure also saving effort from cutting leaves into pieces. A medium containing cytokinins is employed to shorten the time of shoot regeneration. We are able to produce thousands of transgenic alfalfa and *M. truncatula* plants every year with a variety of genes-of-interests from our customers at the Noble Research Institute. In general, regeneration is the most critical step for successful transformation. We rarely have problems inducing calli from either species. The regeneration efficiency reaches over 90% for alfalfa and 60% for *M. truncatula*. A significant amount of effort is saved by using the same set of media for both species.

2 Materials

2.1 Stock Solutions

Always use Milli-Q water to make any solutions used during the whole transformation procedure. Properly store all the chemicals as indicated in the sections below.

1. 20× N6 major salts (*see* **Note 1**, Ref. [13]): 3.7 g of $MgSO_4·7H_2O$, 56.6 g KNO_3, 9.26 g $(NH_4)_2SO_4$, 3.32 g $CaCl_2·2H_2O$, 8.00 g KH_2PO_4, H_2O to 1 L. Store at 4 °C.
2. 1000× SH minor salts (Ref. [13]): 1 g $MnSO_4·H_2O$, 0.5 g H_3BO_3, 0.1 g $ZnSO_4·7H_2O$, 0.1 g KI, 0.01 g $Na_2MoO_4·2H_2O$, 0.02 g $CuSO_4·5H_2O$, 0.01 g $CoCl_2·6H_2O$, H_2O to 100 mL. Store at 4 °C.
3. 1000× SH vitamins (Ref. [13]): 0.5 g nicotinic acid, 0.5 g thiamine HCl (vitamin B_1), 0.5 g pyridoxine HCl (vitamin B_6), H_2O to 100 mL. Store at 4 °C.

4. 1 mg/mL 2,4-dichlorophenoxy acetic acid (2,4-D): store at 4 °C.
5. 1 mg/mL 6-benzylaminopurine (BAP): store at 4 °C.
6. 1 mg/mL kinetin: store at 4 °C.
7. 1 mg/mL indole-3-acetic acid (IAA): store at 4 °C.
8. 400 mg/mL cefotaxime sodium salt: dissolve 10 g powder in 25 mL Milli-Q water, filter sterilize, and store in aliquots at −20 °C.
9. 400 mg/mL ticarcillin disodium salt/potassium clavulanate mixture (15:1) (Timentin): dissolve 10 g powder in 25 mL Milli-Q water, filter sterilize, and store in aliquots at −20 °C.
10. 10 mg/mL Glufosinate-ammonium (phosphinothricin, ppt): dissolve 100 mg powder in 10 mL Milli-Q water, filter sterilize, and store in aliquots at −20 °C.
11. 50 mg/mL hygromycin B: in PBS. Store at 4 °C.
12. 100 mg/mL kanamycin monosulfate: dissolve 1 g powder in 10 mL Milli-Q water, filter sterilize, and store in aliquots at −20 °C.
13. 50 mg/mL rifampicin: dissolve 1 g powder in 20 mL dimethyl sulfoxide (DMSO) and store in aliquots at −20 °C.
14. 100 mM acetosyringone (AS): dissolve 0.196 g powder into 10 mL DMSO, and store in aliquots at −20 °C.

2.2 Media

1. LB liquid medium: add 25 capsules into 1 L Milli-Q water (*see* **Note 2**). Autoclave at 121 °C for 15 min. Store at 4 °C and keep some at room temperature ready for use.
2. LB agar medium: add 40 capsules into 1 L Milli-Q water (*see* **Note 3**). Autoclave at 121 °C for 15 min. Cool to 50 °C, add appropriate antibiotics based on the selection for the plasmid and *Agrobacterium* strain, mix well, pour 25 mL into 100 × 15 mm Petri dishes, and store the plates upside down at 4 °C.
3. SH3a medium (liquid, no Phytagel added; solid, Phytagel added) (Table 1).
4. MSBK medium (*see* **Note 4**, Table 2).
5. SH9 medium (Table 1).
6. MSO medium (Table 2).
7. Phytagel and agar (only used in MSO medium) solidifying agents.

The amount of every component is for 1 L media. Dissolve all the components, adjust pH to 5.8 with KOH or HCl, and then autoclave the media at 121 °C for 15 min. Add appropriate

Table 1
The amount of chemical components in the SH3a and SH9 media

	SH3a	SH9
N6 major salts	50 mL	50 mL
SH minor salts	1 mL	1 mL
SH vitamins	1 mL	1 mL
EDFS[a]	140 mg	140 mg
Myo-inositol	100 mg	100 mg
Sucrose	30 g	20 g
2–4 D (1 mg/mL)	4 mL	–
BAP (1 mg/mL)	0.5 mL	–
Phytagel	3 g	3 g

[a]*EDFS* Ethylenediaminetetraacetic acid, ferric-sodium salt

Table 2
The amount of chemical components in the MSBK and MSO media

	MSBK	MSO
MS basal medium	4.43 g	2.22 g
Sucrose	30 g	10 g
Kinetin	1 mg	–
BAP	0.5 mg	–
	Phytagel 3 g	Agar 8 g or Phytagel 3 g

Table 3
Final concentration of appropriate antibiotics in the media

	Cefotaxime (mg/L)	Timetin (mg/L)	Hygromycin (mg/L)	ppt (mg/L)	Kanamycin (mg/L)
SH3a	200	200	10	3	100
MSBK and SH9	150	150	5	2	50

antibiotics in the media after cooling to 50 °C according to Table 3 (*see* **Note 5**). Pour 25 mL media into the 100 × 15 mm Petri dishes if used before shoot regeneration. After shoots come out, use 100 × 20 mm Petri dishes with thicker medium. Use culture vessels or cups for MSO medium (antibiotics for selection are not necessary) (*see* **Note 6**). Store all media at room temperature if used in a month. Otherwise, keep them at 4 °C for longer storage up to 3 months.

2.3 Plant Materials

Maintain alfalfa (Regen SY4D) plants by clipping the shoots every other week to produce new shoots. To avoid explants from too old plants, use stem cutting for propagation vegetatively every half year. Germinate *M. truncatula* R108 seeds every 2 weeks after sulfuric acid treatment to continuously provide right stage leaf materials ready for use (*see* **Note 7**). Grow both alfalfa and *M. truncatula* plants in Metro-Mix® 360 soil under the conditions of 200 μmol/m^2/s, 16 h day/8 h night at 24 °C/20 °C day/night in the growth chambers (*see* **Note 8**).

2.4 Agrobacterium Strains and T-DNA Vectors

Use *Agrobacterium* strains *AGL1* and *EHA105* in both species. We have not noticed any obvious difference in the transformation procedure due to either of these two *Agrobacterium* strains.

Use binary vectors including pEarleyGate series, pCAMBIA series, pANDA35HK, p*7GWIWG2(II) series, p*7WG2D series, or pMDC series (*see* **Note 9**) containing different selection markers such as phosphinothricin acetyl transferase (*bar*), neomycin phosphotransferase (*nptII*), and hygromycin phosphotransferase (*hpt*) (*see* **Note 10**).

2.5 Lab Equipment and Supplies

1. Growth chambers.
2. Incubators.
3. Laminar flow hood.
4. Autoclave.
5. Shaker/incubator.
6. Branson 2510 Ultrasonic Cleaner with wire mesh basket attachment (40 kHz).
7. Ultrospec 10 Cell Density Meter.
8. Centrifuge with swing rotor.
9. Nalgene transparent polycarbonate desiccator.
10. Steri 250 glass bead sterilizer.
11. Spectrophotometer.
12. 50 mL sterile Falcon tubes.
13. 1.5 mL and 2.0 mL Eppendorf tubes.
14. Autoclavable glass bottles and flasks.
15. Forceps.
16. Whatman filter paper.
17. Sterile plastic petri dishes (100 × 15 mm and 100 × 20 mm).

18. Plastic culture vessels (190 mL with 6.6 cm diameter and 175 mL with 5 cm diameter).
19. Parafilm.
20. Magnetic stirrer and stir bars.

3 Methods

3.1 Agrobacterium Preparation

1. Streak *Agrobacterium EHA105* containing the gene-of-interest from a glycerol stock onto a LB agar plate with appropriate antibiotics for the vector used. Incubate at 28 °C for 2 days (*see* **Note 11**). The plate can be kept at 4 °C for 2 weeks.
2. Dip a single colony from the plate into 4 mL LB medium with 2 μL of rifampicin and appropriate antibiotics for the vector used. Incubate the culture in an incubator/shaker with a speed of 200 rpm at 28 °C overnight.
3. The next afternoon, inoculate 1 mL of the overnight culture into a 30 mL of LB with all antibiotics needed. Continue culturing it in the incubator with shaking at 200 rpm.
4. Check the OD_{600nm} of this overnight culture the following morning. Make sure the culture does not grow over 1.0 (*see* **Note 12**).
5. Centrifuge the *Agrobacterium* culture at 2400 × *g* for 15 min at room temperature. Use 1 mL of room-temperature SH3a liquid medium to resuspend the pellet first and then add more SH3a to make the OD to 0.2–0.3. The *Agrobacterium* suspension is ready to use for infection (*see* **Note 13**).

3.2 Explant Preparation, Infection, and Cocultivation

1. Collect 15–20 leaf trifoliates from healthy alfalfa or *M. truncatula* plants (4–6 weeks old from germination).
2. Sterilize the trifoliates in 10% commercial bleach with a drop of Tween 20 for 10 min on a rotating shaker at 120 rpm at room temperature (*see* **Note 14**). Then wash with sterilized Milli-Q water 3 times.
3. After removing water from the trifoliates, add the *Agrobacterium* suspension into the Magenta box containing the trifoliates.
4. Place these trifoliates in a polycarbonate desiccator and vacuum infiltrate the leaves for 10 min.
5. Sonicate the leaves in the *Agrobacteria* suspension in a Branson 2510 Ultrasonic Cleaner (40 kHz) for 4–5 min at room temperature.
6. Vacuum infiltrate the leaves in the desiccator again for another 10 min.

Fig. 1 Typical major steps in the transformation procedure for alfalfa. (**a**) Intact trifoliate explant on co-cultivation medium after infection with *agrobacterium*. (**b**) Infected leaf explants on selection medium for 2 weeks. (**c**) Six-week old calli on selection medium. (**d**) Calli on regeneration initiation medium for 10–14 days. (**e**) Calli on shoot regeneration medium 4–6 weeks. (**f**) Regenerated shoots are seen on regeneration medium after 6–8 weeks. (**g**) Individual shoot on MSO medium for 2 weeks. (**h**) Confirmed positive transgenic plants ready to transfer into soil

Fig. 2 Typical major steps in the transformation procedure for *M. truncatula*. (**a**) Intact trifoliate explant on co-cultivation medium after infection with *agrobacterium*. (**b**) Infected leaf explants on selection medium for 2 weeks. (**c**) Six-week old calli on selection medium. (**d**) Calli on regeneration initiation medium for 10–14 days. (**e**) Calli on shoot regeneration medium 4–6 weeks. (**f**) Regenerated shoots are seen on regeneration medium after 6–8 weeks. (**g**) Individual shoot on MSO medium for 2 weeks. (**h**) Confirmed positive transgenic plants ready to transfer into soil

7. After removing the *Agrobacterium* suspension, transfer the leaves onto a few layers of sterilized filter paper. Tap dry every trifoliate, and then place it onto SH3a cocultivation medium (Figs. 1a and 2a; *see* **Note 15**).
8. After sealing the plates with parafilm, place them in an incubator at 24 °C in the dark for 24–30 h (*see* **Note 16**).

3.3 Selection, Plant Regeneration, and Rooting

1. Transfer the explants to SH3a selection medium the following day. Continue culturing the explants in the dark incubator at 24 °C.
2. Subculture the explants onto new SH3a selection medium every 2 weeks (Figs. 1b and 2b). At the first transfer, the brittle explants are broken into a few pieces with forceps and then place on the selection medium (*see* **Note 17**). Spots of calli can be visualized within 3 weeks, but the calli will be proliferated in 5–6 weeks (Figs. 1c and 2c) (*see* **Note 18**).
3. Transfer the resistant calli onto MSBK medium when they are big enough (pile diameter larger than 0.5 cm) to initiate regeneration (Figs. 1d and 2d). Culture them in the growth chamber with light intensity of 150 μmol/m^2/s (16 h/8 h day/night) at 24 °C/20 °C for 10–14 days till green embryos are visible (*see* **Note 19**).
4. Transfer the calli with green embryos onto SH9 medium for shoot regeneration (Figs. 1e and 2e). Subculture every 3–4 weeks till shoots grow up (usually after 6–8 weeks on this medium).
5. Transfer the shoot into MSO medium for rooting when it has 2–3 fully expanded trifoliates (Figs. 1f and 2f) (*see* **Note 20**).

3.4 Greenhouse Care

1. Transfer the plantlet with good roots (white and strong) (Figs. 1g and 2g) into Metro-Mix 360 soil, and let it grow in a high humidity environment with light for 1–2 weeks for acclimation (*see* **Note 21**).
2. Transfer the hardened plant to a normal greenhouse. Routinely water and fertilize the plant as needed to ensure vigorous growth.
3. Maintain the transgenic alfalfa plant by trimming every 2–3 weeks or propagate it by stem cutting since it is perennial. Cross a transgenic plant with a wild-type plant to harvest seeds because alfalfa is an outcrossing species. For transgenic *M. truncatula*, it usually takes 2–3 months to flower and set seeds in another month.

4 Notes

1. Sometimes (depending on the batch of order), it is better if $CaCl_2{\cdot}2H_2O$ is completely dissolved before adding other ingredients to help dissolve everything faster.
2. This medium was commercially made with 10 g tryptone-B, 5 g yeast extract-B, and 10 g NaCl per liter by MP Biomedical LLC.

3. This medium was commercially made with 10 g tryptone, 5 g yeast extract, 10 g NaCl, and 15 g agar per liter by MP Biomedical LLC.
4. We noticed using MSBK instead of SH9BK had slightly better response of regeneration. In addition, since MS medium is commercially available, it is simpler and easier to use than SH9 (*see* Table 1 for ingredients) for shoot initiation.
5. Concentration of selection antibiotics is decreased in the MSBK and SH9 media. We usually do not remove selection until we have the plantlet in MSO medium to avoid escapes especially when kanamycin is used. However, when working with some constructs we have difficulty regenerating shoots, we use SH9 medium free of selection after two subcultures on SH9 medium.
6. We sometimes even exclude cefotaxime and Timentin in MSO medium when vigorous growing shoots are obtained after 4 months of selection especially for alfalfa.
7. Seeds are germinated in the dark at room temperature on moistened filter paper after sulfuric acid treatment for 5–8 min depending on how long the seeds are stored. Sand paper can also be used to enhance imbibition by breaking seed coat. The germinated seedlings are transferred into a cell containing MM360 soil 2–3 days later in a tray covered with a transparent plastic dome to keep moisture for their growth. The small plants with 2–3 trifoliates are transferred into pots with MM360 soil after 2 weeks.
8. Chambers are cleaned with bleach and then with heat for a day or two once a while. Plants grown in such growth chambers are free of pests and disease. The quality of plant materials is very critical to ensure the success of the transformation procedure. The plants should be able to grow healthily in a pest-free and optimal growth environment.
9. We have been successful producing transgenic plants with all the vectors listed. We do see variations in transformation efficiency mainly due to individual gene constructs.
10. We have worked with constructs containing *bar*, *nptII*, and *hpt* selection markers for ppt, kanamycin, and hygromycin resistance, respectively. Among all these selection reagents, ppt and hygromycin are more stringent with rare escapes, while 50% escapes were noticed using kanamycin at 75 mg/L as selection for alfalfa transformation. After increasing the concentration of kanamycin to 100 mg/L, we only occasionally see escape higher than 10%.

11. If the glycerol stock of the *Agrobacterium* strain was made from a single colony culture with appropriate antibiotics, we sometimes culture a dip of glycerol stock of *Agrobacteria* into liquid LB medium directly.
12. If the culture did grow over 1.0, subculture 10 mL into 30 mL new LB medium again. Let it grow for another 2 h.
13. In general, acetosyringone is not needed. When working with some difficult constructs for *M. truncatula*, acetosyringone may be helpful for infection.
14. The concentration of bleach and time used for sterilization depends on the cleanness of leaves. If the plants are grown in a dirty environment, more concentrated and longer time may be used for better sterilization. If not sterilized well enough, fungus contamination will be a major problem later on and may ruin the whole transformation procedure.
15. Tap drying is very critical to avoid potential *Agrobacterium* contamination. But make sure the explants are not over-dried to kill the cells.
16. Depending on the time when the transformation is done (in the morning or afternoon), usually 24 h cocultivation is sufficient for the *Agrobacterium* to transfer T-DNA into leaf cells. Shorter time cocultivation also avoids *Agrobacteria* overgrowth later on.
17. Since intact trifoliates are used as explants for infection, breaking the explants into a few pieces at this time could ensure that enough number of independent transformation events are achieved eventually.
18. This also depends on constructs. Sometimes it takes longer to produce resistant calli. We do see calli proliferation 2 months later for some constructs.
19. Pay attention to response of the calli on MSBK medium especially for *M. truncatula*. If the calli turn hard, transfer them onto SH9 medium immediately for better regeneration.
20. We usually do not add any hormones in the rooting medium unless the vigorous shoots have difficulty producing roots in a month. In such situation, IAA is added into MSO medium with the concentration of 1 mg/L.
21. Regenerated plants can be checked for marker gene and gene-of-interest before they are transferred into soil by PCR with genomic DNA extracted from their leaves.

Acknowledgments

We would like to thank our previous and current group members Tim Hernandez, Yewei Wang, Steven Tudor, Mohammad Fereidouni, Guangming Li, Jianfei Yun, and Lishan Yang for their knowledge and suggestions in improving the procedure. We are also grateful to all the researchers and their principal investigators for their trust and support using the service.

References

1. Riday H, Brummer EC (2002) Forage yield heterosis in alfalfa. Crop Sci 42:716–723
2. Wang ZY, Brummer EC (2012) Is genetic engineering ever going to take off in forage, turf and bioenergy crop breeding? Ann Bot 110:1317–1325
3. Deak M, Kiss GB, Koncz C, Dudits D (1986) Transformation of *Medicago* by *Agrobacterium* mediated gene transfer. Plant Cell Rep 5:97–100
4. Shahin EA, Spielmann A, Sukhapinda K, Simpson RB, Yashar M (1986) Transformation of cultivated alfalfa using disarmed *Agrobacterium tumefaciens*. Crop Sci 26:1235–1239
5. Austin S, Bingham ET, Mathews DE, Shahan MN, Will J, Burgess RR (1995) Production and field performance of transgenic alfalfa (*Medicago sativa* L.) expressing alpha-amylase and manganese- dependent lignin peroxidase. Euphytica 85:381–393
6. Samac DA, Temple SJ (2004) Development and utilization of transformation in *Medicago* species. In: Liang GH, Skinner DZ (eds) Genetically modified crops: their development, uses, and risks. Haworth Press, New York, pp 165–202
7. Samac DA, Austin-Phillips S (2006) Alfalfa (Medicago sativa L.). In: Wang K (ed) Methods in molecular biology, Agrobacterium protocols, 2/e, volume 1, vol 343. Humana Press Inc., Totowa, NJ, pp 301–311
8. Bingham ET (1991) Registration of alfalfa hybrid Regen-SY germplasm for tissue culture and transformation research. Crop Sci 31:1098–1118
9. Chen F, Srinivasa Reddy MS, Temple S, Jackson L, Shadle G, Dixon RA (2006) Multi-site genetic modulation of monolignol biosynthesis suggests new routes for formation of syringyl lignin and wall-bound ferulic acid in alfalfa (*Medicago sativa* L.). Plant J 48:113–124
10. Chabaud M, Larsonneau C, Marmouget C, Huguet T (1996) Transformation of barrel medic (*Medicago truncatula* Gaertn) by *Agrobacterium tumefaciens* and regeneration via somatic embryogenesis of transgenic plants with the MtENOD12 nodulin promoter fused to the *gus* reporter gene. Plant Cell Rep 15:305–310
11. Trinh TH, Ratet P, Kondorosi E, Durand P, Kamaté K, Bauer P, Kondorosi A (1998) Rapid and efficient transformation of diploid *Medicago truncatula* and *Medicago sativa* ssp. *falcata* in vitro lines improved in somatic embryogenesis. Plant Cell Rep 17:345–355
12. Chabaud M, de Carvalho-Niebel F, Barker DG (2003) Efficient transformation of *Medicago truncatula* cv. Jemalong using the hypervirulent *Agrobacterium tumefaciens* strain AGL1. Plant Cell Rep 22:46–51
13. Cosson V, Durand P, d'Erfurth I, Kondorosi A, Ratet P (2006) *Medicago truncatula* transformation using leaf explants. In: Wang K (ed) Methods in molecular biology, Agrobacterium protocols, 2/e, volume 1, vol 343. Humana Press Inc., Totowa, NJ, pp 115–127
14. Trieu AT, Harrison MJ (1996) Rapid transformation of *Medicago truncatula*: regeneration via shoot organogenesis. Plant Cell Rep 16:6–11
15. Wright E, Dixon RA, Wang Z-Y (2006) *Medicago truncatula* transformation using cotyledon explants. In: Wang K (ed) Methods in molecular biology, Agrobacterium Protocols, 2/e, volume 1, vol 343. Humana Press Inc., Totowa, NJ, pp 129–136
16. Crane C, Dixon RA, Wang Z-Y (2006) *Medicago truncatula* transformation using root explants. In: Wang K (ed) Methods in molecular biology, Agrobacterium Protocols, 2/e, volume 1, vol 343. Humana Press Inc., Totowa, NJ, pp 137–154
17. Kamaté K, Rodriguez-Llorente ID, Scholte M, Durand P, Ratet P, Kondorosi E, Kondorosi A, Trinh TH (2000) Transformation of floral organs with GFP in *Medicago truncatula*. Plant Cell Rep 19:647–653

Chapter 12

Poplar Transformation

Tobias Bruegmann, Olaf Polak, Khira Deecke, Julia Nietsch, and Matthias Fladung

Abstract

First publications of successful *Agrobacterium*-mediated transformation of tobacco were published more than 30 years ago. Protocols for *Agrobacterium*-based transformation as well as biolistic bombardment and PEG transformation of protoplasts are available for more than 150 plant species from various plant families. Also for many *Populus* species and hybrids, adapted transformation protocols have been published. The standard protocol for *Agrobacterium*-mediated transformation of different *Populus* genotypes is the leaf-disc method. Here, we first describe the transfer of genes into poplar by using the *Agrobacterium*-based leaf disc methods. In addition, alternative basic transformation methods, namely, biolistic bombardment and PEG transformation of protoplasts, are also described. Further, we present improved poplar transformation protocols by simplifying the transformation procedure and optimizing tissue preparation and plant regeneration.

Key words *Populus*, Protoplast, Regenerating callus, Suspension culture, Micro-explant, Plant tissue culture

1 Introduction

Three decades after the discovery of the ability of the plant pathogen *Agrobacterium* to transfer DNA to plant cells, genetic transformation of plant cells is still one of the most frequently used methods in basic molecular biology [1]. Since the first publications of successful *Agrobacterium*-mediated transformation of tobacco that were published almost at the same time [2–4], numerous protocols were invented and subsequently adapted for different plant species. As summarized previously [5], *Agrobacterium*-based transformation protocols are available for more than 150 plant species from various plant families.

Generally, the necessary prerequisite for plant transformation is the availability of an efficient in vitro culture system to regenerate complete plants from isolated tissues or even single cells and protoplasts [6, 7]. Here, parts of any vegetative tissue (leaves, stems, or

Sandeep Kumar et al. (eds.), *Transgenic Plants: Methods and Protocols*, Methods in Molecular Biology, vol. 1864, https://doi.org/10.1007/978-1-4939-8778-8_12,

roots) are placed on an artificial nutrient medium. After a few weeks, shoots are developed at the intersections of the leaf explants, which can be rooted and finally grown into small saplings. Members of the tree genus *Populus* were the first successfully transformed wooden plant [8]. Meanwhile, numerous clones of different poplar species and hybrids of the sections *Populus*, *Aigeiros*, and *Tacamahaca* have also been subject of successful gene transfer by *Agrobacterium* but also by particle bombardment technique [9, 10].

The standard protocol for *Agrobacterium*-mediated transformation of many poplar clones is the leaf-disc method based on [11] as described in [12–15]. To increase yield of transformation, the leaf-disc transformation method has been modified by changing time, age of leaves used, conditions for co-cultivation (light/dark), and regeneration parameters (e.g., hormones) [16–18]. *Agrobacterium*-inoculated aspen seedling hypocotyls followed by direct thidiazuron (TDZ)-mediated shoot regeneration on selective media has been used previously [19]. Nevertheless, despite progress in increasing the transformation efficiency slightly and widening the range of transformable poplar species and hybrids, most of the transformation methods published so far are based on the basic leaf-disc method [11].

Besides *Agrobacterium*, application of two direct DNA transfer approaches, biolistic and protoplast transformation method, has already been described for *Populus* (summarized in [20]). The biolistic method has successfully been applied for stable transformation of different poplar genotypes by using protoplast-derived cells, nodules, and stems [17, 21]. In protoplast transformation, DNA was introduced via polyethylene glycol (PEG) or electroporation in the cells (e.g., [17, 22–24]).

This chapter provides protocols of three basic poplar transformation methods (*Agrobacterium*-based leaf disc, biolistic bombardment, and PEG transformation of protoplasts).The transformation protocols are further improved by simplifying the transformation procedure and optimizing tissue preparation and plant regeneration.

2 Materials

Prepare and store all reagents at room temperature (unless indicated otherwise). Sterilize all media and equipment used by autoclaving or sterile filtration.

Prepare aqueous (dd) stock solutions (unless indicated otherwise) as listed below. Pass aqueous solutions through sterile filter into a small sterile tube. Take appropriate amounts from the listed stock solutions to obtain the respective final media concentrations.

2.1 Hormones, Antibiotics, and Other Stock Solutions

1. 0.0022 mg/L thidiazuron (TDZ).
2. 100 mg/L Pluronic F-68 (Sigma, Heidelberg).
3. 0.25 mg/mL benzylaminopurine (BAP).
4. 0.02 mg/mL 1-naphthylacetic acid (NAA).
5. 50 mg/mL kanamycin monosulfate.
6. 10 mg/mL hygromycin B.
7. 100 mg/mL carbenicillin.
8. 50 mg/mL rifampicin, in 100% methanol.
9. 100 mg/mL spectinomycin.
10. 300 mg/mL streptomycin.
11. 50 mg/mL gentamicin.
12. 50 mg/mL ampicillin.
13. 10 mg/mL tetracycline, in 95% ethanol.
14. 125 mg/mL cefotaxime.
15. 0.1 M 3′5′dimethoxy-4′-hydroxy-acetophenone (acetosyringone): 196 mg in a small amount of 70% EtOH. Bring volume to 10 mL with water (dd).

2.2 Bacterial Strains and Plasmids

Agrobacterium strain GV3101 (C58 family) carrying the Ti-plasmid pMP90RK is an oncogenic nopaline-producing strain with intermediate virulence [25]. Strain LBA4404 is a non-oncogenic octopine-producing derivative of Ach5 with low virulence [26].

For particle bombardment and protoplast PEG transformation, any plasmid carrying the desired gene(s) to be transferred can be used.

2.3 Plant Material

Different poplar clones are used for *Agrobacterium* transformation. In our lab, we started with T89 and Esch5, the latter one being a hybrid between Brauna11 (*P. tremula*) and Tur141 (*P. tremuloides*). Also, pure aspen clones Brauna11 and W52 were successfully subjected to transformation. The hybrid poplar clone INRA 717-1B4 was transformed for the first time by [12]. This clone is a hybrid between *P. tremula* × *P. alba* (*P.* × *canescens*), and until today, it is being used worldwide for transformation. Many transgenic poplar lines based on this clone have been transferred to the field and tested for commercial application [27].

2.4 Culture Media

1. *Woody Plant Medium (WPM)* [28] *or Murashige and Skoog (MS)* [29]: Dissolve the appropriate amount of the powder (depending from the manufactory's product information, e.g., Duchefa, Sigma-Aldrich, Dephyte, HiMedia Laboratories) in 700–800 mL of distilled water. Stir gently until the medium gets completely dissolved. Add 20 g of sucrose and

adjust pH to 5.8 using 1 N NaOH. Depending on application, add 6–7 g of agar-agar (*see* **Note 1**) and heat the medium to boiling until complete dissolution (only for solid medium). Fill up with distilled water to 1 L. Sterilize the medium by autoclaving at 121 °C for 20 min. Cool the autoclaved medium to about 45 °C before adding heat-labile supplements, e.g., antibiotics, hormones, and cefotaxime. Dispense desired amounts of medium under a laminar airflow unit in sterile culture vessels, e.g., petri dishes, Magenta vessels, and glass containers.

2. *Callus-induction medium:* Dissolve half amount of the powder (as given in the manufactory's product information, e.g., Duchefa, Sigma-Aldrich, Dephyte, HiMedia Laboratories) in 700–800 mL of distilled water. Stir gently until the medium gets completely dissolved. Add 20 g of sucrose and adjust pH to 5.8 using 1 N NaOH. Add 6–7 g of agar-agar and heat the medium to boiling until complete dissolution (only for solid medium). Fill up with distilled water to 1 L. Sterilize the medium by autoclaving at 121 °C for 20 min. Cool the autoclaved medium to about 45 °C. Add 1.5 mg/L naphthaleneacetic acid (NAA), 0.25 mg/L benzyl adenine (BAP), and 2.2 μg/L thidiazuron. Dispense desired amounts of medium under a laminar airflow unit in petri dishes.
3. *Yeast extract broth (YEB) medium* [30, 31]*:* Dissolve 5 g of meat extract, 1 g of yeast extract, 5 g of peptone, and 5 g of sucrose in 700 mL distilled water, stir gently until the medium gets completely dissolved, and fill up with distilled water to 1 L. Sterilize the medium by autoclaving at 121 °C for 20 min. Cool the autoclaved YEB medium to about 45 °C. Add 1 mL of 2 M $MgSO_4$ (*see* **Note 2**) and appropriate amounts of antibiotics depending on the *Agrobacterium* resistance genes. To prepare YEB plates, add bacterial agar to 1.5% before autoclaving.

2.5 Protoplast Isolation Solutions [20]

1. Enzyme solution: 2 mL of 0.5 M $CaCl_2$ (10 mM), 1 mL of 20 mM KH_2PO_4 (0.2 mM), 1 mL of 0.1 M KNO_3 (1 mM), 10 mL of 10 mM $MgSO_4 \times 7\ H_2O$ (1 mM), 0.98 g morpholinoethane sulfonic acid (MES) (5 mM), 1 g polyvinylpyrrolidone (1%), 1 g Cellulase Onozuka R10 (1%), 0.24 g macerozyme (0.24%). Adjust pH to 5.6 using 1 M KOH. Adjust the osmolarity to 600 mOsm/kg using mannitol (measure at Osmomat). Filter sterile, prepare 20 mL aliquots, and store at −20 °C.
2. Sucrose solution: 2 mL of 0.5 M $CaCl_2$ (10 mM), 1 mL of 20 mM KH_2PO_4 (0.2 mM), 1 mL of 0.1 M KNO_3 (1 mM), 10 mL of 10 mM $MgSO_4 \times 7\ H_2O$ (1 mM), 0.98 g morpholinoethane sulfonic acid (MES) (5 mM), 1 g polyvinylpyrrolidone 10 (1%), 15 g sucrose. Adjust pH to 5.6 using 1 M KOH. Filter sterile, prepare 20 mL aliquots, and store at −20 °C.

3. Washing solution: 13.88 g of $CaCl_2$ (125 mM), 0.374 g of KCl (5 mM), 0.9 g of glucose (5 mM), 8.9 g of NaCl (155 mM). Adjust pH to 5.6 using 1 M HCl and/or 1 M KOH. Filter sterile.
4. Fluorescein diacetate (FDA) solution: 10 mg of FDA in 2 mL acetone (5 mg/mL).

2.6 PEG Transformation of Protoplasts, Microcalli Formation, and Plant Regeneration [20, 29]

1. MMG solution: 4 mM MES, 0.4 M mannitol, 15 mM $CaCl_2$. Adjust pH to 5.6 using 1 M KOH and osmolarity to 600 mOsm/kg.
2. PEG_{1500} solution: 40% PEG_{1500} in 0.1 M $CaCl_2$, 0.2 M mannitol.
3. Protoplast dividing medium (microcalli formation): Dissolve 2.68 g of MS powder (as given in the manufactory's product information, e.g., PhytoTechnology Laboratories, Shawnee Mission, KS, USA) in 700–800 mL of distilled water, stir gently until the medium gets completely dissolved, and fill up with distilled water to 1 L. Sterilize the medium by autoclaving at 121 °C for 20 min. Cool down the medium to about 45 °C. Add 0.1 mg/L benzyl adenine (BAP), 0.22 mg/L 1-naphthaleneacetic acid (NAA), and a glucose/mannitol mixture (2:1, 600 mOsm/kg).
4. ½ MS medium: Dissolve half amount of the powder (as given in the manufactory's product information, e.g., Duchefa, Sigma-Aldrich, Dephyte, HiMedia Laboratories) in 700–800 mL distilled water. Stir gently until the medium gets completely dissolved. Add 20 g sucrose and adjust pH to 5.8 using 1 N NaOH. Fill up with distilled water to 1 L. Sterilize the medium by autoclaving at 121 °C for 20 min. Cool down the medium to about 45 °C. Add 2.2 µg/L thidiazuron and 0.1 mg/L Pluronic F-68. Dispense desired amounts of medium under a laminar airflow unit in Erlenmeyer flasks.

3 Methods

Carry out all steps under the clean bench unless indicated otherwise. Media are sterilized either by autoclaving or filter sterilization.

3.1 *Agrobacterium*-Based Transformation

For both poplar leaf-disc and micro-explant transformation, *Agrobacterium* suspension has to be prepared properly.

3.1.1 Preparation of *Agrobacterium* Suspension

1. Transfer small spatula tip from frozen bacterial glycerol stock to 4 mL YEB medium containing appropriate antibiotics in correct concentrations.

2. Incubate overnight at 28 °C in the dark by rigorously shaking (800–1000 rpm).
3. Next morning, add 96 mL of fresh YEB medium and incubate for 4 to 5 h at 28 °C in the dark by gently shaking (100–200 rpm). Check OD at 600 nm which should be between 0.3 and 0.8 (optimal about 0.6).
4. 30 min before co-cultivation: add 25 μL of 0.1 M acetosyringone (final concentration 25 μM) and continue shaking in the dark.

3.1.2 Leaf-Disc Transformation (According to Ref. [11])

1. One day before co-cultivation: prepare *Agrobacterium* suspension from glycerol stock. Next morning follow Subheading 3.1.1, **step 3**.
2. Harvest leaves from 3- to 4-week-old axenic shoot cultures and collect in sterile water or in liquid WPM (without hormones). Divide leaves into two halves (one-third base part/two-third upper part) and blot dry on sterile filter paper.
3. 30 min before co-cultivation: add acetosyringone to *Agrobacterium* suspension to be used for co-cultivation.
4. Transfer leaf discs to *Agrobacterium* suspension, and incubate for 10 s up to 30 min at 28 °C in the dark (gently shaking at 100–200 rpm).
5. Blot leaf discs dry again and scatter on sterile filter paper soaked with WPM medium or on solidified WPM medium (both without hormones) in sterile petri dishes.
6. Incubate leaf discs for 2–3 days in growth chamber (24–25 °C) at dimmed light (co-cultivation).
7. After co-cultivation, transfer leaf discs to sterile (metal) tea strainer and wash 3–5 times in sterile aqueous 500 mg/L cefotaxime solution. Between each washing step dip tea strainer shortly (few seconds) on sterile filter paper.
8. Disperse leaf discs on petri dishes containing solid regeneration medium. Cultivate petri dishes in growth chambers at 23–25 °C under convenient light conditions.
9. Microcalli with shoots will appear after 4–8 weeks, depending from the chosen poplar clone and regeneration protocol. Excise shoots and transfer to WPM medium (without hormones, but with antibiotics and cefotaxime) for rooting or on particular rooting medium (if needed).

3.1.3 Transformation of Micro-explants (Fig. 1)

The protocol comprises the simple to harvest complete above-medium-grown part of the poplar plant from the shoot apical meristem until the basal internode. These plants are chopped into micro-explants by hand using scalpel in glass dishes. Obtained micro-explants are then co-cultivated with *Agrobacterium*, washed,

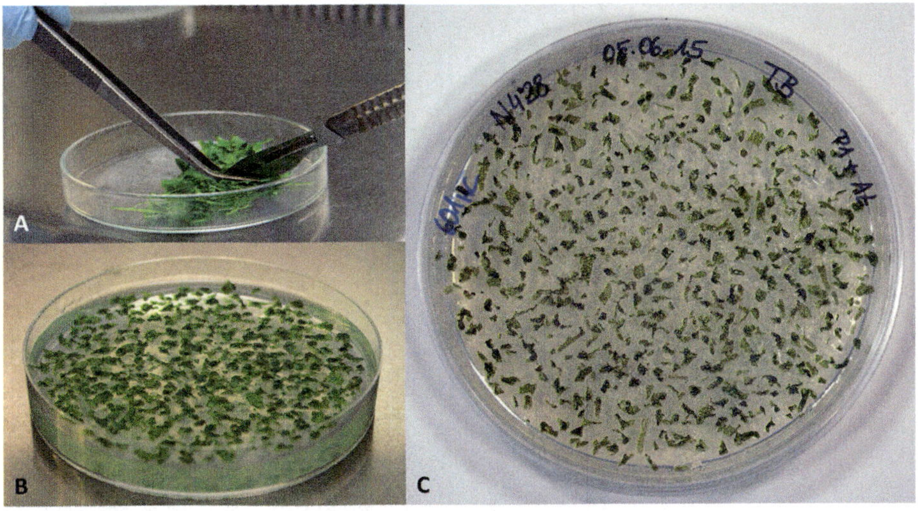

Fig. 1 Handling procedure for leaf-disc transformation. (**a**) The plant material is chopped with a scalpel. (**b**) For two days incubation, the transformation batch is dispersed on WPM without selection antibiotic. (**c**) For regeneration, the plant material is spread reliably on WPM with antibiotics so that every explant had contact with the medium

and transferred either to solid or in liquid culture to initiate regeneration as micro-explants still have the potential to form callus and regenerate into complete poplar plants.

1. One day before co-cultivation: prepare *Agrobacterium* suspension from glycerol stock. Next morning follow Subheading 3.1.1, **step 3**.
2. Cut 3–4-week-old axenic grown plants (with leaves) at the shoot basis (*see* **Note 3**), and collect in liquid WPM (without hormones) or sterile tap water (*see* **Note 4**).
3. Approximately 60–90 min before co-cultivation: chop plants by hand using sterile scalpel in sterilized glass dishes.
4. 30 min before co-cultivation: add acetosyringone to *Agrobacterium* suspension (*see* Subheading 3.1.1, **step 4**) to be used for co-cultivation.
5. Transfer micro-explants to *Agrobacterium* suspension, and incubate for 30 min at 28 °C in the dark (gently shaking at 100–200 rpm).
6. Sieve micro-explants/*Agrobacterium* suspension in sterile (metal) tea strainer and dip tea strainer shortly (few seconds) on sterile filter paper.
7. Scatter micro-explants on solid WPM (without hormones, in petri dishes). Incubate petri dishes for 2–3 days at 23 °C in the dark (co-cultivation).

Fig. 2 Putative transgenic regenerates in liquid culture. After 8 weeks, numerous small regenerates (**a**) and one large regenerate (**b**) had been formed. Since the stirring was interrupted for documentation, the regenerates were settled at the bottom of the flask bottom

8. Washing: collect all micro-explants in Erlenmeyer flask (250 mL). Add 150–200 mL aqueous cefotaxime solution (500 mg/L, sterile water) and incubate for 2 min by gently hand shaking (100–200 rpm). Sieve micro-explants in sterile (metal) tea strainer, and dip tea strainer shortly (few seconds) on sterile filter paper (*see* **Note 5**).
9. Repeat washing of micro-explants in aqueous cefotaxime solution three times as described before, but incubate only for 2 min.
10. Finally wash micro-explants in Erlenmeyer flask for 2 min by gently hand shaking (100–200 rpm) in sterile tap water (*see* **Note 4**). Sieve in sterile (metal) tea strainer and dip tea strainer shortly (few seconds) on sterile filter paper.
11. Transfer micro-explants into Erlenmeyer flask with baffles (250 mL) containing 200 mL liquid regeneration medium. Incubate baffled Erlenmeyer flask in growth chambers at dim light and 23 °C by gently shaking (100–200 rpm); *see* **Note 6**.
12. Change regeneration medium every week.
13. Microcalli with shoots will appear after 4–8 weeks (Fig. 2), depending from the chosen poplar clone and regeneration protocol. Collect micro callus-containing shoots and transfer to petri dishes with solidified regeneration medium. Excise shoots and transfer to WPM medium (without hormones, but with antibiotics and cefotaxime) for rooting or on particular rooting medium (if needed).

3.2 Particle Gun Transformation of Callus

3.2.1 Callus Induction

1. Cut stems from axenic grown plantlets and isolate internodal stem segments of 0.5–1.0 cm in size.
2. Divide internodal stem segments with a scalpel in the longitudinal direction. Place halved stem segments with the cut face up on callus induction medium.
3. Callus is induced within 10 days in a growth chamber under convenient light conditions.

3.2.2 Biolistic Transformation

1. Harvest callus from stems cultivated for 10 days for callus induction.
2. Place callus in the middle of a 6 cm petri dish, filled with MS medium supplemented with 200 g sucrose per liter (525 mOsm/kg). Incubate in the dark at 26 °C for 4 h before bombardment.
3. For biolistic bombardment, use 120 μg gold particles (0.4–0.8 μm in diameter [Heraeus, Frankfurt/Main, Germany]) coated with equal amounts of plasmids which were solved in 200 μL 98% ethanol.
4. Perform bombardment with the Particle Delivery System PDS-1000 (Bio-Rad, Munich, Germany) using a partial vacuum of 0.9 bar and a pressure of 107 bar. Adjust the distance between the macrocarrier and the target tissue to 45 mm.
5. After biolistic bombardment: Store callus in the dark in callus induction medium supplemented with 200 g sucrose per liter (525 mOsm/kg) at 26 °C for 24 h.
6. After 24 h incubation, transfer bombarded callus to callus inducing medium (with standard sucrose concentration), supplemented with the antibiotic (dependent from the chosen plant selection marker).

3.2.3 Plant Regeneration

1. Cultivate bombarded callus on callus-inducing media (with antibiotics) in a growth chamber in the dark at 26 °C for 4–6 weeks.
2. Separate putative transgenic callus tissue (after reaching a size of 5 mm in diameter) and transfer to regeneration medium.
3. Transfer callus with formed regenerative structures to regeneration media based either on WPM or MS (depending of the poplar clone), supplemented with 2.2 μg/L thidiazuron and 0.1 mg/L Pluronic F-68, 20 g/L sucrose, and 7 g/L agar.
4. Subculture transferred regenerating callus cultures in a growth chamber at 26 °C, with a 16/8 h light/dark cycle.
5. Six to ten weeks later: transfer putative independent transgenic plantlets to magenta boxes containing WPM medium (without hormones) for further development.

3.3 PEG Transformation of Protoplasts

3.3.1 Protoplast Isolation

1. Harvest leaves from leaves of 4–6-week-old aseptic-grown shoot cultures from clone INRA 717-1B4.
2. Finely chop isolated leaves (1 g fresh weight) with two scalpels. Add 20 mL of filter-sterilized enzyme solution, infiltrate under vacuum (6 × 10 s, 0.9 bar), and incubate in the dark for 3 h with gentle shaking (40 rpm).
3. Enrich obtained protoplasts by sieving through a 100 μm net with W5 solution, and collect by careful centrifugation at 150 × *g* for 5 min (e.g., Hettich, Tuttlingen, Germany, or any other centrifuge with slow acceleration and braking).
4. Resuspend pellet in 2 mL W5 solution. Further purify protoplasts by sucrose density gradient centrifugation, and collect at the interface of sucrose solution and W5 buffer after careful centrifugation for 10 min at 300 × *g* (e.g., Hettich, Tuttlingen, Germany, or any other centrifuge with slow acceleration and braking).
5. Wash protoplasts twice in W5. Pellet by centrifugation at 150 × *g* for 5 min and resuspend in 10 mL W5 solution.
6. Count number of protoplasts in a hemocytometer. Store protoplasts in W5 buffer over night at 4 °C until use.

3.3.2 Protoplast Yield and Viability

1. Mix suspended protoplasts (500 μL) with 12 μL fluorescein diacetate (FDA) solution, and subsequently place 20 μL of suspended protoplasts in a hemocytometer slide.
2. After 5 min, count protoplasts under white light. Express results as protoplast yield per 1 g fresh weight.
3. Examine viability of FDA-stained protoplasts under UV illumination.
4. Express protoplast viability as percentage of protoplasts showing fluorescence under UV light out of the total number of isolated protoplasts observed in the same microscopic field under white light.

3.3.3 PEG Transformation

1. Resuspend 5×10^5 purified protoplasts in 100 μL MMG solution.
2. Add 10 μL of desired plasmid (5 μg/μL) and mix gently by pipetting slowly up and down.
3. Then add 110 μL PEG_{1500} solution and mix completely by gently tapping the tube.
4. Incubate the assay in the dark at room temperature for 10 min. Dilute with 5 mL W5 solution and mix carefully by inverting the tube.
5. Centrifuge at 150 × *g* for 5 min. Remove the supernatant and culture the protoplasts at a density of 2×10^5 per mL in protoplast dividing medium.

3.3.4 Formation of Microcalli

1. Culture protoplasts in protoplast dividing medium (without NH_4NO_3) in the dark at 26 °C for 4–6 weeks.
2. Transfer microcalli in the scale of 5 mm in diameter to solid ½MS or ½WPM plant regeneration medium (depending from the poplar clone) and incubate in a growth chamber at 26 °C with a 16/8 h light/dark cycle. Carefully adapt microcalli to ambient light by weekly increase of light intensity.
3. Start selection (kanamycin or hygromycin; depending from the plant selection marker) 10 days after transfer of microcalli onto solid media.

4 Notes

1. The amount of used agar-agar depends on the intended application. High agar concentrations (up to 7 g) are feasible for petri dishes. Lower agar content hampers transfer of leaf explants to new medium during leaf-disc cultivation. For standard plant cultivation, e.g., in Magenta boxes or glass vessels, lower amounts of agar (5–6 g) are suitable. Especially if plants should be transferred to soil completely, softer medium avoids root rupture.
2. The solubility of $MgSO_4$ decreases with increasing temperatures. Thus we recommend to sterilize $MgSO_4$ solution using sterile filters and to add the solution after autoclaving the medium under a laminar airflow unit.
3. For transformation of micro-explants, the complete above-medium plants are harvested and processed. Especially stem tissue is suitable for high regeneration rates. Regarding the explant size, you might face a discrepancy. Too small plant explants are hard to regenerate, whereas too large explants increase the risk of multiple transformation events in one explant. Albeit, multiple transformation events could occur even in small explants. The cultivation of chimeric tissue could be avoided by transfer of only one regenerated stem from the callus with multiple stem formation. To determine the absolutely certain T-DNA integration locus in your regeneration line, TAIL-PCR is recommended [32, 33].
4. Tap water has proved of value. Since tap water contains minerals in contrast to distilled water, osmotic stress to the harvested plant tissue will be avoided.
5. Plant explants tend to stick to the metal strainer, especially if the tissue gets too dry. If the tissue remains wet, it is sticky to other explants thus it is easier to unhinge from the strainer with tweezers.

6. Here, baffled flasks are applied for improved venting. Adjust the stirring speed to your baffled flask. It should not be too high to allow regeneration processes and later on to avoid the breakup of regenerates. Otherwise you obtain a lot of regenerates, which developed from one transformation event and are genetically identical. Nevertheless, the regenerates should overcome the baffles for optimal mixing.

Acknowledgments

We would like to thank all technical assistants of the genome research group for valuable input, namely, Susanne Jelkmann, Denise Lehnhardt, Jennifer Lüneburg, Anke Schellhorn, and Doris Ebbinghaus.

References

1. Van Montagu M (2011) It is a long way to GM agriculture. Ann Rev Plant Biol 62:1–23
2. Bevan M (1984) Binary *Agrobacterium* vectors for plant transformation. Nucl Acid Res 12:8711–8721
3. Horsch RB, Fry JE, Hoffman NL et al (1985) A simple and general method for transferring genes into plants. Science 227:1229–1232
4. Herrera-Estrella L, De Block M, Messens EHJP et al (1983) Chimeric genes as dominant selectable markers in plant cells. EMBO J 2:987
5. Kempken F, Kempken R (2012) Gentechnik bei Pflanzen: Chancen und Risiken. Springer, Heidelberg
6. De Block M (1990) Factors influencing the tissue culture and the *Agrobacterium tumefaciens*-mediated transformation of hybrid aspen and poplar clones. Plant Physiol 93:1110–1116
7. Ahuja MR (1993) Regeneration and germplasm preservation in aspen-populus. In: Ahuja MR (ed) Micropropagation of woody plants. Springer, Netherlands, pp 187–194
8. Fillatti JJ, Sellmer J, McCown B et al (1987) *Agrobacterium* mediated transformation and regeneration of *Populus*. Mol Gen Genet 206:192–199
9. Peña L, Séguin A (2001) Recent advances in the genetic transformation of trees. Trends Biotechnol 19:500–506
10. Busov V, Strauss S, Pilate G (2010) Transformation as a tool for genetic analysis in *Populus*. In: Jansson S et al. (eds) Genetics and genomics of *Populus,* vol 8. Plant Genetics and Genomics: Crops and Models. pp 113–128
11. Horsch RB, Fry JE, Hoffmann NL et al (1985) A simple and general method for transferring genes into plants. Science 227:1229–1231
12. Leple JC, Brasileiro ACM, Michel MF, Delmotte F, Jouanin L (1992) Transgenic poplars: expression of chimeric genes using four different constructs. Plant Cell Rep 11:137–141
13. Fladung M, Muhs HJ, Ahuja MR (1996) Morphological changes observed in transgenic *Populus* carrying the *rolC* gene from *Agrobacterium rhizogenes*. Silv Genet 45:349–354
14. Fladung M, Kumar S, Ahuja MR (1997) Genetic transformation of Populus genotypes with different chimaeric gene constructs: transformation efficiency and molecular analysis. Trans Res 6:111–121
15. Kumar S, Fladung M (2001) Gene stability in transgenic aspen (*Populus*). II. Molecular characterization of variable expression of transgene in wild and hybrid aspen. Planta 213:731–740
16. Yevtushenko DP, Misra S (2010) Efficient *Agrobacterium*-mediated transformation of commercial hybrid poplar *Populus nigra* L. × *P. maximowiczii* A. Henry. Plant Cell Rep 29:211–221
17. Wang H, Wang C, Liu H et al (2011) An efficient *Agrobacterium*-mediated transformation and regeneration system for leaf explants of two elite aspen hybrid clones *Populus alba* × *P. berolinensis* and *Populus davidiana* × *P. bolleana*. Plant Cell Rep 30:2037–2044

18. Maheshwari P, Kovalchuk I (2016) *Agrobacterium*-mediated stable genetic transformation of *Populus angustifolia* and *Populus balsamifera*. Front Plant Sci 7:296
19. Cseke LJ, Cseke SB, Podila GK (2007) High efficiency poplar transformation. Plant Cell Rep 26:1529–1538
20. Nietsch J, Brügmann T, Becker D, Fladung M (2017) Old methods rediscovered: application and improvement of two direct transformation methods to hybrid poplar (*Populus tremula* × *P. alba*). Plant Cell Tissue Organ Cult 130:183–196
21. McCown BH, McCabe D, Russell D, Robinson D, Barton K, Raffa K (1991) Stable transformation of *Populus* and incorporation of pest resistance by electric discharge particle acceleration. Plant Cell Rep 9:590–594
22. Russell JA, McCown BH (1986) Culture and regeneration of *Populus* leaf protoplasts isolated from non-seedling tissue. Plant Sci 46:133–142
23. Chupeau MC, Pautot V, Chupeau Y (1994) Recovery of transgenic trees after electroporation of poplar protoplasts. Trans Res 3:13–19
24. Yoo SD, Cho YH, Sheen J (2007) Arabidopsis mesophyll protoplasts: a versatile cell system for transient gene expression analysis. Nat Protocol 2:1565–1572
25. Hamilton RH, Fall MZ (1971) The loss of tumor-initiating ability in *Agrobacterium tumefaciens* by incubation at high temperature. Cell Mol Life Sci 27:229–230
26. Hoeckema A, Hirsch PR, Hooykaas PJJ, Schilperoort RA (1983) A binary plant vector strategy based on separation of vir- and T-region of the *Agrobacterium tumefaciens* Ti-plasmid. Nature 303:179–180
27. Mader M, Le Paslier MC, Bounon R et al (2016) Whole-genome draft assembly of *Populus tremula* × *Populus alba* clone INRA 717-1B4. Silvae Genetica 65:2
28. Lloyd G, McCown B (1980) Commercially-feasible micro-propagation of mountain laurel, Kalmia latifolia, by use of shoot-tip culture. Int Plant Prop Soc Proc 30:421–426
29. Murashige T, Skoog F (1962) A revised medium for rapid growth and bioassays with tobacco tissue cultures. Physiol Plant 15:473–497
30. Bridson EY, Brecker A (1970) Chapter III Design and formulation of microbial culture media. Meth Microbiol 3:229–295
31. Wise AA, Liu Z, Binns AN (2006) Culture and maintenance of *Agrobacterium* strains. In: Methods in molecular biology, vol 343, *Agrobacterium* protocols. pp. 3–14
32. Liu YG, Mitsukawa N, Oosumi T, Whittier RF (1995) Efficient isolation and mapping of *Arabidopsis thaliana* T-DNA insert junctions by thermal asymmetric interlaced PCR. Plant J 8:457–463
33. Fladung M, Polak O (2012) *Ac*/Ds-transposon activation tagging in poplar: a powerful tool for gene discovery. BMC Genomics 13:61

Chapter 13

The Genetic Transformation of Sweet Orange (*Citrus sinensis* L. Osbeck) for Enhanced Resistance to Citrus Canker

Lorena Noelia Sendin and María Paula Filippone

Abstract

Developing disease resistance is one of the most important components of any plant breeding program. Citrus traditional breeding methods (bud sport selection, crossbreeding, and other breeding channels) are a laborious task and often hampered by long juvenility, a high degree of heterozygosity, polyembryony, self-incompatibility, and abortion of reproductive organs. An interesting alternative to the classical breeding approach is the use of genetic transformation, which provides the means for adding a single agronomic trait to a plant without otherwise altering its phenotype. *Agrobacterium tumefaciens*-mediated transformation has been carried out with numerous hybrids and citrus species. This technique allowed us to introduce the *Bs2* gene in *Citrus*, as well as to increase citrus canker resistance in transgenic *Bs2* gene-expressing lines.

Key words Citrus breeding, *A. tumefaciens*, Shoot regeneration, In vitro grafting, Acclimatization

1 Introduction

Plants have to deal with various and complex types of interactions involving numerous environmental factors. In the course of evolution, they have evolved specific mechanisms allowing them to adapt and survive stressful events [1]. Biotic stress is one of the major environmental factors affecting plants [2]. In this sense, *Citrus* crop is affected by a variety of pathogens: viroids, viruses, bacteria, fungi, and nematodes. Some of these are very destructive and have had a great economic impact in several regions of the world; examples include *Citrus tristeza virus* (which causes tristeza decline), *Xanthomonas citri* subsp. *citri* (which causes citrus canker), *Phyllosticta citricarpa* (*Guignardia citricarpa*, which produces black spot disease), and *Candidatus liberibacter* spp., the latter one being the causal agent of Huanglongbing (HLB), a devastating and irreversible disease leading to the death of the plant, which has produced a crisis in the global citrus industry [3].

Sandeep Kumar et al. (eds.), *Transgenic Plants: Methods and Protocols*, Methods in Molecular Biology, vol. 1864,
https://doi.org/10.1007/978-1-4939-8778-8_13,

Specifically, citrus canker is a serious disease caused by the Gram-negative plant pathogenic bacterium *Xanthomonas citri* subsp. *citri* (*Xcc*) [4], which affects most commercial citrus species and varieties. The first symptom is an oily circular spot that becomes white or yellow with spongy pustules [5]. This oily spot later darkens into a brown corky canker with a yellow halo that can be easily seen on leaves, stems, and fruit [5]. Severe infections result in defoliation and blemished fruit, and premature abscission may also constrain the commercialization of citrus fruit from canker-affected to disease-free areas [6].

Developing disease resistance is one of most important components of any plant breeding program. Citrus traditional breeding methods (bud sport selection, crossbreeding, and other breeding channels) are a laborious task and often hampered by long juvenility, a high degree of heterozygosity, polyembryony, self-incompatibility, and abortion of reproductive organs [7]. An interesting alternative to the classical breeding approach is the use of genetic transformation, which provides the means for adding a single agronomic trait to a plant without otherwise altering its phenotype [8].

Citrus genetic transformation is being increasingly used as an improvement tool; in fact, in some countries it has already been incorporated into breeding programs [9]. *Agrobacterium tumefaciens*-mediated transformation has been carried out with numerous hybrids and citrus species, such as grapefruit, sour orange, sweet orange, Carrizo citrange, Mexican lime, citrumelo, and mandarin [10–12]. Resistance to diseases is one of the main objectives of transgenesis. With respect to HLB, transgenic lines of *C. sinensis* expressing the *Arabidopsis* Npr1 gene [13] were obtained. Also, transgenic lines with increased resistance to citrus tristeza virus (CTV) and psoriasis virus have been obtained through different strategies: pathogen-derived resistance [14, 15], introduction of candidate resistance genes [16], and host-mediated gene silencing of pathogen. In order to improve resistance to bacterial diseases in plants through genetic engineering, different strategies have been proposed. To contribute to the control of citrus canker produced by the bacterium *Xanthomonas citri* subsp. *citri*, the strategies include the overexpression of genes that code for antibacterial peptides [17], disease resistance proteins [18], and transcription factors from non-plant origin that enhance natural plant defenses or are involved in the induction of programmed cell death at the infection site [13, 19, 20].

This chapter describes the methodology used in our laboratory to introduce foreign genes into *Citrus* genome by transformation mediated by *A. tumefaciens*. We successfully introduced the *Bs2* resistance gene from pepper [21] in *Citrus sinensis* plants [22]. A reduction of canker symptoms of up to 70% was observed in transgenic lines expressing *Bs2* with respect to non-transformed control plants. It was proposed that *Bs2* gene had restricted taxonomic functionality (RTF) and only conferred resistance to species belonging to Solanaceae [21]. However, our results indicate that pepper *Bs2* resistance gene is also functional in families other than the Solanaceae and could be considered for canker control and other diseases produced by *Xanthomas.*

2 Materials

2.1 Plant Material

1. Source of explants: seedlings of *Citrus sinensis* L. Osbeck cv. Pineapple obtained by in vitro germination of uncovered and disinfected seeds (*see* **Note 1**) and grown in controlled conditions for 6 weeks (Fig. 1a; *see* **Notes 2** and **3**)
2. Rootstock for in vitro grafting: seedling of Troyer citrange (*C. sinensis* L. Osbeck x *Poncirus trifoliata* L. Raf.) germinated in vitro and grown in the dark for 2 weeks

2.2 Strains and Plasmid

1. Bacterial strain: *A. tumefaciens* super-virulent strain EHA 105 [23] which has shown to be efficient in citrus.
2. Plasmid constructions: binary plasmids may contain one selectable marker gene, such as neomycin phosphotransferase II (*nptII*), which confers kanamycin resistance to transformed cells, and a reporter gene marker which leads to differentiate the transformed cells. The two marker genes mostly used are β-D-glucuronidase (*gus A*), which confers blue coloration after a histochemical coloration, and green fluorescein protein (*gfp*) which confers green coloration under UV light. Both types of marker genes (selectable and reporter) should be driven by both promoter and terminator sequences. In case of transgenic *Bs2* lines of *C. sinensis*, we constructed the plasmid pGST1-BS2, which carries the *Bs2* gene under the control of the *gst1* promoter [24] from potato (inducible for bacterial and wounds, kindly sent by Dr. Beer from Cornell University, USA) and *T-nos* terminator between the 35SCaMV-promoter/*npt*II /35S-ter and the 35SCaMV-promoter/*gus A*/ *nos*-ter cassettes. The plasmid vector is introduced into an *A. tumefaciens* super-virulent strain EHA 105 [23] through electroporation.
3. *Xanthomonas* strain: To reproduce canker disease under controlled growth conditions, plant leaves are inoculated with a *Xanthomonas citri* subsp. *citri* strain isolated from lemon leaves with canker symptoms and identified in the Phytopathology laboratory at EEAOC. Xcc is cultured at 28 °C for 24 h in PYM medium.

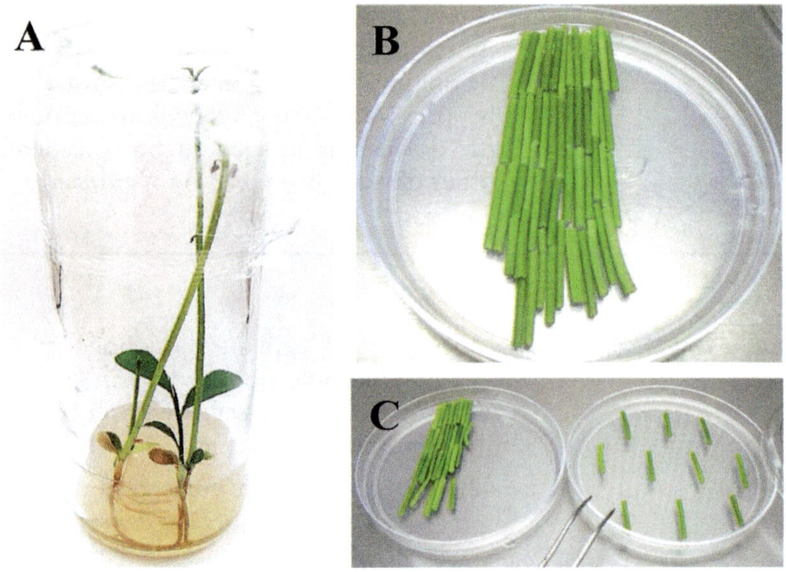

Fig. 1 Plant material. (**a**) Seedlings of *C. sinensis* germinated in vitro on germination medium and grown for 6 weeks. (**b**) Co-cultivation of internodal stem segments and *A. tumefaciens*. (**c**) Transference of explants from co-cultivate medium to regeneration and selection medium

2.3 Specific Primers

To verify the presence of transgene in regenerated in vitro shoots, specific primers are required. In this work following primers amplifying *Bs2* gene are used: *Bs*2 L1 (5′-CTGCTTCACCAATCA TCTTAAC-3′) and *Bs*2 R1 (5′-TGAGACTAACAGGAACTGTA CT-3′).

2.4 Culture Media and Solutions

1. Seed germination medium: 4.3 g/L salts of the basal medium Murashige and Skoog (MS) [25], 30 g/L sucrose, and 10 g/L agar-agar.
2. Inoculation medium: 4.3 g/L MS salts, vitamins (100 mg/L myoinositol, 0.2 mg/L thiamine-HCl, 1 mg/L pyridoxine-HCl, 1 mg/L nicotinic acid), and 30 g/L sucrose.
3. Co-cultivation medium: 4.3 g/L MS salts, 30 g/L sucrose, vitamins (the same on 2.3.2), 2 mg/L 2,4-Dichlorophenoxyacetic acid (2,4-D), 2 mg/L indole-3-acetic acid (AIA), 1 mg/l 2-isopentenil adenine (2,i-P), and 8 g/L agar-agar.
4. Regeneration medium: 4.3 g/L MS salts, 30 g/L sucrose, vitamins (the same on 2.3.2), 3 mg/L 6-benzylaminopurine (BAP), and 7 g/L agar-agar (Sigma). After sterilization, 750 mg/L the antibiotic cefotaxime for eliminate *A. tumefaciens* and 100 mg/L of kanamycin (selection agent of plasmid construction used) are added.
5. Liquid grafting medium: 4.3 g/L MS salts and 45 g/L sucrose.

The pH of all culture media was adjusted to 5.7 prior to the addition of the gelling agent and autoclaved at 121 °C for 20 minutes at a pressure of 1 atm (*see* **Note 4**).

6. Luria broth (LB) medium: 10 g/L tryptone, 10 g/L NaCl, and 5 g/L yeast extract. pH 7.5. After autoclaving adequate antibiotics must be added. In case of *A. tumefaciens*, this strain has chromosomal resistance to nalidixic acid (25 mg/L). Moreover, it must to be added the selective agent codified by the plasmid construction. In case of plasmid GST1-Bs2 which carries nptII gene, 50 mg/L Kanamycin are added.
7. PYM medium: peptone 0.5%, yeast extract 0.3%, malt extract 0.3%, supplemented with D-glucose at 1.5% (w /v) final concentration.
8. GUS staining solution: 1 mM EDTA (pH 8), 5 mM potassium ferricyanide, 5 mM potassium ferrocyanide, 100 mM sodium phosphate (pH 7.0), 1% Triton X-100, and 2 mM X-Gluc (5-bromo-4-chloro-3-indolyl-beta-D-glucuronide, Sigma). Dissolve the X-Gluc in a drop of *N*,*N*-dimethylformamide immediately prior to use. Staining solution can be stored at −20 °C.
9. Disinfectant solution: 2% (v/v) sodium hypochlorite.

2.5 Equipment

1. Culture chambers allowing temperature, humidity, and illumination control
2. Laminar flow cabinets
3. Shakers
4. Centrifuge
5. Spectrophotometer

3 Methods

3.1 *C. sinensis* Transformation and Plant Regeneration

Experiments are carried out following the established protocol for citrus transformation [26].

3.1.1 Plant Material Preparation

1. Select the seedlings of uniform size.
2. Cut transverse internodal segments of approximately 1 cm long by using a sterile scalpel.
3. Keep the explants in a petri dish with paper moistened with sterile distilled water until the inoculation is initiated (*see* **Note 5**).

3.1.2 *Agrobacterium* Inoculum Preparation

1. To prepare the pre-inoculum of *A. tumefaciens* for the infection of the explants, take an isolated colony from a fresh plate of the bacterium.
2. Inoculate the colony into 5 mL of liquid LB medium, supplemented with 25 mg/L of nalidixic acid and 50 mg/L of kanamycin, into 15 mL glass assay tubes. Incubate over night at 28 °C on an orbital shaker (200 rpm).
3. Take 100 μL of the pre-inoculum obtained in **step 2**, and inoculate 100 mL of liquid LB medium, supplemented with 25 mg/L of nalidixic acid and 50 mg/L of kanamycin, into 500 mL glass Erlenmeyer. Incubate at 28 °C on an orbital shaker (200 rpm) until the culture reached 0.25 to 0.4 optical density (OD) at 600 nm (16–20 h approximately).
4. Centrifuge the culture at 2150 × *g* for 10 min.
5. Discard the supernatant, and suspend the bacterial pellet delicately in the same volume by using liquid inoculating medium.
6. Perform a dilution of the bacterial culture to obtain a suspension of 4 × 10^7 cells/mL (*see* **Note 6**).

3.1.3 Inoculation and Culture of Explants

1. Inoculate epicotyl segments by immersion (15 min) in the suspension of *A. tumefaciens* EHA 105 (10^7 cell/mL) (*see* **Note 7**).
2. Blot dry explants with tissue (*see* **Note 8**) and transfer to co-cultivation medium for 3 days at 26 °C at low light intensity (10 μE/m^2/s, 16 h photoperiod) (*see* **Note 9** and Fig. 1b).
3. Transfer explants to regeneration and selection medium (Fig. 1c).
4. Incubate cultures in the dark for 2 weeks, at 26 °C, and then transfer to a 16-h photoperiod, 45 μE/m^2/s, at 26 °C (*see* **Note 10**).
5. Subculture explants every 4 weeks.

3.2 Reporter Gene Activity Analysis

1. Once shoots are formed on the ends of explants, check the shoots for transgenic nature through expression analysis of reporter marker genes (*see* **Note 11**). In case of GST1-*Bs2*, perform the histochemical assay for reporter gene *gus A*.
2. Excise kanamycin-resistant shoots from the explants, and transfer small basal tissue to the GUS staining solution and incubate at 37 °C overnight.
3. Clear the solution by incubation in 70% ethanol for 24 h, and observe for *gus* stain using microscope or magnifying glasses.

3.3 Recovery of Transgenic Shoots

1. Recover *gus*-positive shoots through an in vitro grafting of apical portions on in vitro-grown rootstocks [26].

Fig. 2 Recovery of whole transgenic plants (**a**) Shoot regenerated from *Agrobacterium*-inoculated explants. (**b**) Transgenic shoot grafted in vitro on a decapitated rootstock. (**c**) Seedling obtained 2 months after grafting

2. Decapitate seedlings of Troyer citrange and remove the cotyledons and the secondary roots.
3. Make a wedge-shaped cut at the base of the putative transgenic shoots, and then introduce into a small vertical incision that is made on the apical end of rootstock (Fig. 2a, b).
4. Carefully, place the grafting inside a tube with liquid graft medium which has a sterile paper inside as a support. The radicular system passes through one small hole into the paper, and the area part remains on upper the support (Fig. 2c).
5. Transfer well-developed plantlets to container pots, and keep in controlled conditions or may be grafted in a rootstock in greenhouse.

3.4 Detection of Transgene Presence

1. Extract genomic DNA from 200 mg of fresh leaves sampled from 2-month-old grafted shoots of transgenic and non-transformed control plants, by using the CTAB method [27].
2. Amplify a fragment of 0.5 kb of *Bs2* gene is by PCR using specific primers *Bs2* L1 and *Bs2* R1 to detect genetically transformed lines.
3. Perform PCR amplification of the *Bs2* gene using 30 cycles according to the following conditions: denaturation at 95 °C for 30 s, annealing at 58 °C for 30 s, and extension at 72 °C for 60 s.
4. Detect PCR products by electrophoresis using 1% agarose gels.

3.5 Evaluation of Transgenic Lines for Resistance to Citrus Canker Disease

1. Inoculate ten leaves from new flushes of each of the transgenic lines and non-transformed plants by infiltration of 10^5 cells/mL *Xcc* suspension in 10 mM Cl_2Mg by using a syringe without a needle.
2. Keep the inoculated plants at 28 °C and 70% relative humidity. Bacterial population in the inoculated sites of leaves from transgenic plants is analyzed 14 days post inoculation.
3. Disinfect a total of ten 1 cm^2 leaf discs from inoculated sites of each plant with 2% (v/v) sodium hypochlorite for 1 min, and wash three times with sterile distilled water.
4. Macerate discs in 1 mL of sterile 10 mM Cl_2Mg solution, and later dilute and spread on a PYM nutrient medium.
5. After incubation for 2 days at 28 °C, count the number of colonies to calculate the colony-forming units (CFU), expressed as CFU/cm^2 of infected leaves. The samples must be taken from three plants and the experiment must be done in triplicate.

4 Notes

1. Seeds are extracted from mature fruits of *C. sinensis* cv. Pineapple, placed inside a vessel with distilled water, and shaken vigorously. We rinse the seeds three times to make sure that there is no mucilage around the seed so as to prevent fungal development. After that, the seeds are dried at room temperature for 24 h and their coats removed. There are two seed coats: the external coat, white and thick, and the internal coat, brown and thin. Both of them must be carefully removed to prevent injuring the seed embryo. The brown coat is usually more difficult to remove, so we strip the outer coat and then soak the seed in distilled water for 30 min. Then, we make a small wound with a needle only in the internal coat to facilitate

its removal. It is important not to break or disarm the seed, which could affect the percentage of germination. After the seed coats are removed, the seeds are treated with a solution of sodium hypochlorite (0.5%) for 15 min, followed by three rinses with sterile distilled water.

2. For germination, seeds can be grown in 2.5 cm × 15 cm tubes with 10 mL germination medium (in this case, only one seed is placed into each tube) or in 7 cm × 12 cm glass flasks with 70 mL germination medium in which two or three seeds can be placed. The seeds are incubated in the dark for 2 weeks and then transferred to a 16-h photoperiod, 45 $\mu E/m^2/s$, at 26 °C for another 4 weeks.
3. One of the difficulties to work with citrus seeds is that they are not available all the time; the fruits are ripe in a specific period of the year. Therefore, when fruits are in season, all the seeds must be collected and conserved for their further use. The seeds that will not be immediately used are immersed into 1% (p/v) solution of oxyquinoline sulfate (almacigol) for 1 min and then allowed to dry at room temperature for 24 h. Once the seeds are dry, cover them with paper napkins, and store in paper bags at 6–8 °C. The seeds could be stored for 3–4 months without a viability decrement.
4. Shoot regeneration and selection media must be autoclaved, dispensed in sterile Petri dishes (25 mL), and kept at room temperature for 3–5 days to check fungal development. After this period, uncontaminated media should be stored at 7 °C to avoid evaporation. This media is preferably used within 15 days after preparation to ensure the composition is not altered.
5. To obtain the explants, first the seedlings with uniform growth must be selected; i.e., the smallest ones or those with a thin stem are eliminated. Subsequently, the seedlings are taken with a clamp from the apical end and removed gently to avoid damaging them. Once the roots are released from the culture medium, a cut at 0.5–1 cm above the cotyledons is made, dropping the root portion into the tube or flask. It ensures that the radicular system remains inside the container where the seedling grew. This decreases the probability of contamination in the assay. Subsequently, the aerial part of the seedling is placed on sterile filter paper, and the upper ends (1–2 cm) are discarded since the stem in this area is usually thinner and does not outlive the posterior infection.
6. It is important to draw a growth curve for the *Agrobacterium* strain that will be used to determine the number of bacteria as function of optical density at 600 nm. From this curve, the dilution needed to prepare a suspension of 4×10^7 cells/mL will be calculated. It is possible to determine one formula to

linear regression of OD_{600} versus bacterial number. For example, for *A. tumefaciens* EHA 105, the formula is as follows: Bacterial number = $((137.16 \times OD600) - 5.56) \times 10^7$ cells/mL [26], which means that for culture OD = 0.37, the suspension has 4.5×10^8 cells/mL. So, a dilution of 1:10 proximally leads to the appropriate concentration for inoculation.

7. We use sterile 50 mL Falcon tubes to perform the inoculation of the explants. We place 30 mL of the bacterial dilution and a total of 100 explants in each tube and gently shake for 15 min. To optimize the time, it is possible to start an inoculation tube every 3–5 min depending on the operator's ability. A transformation experiment must have at least 500 explants; therefore, five transformation tubes will be used, which must be numbered to identify the order in which they will be stopped.
8. After inoculation, it is important to dry the explants carefully onto sterile filter paper and to absorb the excess of bacterial suspension with sterile paper napkins. Once this process is concluded, the explants are immediately transferred to co-culture medium to prevent the excessive drying of plant material due to the air from the laminar flow; this process can stress the tissue and decrease the transformation efficiency.
9. The explants inoculated with *A. tumefaciens* are placed side by side in the co-culture medium (50 explants per plate) in order for the ends of one explant to be in contact with the end of the next (Fig. 1b). This procedure helps phenolic compounds that are released from plant tissues in wound response to concentrate around the tissue. These compounds facilitate the transfer of *Agrobacterium* T-DNA to the plant, thus leading to an increment of genetic transformation efficiency.
10. The explants must be cultivated in the dark until the formation of a white callus. In the case of *C. sinensis*, this callus can be observed from the seventh to the tenth day after inoculation with *A. tumefaciens*. Once the callus is formed, the plates are transferred to a 16-h photoperiod, 45 $\mu E/m^2/s$, at 26 °C to induce bud formation from the transformed callus. Darkness period is variable for each citric genotype and should be optimized in each case since a large callus formation can lead to a decrement of shoot regeneration efficiency.
11. A practical way to select the shoots expressing the reporter gene is to detach them from the ends of the explants; to cut a basal portion, which will be placed in numbered 0.06 mL Eppendorf tubes to determine their GUS activity; and to insert them vertically into a plate with regeneration medium following the same order of the Eppendorf tubes. This allows the transgenic shoots to be preserved at 26 °C for several days until they are grafted.

Acknowledgments

This project was supported by Estación Experimental Agroindustrial Obispo Colombres (EEAOC) and Consejo Nacional de Investigaciones Científicas y Tecnológicas (CONICET).

References

1. Rejeb I, Pastor V, Mauch-Mani B (2014) Plant responses to simultaneous biotic and abiotic stress: molecular mechanisms. Plants 3:458–475. https://doi.org/10.3390/plants3040458
2. Sultanova N, Huseynova I, Mammadov A et al (2014) Biotic stress and crop improvement. Improv Crop Era Clim Chang. https://doi.org/10.1007/978-1-4614-8824-8_4
3. Bové JM (2006) Huanglongbing: a destructive, newly-emerging, century-old disease of citrus. J Plant Pathol 88:7–37. https://doi.org/10.4454/jpp.v88i1.828
4. Schaad NW, Postnikova E, Lacy GH et al (2005) Reclassification of Xanthomonas campestris pv. citri (ex Hasse 1915) Dye 1978 forms A, B/C/D, and E. Syst Appl Microbiol 28:494–518. https://doi.org/10.1016/j.syapm.2005.03.017
5. Brunings AM, Gabriel DW (2003) Xanthomonas citri: breaking the surface. Mol Plant Pathol 4:141–157. https://doi.org/10.1046/j.1364-3703.2003.00163.x
6. Koh E-J, Zhou L, Williams DS et al (2011) Callose deposition in the phloem plasmodesmata and inhibition of phloem transport in citrus leaves infected with "Candidatus Liberibacter asiaticus.". Protoplasma 249:687. https://doi.org/10.1007/s00709-011-0312-3
7. Grosser JW, Gmitter FG Jr (1990) Protoplast fusion and citrus improvement. Plant Breed Rev 8:339–374
8. Deng X, Duan Y (2006) Modification of perennial fruit trees BT – tree transgenesis: recent developments. In: Fladung M, Ewald D (eds) . Springer, Berlin, Heidelberg, pp 47–66
9. Donmez D, Simsek O, Izgu T et al (2013) Genetic transformation in citrus. Sci World J 2013:1. https://doi.org/10.1155/2013/491207
10. Luth D, Moore G (1999) Transgenic grapefruit plants obtained by Agrobacterium tumefaciens-mediated transformation. Plant Cell Tissue Organ Cult 57:219–222. https://doi.org/10.1023/A:1006387900496
11. Dominguez A, Guerri J, Cambra M et al (2000) Efficient production of transgenic citrus plants expressing the coat protein gene of citrus tristeza virus. Plant Cell Rep 19:427–433
12. Peña L, Pérez RM, Cervera M et al (2004) Early events in agrobacterium-mediated genetic transformation of citrus explants. Ann Bot 94:67–74. https://doi.org/10.1093/aob/mch117
13. Dutt M, Barthe G, Irey M, Grosser J (2015) Transgenic citrus expressing an arabidopsis NPR1 gene exhibit enhanced resistance against Huanglongbing (HLB; Citrus greening). PLoS One 10:e0137134. https://doi.org/10.1371/journal.pone.0137134
14. Fagoaga C, López C, De Mendoza AH et al (2006) Post-transcriptional gene silencing of the p23 silencing suppressor of Citrus tristeza virus confers resistance to the virus in transgenic Mexican lime. Plant Mol Biol 60:153–165. https://doi.org/10.1007/s11103-005-3129-7
15. Zanek MC, Reyes CA, Cervera M et al (2008) Genetic transformation of sweet orange with the coat protein gene of Citrus psorosis virus and evaluation of resistance against the virus. Plant Cell Rep 27:57–66. https://doi.org/10.1007/s00299-007-0422-8
16. Çevik B, Lee RF, Niblett CL (2006) Genetic transformation of Citrus paradisi with antisense and untranslatable RNA-dependent RNA polymerase genes of Citrus tristeza closterovirus. Turkish J Agric For 30:173–182
17. Boscariol RL, Monteiro M, Takahashi EK et al (2006) Attaching A gene from trichoplusia ni reduces susceptibility to Xanthomonas axonopodis pv. citri in Transgenic Citrus sinensis 'Hamlin'. J Am Soc Hortic Sci 131:530–536
18. Omar AA, Song WY, Grosser JW (2007) Introduction of Xa21, a Xanthomonas-resistance, gene from rice, into "Hamlin" sweet orange [Citrus sinensis (L.) Osbeck] using protoplast-GFP co-transformation or single plasmid transformation. J Hortic Sci Biotechnol 82:914–923
19. Yang L, Hu C, Li N et al (2011) Transformation of sweet orange [Citrus sinensis (L.)

Osbeck] with pthA-nls for acquiring resistance to citrus canker disease. Plant Mol Biol 75:11–23

20. Zhang X, Francis MI, Dawson WO et al (2010) Over-expression of the Arabidopsis NPR1 gene in citrus increases resistance to citrus canker. Eur J Plant Pathol 128:91–100. https://doi.org/10.1007/s10658-010-9633-x
21. Tai TH, Dahlbeck D, Clark ET et al (1999) Expression of the Bs2 pepper gene confers resistance to bacterial spot disease in tomato. Proc Natl Acad Sci U S A 96:14153–14158. https://doi.org/10.1073/pnas.96.24.14153
22. Sendín LN, Orce IG, Gómez RL et al (2017) Inducible expression of Bs2 R gene from Capsicum chacoense in sweet orange (Citrus sinensis L. Osbeck) confers enhanced resistance to citrus canker disease. Plant Mol Biol 93:607–621. https://doi.org/10.1007/s11103-017-0586-8
23. Hood EE, Gelvin SB, Melchers LS, Hoekema A (1993) New Agrobacterium helper plasmids for gene transfer to plants. Transgenic Res 2:208–218. https://doi.org/10.1007/BF01977351
24. Barbosa-Mendes JM, de Filho F AAM, Filho AB et al (2009) Genetic transformation of Citrus sinensis cv. Hamlin with hrpN gene from Erwinia amylovora and evaluation of the transgenic lines for resistance to citrus canker. Sci Hortic (Amsterdam) 122:109–115. https://doi.org/10.1016/j.scienta.2009.04.001
25. Murashige T, Skoog F (1962) A revised medium for rapid growth and bioassays with tobacco tissue cultures. Physiol Plant 15:473–497. https://doi.org/10.1111/j.1399-3054.1962.tb08052.x
26. Pena L, Cervera M, Juárez J et al (1995) High efficiency Agrobacterium-mediated transformation and regeneration of citrus. Plant Sci 104:183–191. https://doi.org/10.1016/0168-9452(94)04021-8
27. McGarvey P, Kaper JM (1991) A simple and rapid method for screening transgenic plants using the PCR. BioTechniques 11:428–432

Chapter 14

Genetic Modification of Grapevine Embryogenic Cultures

S. A. Dhekney, S. K. Sessions, M. Brungart-Rosenberg, C. Claflin, Z. T. Li, and D. J. Gray

Abstract

Precision breeding is an approach to grapevine genetic improvement that transfers only specific traits among sexually compatible species via the relatively stable mitotic cell division pathway in order to avoid the significant disruption imposed upon conventional breeding by meiosis. Factors enabling precision breeding include the availability of the *Vitis* genome sequence combined with highly optimized gene insertion and plant regeneration protocols. A protocol for the production of grapevine embryogenic cultures and their genetic transformation is described. Embryogenic cultures are produced from either leaf or floral explants. Somatic embryos at the cotyledonary stage of development are used for *Agrobacterium*-mediated transformation. Following co-cultivation with *Agrobacterium* containing the genes of interest, modified embryos are selected on the basis of anthocyanin pigmentation and antibiotic resistance. Somatic embryos are then germinated to produce modified plants that are hardened and transferred to a greenhouse. The presence of the genes of interest is confirmed by PCR.

Key words *Vitis*, Somatic embryogenesis, *Agrobacterium*, Transgenic plants, Transgenes, Culture medium, Growth regulators, Plant tissue culture

1 Introduction

Grape is among the most ancient cultivated crops and is valued for its multiple uses as a fresh fruit and processed food product (jelly, juice, raisins, and wine). Grape and its products are well known for its health attributes due to the presence of antioxidants, vitamins, and other health-beneficial compounds. Grapevine is cultivated on 1.04 million acres in the United States with an annual production of 74.5 million tons [1]. The viticulture industry annually generates a large number of jobs and makes a substantial contribution to the US economy [2].

Grapevine improvement and development of new cultivars have traditionally been achieved using clonal selection of bud sports and breeding for the incorporation of biotic and abiotic stress tolerance. Breeding has limited applications in grape improvement

Sandeep Kumar et al. (eds.), *Transgenic Plants: Methods and Protocols*, Methods in Molecular Biology, vol. 1864, https://doi.org/10.1007/978-1-4939-8778-8_14,

due to the heterozygous nature of the grape genome, which results in the production of F1 hybrids with intermediate qualities between two parents [3]. Severe inbreeding depression observed in grapevines results in the application of techniques such as "pseudo-backcrossing" that involves crossing the F1 hybrid to a different parent that was not involved in the original cross [4, 5].

Precision breeding of grapevine for genetic improvement involves utilizing DNA sequences found solely in the genome of sexually compatible *Vitis* species and is a logical refinement of conventional breeding procedures [6, 7]. Factors enabling the successful development of precision breeding technology involve the availability of sequence information on the genes and regulatory elements that confer specific traits of interest along with efficient cell culture and plant regeneration protocols [8].

Grapevine embryogenic cultures have been very frequently used for transgene insertion and recovery of transgenic plant lines [9–14]. We have optimized protocols for initiation and maintenance of embryogenic cultures for several *Vitis* species and varieties [15, 16]. Cultures can be initiated from leaves or floral explants and can be maintained on hormone-free embryo development medium for extended periods of time. Somatic embryos at the cotyledonary stage of development are used as target tissues for genetic transformation and plant regeneration. Additionally, a wide array of regulatory elements and functional gene sequences have been isolated and characterized from a number of sexually compatible *Vitis* species [17].

The success of a transformation system depends on the ability to successfully recover transformants after co-cultivation. The ability to recover transformants following *Agrobacterium*-mediated transformation can be greatly enhanced by the use of reporter genes that provide visual indication of the T-DNA insertion process and selectable marker genes that provide a selective advantage for the growth and development of cells harboring the inserted T-DNA. Earlier studies involving genetic modification of plant species widely used the green fluorescent protein (GFP) derived from the Pacific jellyfish *Aequorea victoria* and the β-glucuronidase (GUS) gene, derived from *E. coli*, both of which are transgenic in origin [11]. A grapevine transcription factor *Vv*MybA1 has been extensively studied for its use as a precision-bred reporter gene in *Vitis* [18, 19]. The *Vv*MybA1 gene targets the final step in the anthocyanin biosynthesis pathway by activating the UFGT enzyme, which in turn catalyzes the formation of stable anthocyanin. Thus, any modified cells and plants can be readily identified on the basis of anthocyanin pigmentation (Fig. 1). The *Vv*MybA1 gene thus provides for convenient screening of genetically modified events in a nondestructive fashion using a simple dissecting microscope compared to GFP that requires a microscope equipped with

Fig. 1 Expression of the *Vv*MybA1 reporter system in *Vitis*. Transient expression in cotyledonary-stage SE explants after 3 days of co-cultivation (**a**). Stable callus formation (**b**), secondary embryo production (**c**), and plant regeneration from MybA1-expressing somatic embryos (**d**). Note that the intense red pigmentation occurring from anthocyanin accumulation is uniformly observed in all plant organs

epifluorescence illumination and GUS that necessitates destructive screening of putative transformants.

Selectable markers such as the neomycin phosphotransferase II and hygromycin phosphotransferase (hpt) genes are frequently used along with reporter genes in genetic transformation [20–22]. Genetically modified cells that express the enzymes coded by the selectable marker genes can grow on medium containing kanamycin or hygromycin antibiotics while inhibiting the growth of non-transformed cells.

This chapter presents the latest updates in methods to genetically modify grapevine. Grapevine embryogenic cultures are co-cultivated with *Agrobacterium* containing a binary plasmid with the genes of interest. Transformed somatic embryo lines obtained through the process of secondary embryogenesis are identified based on anthocyanin pigmentation and antibiotic resistance. Modified embryos are then germinated to obtain plants, which are hardened in a growth room and transferred to a greenhouse. Presence of transgenes in regenerated plants is estimated using polymerase chain reaction (PCR).

2 Materials

2.1 Supplies and Equipment

1. Laminar airflow hood.
2. 100 × 15 mm plastic Petri dishes.
3. 50-mL Centrifuge tube.
4. 125-mL conical flasks.
5. Scalpels.
6. Forceps.
7. Sterile filter paper.
8. Orbital shaker.
9. Centrifuge.
10. Clorox®.
11. Bead sterilizer.
12. Micropipettors and micropipette tips.
13. 70% Ethanol.
14. Tween 20.
15. 96-Well PTC 100 Programmable Thermal Controller (Techne Prime, Bibby Scientific).
16. 1.5 mL Microcentrifuge tubes.
17. Growth/culture room.

2.2 Plant Tissue Culture Media

1. NB2 medium: [23] salts and vitamins, g/L sucrose, 0.1 g/L myoinositol 1.0 μM BAP, 5.0 μM 2,4-D, 7.0 g/L TC agar, pH 6.0 (*see* **Note 1**).
2. MSI medium: [24] salts and vitamins, g/L sucrose, 0.1 g/L myoinositol, 4.5 μM BAP, 5.0 μM 2,4-D, 7 g/L TC agar, pH 6.0 (*see* **Note 2**).
3. PIV medium: [23] salts, B5 vitamins, 60 g/L sucrose, 8.9 μM BAP, 4.5 μM 2,4-D, 3.0 g/L Phytagel, pH 5.7 (*see* **Note 3**).
4. X6 Medium: MS medium lacking glycine and supplemented with 3.033 g/L KNO_3 and 0.364 g/L NH_4Cl (as the sole nitrogen source), 60.0 g/L sucrose, 1.0 g/L *myo*-inositol, 7.0 g/L TC agar, 0.5 g/L activated charcoal, pH 5.8 (*see* **Note 4**).
5. DM medium: DKW basal salts [25], 0.3 g/L KNO_3, 1.0 g/L *myo*-inositol, 2.0 mg/L each of thiamine-HCl and glycine, 1.0 mg/L nicotinic acid, 30 g/L sucrose, 5.0 μM BA, 2.5 μM each NOA and 2,4-D, pH 5.7.
6. DMcck medium: DKW basal salts [25], 0.3 g/L KNO_3, 1.0 g/L *myo*-inositol, 2.0 mg/L each of thiamine-HCl and glycine, 1.0 mg/L nicotinic acid, 30 g/L sucrose, 5.0 μM BA, 2.5 μM

each NOA, 2,4-D, 7.0 g/L TC agar, 200 mg L^{-1} each of carbenicillin and cefotaxime, and 100 mg/L kanamycin.

7. MS1B medium, pH 5.8: MS salts and vitamins, 20.0 g/L sucrose, 0.1 g/L *myo*-inositol, 1.0 μM BAP, 7.0 g/L TC agar, pH 5.7.
8. Yeast extract peptone (YEP) medium: 10 g/L Peptone, 10 g/L yeast extract, 5 g/L NaCl, supplemented with 15 g/L Bactoagar, pH 7.0.
9. Liquid mg/L medium : 5.0 g/L Mannitol, 1.0 g/L L-glutamate, – 5.0 g/L tryptone, 2.5 g/L yeast extract, 5.0 g/L NaCl, 150.0 mg/L KH_2PO_4, 100.0 mg/L $MgSO_4{\cdot}7H_2O$, 2.5 mL L^{-1} Fe-EDTA (to make a stock solution of Fe-EDTA, dissolve 7.44 g of $Na_2EDTA{\cdot}2H_2O$ and 1.86 g $FeSO_4.7H_2O$ in sterile distilled water and make final volume to 1 L), pH 7.0 (*see* **Note 5**).
10. Liquid ×2 medium: ×6 Medium modified to contain 20.0 g/L sucrose without TC agar and activated charcoal, pH 5.8.

2.3 In Vitro Plant Materials

1. Micropropagation/in vitro shoot cultures (*see* **Note 6**).
2. Dormant vine cuttings (*see* **Note 7**).

2.4 Antibiotic Stock Solutions

1. Kanamycin sulfate: 100 mg of antibiotic dissolved in 1 mL of water and filter-sterilized using a 0.2 μm membrane.
2. Carbenicillin and cefotaxime: 200 mg of each antibiotic dissolved in 1 mL of water and filter-sterilized using a 0.2 μm membrane.
3. Rifampicin: 20 mg of antibiotics dissolved in 1.0 mL water and filter-sterilized using a 0.2 μm membrane (*see* **Note 8**).

2.5 Agrobacterium Culture

1. Binary vector containing the VvMybA1 and npt II genes under the control of a double-cassava vein mosaic virus (CsVMV) or double-cauliflower mosaic virus (CaMV 35S) promoter.
2. *Agrobacterium* culture stock containing the binary vector (stored in glycerol at −70 °C).

2.6 PCR Reagents

1. *Taq* DNA Polymerase.
2. 10× PCR buffer with Mg++.
3. dNTP mixture.
4. Primers:

   myba1- and npt II gene-specific oligonucleotide primer pairs:

   (a) GAA-51 (5′- GGG GAC AAG TTT GTA CAA AAA AGC AGG CTC CGC CAC CAT GGA GAG CTT AGG AGT TAG AAA-3′).

(b) GAA-32 (5′- GGG GAC CAC TTT GTA CAA GAA AGC TGG GTA CTA GTT CAG ATC AAG TGA TTT ACT TGT GTG-3′).

(c) NP-53 (5′- CGGCCGCTTGGGTGGAGAGG CTATT-3′).

(d) NP-34 (5′- TCAGAAGAACTC GTCAAGAAGGCGA -3′).

3 Methods

3.1 Production of Embryogenic Cultures from Leaf Explants

1. Excise 1.5–5.0 mm long unopened leaves from micropropagation cultures and place leaves on a Petri dish containing NB2 medium (*see* **Note 9**).
2. Cover Petri dishes in aluminum foil and incubate in darkness at 26 °C for 4 weeks.
3. After 5 weeks, transfer Petri dishes containing the callus cultures to cool white fluorescent light (65 μm m^{-2} s^{-1} and 16-h photoperiod) at 26 °C for 5 weeks.
4. Screen callus cultures at weekly intervals to identify compact, cream-colored embryogenic callus that is interspersed with loose, brown-colored non-embryogenic callus (*see* **Note 10**).
5. Transfer embryogenic callus to growth regulator-free X6 medium for the development and proliferation of proembryonic masses (PEM) and somatic embryos (SE).
6. Maintain embryogenic cultures by transferring gray-colored PEM to fresh X6 medium at 4–6-week intervals.
7. Use cotyledonary-stage SE for *Agrobacterium*-mediated transformation.

3.2 Production of Embryogenic Cultures from Floral Explants

1. Obtain inflorescences from either field-grown grapevines or dormant vine cuttings (*see* **Note 11**).
2. Screen floral explants for the various developmental stages and select the optimum stage that produces an embryogenic response (*see* **Note 12**).
3. Immerse inflorescences in 70% ethanol for 30 s followed by washing them in distilled water for 30 s.
4. Wash inflorescences in 25% Clorox® solution containing one drop of Triton X-100 with constant agitation for 5 min, followed by three, 5-min washes in sterile distilled water.
5. Carefully separate stamens from the calyptra and the pistil before placing them on embryogenic culture medium. Also place the pistil with the remaining filament stubs on the medium.
6. Seal Petri dishes with Parafilm® and place in darkness at 26 °C for 4 weeks.

7. After 4 weeks, transfer Petri dishes under cool white fluorescent lights (65 μmoles m^{-2} s^{-1} and 16-h photoperiod). Screen developing cultures using a microscope for the presence of embryogenic callus at weekly intervals for 16 weeks.
8. Once production of embryogenic callus is observed transfer it to ×6 medium for development and proliferation of SE. Maintain embryogenic cultures by transferring gray-colored PEM to fresh ×6 medium at 4–6-week intervals (*see* **Note 13**).

3.3 Initiation of Agrobacterium Culture

1. Transfer *Agrobacterium* culture containing the binary plasmid from −80 °C to room temperature and thaw.
2. Spread approximately 20 μL of bacterial culture on a Petri dish containing solid YEP medium with 20 mg/L rifampicin and 100 mg/L kanamycin. Incubate dishes in the dark at 26 °C for 3 days.
3. Isolate a single colony growing on YEP medium and transfer it to a 125 mL conical flask containing 30 mL mg/L medium with 20 mg/L rifampicin and 100 mg/L kanamycin.
4. Incubate on a rotary shaker at 180 rpm at 26 °C for 16–20 h. The bacterial culture should appear cloudy at the end of the culture period.
5. Transfer the culture to a 50 mL centrifuge tube and spin at 4200 × *g* for 8 min at room temperature. Discard the supernatant and resuspend the pellet in 30 mL liquid X2 medium. Adjust OD value to 0.6 using liquid X2 medium.
6. Transfer the contents of the tube to a 125 mL conical flask and incubate for an additional 4 h on a rotary shaker under the same conditions as above. Use this culture for co-cultivation.

3.4 Transformation of Somatic Embryos

1. Transfer cotyledonary-stage SE to sterile Petri dishes. Avoid wounding SE during transfer to prevent browning (*see* **Note 14**).
2. Add 5.0 mL *Agrobacterium* culture to the SE and mix thoroughly by swirling. Incubate for 7–10 min and then remove bacterial culture.
3. Transfer SE to a Petri dish containing two layers of filter paper that are thoroughly soaked in liquid DM medium (*see* **Note 15**).
4. Seal the Petri dish with Parafilm® and co-cultivate in darkness at 26 °C for 3–4 days.
5. After 5 days observe SE for transient anthocyanin expression using a stereomicroscope.

6. Transfer co-cultivated SE to a 125 mL conical flask containing liquid DM medium with 200 mg/L each of carbenicillin and cefotaxime.
7. Transfer the flask to a rotary shaker at 110 rpm and wash SE for 3–4 days to eliminate bacterial cells.
8. Transfer washed SE to each 100 × 15 mm Petri dish containing solid DM medium with 200 mg/L each of carbenicillin and cefotaxime and 100 mg/L kanamycin.
9. Place Petri dishes in the dark at 26 °C for 4 weeks to permit callus development and proliferation.
10. After 4 weeks, transfer callus cultures to 100 × 15 mm Petri dishes containing X6 medium with 200 mg/L each of carbenicillin and cefotaxime and 70 mg/L kanamycin for secondary embryo development. Place Petri dishes in the dark and screen at weekly intervals for the presence of transgenic SE lines.
11. Independent SE lines can be identified by anthocyanin pigmentation and can be used to separate transgenic lines from non-transformed SE.
12. Designate each transgenic SE line as an independent event and transfer to fresh X6 medium with 200 mg/L each of carbenicillin and cefotaxime and 70 mg/L kanamycin.

3.5 Germination of SE Lines and Plant Regeneration

1. Transfer 5 SE at the late cotyledonary stage from each independent transgenic line to 100 × 15 mm Petri dishes containing 25 mL MS1B medium. Place Petri dishes under cool white fluorescent lights at 65 μmoles m^{-2} s^{-1} and 16-h photoperiod.
2. After 2 weeks of culture, when SE becomes enlarged and pigmented, excise the enlarged cotyledons to enhance subsequent shoot development (*see* **Note 16**). Transfer resulting plants having a robust root and shoot system plants to 7 cm plastic pots containing Pro-Mix BX potting mix (Premier Horticulture Inc., Red Hill, PA) and acclimate in a growth room for 2 weeks before transfer to a greenhouse (*see* **Note 17**).

3.6 PCR Analysis of Transgenic Plants

1. Isolate genomic DNA from leaves using CTAB protocol [26].
2. Make a 10 μL reaction mixture containing 1.0 μL each of the forward and reverse primers, 1.0 μL genomic DNA (concentration 20 ng μL^{-1}), 1.0 μL 10× buffer, 0.6 μL MgCl2, 1.6 μL dNTP (concentration 1.25 mM), 0.1 μL Taq polymerase, and 3.7 μL water.
3. Conduct PCR reactions with 1 cycle at 95 °C for 4 min; 39 cycles at 94 °C for 1 min, 58 °C for 1 min, and 72 °C for 1 min; and a final cycle at 72 °C for 4 min.
4. Use plasmid DNA as a positive control and DNA from non-transformed plants as a negative control.

5. Load PCR products on a 0.6% agarose gel containing 1.0 μL of ethidium bromide (1.0 mg mL^{-1}). Run the PCR products at 100 volt current for 30 min in a gel electrophoresis unit and observe separated DNA bands with a UV transilluminator.

4 Notes

1. Media are autoclaved at 121 °C and 15 psi pressure for 20 min. pH was adjusted using KOH prior to autoclaving. Media are cooled at 55 °C after autoclaving and then poured in Petri dishes.
2. Embryogenic response from leaves and floral explants is genotype dependent. In general, a greater number of varieties produce embryogenic cultures from floral explants than leaves. This factor needs to be considered prior to culture initiation from a specific explant.
3. Production of embryogenic response on MSI and PIV medium is genotype dependent and needs to be standardized for the genotype being studied.
4. An important factor affecting maintenance of embryogenic cultures is the use of TC agar as a gelling agent. TC agar produces higher quality embryogenic cultures compared to other agar sources such as Bactoagar or Phytagel.
5. Although several bacterial media have been used for culture of *Agrobacterium*, MG/L tends to yield better cell growth and a consistent culture quality. Therefore, we recommend this medium for routine use in transformation of grapevine.
6. In vitro micropropagation cultures can be established from shoot tips of field or greenhouse-grown grapevines following the protocol established by Gray and Benton [27]. Once established, explants can be obtained from these cultures.
7. Dormant vine cuttings can be obtained after pruning of grapevines during the winter season. Alternatively, cuttings can be obtained from the Foundation Plant Services, University of California, Davis, CA, or the USDA clonal germplasm repository in Geneva, NY.
8. Antibiotic stock solutions are filter-sterilized and stored at −20 °C. Solutions are thawed just prior to use. Antibiotics are added to culture medium after autoclaving and cooling the medium to 55 °C. Rifampicin is light sensitive and should be stored and used away from light.
9. We use *V. vinifera* "Thompson Seedless" in the majority of our experiments because of its ability to produce embryogenic cultures from leaves and higher transformation efficiency.

10. A positive embryogenic response can be obtained only from unopened leaves, 1.5–5.0 mm in size. Larger size leaves will produce a non-embryogenic callus and should not be used to initiate cultures.
11. Dormant vine cuttings are forced to flower by placing them in 500 mL Erlenmeyer flasks containing 250 mL distilled water and placing flasks under light (65 μmoles m^{-2} s^{-1} and 16-h photoperiod), at 26 °C for 3 weeks.
12. The developmental stage of stamens and pistils suitable for obtaining an embryogenic response is variety dependent. In general stage II and stage III stamens and pistils [15] produce the best embryogenic response among a large number of varieties.
13. It is critical to transfer embryogenic cultures to fresh ×6 medium at 4–6-week intervals. Failure to do so will result in a decrease in regeneration potential of SE and subsequent reduction in transformation efficiency and plant regeneration.
14. Grapevine SE proliferates by direct secondary embryogenesis with new embryos emerging from epidermal or subepidermal cells of primary SE. Thus, surface cells of cotyledonary-stage SE are the best target tissues for *Agrobacterium*-mediated transformation.
15. Co-cultivation of SE on filter paper thoroughly soaked in DM medium dramatically increases the transformation efficiency while reducing browning and subsequent necrosis of embryogenic cultures.
16. Cotyledon excision to improve shoot recovery from germinated SE is variety dependent and needs to be tested for each variety being studied.
17. Transformation efficiency varies among *Vitis* species and varieties. The highest recovery of transgenic plant lines has been recorded in *V. vinifera* "Thompson Seedless."

Acknowledgments

S. A. Dhekney holds the E.A Whitney Endowed Professorship in the Department of Plant Sciences. M. Brungart-Rosenberg and C. Claflin were supported in part by a grant from the National Institute of General Medical Sciences (2P20GM103432) from the National Institute of Health. The research was supported in part by the Wyoming Agricultural Experiment Station and grants from the USDA/USAID US-Egypt Science and Technology Development Funds program (Grant No. 0210-22310-005-24G).

References

1. FAOSTAT (2014) Faostat.fao.org
2. Grapes, wine fuel US economy (2007) Food and Drink Weekly 1
3. Olmo HP (1942) Choice of parents as influencing seed germination in fruits. J Am Soc Hortic Sci 41:171–175
4. Einset J, Pratt C (1975) Grapes. In: Moore JN, Janick J (eds) Advances in fruit breeding. Purdue University Press, Lafayette, IN
5. Swaler J, Reisch B, Aradhya MK, Prins B, Zhong GY, Schwaninger H, Simon C, Buckler E, Myles S (2013) Genomics assisted ancestry deconvolution in grape. PLoS One 8 (11):e80791. https://doi.org/10.1371/journal.pone.0080791
6. Gray DJ, Li ZT, Grant NL, Dean DA, Trigano RN, Dhekney SA (2016) The application of precision breeding (PB) for crop improvement is consistent with the plant life cycle: the utility of PB for grapevine. Acta Hortic 1115:49–56
7. Gray DJ, Li ZT, Grant TNL, Dean DA, Dhekney SA (2016) The new biology of precision breeding. P-20. In Vitro Cell Dev Biol Anim. https://doi.org/10.1007/s11626-016-0034-0
8. Gray DJ, Li ZT, Dhekney SA (2014) Precision breeding of grapevine for improved traits. Plant Sci 228:3–10
9. Dhekney SA, Li ZT, Dutt M, Gray DJ (2008) *Agrobacterium*-mediated transformation of embryogenic cultures and regeneration of transgenic plants in *Vitis routundifolia* Michx. (muscadine grape). Plant Cell Rep 27:865–872
10. Li Z, Jayasankar S, Gray DJ (2001) Expression of a bifunctional green fluorescent protein (GFP) fusion marker under the control of three constitutive promoters and enhanced derivatives in transgenic grape (*Vitis vinifera*). Plant Sci 160:877–887
11. Li Z, Jayasankar S, Gray DJ (2004) Bi-directional duplex promoters with duplicated enhancers significantly increase transgene expression in grape and tobacco. Transgenic Res 13:143–154
12. Li ZT, Dhekney SA, Dutt M, Van Aman M, Tattersall J, Kelley KT, Gray DJ (2006) Optimizing *Agrobacterium*-mediated transformation of grapevine. In Vitro Cell Dev Biol Plant 42:220–227
13. Li ZT, Dhekney SA, Dutt M, Gray DJ (2008) An improved protocol for *Agrobacterium*-mediated transformation of grapevine. Plant Cell Tissue Organ Cult 93:311–321
14. Dutt M, Li ZT, Dhekney SA, Gray DJ (2008) A co-transformation system to produce transgenic grapevines free of marker genes. Plant Sci 175:423–430
15. Dhekney SA, Li ZT, Compton ME, Gray DJ (2009) Optimizing initiation and maintenance of *Vitis* embryogenic cultures. Hortscience 44:1400–1406
16. Dhekney SA, Li ZT, Gray DJ (2010) Factors influencing induction and maintenance of *Vitis rotundifolia* Michx. embryogenic cultures. Plant Cell Tissue Organ Cult 105:175–180
17. Li ZT, Kim KH, Jasinski JR, Creech MR, Gray DJ (2012) Large-scale characterization of promoters from grapevine (*Vitis* spp.) using quantitative anthocyanin and GUS assay. Plant Sci 196:132–142
18. Li ZT, Dhekney SA, Gray DJ (2011) Use of the VvMybA1 gene for non-destructive quantification of promoter activity via color histogram analysis in grapevine (*Vitis vinifera*) and tobacco. Transgenic Res 20:1087–1097
19. Kandel R, Dutt M, Gray DJ, Li ZT, Bergey DR, Sitther V, Dhekney SA (2016) Evaluation of a grapevine-derived reporter gene system for precision breeding of *Vitis*. Plant Cell Tissue Organ Cult 124:599–609
20. Dhekney SA, Li ZT, Zimmerman TW, Gray DJ (2009b) Factors influencing genetic transformation and plant regeneration of *Vitis*. Am J Enol Vitic 60:285–292
21. Benveniste R, Davies J (1973) Mechanisms of antibiotic resistance in bacteria. Annu Rev Biochem 42:471–506
22. van den Elzen PJM, Townsend J, Lee KY, Bedbrook JR (1984) A chimeric hygromycin resistant gene as a selectable marker in plant cells. Plant Mol Biol 5:299–302
23. Nitsch JP, Nitsch C (1969) Haploid plants from pollen grains. Science 163:85–87
24. Murashige T, Skoog F (1962) A revised medium for rapid growth and bioassay with tobacco tissue cultures. Physiol Plant 15:473–497
25. Driver JA, Kuniyuki AH (1984) In vitro propagation of Paradox walnut rootstock. HortScience 19:507–509
26. Lodhi MA, Ye GN, Weeden NF, Reisch BJ (1994) A simple and efficient method for DNA extraction from grapevine cultivars and *Vitis* species. Plant Mol Biol Rep 12:6–13
27. Gray DJ, Benton CM (1991) In vitro micropropagation and plant establishment of muscadine grape cultivars (*Vitis rotundifolia*). Plant Cell Tissue Organ Cult 27:7–14

Chapter 15

Agrobacterium-Mediated Transformation of *Solanum tuberosum* L., Potato

Myron A. Bruce and Jessica L. Shoup Rupp

Abstract

Potato is considered the fourth most important food crop in the world, and the most important non-cereal crop. Potato is transformed using *Agrobacterium tumefaciens* with relative ease. Several improvements have been made in the last 20 years with respect to tissue culture, transformation, and regeneration of potato. This chapter describes a reliable and efficient potato transformation system using internodal explants. Plasmid preparation, *Agrobacterium* transformation, potato transformation, regeneration, and recovery are described in detail, as well as molecular characterization of resulting putative transgenic plants.

Key words Potato, *Solanum tuberosum* L., Internodal explant, Tuber, Antibiotic selection

1 Introduction

Potato is the most important vegetable root crop in the world with the tuber being the edible portion of the plant. Approximately 400 million tons of potatoes are produced annually (Halterman et al. [1]). Potatoes are produced on a wide scale and are historically important due to the Irish Potato Famine beginning in 1845 [2]. Potato is also a relatively easy plant to transform. It has played a significant role in the recent application of biotechnology with several different transgenic lines approved for human consumption. While the traits introduced (anti-browning, disease tolerance) function as intended, their economic success remains unclear at this time. Potato was one of the first crops to be successfully transformed, first reported in 1986 [3]. Early studies used *Agrobacterium rhizogenes* [3], shifting quickly to *Agrobacterium tumefaciens* for enhanced transformation ease and success [4, 5]. Varied success has been attained based on cultivar and explant used, with the greatest transformation efficiency reported using internodal explants. This level of success has led to the development of transgenics targeting a variety of characteristics

Sandeep Kumar et al. (eds.), *Transgenic Plants: Methods and Protocols*, Methods in Molecular Biology, vol. 1864,
https://doi.org/10.1007/978-1-4939-8778-8_15,

including tuber quality, modified starch, anti-browning, and pest and disease resistance [1].

The cultivars 'Desiree' and 'Bintjee' were early favorites due to their amenability for transformation [4], but most potato varieties respond well in tissue culture. At present, most commercially produced cultivars have been successfully transformed with only minor modifications to standard protocols, similar to the one described here. These fall under the category of a two-step regeneration process using shoot tissue as the explant [6]. Potato has a high likelihood of somaclonal variation [7], which is the reason for the abbreviated callus induction step found in this protocol.

There are many types of plant tissue culture media, most based on Murashige and Skoog's MS Media [8]. They fall under three basic types: callus induction media, shoot/root induction, and propagation media. Each of these has their basis in MS media. MS media contains the major salts ammonium nitrate, calcium chloride, magnesium sulfate, monopotassium nitrate, and potassium nitrate. There are a variety of minor salts as well including boric acid, cobalt chloride, ferrous sulfate, manganese (II) sulfate, potassium iodide, sodium molybdate, zinc sulfate, and ethylenediaminetetraacetic acid ferric sodium. Finally, this base media is often supplemented with sucrose. Vitamins, such as the B5 vitamins [9], and plant growth regulators are also required.

Plant growth regulators (PGR) are an important component of potato transformation. PGRs, also known as hormones, include cytokinins, auxins, gibberellins, abscisic acid, and ethylene. Cytokinins, such as 6-benzylaminopurine (BAP), kinetin, and zeatin used in potato tissue culture, stimulate cell division and induce shoot bud formation. The auxins, such as 3-indole acetic acid (IAA) and 1-naphthaleneacetic acid (NAA), serve to stimulate cell elongation and influence root initiation, vascular development, and apical dominance. In potato tissue culture, auxins are used to induce callus tissue formation from explants and initiate shoot and root production. The PGRs described here can be found in both the stock plant media (PHP NHK) and the 3C5ZR media. These PGRs can be added to media as described following as either a dry powder or a liquid stock solution. Liquid stock solutions provide a time-saving element, but consideration should be given to shelf life of liquid stock solutions, which is often limited. Prepare only amounts required to complete your transformation experiment.

In cloning, transformation, and tissue culture, selectable markers are often used. The selectable markers commonly used in cloning vectors (plasmids) are often antibiotics. For example, under selective conditions, only cells that contain plasmids with the appropriate selectable marker will survive on media containing that specific antibiotic. That means when an antibiotic resistance gene is included in the construct (plasmid with integrated target), only the successfully transformed cells will grow. In every case, there are those cells which are not transgenic that are able to grow. Those cells are termed "escapes," meaning those cells have escaped the selection process. Herbicides, such as glufosinate, are also often used as a selectable marker. Once again, it is important to choose your vector first, giving consideration to your transgene and match your selectable marker usage.

Several *Escherichia coli* strains are available for use, not limited to DH5a, which are readily available at most institutions, but also JM109, DB3.1, and Top10. Whether you obtain a stock from your institution or order one, there are a few considerations to keep in mind. The foremost, does the plasmid vector you have chosen have a recommendation? While many vectors are capable of working with many strains, elements within those vectors may require a specific strain to work to their full advantage. It is wise to begin by choosing a vector that is ideal for your experiment, and then match your *E. coli* strain. For example, this includes selecting a strain that does not carry antibiotic resistance to streptomycin, if your plasmid also carries resistance to streptomycin. A decision such as this would make identifying transformed colonies impossible since all bacteria would grow, not just those transformed.

Binary vectors used in potato should have a variety of features that are suitable for this dicot system. Avoid monocot promoters driving either plant selectable markers or transgene. Instead, dicot promoters are preferred. Commonly, the CaMV35s promoter has been successfully used to express transgenes in potato [10]. The CaMV35S promoter is a constitutive promoter, and due to its ability to drive expression in potato it has remained very prominent, continuing to be used for newer techniques including CRISPR/Cas9. The patatin promoters, i.e. B33, have been successful at inducing tuber expression. The arrangement of elements in the plasmid should also be considered. *A. tumefaciens* transfers T-DNA from right border to left border. To aid identification of events that contain the entire cassette, selectable/screenable markers near the left border are desirable for expression. In some instances, a second selectable marker can make the process easier. In our example shown below kanamycin and hygromycin serve as dual-selectable markers.

Vector confirmation is necessary at every step. Confirmation must take place at first construction, *E. coli* transformation, *Agrobacterium* transformation, potato transformation, and plantlet regeneration. Most commonly, this is done through DNA sequencing and PCR analysis. Transgenic plants should undergo PCR and sequencing to confirm transgene presence, RT-PCR for transgene expression, and Southern blot analysis for transgene integration, as well as other experimental techniques as appropriate. Each experiment within transgenic plantlet confirmation requires a specific set of controls to be used throughout. For example, in PCR, a water-only control, a non-transgenic control (wild-type), a transgenic, non-target control, or a vector-only (no target addition) should be used.

2 Materials

2.1 Plant Material

Potato cv. Desiree, Ranger Russet, Umatilla Russet, Alturas, and Yukon Gold all perform well as in vitro cultures (obtained from the Montana State University Seed Potato Certification Laboratory, montanaspud.org, or other certified sources). With minor modifications during transformation and regeneration, most potato cultivars are amendable to transformation. These plantlets are grown in sterile culture from virus-free clonal cuttings. It is critical to ensure that plantlets are virus-free prior to transformation. Plantlets should be kept in Magenta GA-7 vessels (Sigma V8505), PTL-100 vessels (PhytoTechnology Laboratories, C2100) or glass jars with lids (PhytoTechnology Laboratories, C1770, ideally in a growth chamber at 18 °C and 16-h/8-h light cycle (80–110 $\mu E/m^2/s$)). The easiest method of establishing virus-free in vitro cultures is through the sterile propagation of nodal stem cuttings (Subheading 3.1.3). Cultures can also be started through the use of greenhouse-grown plants or sprouting tubers, but in this instance it is appropriate to allow time for thorough testing to ensure a pathogen-free plant stock. Potatoes may be grown to maturity in a growth chamber or greenhouse at 15–20 °C (150 $\mu E/m^2/s$ light intensity and 16-h light photo period). Harvest tubers and store in a dry, dark environment at 5 °C a minimum of 8 weeks until ready to use. Tubers moved to room temperature will begin sprouting following cold storage. Cultures may then be started by using sprouting tubers (described in Subheading 3.1.1) or growing those tubers in growth chambers or greenhouses at 15–20 °C (150 $\mu E/m^2/s$ light intensity and 16-h light photo period for stem cuttings (described in Subheading 3.1.2)).

2.2 Stock Solutions

1. 10 mM Kinetin: Dissolve 21.5 mg of kinetin in 1 mL of 1 N NaOH. Add molecular biology-grade water to a final volume of 10 mL. Filter sterilize through a 0.22 μm filter. Aliquot and store at −20 °C.
2. 1 mg/mL Naphthaleneacetic acid (NAA): Dissolve 10 mg of NAA in 200 μL of 1 N NaOH. If powder is reluctant to dissolve, heat in a heating block at 50 °C. Add molecular biology-grade water to a total volume of 10 mL. Filter sterilize through a 0.22 μm filter. Aliquot and store at 4 °C.
3. 1 mg/mL 6-Benzylaminopurine (BAP): Dissolve 10 mg of BAP in 400–1000 μL of 1 N NaOH. Slowly add molecular

biology-grade water while stirring to a total volume of 10 mL. Filter sterilize through a 0.22 μm filter. Aliquot and store at 4 °C.

4. 0.5 mg/mL Indole-3-acetic acid (IAA): Dissolve 5 mg of IAA in 200–500 μL of 1 N NaOH. Mix to ensure complete dissolution of the powder. Add molecular biology-grade water to a total volume of 10 mL. Filter sterilize through a 0.22 μm filter. Aliquot and store at −20 °C.
5. 1 mg/mL trans-Zeatin riboside: Dissolve 10 mg of trans-zeatin riboside in 200–500 μL of 1 N NaOH, Slowly add molecular biology-grade water while stirring to a total volume of 10 mL. Filter sterilize through a 0.22 μm filter. Aliquot and store at −20 °C.
6. 50 mg/mL Kanamycin (Kan): Dissolve 0.5 g of kanamycin sulfate in 10 mL of molecular biology-grade water. Filter sterilize through a 0.22 μm filter (*see* **Note 1**). Aliquot and store at −20 °C.
7. 10 mg/mL Rifampicin (Rif): Dissolve 0.1 g of rifampicin in 10 mL of 100% methanol. Aliquot and store at −20 °C.
8. 100 mg/mL Ampicillin (Amp): Dissolve 1.0 g of ampicillin in 10 mL of molecular biology-grade water. Filter sterilize through a 0.22 μm filter (*see* **Note 1**). Aliquot and store at −20 °C.
9. 50 mg/mL Hygromycin B (Hyg): Dissolve 0.5 g of hygromycin B in 10 mL of molecular biology-grade water. Filter sterilize through a 0.22 μmn filter (*see* **Note 1**). Aliquot and store at −20 °C.
10. 100 mg/mL Timentin™ (Ticarcillin Disodium and Potassium Clavulanate): Dissolve 1.0 g of Timentin in 10 mL of molecular biology-grade water. Filter sterilize through a 0.2 μm filter (*see* **Note 1**). Aliquot and store at −20 °C.
11. MSVI vitamins: Add 200 mg of glycine, 50 mg of nicotinic acid, 50 mg of pyridoxine HCl, and 40 mg of thiamine HCl in 100 mL of ddH_2O. Mix thoroughly until fully dissolved. 1 L of media requires 1 mL of MSVI vitamins. Aliquot 1 mL of MSVI vitamins into 1.5 mL microfuge tubes and store at −20 °C.
12. JHMS vitamins: Add 25 mg of folic acid and 5 mg of biotin in 100 mL of ddH_2O. Mix thoroughly until fully dissolved. 1 L of media requires 1 mL of JHMS vitamins. Aliquot 1 mL of JHMS vitamins into 1.5 mL microfuge tubes and store at −20 °C.
13. 3R vitamins: Add 50 mg of thiamine HCl, 25 mg of nicotinic acid, and 25 mg of pyridoxine HCl, in 100 mL of ddH_2O. Mix thoroughly until fully dissolved. 1 L of media requires 1 mL of 3R vitamins. Aliquot 1 mL of 3R vitamins into 1.5 mL microfuge tubes and store at −20 °C.

14. 74 mM Acetosyringone (3′,5′-Dimethoxy-4′-hydroxyacetophenone): Add 145.18 mg of acetosyringone to 10 mL of dimethyl sulfoxide (DMSO). Aliquot into 0.2 mL tubes and store at −20 °C.

2.3 Media

All media is made in 1 L volume and distributed either directly into culture vessels or into two 1 L glass bottles, each containing 500 mL, and autoclaved for 20 min at 120 °C (*see* **Notes 2** and **3**).

1. Bacterial propagation media: Lysogeny broth or Luria-Bertani broth (LB)—1 L. Add 10 g of tryptone, 5 g of yeast extract, and 5 g of NaCl to approximately 800 mL of ddH_2O. For solid media, also add a solidifying agent, such as agar, at an appropriate rate (8–15 g/L, *see* **Note 4**). Mix well to ensure dissolution of all components. Increase volume to 1 L. Autoclave. Cool below 50 °C before adding selective antibiotics (*see* **Note 5**). Store at 4 °C.
2. Yeast mannitol broth (YM): 1 L. Add 0.4 g of yeast extract, 10 g of mannitol, 0.1 g of NaCl, 0.2 g of $MgSO_4{\cdot}7H_2O$, and 0.5 g of KH_2PO_4 to approximately 800 mL of ddH_2O. For solid media, also add solidifying agent at an appropriate rate (8–15 g/L, *see* **Note 4**). Mix well to ensure dissolution of all components. Increase volume to 1 L. Autoclave. Cool below 50 °C before adding selective antibiotics (*see* **Note 5**). Store at 4 °C.
3. Stock plant material media (PPM NHK): Add 800 mL of ddH_2O to a 1 L glass bottle containing a magnetic stir bar. Place on a stir plate. Add 20 g of sucrose and dissolve. Add 100 μL of kinetin (10 mM), 4.3 g of MS Salts + Gamborg's vitamins [PhytoTechnology M404, Caisson MSP006, Sigma M0404], and 0.5 mL of PPM (Plant Cell Technology). Volume up to 1 L with ddH_2O. pH to 5.8 ± 0.05 using 2% NaOH to raise the pH or 0.37% HCl to lower pH. If pouring directly into 36 magenta boxes GA-7 vessels, 6 oz baby food jars, add 5 g of plant micropropagation-grade agar (PhytoTechnology A111, Caisson A038) and microwave to melt agar. Ensure that the media boils for 30–60 s. Pour into 36 vessels, cover with lids, and autoclave (*see* **Note 6**). If using two 1 L glass bottles, you may omit microwaving, divide media, and add ½ of the agar to each bottle. Screw glass bottle lids on loosely to allow ventilation during the autoclave process. Autoclave for 20 min. Store autoclaved materials at 4 °C.
4. Callus induction media (CIM): Add 4.3 g of MS salt, 1 mL of MSVI vitamins, 1 mL of JHMS vitamins, 0.1 g of inositol, 30 g of sucrose, 1 mL of BAP stock solution (1 mg/mL), 2 mL of NAA stock solution (1 mg/mL), and 10 g of micropropagation-grade agar. Add 900 mL of distilled water.

pH to 5.6 with 2% NaOH. Volume up to 1 L with ddH_2O. Autoclave. Use petri plates.

5. Shoot Induction Media (3C5ZR): Add 4.3 g of MS salt, 1 mL of 3R vitamins, 0.1 g of inositol, 30 g of sucrose, and 10 g of agar. Add 900 mL of distilled water. Adjust pH to 5.9 with 2% NaOH. Volume up to 1 L with ddH_2O. Autoclave. Allow media to slightly cool below 50 °C. Add 3 mL of trans-zeatin riboside stock (1 mg/mL), 500 μL of Timentin stock, and 1 mL of IAA stock solutions. Add selection agent (e.g., kanamycin, hygromycin) as appropriate for the vector in use. Use petri plates.
6. Propagation media: Add 4.3 g of MS salt, 0.17 g of $NaH_2PO_4H_2O$, 0.1 g of inositol, 0.4 g of thiamine HCl, 30 g of sucrose, 8 g of micropropagation-grade agar, and 900 mL of distilled water. Adjust to pH 5.7 with 2% NaOH. Volume up to 1 L with ddH_2O. Autoclave. Allow media to cool below 50 °C. Add selection agent (e.g., kanamycin, hygromycin) as appropriate for the vector in use. Supplement with 100 μL of Timentin stock to remove any remaining *Agrobacterium*. Use 25 mL culture tubes.

2.4 Bacterial Strains, Vectors, and Supplies

1. *Escherichia coli*—DH5α or equivalent standard cloning strain.
2. *Agrobacterium tumefaciens*—GV3101 or LBA4404.
3. Entry vector—pENTR-D-TOPO (Invitrogen-Thermo Fisher Scientific).
4. RNAi destination vector—pANDA-35HK [11].
5. Gateway LR Clonase kit (Invitrogen-Thermo Fisher Scientific).
6. Gel extraction kit (e.g.: Qiagen, QIAquick Gel Extraction Kit).
7. Plasmid miniprep kit (e.g.: Qiagen Qiaprep Spin Miniprep Kit, Cat).
8. DNA Extraction kit (e.g.: Qiagen DNeasy Plant Mini Kit).
9. M13 F and M13R primers:

   M13 F: GTA AAA CGA CGG CCA GT.
   M13 R: CAG GAA ACA GCT ATG AC.
10. Gene/fragment-specific primers.
11. Tween-20.

3 Methods

3.1 Establishment of In Vitro Stock Plants (See Note 7)

3.1.1 Establish In Vitro Plants from Sprouting Tubers

1. Excise shoots from spouting tubers that are 2.5–5 cm in length. Place in a 50 mL conical tube and fill with ddH_2O. Do not overfill the tube with excised shoots. Ensure that at least 25 mL of ddH_2O can be added for thorough rinsing to take place. Place tube on an orbital shaker, and agitate for 15 min at 200 rpm. Discard ddH_2O and apply subsequent rinse of ddH_2O. Discard water.
2. Immerse shoots in 25 mL of 10% sodium hypochlorite with two drops of Tween-20. Place again on shaker and agitate for 15 min at 250 rpm.
3. Discard disinfectant, add 25 mL of 70% ethanol, and rinse for 1–2 min by gently rocking the tube.
4. Rinse shoots from sprouting tubers five times in ddH_2O to remove all traces of ethanol.
5. Trim the sterilized shoots with sterilized scalpel to remove any superficial areas damaged by sterilization (*see* **Note 8**).
6. Place trimmed pieces into 6 oz baby food jars or Magenta GA-7 vessels with lids containing stock plant material media. Seal culture vessels using Glad Wrap cut to size (*see* **Note 9**) or breathable plate-sealing film (PhytoTechnology A003).
7. Place in growth chambers at 18 °C–24 °C and 16-h/8-h light cycle (80–110 $\mu E/m^2/s$). Cultures will establish in 4–6 weeks.

3.1.2 Establish In Vitro Plants from Greenhouse-Grown Plants

1. Ensure that source plants are free from pathogens or pests. Excise stem cutting with internode and/or shoot tips from actively growing plants. Place in a 50 mL conical tube and fill with ddH_2O. Do not overfill the tube with cuttings. Ensure that at least 25 mL of ddH_2O can be added in order for thorough rinsing to take place. Place tube on an orbital shaker, and agitate for 15 min. Drain ddH_2O and apply subsequent rinse of ddH_2O. Drain tube.
2. Immerse stem cuttings and shoots in 25 mL of 10% sodium hypochlorite with two drops of Tween-20. Place again on shaker and agitate for 15 min.
3. Discard disinfectant, add 25 mL of 70% ethanol, and rinse for 1–2 min by gently rocking the tube.
4. Rinse shoots from sprouting tubers five times in ddH_2O to remove all traces of ethanol.
5. Trim the sterilized shoots with sterilized scalpel to remove any superficial areas damaged by sterilization.
6. Place trimmed pieces into 6 oz baby food jars or Magenta GA-7 vessels with lids containing stock plant material media. Seal

Fig. 1 Stock plants inside Magenta GA-7 boxes or 6 oz baby food jars in growth chamber set at 16-h/8-h light cycle (80–110 μE/m^2/s)

culture vessels using Glad Wrap cut to size or breathable plate-sealing film.

7. Place in growth chambers at 18 °C–24 °C and 16-h/8-h light cycle (80–110 μE/m^2/s). Cultures will establish in 4–6 weeks (Fig. 1).

3.1.3 Establish In Vitro Plants from In Vitro Stock Plantlets

1. Obtain sterile cultures of actively growing plants.
2. Using sterile forceps, remove plant from culture vessel and place on a sterile petri plate.
3. Use sterile scissors to remove leaf tissue.
4. Using a sterile scalpel or sterile scissors, cut nodal stem sections approximately 1 cm in length to produce multiple stem cuttings.
5. Using sterile forceps place stem cuttings into culture vessels containing stock plant material media while leaving the top part of the stem exposed. A 6 oz baby food jar with lid is best for up to four cuttings, while a Magenta GA-7 vessel can hold up to five cuttings. Seal culture vessels using Glad Wrap cut to size or breathable plate-sealing film.
6. Place in growth chambers at 18 °C–24 °C and 16-h/8-h light cycle (80–110 μE/m^2/s). Cultures will establish in 4–6 weeks (Fig. 2).

Fig. 2 Established culture in Magenta GA-7 box holding five plantlets in stock plant medium

3.2 Plasmid Vector Construction and *E. coli* Transformation

3.2.1 Choosing a Plasmid

For this example, cloning will take place using Invitrogen's Gateway system and a Gateway-compatible RNAi binary vector. However, alternative systems are available. The most important ideas to retain from this section are not the specific technologies used to generate the binary vector, rather the function of each component of the system to achieve the desired aim in potato.

3.2.2 Generating an Entry Vector Using pENTR-D-TOPO

1. Primers to amplify the region of interest should be designed to be specific (*see* **Note 10**). In the case of RNAi, a 100–400 bp fragment with no more than 20 contiguous bases matching elsewhere in the genome is sufficient for gene silencing. Directional cloning in pENTR is enhanced by the addition of "CACC" to the 5′ end of the forward primer. No modification is required for the reverse primer.
2. Amplify the fragment and analyze by gel electrophoresis to ensure that the fragment is of the expected size, and that amplification was specific.
3. Extract the band from the gel, and purify it using a gel purification kit.
4. Prepare the entry vector and PCR fragment as recommended by the manufacturer.

3.2.3 Heat-Shock Transformation of Chemically Competent *E. coli*

1. Remove cells from −80 °C and thaw on ice.
2. Add 2 μL of entry vector/PCR fragment reaction to 50–100 μL of cells on ice. Flick tube to mix.
3. Incubate on ice for 5 min.
4. Heat shock cells at 42–45 °C for 45 s–1 min.
5. Return to ice for 10 min.
6. Add 900 μL of LB (or SOC, if provided), and shake at 37 °C, 250 rpm, for 30–45 min.
7. Plate 10 and 100 μL of each transformation on LB agar plates amended with 50 mg/L Kan (*see* **Note 13**).
8. Incubate plates inverted (*see* **Note 11**) at 37 °C overnight.
9. Inoculate 3–6 well-separated colonies into 2 mL of LB amended with 50 mg/L Kan.
10. Incubate overnight at 37 °C, 250 rpm.

3.2.4 Confirming Presence of the Insert in the Plasmid

1. Prepare plasmids from overnight cultures with a plasmid miniprep kit.
2. Confirm the presence of the insert by PCR using M13F/R (*see* **Note 12**), or the primers used to generate the fragment.
3. Sequence inserts using M13F and M13R to confirm the orientation and sequence of the fragment.

3.2.5 Moving the Fragment from the Entry Vector to the Destination Vector

1. Perform the LR clonase recombination reaction as described by the manufacturer bearing in mind the number of recombination sites. In the case of pANDA-35HK [11], there are two recombination sites. Use twice the recommended molar amount of the entry vector to compensate.
2. Transform DH5α as described in "Heat-shock transformation" with the exception that LB plates and liquid media should be amended with 100 mg/L Kan and 100 mg/L Hyg (*see* **Note 5**).

3.2.6 Confirming the Presence of the Insert (s) in the Destination Vector

1. Prepare plasmids from overnight cultures with a plasmid miniprep kit.
2. Use an appropriate combination of primers specific to the vector and the insert to confirm the presence of insert(s) by PCR and gel electrophoresis (*see* **Note 13**).
3. Sequence using a combination of appropriate vector primers and insert primers to ensure the presence and proper orientation of insert(s). Examine the sequence at the recombination junctions to establish proper recombination and reading frame if appropriate.
4. Prepare confirmed plasmid for transformation into *A. tumefaciens*.

3.3 Agrobacterium Preparation and Transformation

The following protocols for *Agrobacterium* preparation and transformation were adapted in 2013 by David Neece from Höfgen and Willmitzer [12]. The protocols regarding inoculation, co-cultivation, and regeneration were adapted from Chronis et al. [13].

3.3.1 Preparation of A. tumefaciens-Competent Cells

1. Start overnight culture of bacteria in 2 mL YM from glycerol stock. Incubate at 28 °C, 250 rpm.
2. Inoculate 200 mL YM with 1 mL of overnight culture, and incubate on an incubator shaker at 28 °C, 250 rpm, to log-phase OD550 density of 0.5–0.8 (overnight).
3. Harvest cells in centrifuge at 5000 × *g* for 10 min, room temperature.
4. Discard media and wash cells in 100 mL sterile 1× TE.
5. Pellet cells in centrifuge at 5000 × *g* for 10 min, room temperature.
6. Discard TE.
7. Resuspend cells in 20 mL (0.1× original media volume), and aliquot into 250 μL fractions.
8. Snap freeze in liquid nitrogen, and store at −80 °C.

3.3.2 Transformation of A. tumefaciens GV3101 or LBA4404

1. Thaw cells on ice. Add 100–1000 ng of vector, and flick tube gently to mix.
2. Incubate on ice 5 min, and transfer to liquid N2 for 5 min.
3. Transfer cells to 15 mL tube with 1 mL YM and incubate for 2 h at 28 °C, 250 rpm.
4. Collect cells by centrifuge at 4500 × *g*, for 2 min, at room temperature. Discard liquid.
5. Resuspend cells in 100–200 μL YM. Split the transformation between two YM plates containing appropriate antibiotics (for pANDA-35HK, 50 mg/L Kan, 50 mg/L Hyg).
6. Incubate plates inverted for 2 days at 28 °C.
7. Pick well-isolated colonies to YM broth amended with appropriate antibiotics and proceed with vector confirmation as described previously, omitting the sequencing step (Fig. 3).
8. Store cultures of confirmed clones in sterile 30% glycerol at −80 °C.

3.4 Inoculation, Co-cultivation, and Regeneration

1. The day prior to transformation, inoculate 10 mL of YM broth with *Agrobacterium* stock culture with confirmed clones and culture at 28 °C on an incubator shaker at 200 rpm until the OD_{00} reaches 0.7 (overnight culturing: 12–15 h).

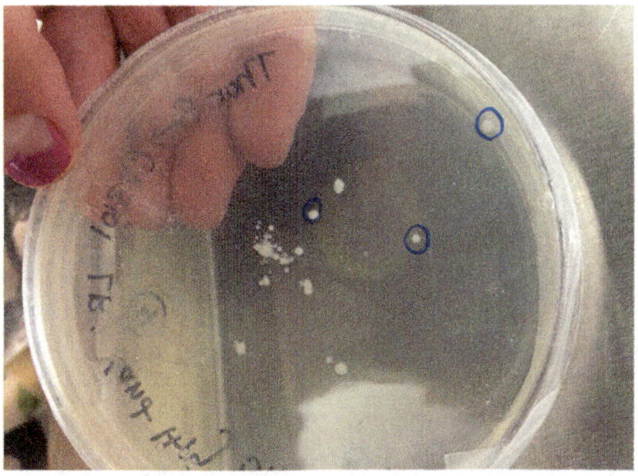

Fig. 3 Single colonies of *Agrobacterium* strain GV3101 harboring plasmid. Choose well-isolated colonies for Subheading 3.3.2, **step 7**

2. On the day of inoculation, centrifuge the *Agrobacterium* culture at 9000 × *g* for 15 min at room temperature and resuspend the pellet in 10 mL CIM liquid medium.

   Add 50 μL of 74 mM acetosyringone to the *Agrobacterium* suspension. Use within 1 h, as cells will begin to aggregate.
3. Remove a 4-week-old stock plant from its culture vessel and hold with sterile forceps. Using sterile scissors, trim off excess leaf tissue. Trim internodal segments to approximately 1 cm in length.
4. Up to 100 explants (stem cuttings) fit easily on a petri plate containing CIM media. This ensures that cuttings do not dry out (Fig. 4).
5. When all cuttings are prepared, move them to a petri dish containing up to 10 mL of *Agrobacterium* suspension from morning culture (explained in Subheading 3.2.6). Wrap plates in plate-sealing film and place on an orbital shaker. Plates should remain on the orbital shaker at 50–90 rpm for 45 min at room temperature (22–25 °C). Ensure that the *Agrobacterium* solution is washing over explants.
6. Pour off *Agrobacterium* suspension into appropriate disposal vessel, such as 10% bleach solution.
7. Using sterile forceps, gently blot cuttings on a sterile filter paper (*see* **Note 14**) placed into a petri dish (optional). Transfer approximately 25 blotted cuttings to a fresh CIM media plate, wrap plate in sealing film, and place in the dark at room temperature for 3 days for co-cultivation. A drawer in the laboratory can serve this purpose (Fig. 5).
8. At the conclusion of the third day, transfer the stem cuttings to 3C5ZR media plates containing the appropriate antibiotics for

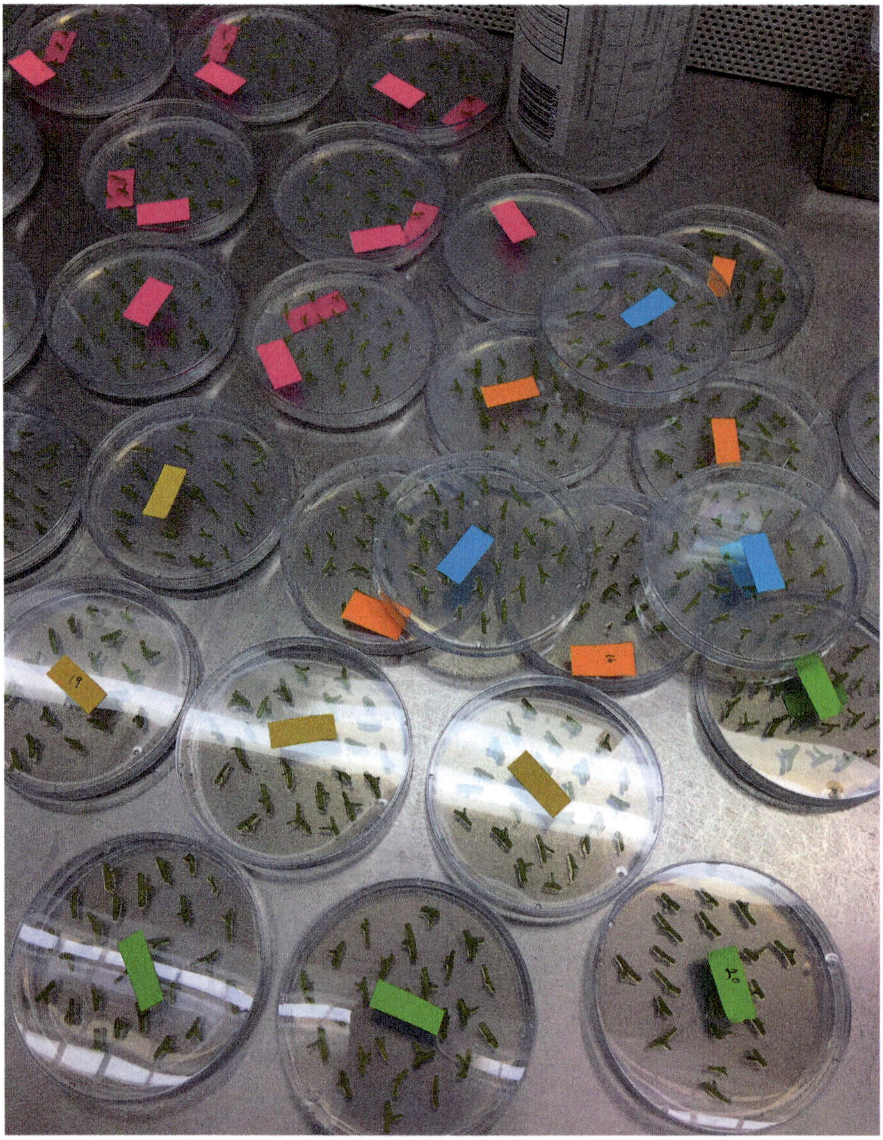

Fig. 4 Stem cuttings awaiting transformation on CIM media to prevent drying out of plant material

use with your plasmid of choice and 500 μg/mL of Timentin to eliminate *Agrobacterium* from the culture.

9. Place new putative transformants in growth chambers at 18 °C–24 °C and 16-h/8-h light cycle (80–110 μE/m^2/s).
10. Explants are subcultured onto freshly prepared plates of 3C5ZR containing selective agents every 7–10 days (Fig. 6).
11. Plantlet emergence should occur 4–6 weeks later (Fig. 7). Excise developing shoots (5–10 mm in size) and transfer into 25 mL culture tubes containing 3C5ZR media with selective

Fig. 5 Transfer of stem cuttings to fresh CIM media for co-culture in the dark

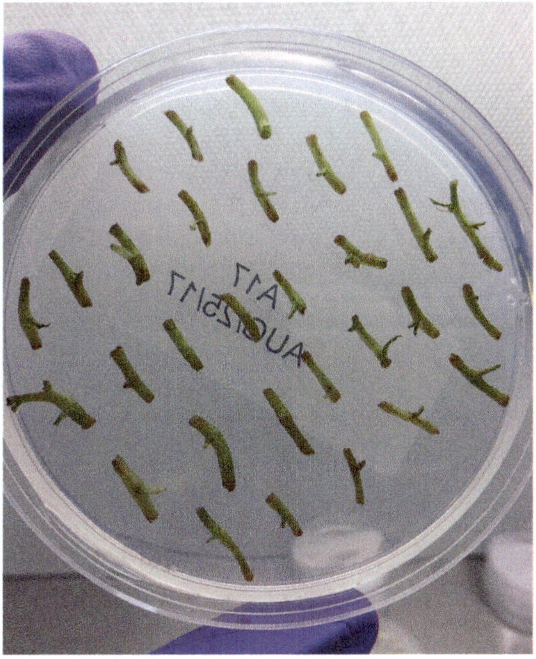

Fig. 6 Stem cuttings 10 days after transformation to be transferred to fresh 3C5ZR media

agents. Use forceps to ensure that the plantlet is securely placed in the media.

12. Continue transferring explants every 14 days and excising developed shoots. Ensure that newly excised plantlets are considered an independent event.

Fig. 7 Plantlet emerging from stem cutting

3.5 Selection and Further Growth of Transgenic Shoots

1. If desired, a second round of selection may be performed. If so, excise the shoot tip from the 4-week-old plantlet and subculture into a fresh 3C5ZR tube. Any plantlet that results in rooting from the cut end can be considered transgenic.
2. After 4 weeks, roots should be established. Discard any events that failed to root after 6 weeks.
3. Well-rooted plantlets at the 4–5 node stage can be transferred into a peat pot. A sleeve of peat pots should be wrapped in foil, and autoclaved. Immediately before planting, peat pots should be soaked in ddH_2O. Following imbibing of peat pot, a moist potting soil mix containing a slow-release fertilizer should be added and gently pressed down to avoid compaction.
4. Gently remove plantlets from tubes and rinse under running ddH_2O to remove the rooting media.
5. Mix potting soil in a 1 gal bucket with water to create a moist mix. Add one tablespoon Osmocote Slow Release 14-14-14 N:P:K to bucket and mix thoroughly.
6. Press plantlet into soil-filled peat pot. Gently add any soil required to cover the plantlet to the level that was previously present in the tube. Ensure that the peat pot and soil is quite moist, but not soaking wet.

Fig. 8 Peat pots in plastic bin with wet paper towels in bottom following removal of lid

7. Place peat pots into a clear plastic container with a clear lid. Add a small stack of ddH_2O-soaked paper towels to the side of the plastic bin. Cover the bin with the lid, but do not seal. Place in a greenhouse or growth chamber at 15–20 °C (150 μE/m^2/s light intensity and 16-h light photo period).
8. After 3 days, slowly begin moving the lid off of the bin to slowly acclimatize transgenic plants. Water plantlets as needed. Remove lid completely after 3 days (Fig. 8).
9. Once plants have filled the peat pots (Fig. 9), place peat pot within a 10″ pot filled with potting soil. Fertilize plants once a week with a liquid fertilizer (1:1:1 N:P:K) (Fig. 10).
10. Transgenic lines can be maintained in vitro as microplants, propagated as described above, or tuber stocks in dark plastic containers and placed at 4 °C.

3.6 Ensuring Transgenic Plants

Two main steps are commonly used to ensure that the events recovered are transgenic. These are herbicide painting and PCR

Fig. 9 Potato plant fully acclimatized to peat pot and large enough to transfer to a 10″ pot for further growth

screening. Herbicide painting is only used when your plasmid of choice has a herbicide resistance gene. This will have been the same selection agent used in your 3C5ZR media. In this case, mix up the appropriate amount of herbicide in a 15 mL conical tube. Use a cotton-tipped applicator to dip into the solution and gently stroke onto leaf tissue. If the leaf dies, two possibilities may be occurring. The first is that the potato plantlet is an escape, meaning it is not transgenic. The second involves the possibility that the plantlet is transgenic, but not expressing the transgene. At this stage of analysis, both of these possibilities would result in discarding the plantlet. Once a plantlet has survived herbicide painting, PCR analysis is appropriate. PCR primers must be designed to work well with your insert. Perform DNA extraction using a Qiagen Plant DNA extraction kit.

Fig. 10 Potato plants in greenhouse

4 Notes

1. Pre-wet the 0.22 μm filter by drawing through 5–10 mL sterile ddH_2O and discard water.
2. Use slow exhaust on autoclave for all media.
3. All media should be poured into culture vessels or petri plates in a laminar flow hood.
4. 8 g/L Agar is sufficient for solidification on LB media on plates and is the most economical, though care should be taken while streaking to avoid ripping media surface.
5. Addition of antibiotics in media that has not cooled below 50 °C can result in inactivation of selective agent.
6. 1 L divides into 36 6 oz jars or 36 magenta vessels at 27 mL per vessel.
7. Any step of tissue culture that requires plant material to be exposed to air should be done within the laminar flow hood.

8. It is helpful to keep a Bunsen burner and a 25 mL culture tube filled with 95% ethanol in the laminar flow hood to facilitate sterilization of tools used for cutting, trimming, and movement.
9. Use a vice and a serrated knife to trim Glad Wrap into 1–1/2′ rolls for wrapping plates.
10. For primer design assistance, many Web-based tools are available. Integrated DNA Technologies, idtdna.com, has a variety of tools to assist with primer design, synthesis, resuspension, dilution, and more.
11. Inverted plates prevent condensation from dripping onto your bacterial colonies.
12. Consult your plasmid vector information to determine if M13 F/R will work for your insert. A variety of other standard primers may work, but for nonstandard plasmids/targets, special primer design for confirmation may be required.
13. Use proper controls at every step of all bacterial transformation steps, often recommended by the manufacturers. These controls should carry forward into PCR reactions as well to ensure transformants.
14. Autoclave a stack of Whatman filter papers, petri dish-sized prior to experiment. Remove a single filter paper with sterilized forceps and place into an empty petri dish. For this purpose you may use both the base and lid of the petri plate.

Acknowledgments

A sincere thank you to Whitney Harchenko, for providing photos of her work in transforming potato.

References

1. Halterman D, Guenthner J, Collinge S, Butler N, Douches D (2016) Biotech potatoes in the 21st century: 20 years since the first biotech potato. Am J Potato Res 93(1):1–20
2. Yoshida K, Schuenemann VJ, Cano LM, Pais M, Mishra B, Sharma R, Lanz C, Martin FN, Kamoun S, Krause J, Thines M, Weigel D, Burbano HA (2013) The rise and fall of the Phytophthora infestans lineage that triggered the Irish potato famine. elife 2:e00731. https://doi.org/10.7554/eLife.00731
3. Ooms G, Bossen M, Burrell M, Karp A (1986) Genetic manipulation in potato with Agrobacterium rhizogenes. Potato Res 29(3):367–379
4. Stiekema WJ, Heidekamp F, Louwerse JD, Verhoeven HA, Dijkhuis P (1988) Introduction of foreign genes into potato cultivars Bintje and Desiree using an Agrobacterium tumefaciens binary vector. Plant Cell Rep 7(1):47–50
5. De Block M (1988) Genotype-independent leaf disc transformation of potato (Solanum tuberosum) using Agrobacterium tumefaciens. Theor Appl Genet 76(5):767–774
6. Visser RGF, Jacobsen E, Hesseling-Meinders-A, Schans MJ, Witholt B, Feenstra WJ (1989) Transformation of homozygous diploid potato with an Agrobacterium tumefaciens binary vector system by adventitious shoot regeneration on leaf and stem segments. Plant Mol Biol 12

(3):329–337. https://doi.org/10.1007/bf00043210

7. Heeres P, Schippers-Rozenboom M, Jacobsen E, Visser RGF (2002) Transformation of a large number of potato varieties: genotype-dependent variation in efficiency and somaclonal variability. Euphytica 124 (1):13–22. https://doi.org/10.1023/a:1015689112703
8. Murashige T, Skoog F (1962) A revised medium for rapid growth and bio assays with tobacco tissue cultures. Physiol Plant 15 (3):473–497. https://doi.org/10.1111/j.1399-3054.1962.tb08052.x
9. Gamborg OL, Miller RA, Ojima K (1968) Nutrient requirements of suspension cultures of soybean root cells. Exp Cell Res 50 (1):151–158. https://doi.org/10.1016/0014-4827(68)90403-5
10. Matsuda F, Yamada T, Miyazawa H, Miyagawa H, Wakasa K (2005) Characterization of tryptophan-overproducing potato transgenic for a mutant rice anthranilate synthase α-subunit gene (OASA1D). Planta 222(3):535–545. https://doi.org/10.1007/s00425-005-1565-x
11. Miki D, Shimamoto K (2004) Simple RNAi vectors for stable and transient suppression of gene function in rice. Plant Cell Physiol 45(4). doi:https://doi.org/10.1093/pcp/pch048
12. Höfgen R, Willmitzer L (1988) Storage of competent cells for Agrobacterium transformation. Nucleic Acids Res 16(20):9877
13. Chronis D, Chen S, Lang P, Tran T, Thurston D, Wang X (2014) Potato transformation. bio-Protocol 4(1). doi:https://doi.org/10.21769/BioProtoc.1017

Chapter 16

Agrobacterium tumefaciens-Mediated Transformation of Tomato

Joyce Van Eck, Patricia Keen, and Michelle Tjahjadi

Abstract

Tomato is both an important food crop and serves as a model plant species that is used for various research investigations including understanding gene function. Transformation is commonly utilized to facilitate these investigations in combination with all the extensive genetic and genomic resources available for tomato. The transformation protocol routinely used in our laboratory has been applied to many different tomato genotypes and relies on *Agrobacterium tumefaciens* infection of young cotyledon sections. We have used vector systems for overexpression, RNA interference for gene silencing, and CRISPR/Cas9 for genome editing. Vectors used to design gene constructs contained selectable marker genes that conferred resistance to kanamycin, hygromycin, and the herbicide component, bialaphos. The protocol we follow for *Agrobacterium*-mediated transformation of both cultivated and wild species of tomato is detailed in this chapter.

Key words AGL1, *Agrobacterium tumefaciens*, LBA4404, *Solanaceae*, *Solanum lycopersicum*, *Solanum pimpinellifolium*

1 Introduction

Tomato, a member of the *Solanaceae* family, is native to South America and was brought to Europe in the 1500s and then to North America in the 1800s [1]. It is a perennial plant that has determinate and indeterminate plant architecture where determinate is a compact growth habit, and indeterminate is a sprawling type of habit. There are two different market types of tomatoes, fresh market and processing. According to the Agricultural Marketing Resource Center, in 2014 the US dollar value for fresh market tomatoes was 1.14 billion and 1.325 billion for processing types, which are used to make products such as juice, sauces, and ketchup [2].

In addition to being an important food crop, tomato has also been exploited as a model species for various studies. Over the years, utilization of tomato as a model plant species has increased

Sandeep Kumar et al. (eds.), *Transgenic Plants: Methods and Protocols*, Methods in Molecular Biology, vol. 1864, https://doi.org/10.1007/978-1-4939-8778-8_16, © Springer Science+Business Media, LLC, part of Springer Nature 2019

because of readily available resources such as mutant populations, bioinformatics tools [3], a high-quality reference genome [4], and efforts to improve *Agrobacterium*-mediated transformation efficiency and recovery time of transgenic lines [5, 6]. The first report of *Agrobacterium tumefaciens*-mediated transformation of tomato (*Solanum lycopersicum*) was by McCormick et al. [7] more than 30 years ago. In the intervening years, there have been reports of transformation of different tomato genotypes and different explant types such as cotyledons, hypocotyls, and leaves [5, 8–13]. Availability of efficient transformation methods is especially critical with the rapid development of genome editing technologies, which will result in an increased demand for generation of transgenic lines for basic research studies that can lead to crop improvement.

The methodology reported in this chapter is based on a modified version reported by Fillatti et al. [9] for infection of tomato cotyledons with *Agrobacterium* (Fig. 1). Briefly, our method starts with disinfected seeds cultured on a medium for germination. Cotyledons collected before the first true leaves emerge are sectioned and infected with *Agrobacterium tumefaciens* followed by a cocultivation period of 2 days. The cotyledon sections are transferred to a selective regeneration medium that contains selection

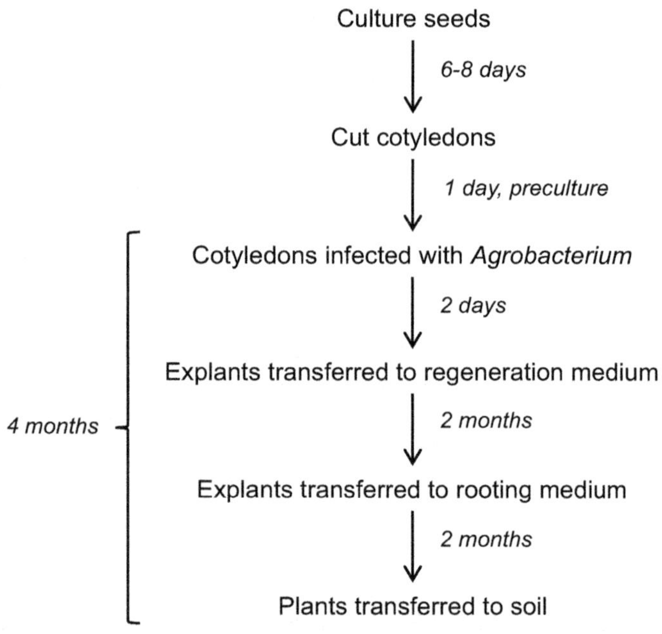

Fig. 1 Tomato transformation workflow. Estimated time is for best-case scenario to recover the first rooted plant. Duration and efficiency of transformation are dependent on the gene of interest and genotype. Procedure from time of infection to soil transfer can take up to 6 months to recover ten rooted plants as plants regenerate at different time points during the process of generating transgenic lines

agents specific for the selectable marker gene in the vector used to build a construct. We have continued to build and optimize our method since the first report in 2006 [13], including the removal of a *Nicotiana tabacum* cell suspension (NT1) feeder layer during cocultivation. In a recent study, we showed that the time for recovery of transgenic lines from the tomato cultivar M82 can be significantly decreased by the addition of indole-3-acetic acid (IAA) to the regeneration medium [5]. However, we have not systematically tested this method on other genotypes; therefore, our recommendation is that it be tested in comparison with medium not containing IAA before adopting this as a standard medium component. Revisiting protocols for generation of stable transgenic lines to determine if methods could be made more efficient or less genotype dependent should always be considered in order to alleviate bottlenecks that prevent progress for investigations that rely on transformation [14].

2 Materials

2.1 Plant Material, *Agrobacterium tumefaciens*, and Vectors

1. One gram of seed of tomato genotype of interest (*see* **Note 1**).
2. *Agrobacterium tumefaciens* strain AGL1 or LBA4404 (*see* **Note 2**).
3. Vectors commonly used for our work include but are not limited to p201N:Cas9 and pAGM4723 for CRISPR/Cas9-mediated gene editing, pHellsgate8 for RNAi-mediated silencing, and pBTEX for overexpression.

2.2 Media Components and Stock Solutions (See Note 3)

1. 1 mg/mL Thiamine HCl: Dissolve 50 mg of thiamine HCl in 50 mL deionized H_2O. Wrap in foil and store at 4 °C.
2. 0.5 mg/mL Pyridoxine HCl: Dissolve 25 mg of pyridoxine HCl in 50 mL deionized H_2O. Store in 1 mL aliquots at −20 °C.
3. 0.5 mg/mL Nicotinic acid: Dissolve 25 mg of nicotinic acid in 50 mL deionized H_2O. Store in 1 mL aliquots at −20 °C.
4. 1000× Modified Nitsch vitamins: Dissolve 0.1 g of glycine, 0.5 g of nicotinic acid, 0.025 g of pyridoxine HCl, 0.025 g of thiamine HCl, 0.025 g folic acid, and 0.002 g of d-biotin in 50 mL deionized H_2O. Adjust the pH to 7.00. Store in 1 mL aliquots at −20 °C.
5. 1 mg/mL trans-Zeatin: Dissolve 50 mg of trans-Zeatin in a few drops of 0.5 M HCl. Add deionized H_2O to a total volume of 50 mL. Filter sterilize and store in 1 mL aliquots at −20 °C. Add after autoclaving.
6. 100 mg/mL Timentin: Dissolve 5.17 g of timentin in 50 mL of deionized H_2O (*see* **Note 4**). Filter sterilize and store in aliquots at −20 °C. Add after autoclaving.

7. 100 mg/mL Kanamycin: Dissolve 5 g of kanamycin (Caisson Labs, Smithfield, UT, USA, product #K003) in 50 mL of deionized H2O. Filter sterilize and store in 1 mL aliquots at −20 °C. Add after autoclaving.
8. 50 mg/mL Spectinomycin: Dissolve 2.5 g of spectinomycin in 50 mL of deionized H_2O. Filter sterilize and store in 1 mL aliquots at −20 °C. Add after autoclaving.
9. 100 mg/mL Carbenicillin: Dissolve 5 g of carbenicillin in 50 mL of deionized H_2O. Filter sterilize and store in 1 mL aliquots at −20 °C. Add after autoclaving.
10. 12 mg/mL Rifampicin: Dissolve 120 mg of rifampicin in 10 mL of methanol. Wrap in foil and store at −20 °C.

2.3 Culture Medium

The media components listed are on a per-liter basis. Cool the medium in a water bath to 55–60 °C. This preserves the integrity of the hormones and antibiotics added after autoclaving while keeping the media from solidifying before poured. Dispense 24.5 mL of medium per Petri plate. For Magenta™ GA-7 boxes (referred to as Magenta boxes throughout the remainder of the chapter), we dispense 62.5 mL per Magenta box.

2.3.1 For *Agrobacterium* AGL1

1. MG/L bacterial medium: 5 g of tryptone, 2.5 g of yeast extract, 5 g NaCl, 5 g of mannitol, 0.1 g of $MgSO_4$, 0.25 g of K_2HPO_4, 1.2 g of glutamic acid, 15 g of sucrose, pH to 7.2, 15 g Bacto™ Agar. After autoclaving, add 0.5 mL of carbenicillin stock and antibiotics appropriate for vector. Dispense into 100 × 15 mm Petri plates.
2. Luria Broth (LB): 10 g of tryptone, 5 g of yeast extract, 10 g of NaCl, and antibiotics appropriate for vector.

2.3.2 For *Agrobacterium* LBA4404

1. LB bacterial medium: 10 g of tryptone, 5 g of yeast extract, 10 g of NaCl, 15 g of Bacto™ Agar. After autoclaving, add 1.25 mL rifampicin stock and antibiotics appropriate for vector. Dispense into 100 × 15 mm Petri plates.
2. YEP liquid: 10 g of yeast extract, 10 g of peptone, 5 g of NaCl, and antibiotics appropriate for vector.

2.3.3 For Seed Germination and Transformation

1. Seed germination medium (½ MSO): 2.15 g of Murashige and Skoog (MS) salts, 100 mg of myoinositol, 2 mL of thiamine HCl stock, 1 mL of pyridoxine HCl stock, 1 mL of nicotinic acid stock, 10 g of sucrose, pH to 5.8, 8 g agar (Millipore Sigma, Burlington, MA, USA, product #A1296). Dispense 50 mL per Magenta box.
2. Liquid coculture medium (2% MSO): 4.3 g of MS salts, 100 mg of myoinositol, 0.4 mL of thiamine HCl stock, 20 g of sucrose, pH to 5.8.

3. Preculture/coculture medium (2Z preculture): 4.3 g of MS salts, 100 mg of myoinositol, 1 mL of modified Nitsch vitamin stock, 20 g of sucrose, pH to 6.0, 5.2 g of TC Gel (Caisson Labs, Smithfield, UT, USA, product #PTP02). After autoclaving, add 2 mL of trans-zeatin stock. Dispense into 100 × 20 mm Petri plates (*see* **Note 5**).

2.3.4 For Plant Regeneration and Rooting

1. Primary regeneration medium (2Z): 4.3 g of MS salts, 100 mg of myoinositol, 1 mL of modified Nitsch vitamin stock, 20 g of sucrose, pH to 6.0, 5.2 g of TC Gel. After autoclaving, add 2 mL of trans-zeatin stock, 3.5 mL of timentin stock, and antibiotics appropriate for vector. Dispense into 100 × 20 mm Petri plates (*see* **Note 5**).
2. Secondary regeneration medium (1Z): Same as 2Z but with only 1 mL of trans-zeatin stock. Dispense into 100 × 20 mm Petri plates (*see* **Note 5**) and Magenta boxes.
3. Rooting medium (RM): 4.3 g of MS salts, 1 mL of modified Nitsch vitamin stock, 30 g of sucrose, pH to 6.0, 8 g Bacto™ Agar. After autoclaving, add 3.5 mL of timentin stock and antibiotics appropriate for vector. Dispense into Magenta boxes.

2.4 Other Supplies and Reagents

1. Sterile deionized water.
2. Sterilization solution (20% bleach plus one drop Tween 20).
3. Sterile Whatman filter paper, 7 cm in diameter.
4. Pipetman and sterile tips.
5. Sterile paper towels.
6. 100 × 15 mm and 100 × 20 mm sterile Petri plates.
7. 0.5 in Micropore tape (Fisher Scientific, Birmingham, AL, USA, product #19-061655).
8. Parafilm.
9. 50 mL Falcon disposable centrifuge tube, or similar.
10. Magenta™ GA-7 boxes (referred to as Magenta boxes in this chapter).

3 Methods

3.1 Preparation of Plant Material

1. Sterilize seed by immersing 0.9–1.0 g seeds in 25 mL sterilization solution and agitate for 20 min at 250 rpm on an orbital shaker. Rinse three times with sterile deionized H_2O.
2. Place 25–30 seeds per Magenta box containing ½ MSO (Fig. 2a).

Fig. 2 Generation of transgenic tomato lines from cotyledon explants. Explants were infected with *Agrobacterium tumefaciens* that contained a plasmid harboring the selectable marker gene neomycin phosphotransferase II (*nptII*) gene which confers resistance to kanamycin. (**a**) Tomato seeds placed in Magenta boxes containing ½ MSO. (**b**) 7-Day-old seedlings grown on ½ MSO medium. (**c**) Cotyledons cut and placed adaxial (top) side down on sterile filter paper in Petri plates containing 2Z preculture medium. (**d**) Explants on 1Z selective medium containing kanamycin. (**e**) Putative transgenic lines on first phase of RM selective medium. (**f**) Close-up of nodal region ready for excision and transfer into test tube containing RM. (**g**) Excised shoot in test tube containing RM selective medium. (**h**) Rooted shoot, 2 weeks after transfer into tube containing RM selective medium

3. Cut cotyledons from 6- to 8-day-old seedlings (Fig. 2b) (*see* **Note 6**), before the first true leaves have appeared (*see* **Note 7**). Place seedling on a sterile paper towel moistened with sterile deionized H_2O. Excise cotyledons at petiole and cut off tips. Cut cotyledons into 1 cm sections.

4. Place cut cotyledons adaxial (top) side down on plates of 2Z preculture medium, 50 pieces per plate (Fig. 2c). A piece of sterile Whatman filter paper is laid on top of the preculture medium and moistened with sterile water. For controls, ten pieces are used (five for positive controls, five for negative controls).
5. Seal plates with Parafilm. Culture for 1 day at 24 °C ± 2 °C with a 16-h photoperiod.

3.2 Transformation and Selection of Transgenic Lines

1. Two to three days before transformation, streak *Agrobacterium* containing the construct of interest onto selective MG/L or LB bacterial medium. Incubate for 36–48 h at 28 °C until colonies have developed.
2. Select four single, well-formed colonies from the plate and transfer to 50 mL LB or YEP medium containing antibiotic selection appropriate for the vector. Culture in a shaking incubator at 250 rpm at 28 °C overnight.
3. Check the OD_{600}. Optimum OD_{600} is 0.6. If OD_{600} is more than 0.65, dilute the culture until the OD_{600} reading is below 0.5 and grow for another hour. Check the OD_{600} reading periodically.
4. Centrifuge at 8228 × *g* for 10 min at 20 °C.
5. Note and record the volume of the supernatant. Discard the supernatant. Add 10 mL 2% MSO.
6. Resuspend the pellet by vortexing three times for 3–4 s each time. Then bring up to the volume recorded in **step 5** with 2% MSO.
7. Pipette 25 mL *Agrobacterium* culture into a sterile Magenta box. Transfer explants from 2 plates into inoculum (100 explants per 25 mL culture). Incubate explants for 5 min with occasional shaking. For controls, 25 mL of 2% MSO is used.
8. Remove explants to a sterile paper towel. Return explants adaxial side down to plates containing 2Z preculture medium. Seal plates with Parafilm.
9. Maintain plates under dark conditions at 19 °C for 48 h.
10. Transfer 25 explants to each plate of 2Z selective medium, adaxial side up. For controls, transfer five explants to 2Z medium without a selective agent (positive controls), and the other five explants to 2Z medium with selective agent (negative controls). Seal plates with Micropore tape. Culture at 24 °C ± 2 °C, with a 16-h photoperiod for 2 weeks.

11. Transfer to fresh 1Z selective medium (15 explants per plate), and to freshly prepared medium every 2 weeks (Fig. 2d) (*see* **Note 8**). When shoots begin to appear and touch the lid of the plate, transfer explants to 1Z selective medium in Magenta boxes, five explants per Magenta box (*see* **Note 9**).

3.3 Regeneration and Rooting

1. Initial shoots should appear within 4–6 weeks (Fig. 2e).
2. Excise shoots from explants when shoots are at least 2 cm and include at least one node (Fig. 2f). Place in Magenta boxes containing RM selective medium, four shoot cuttings per Magenta box (*see* **Note 9**). Tubes containing RM selective medium can also be used, with one shoot cutting per tube (Fig. 2g) (*see* **Note 10**).
3. Roots should begin to appear in 5–9 days (Fig. 2h).

4 Notes

1. Various tomato genotypes have been transformed with this protocol including M82, Rio Grande, Castlemart, cherry tomatoes including Sweet100, and the wild species *S. pimpinellifolium*.
2. There are some differences between *Agrobacterium tumefaciens* strains LBA4404 and AGL1. Transgenics generated with either strain generally contain a low copy number of the introduced transgene as compared to other *Agrobacterium* strains. AGL1 is a hypervirulent strain, and we have observed faster growth compared to LBA4404 with both solidified and liquid media. There are also more incidences of *Agrobacterium* AGL1 overgrowth in tissue culture postinfection as compared to LBA4404. Therefore, we recommend using 350 mg/L timentin in all culture media for AGL1 while 300 mg/L timentin is sufficient when using LBA4404. Overall, we see higher transformation efficiency with AGL1 (unpublished data), possibly because of its hypervirulence. Also, being that AGL1 is *recA*$^-$, there is less chance of the occurrence of recombination of introduced plasmid DNA [15].
3. In general, we keep refrigerated stock solutions for up to 3 months and frozen stock solutions for up to 6 months; however, this can be dependent on the stability of individual compounds. For frozen stock solutions, do not exceed two freeze/thaw cycles, as this impacts efficacy of the solution.
4. We have used timentin from bioWORLD, and GoldBio with similar effectiveness. Bioworld gives slightly higher transformation efficiency but results are not significant (unpublished data).

5. We use 100 × 20 mm Petri plates because they provide more space compared to 100 × 15 mm plates, allowing more room for shoots to expand. This additional space of the deeper plates has the potential to lessen the negative effects of accumulated ethylene, which can inhibit growth.
6. Time of germination, synchrony, and percentage vary depending on the tomato genotype.
7. Using cotyledons from seedlings after the first true leaves have appeared dramatically decreases transformation efficiency.
8. Media in Petri plates and Magenta boxes kept at room temperature are good for 2 weeks. If stored at 4 °C, they can be used for up to a month. Decreased timentin efficacy in older media allows *Agrobacterium* overgrowth, leading to losses of plant material.
9. It is important to not exceed five shoots per Magenta box containing 1Z selective medium and only four shoots per Magenta box containing RM selective medium. We have observed significant negative effects on growth when these numbers are exceeded. Excised shoots can be rooted on medium in either Magenta boxes or test tubes. For shipping, we recommend test tubes because the medium is more secure and will not detach from the bottom as it does in Magenta boxes.
10. Excised shoots can be rooted in Magenta boxes or test tubes. For shipping, we recommend test tubes because the medium will not detach from the bottom as it does in Magenta boxes.

Acknowledgments

Van Eck acknowledges the National Science Foundation Plant Genome Research Program (IOS-1732253) for support related to her research on genetic engineering and genome editing of tomato.

References

1. Jones JBJ (1998) Tomato plant culture: in the field, greenhouse, and home garden. CRC Press LLC, Boca Raton, FL
2. Center AMR (2017) http://www.agmrc.org/commodities-products/vegetables/tomatoes/
3. Bombarely A, Menda N, Tecle IY, Buels RM, Strickler S, Fischer-York T, Pujar A, Leto J, Gosselin J, Mueller LA (2011) The sol genomics network (solgenomics.net): growing tomatoes using perl. Nucleic Acids Res 39: D1149–D1155
4. Consortium TTG (2012) The tomato genome sequence provides insights into fleshy fruit evolution. Nature 485:635–641
5. Gupta S, Van Eck J (2016) Modification of plant regeneration medium decreases the time for recovery of *Solanum lycopersicum* cultivar M82 stable transgenic lines. Plant Cell Tissue Organ Cult 127:417–423
6. Yasmeen A (2009) An improved protocol for the regeneration and transformation of tomato (cv Rio Grande). Acta Physiol Plant 31:1271–1277

7. McCormick S, Niedermeyer J, Fry J, Barnason A, Horsch R, Fraley R (1986) Leaf disc transformation of cultivated tomato (*L. esculentum*) using *Agrobacterium tumefaciens.* Plant Cell Rep 5:81–84
8. Chyi YS, Phillips GC (1987) High efficiency *Agrobacterium*-mediated transformation of *Lycopersicon* based on conditions favorable for regeneration. Plant Cell Rep 6:105–108
9. Fillatti JJ, Kiser J, Rose R, Comai L (1987) Efficient transfer of a glyphosate tolerance gene into tomato using a binary *Agrobacterium tumefaciens* vector. Bio/Technology 5:726–730
10. Frary A, Earle ED (1996) An examination of factors affecting the efficiency of *Agrobacterium*-mediated transformation of tomato. Plant Cell Rep 16:235–240
11. Park SH, Morris JL, Park JE, Hirschi KD, Smith RH (2003) Efficient and genotype-independent *Agrobacterium*-mediated tomato transformation. J Plant Physiol 160:1253–1257
12. Sun HJ, Uchii S, Watanabe S, Ezura H (2006) A highly efficient transformation protocol for Micro-Tom, a model cultivar for tomato functional genomics. Plant Cell Physiol 47:426–431
13. Van Eck J, Kirk DD, Walmsley AM (2006) Tomato (*Lycopersicum esculentum*). In: Wang K (ed) Methods in molecular biology, *Agrobacterium* protocols, vol 343, vol 1. Humana Press Inc., Totowa, NJ, pp 459–473
14. Altpeter F, Springer NM, Bartley LE, Blechl AE, Brutnell TP, Citovsky V, Conrad LJ, Gelvin SB, Jackson DP, Kausch AP, Lemaux PG, Medford JI, Orozco-Cardenas ML, Tricoli DM, Van Eck J, Voytas DF, Walbot V, Wang K, Zhang ZJ, Stewart CN Jr (2016) Advancing crop transformation in the era of genome editing. Plant Cell 28:1510–1520
15. Lazo GR, Stein PA, Ludwig RA (1991) A DNA transformation-competent *Arabidopsis* genomic library in *Agrobacterium.* Nat Biotechnol 9:963–967

Part II

Genome Editing

Chapter 17

DNA Break Repair in Plants and Its Application for Genome Engineering

Carla Schmidt, Michael Pacher, and Holger Puchta

Abstract

Genome engineering is a biotechnological approach to precisely modify the genetic code of a given organism in order to change the context of an existing sequence or to create new genetic resources, e.g., for obtaining improved traits or performance. Efficient targeted genome alterations are mainly based on the induction of DNA double-strand breaks (DSBs) or adjacent single-strand breaks (SSBs). Naturally, all organisms continuously have to deal with DNA-damaging factors challenging the genetic integrity, and therefore a wide range of DNA repair mechanisms have evolved. A profound understanding of the different repair pathways is a prerequisite to control and enhance targeted gene modifications. DSB repair can take place by nonhomologous end joining (NHEJ) or homology-dependent repair (HDR). As the main outcome of NHEJ-mediated repair is accompanied by small insertions and deletions, it is applicable to specifically knock out genes or to rearrange linkage groups or whole chromosomes. The basic requirement for HDR is the presence of a homologous template; thus this process can be exploited for targeted integration of ectopic sequences into the plant genome. The development of different types of artificial site-specific nucleases allows for targeted DSB induction in the plant genome. Such synthetic nucleases have been used for both qualitatively studying DSB repair in vivo with respect to mechanistic differences and quantitatively in order to determine the role of key factors for NHEJ and HR, respectively. The conclusions drawn from these studies allow for a better understanding of genome evolution and help identifying synergistic or antagonistic genetic interactions while supporting biotechnological applications for transiently modifying the plant DNA repair machinery in favor of targeted genome engineering.

Key words DSB repair, HDR, NHEJ, Programmable nucleases, Gene targeting, Genome engineering

1 Randomly Induced DNA Damage

Different endogenous and exogenous factors continuously damage genetic information, causing various kinds of DNA lesions (Fig. 1). As photoautotrophic organisms, plants use light energy for photosynthesis and thus are also exposed to ultraviolet radiation (UV). This radiation ranges from 100 to 400 nm in wavelength. Three types of UV radiation are described, namely UV-A, UV-B, and UV-C, with UV-C being directly absorbed by the DNA and thus

Sandeep Kumar et al. (eds.), *Transgenic Plants: Methods and Protocols*, Methods in Molecular Biology, vol. 1864, https://doi.org/10.1007/978-1-4939-8778-8_17, © Springer Science+Business Media, LLC, part of Springer Nature 2019

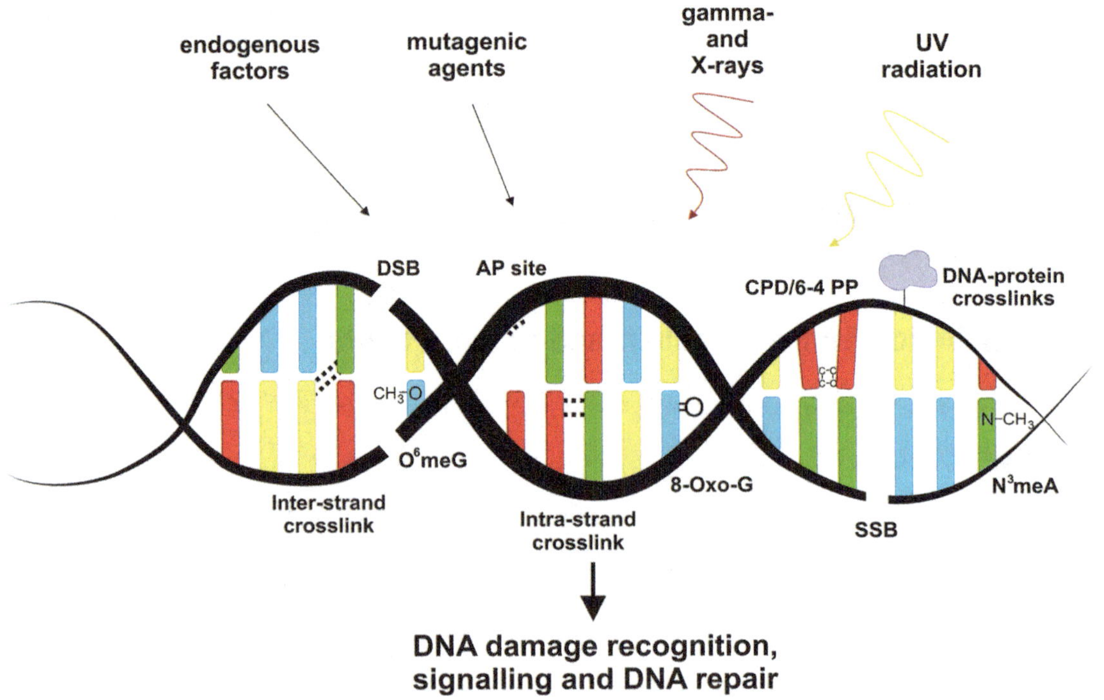

Fig. 1 Possible DNA lesions induced by various factors. The DNA double helix is exposed to different endogenous and exogenous factors, resulting in different DNA damages, followed by damage recognition and DNA repair. CPDs and 6–4 PPs are mainly produced via UV radiation, while gamma- and X-rays directly ionize the DNA, resulting in DSBs and SSBs. Both UV and ionizing radiations are also involved in the formation of ROS. This highly reactive species can cause several kinds of base damages like 8-oxo-guanine, O^6-meG, N^3meA, and AP sites. ROS are formed endogenously due to the presence of free radicals. Mutagenic agents can cause cross-links between adjacent and paired bases, as well as DNA-protein cross-links

being most harmful for genetic material. UV radiation mainly causes the formation of cyclobutane pyrimidine dimers (CPDs) and 6–4 photoproducts (6–4 PP) by covalent binding of C_5 and C_6 carbon atoms of the deoxyribose backbone. Additionally, oxidative DNA damage can be caused by UV irradiation, as the high-energy rays produce reactive oxygen species (ROS) by interactions with molecules within a living cell [1]. Free radicals can also appear in plant cells resulting in ROS formation, due to an inevitable leakage of electrons onto O_2 molecules during electron transport in chloroplasts, mitochondria, and plasma membranes, or as a by-product of different metabolic pathways [2]. ROS can cause alkylation or oxidation of bases (e.g., 8-oxo-guanine), as well as hydrolysis of the N-glycosidic bond of bases, both resulting in an apurinic/apyrimidinic site (AP site). Besides these endogenous factors, agents like methyl-methanesulfonate (MMS) and ethyl-methanesulfonate (EMS) can alkylate bases at their O/N-positions, resulting, for example, in O^6-methylguanine (O^6-meG) or N^7-methylguanine/N^3-methyladenine (N^7meG/N^3meA). During replication, the O^6-meG mispairs with thymine instead of cytosine, leading to a point mutation. The mutagenic effect of EMS is widely used for random mutagenesis to perform crop improvements and execute forward genetic studies [3]. The most noted procedure is called targeting-induced local lesions in genomes (TILLING), in which traditional chemical mutagenesis is combined with a high-throughput genome-wide screening for point mutations [4, 5]. Aside from methylation, DNA bases can covalently bind to an adjacent base on the same strand, leading to intra-strand cross-links, or mispairing with bases of the complementary strand can result in inter-strand cross-links. Agents like mitomycin C and cisplatin, typically used as cytostatic agents, create inter- and intra-strand cross-links and are also applied for mutagenic treatments of plants. Cross-links can occur not only between two DNA strands, but also between the DNA backbone and proteins. Protein-DNA interactions are required for nearly all essential DNA transactions and need to be highly dynamic; however, such transient interactions can also result in covalent binding of proteins to the DNA. Such DNA-protein cross-links are highly toxic, as they interfere with transcription and replication processes [6]. These DNA damages can occur naturally or can be induced by camptothecin, leading to covalent binding of topoisomerase I to DNA. Besides base damages and cross-linking, the induction of DSBs into the DNA backbone is highly mutagenic due to the absence of a complementary strand, usually serving as repair template. DSBs can arise endogenously during processes involved in transmission, expression, and maintenance of the genetic information, as well as by exogenous factors like ionizing radiation or radiomimetic agents, e.g., bleomycin. Ionizing radiation like X- and gamma rays was discovered in the late nineteenth century. Its highly mutagenic potential is caused by its ability to directly ionize DNA molecules or to produce highly reactive radicals and free electrons. Ionizing radiation can lead to DNA fragmentation and especially to modifications at the break sites, like the formation of 3′-phosphate, 3′-phosphoglycolate, and 3′-phosphoglycaldehyde ends, which need to be processed before ligation [7]. AP sites, base oxidations, and SSBs can also occur via secondary DNA ionization by highly reactive radicals or free electrons. The first mutagenesis studies applying X-rays in eukaryotes were made in maize, barley, wheat and *Drosophila* [8, 9]. Since then, ionizing radiation has been applied as a useful tool to create mutants and novel traits and to study DNA repair mechanisms. Large-scale mutagenesis of plants using gamma rays was performed on the gamma field of the Japanese Institute of Radiation Breeding [10]. The number of mutations arising in plant cells can differ depending on the exposure time and the distance to the radiation source. Subsequently, laborious screening of plants carrying mutations associated to desired phenotypes or to the production of greater yields is indispensable.

Mutagens like bleomycin are also able to induce DSBs by mimicking the effects of ionizing radiation. Due to its cytostatic effect, this family of glycopeptides are commonly known from cancer treatments. The DSB-inducing property of bleomycin is exploited for studying DNA repair in plants, while it only rarely plays a role in mutagenesis [11, 12].

Plants continuously face the challenge to repair all of these different base lesions, cross-links, SSBs, and DSBs (reviewed in [13]). Therefore, a complex network of various damage surveillance and repair mechanisms is in place, acting through highly specific proteins and signaling cascades. To get a deeper insight into how these mechanisms are regulated and which protein functions are required for adequate repair, plant mutant lines lacking certain protein functions are required. For the model plant *Arabidopsis thaliana*, many mutant lines are available for such studies that have been produced via Agrobacterium-mediated T-DNA transformation [14]. Based on the elucidation of repair mechanisms in plants, different DNA repair models have been developed, which now provide the basis for controlled sequence modifications, mediated by targeted DNA damage. The discovery of synthetic nucleases with specific or adaptable target sequences led to a revolution of plant genome engineering approaches [15].

2 Site-Specific Endonucleases for Targeted Genome Manipulations

The first experiments in plants to induce site-specific breaks in the genome were performed with the homing endonuclease I-*Sce*I, which was discovered in *Saccharomyces cerevisiae* [16, 17]. Due to its non-palindromic 18mer recognition sequence [18, 19], I-*Sce*I is also referred to as megaendonuclease. The probability of the I-*Sce*I target site occurring randomly in the genome of a plant species is very low and correlated to the respective genome size. Obviously, this discovery and the characterization of this unique enzyme provided an unprecedented tool for targeted DSB induction in vivo, by I-*Sce*I expression in plants harboring transgenes with the 18 nt recognition site. It was demonstrated by Puchta et al. that I-*Sce*I can induce DSBs in plant cells and can therefore be used for precise genome alterations and to study basic mechanisms of DSB repair [20]. A disadvantage of I-*Sce*I, however, is its static intrinsic 18 nt recognition motif, as it is desirable to target any sequence in the genome. For that reason, different types of artificial nucleases with adaptive recognition sites have been investigated. As the characteristics of these nucleases and their applications are discussed in detail in the following chapters, we will provide only a short overview here.

First attempts to create site-specific nucleases were based on the combination of a DNA-binding domain with a nuclease domain.

This was initially realized by a combination of a highly variable class of zinc-finger transcription factors and a cleavage domain of the restriction enzyme *Fok*I [21]. Different combinations of these zinc fingers, with one zinc finger binding three adjacent bases, implemented sequence-specific recognition of the target sequence. However, sometimes interactions between zinc fingers and proximal nucleotides led to unpredictable binding; thus each nuclease required thorough testing. As cloning and testing of these zinc-finger nucleases (ZFNs) are quite time consuming, a similar approach was investigated using transcription activator-like effectors (TALEs), which were discovered in the plant pathogen species *Xanthomonas*. The DNA-binding domain of these TALEs consists of tandem repeats, varying only in amino acids 12 and 13, referred to as repeat-variable diresidue. The variable region of the tandem repeat was found to provide the specificity for all four DNA bases and thus a combination of the TALEs with a *Fok*I nuclease domain, called a TALE nuclease (TALEN), can be applied for targeting a DSB. Both ZFNs and TALENs were successfully used in different plant species for genome engineering approaches (reviewed in [22]).

The programmable nuclease discovered most recently was originally described as an adaptive immune system in bacteria and archaea, consisting of a clustered regularly interspaced short palindromic repeat (CRISPR) locus and Cas proteins [23]. This CRISPR/Cas system can be used for site-specific break induction by combining the Cas9 nuclease with a sgRNA complementary to the target sequence [24]. Besides the 20 nt spanning target sequence, a protospacer adjacent motif (PAM) is required at the target locus for break induction via Cas9. The adaption of the sgRNA to the target site and cloning is simple and can be conducted very quickly in any molecular biology laboratory. This technique is currently used in many eukaryotic systems and has proved successful as a tool for genome engineering in plants (reviewed in [25–27]).

The availability of these programmable site-specific nucleases provides the basis of targeted genome modifications and tremendously broadens the possibilities of genome engineering. However, these new biotechnological opportunities are limited by the properties of the intrinsic DNA repair machinery. Thus, a thorough understanding of the plant DSB repair mechanisms and its regulation is a mandatory prerequisite for knowledgeable optimization of genome editing approaches.

3 DNA Repair Mechanisms in Plants and Their Application for Genome Engineering

Plants have to protect their genome against several exogenous and endogenous factors. Mostly, only one DNA strand is affected, but the more severe damages are DSBs, due to the absence of a complementary strand, normally serving as a repair template. Therefore, eukaryotic organisms evolved different pathways to repair these lesions and to preserve genomic integrity. Some of these pathways are error prone and thus cause mutations; however, there are also repair pathways resulting in an accurate restoration of the genome. Most applications of genome engineering in living organisms are based on the exploitation of repair pathways, e.g., mutagenesis, induced chromosomal rearrangements, and gene targeting (GT). In particular, NHEJ as well as HDR are the bases for most applications of genome engineering [28]. While base editing and introduction of epigenetic modifications are indirect ways of either exchanging single nucleotides or modulating gene expression activity, any targeted genetic rearrangement requires the transient opening of the genetic code itself with a site-specific cut, followed by changes therein catalyzed by the intrinsic DNA repair machinery. Thus, numerous studies have been conducted over the last 25 years, in order to understand how DNA double-strand breaks (DSBs) are repaired in plants.

3.1 HDR

In somatic plant cells, HDR is a minor pathway occurring mainly during S and G_2 phases of the cell cycle [29]. In somatic cells, two different HDR mechanisms have been described, the single-strand annealing (SSA) and the synthesis-dependent strand annealing (SDSA), which are the mechanisms employed for the majority of HDR repair outcomes, while the double-strand break repair (DSBR) model [30] and the dissolution pathway play only a minor role (for overview, *see* Fig. 2). DSBR and dissolution occur most frequently during meiosis, resulting in crossover (CO) and non-crossover (NCO) products, and thus DSBR is the most prominent mechanism for explaining the observed meiotic recombination events [31].

All HR pathways have one common step, mediated by the MRN (Mre11, Rad50, Nbs1) complex, leading to the formation of 3′ single-stranded overhangs. To protect the produced DNA single strands from exonuclease-mediated degradation, a heterodimeric protein complex is formed by the three RPA subunits RPA1, RPA2, and RPA3 in plants [32].

In the presence of homologies on both ends of the break site, the nonconservative SSA pathway can be utilized for repair (Fig. 2a). Following end resection by the MRN complex, the protein RAD52 is recruited to the RPA-bound single-stranded DNA nucleoprotein complex in yeast [33]. In *Arabidopsis*, homologs of RAD52 have been identified; however, they display only a moderate effect on homologous recombination [34]. Instead, it was demonstrated that the *Arabidopsis* RAD51 paralogues XRCC2, RAD51B, and RAD51D are involved in SSA [35]. In contrast to the MMEJ pathway, it was shown in yeast that the homology length

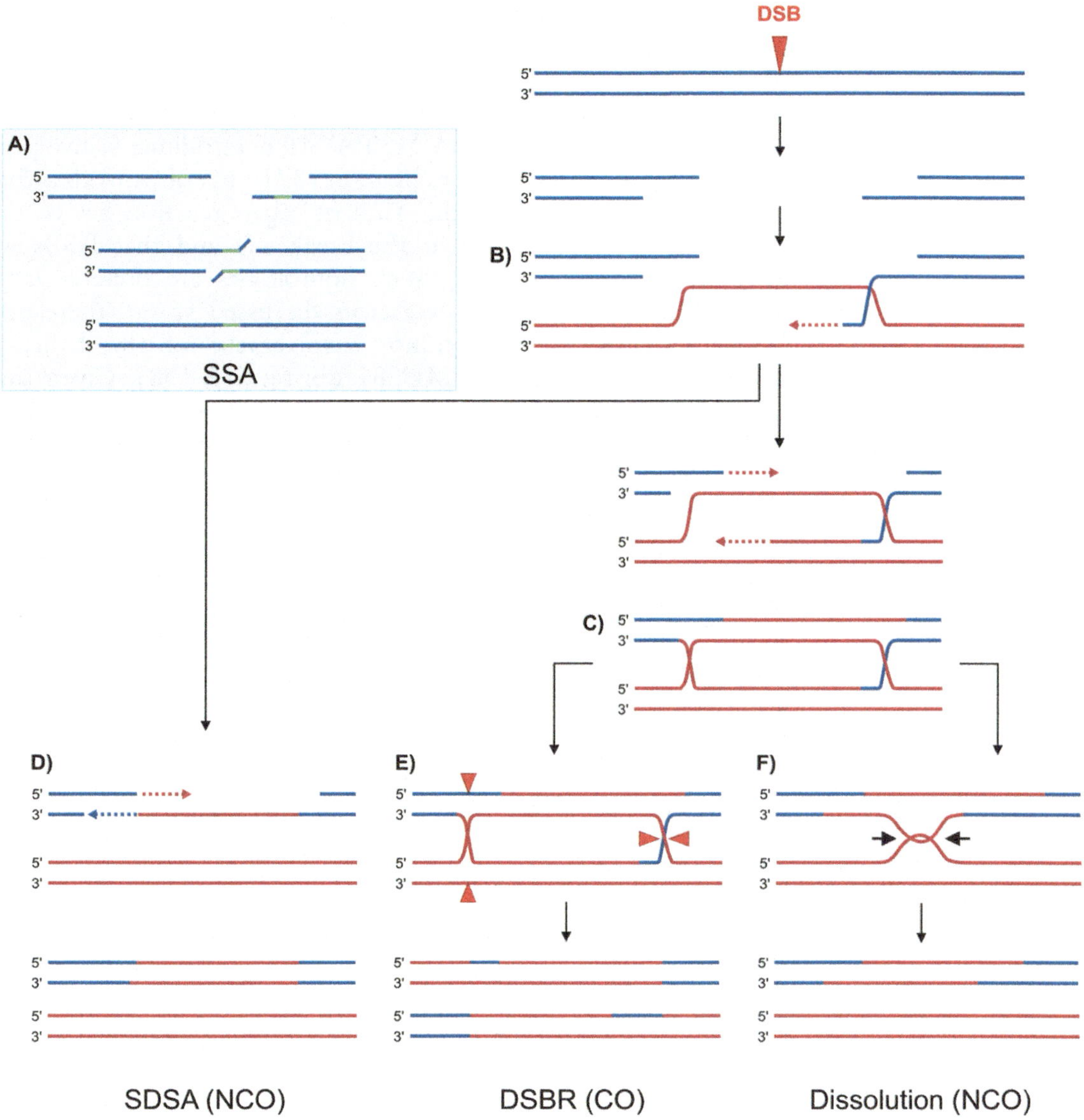

Fig. 2 Overview of HR-related DSB repair mechanisms. After DSB induction, single-stranded 3′ overhangs are produced. (**a**) In the case of homologies (green) flanking the break site, the repair reaction occurs via SSA. Subsequent to annealing of homologous sequences, the 3′ overhanging flaps are removed by an endonuclease and the gaps are filled by a polymerase. (**b**) When a homologous template is accessible, the 3′ end is able to invade into the donor molecule (red), forming a D-loop. Subsequently, elongation of the invaded single strand can take place, using the complementary donor strand as a template. If the displaced strand realigns with the other end of the break and DNA synthesis occurs, a double Holliday junction (dHJ) is formed (**c**). In the case of SDSA, the invading strand is released after elongation and realigns with the original break site, resulting in a NCO event, while the information is only copied to one strand (**d**). The dHJ can be resolved via HJ-processing endonucleases, inducing orthogonal orientated cuts (red triangles), resulting in a CO event (**e**). Alternatively, the dHJ can be dissolved by the formation of a hemicatenane involving a helicase followed by strand separation by a topoisomerase, resulting in a NCO event (**f**)

should add up to a minimum size of between 63 and 89 bp, but in the absence of competition SSA can utilize repeats as small as 29 bp [36]. During subsequent strand annealing, overhanging flaps are removed by an endonuclease, resulting in a deletion between the homologies. In plants, the RAD1/RAD10 heterodimer is involved in trimming these flaps before ligation [37]. It is obvious that the homologous template for this kind of repair reaction has to be located in close proximity to the break site and that DSBs in genomic regions containing such homologies are preferentially repaired via SSA. Besides the factors discussed so far, there are indications hinting to a minor involvement of the helicase FANCM and the nuclease MUS81 during SSA [38]. Given the current limited knowledge regarding the factors involved in SSA, it is likely that additional helicases and nucleases can substitute each other, hampering the detection of SSA key factors.

In the case of a DSB occurring in a genomic region lacking repetitive homologies, a homologous double-stranded template elsewhere can serve as the repair template. Following the initial MRN-mediated formation of 3′ single-stranded ends, these overhangs are able to invade into the donor molecule, forming a structure called a D-loop (Fig. 2b). The formation of such a D-loop can lead to either repair via SDSA or processing of the double Holliday junction (dHJ) as an intermediate, resulting in repair via DSBR or dissolution. A dHJ is formed when the displaced strand aligns with the original break site, so that DNA synthesis can also take place at the other broken end (Fig. 2c). This process results in a structure in which information is copied from both strands of the matrix and separation of the homologous molecules can only be achieved by resolution or dissolution of the dHJ. For resolution, specific HJ-processing endonucleases are essential which can bind and introduce symmetric cuts with opposite orientation at the junction (Fig. 2e). Resolution is mainly observed during meiosis and results in CO events leading to an exchange of chromosome arms and is required for mixing of parental genomes. Dissolution occurs via the RTR complex, composed of a RECQ helicase, type 1 topoisomerase, and a structural protein. The helicase forms the hemicatenane in which the topoisomerase cleaves one strand, enabling strand separation (Fig. 2f).

During SDSA, sequence information of the donor strand is replicated to close the gap, with the invading strand being released to realign with the original break site. Thus, the repair reaction always results in a NCO, while the information of the matrix is only copied to one strand (Fig. 2d). Strand invasion of the 3′ ends is mediated by RAD51 and its paralogs [39, 40]. Subsequently, the repair reaction is performed utilizing the complementary donor strand as a template. The binding of the RAD51-ssDNA complex to the homologous strand is stabilized by the SWI2/SNF2 class chromatin remodeler RAD54, which harbors a DNA-binding domain, mediating homology search [41, 42]. It was shown before that RAD54, as well as RAD51 and its paralogs RAD51C and XRCC3, is essential for SDSA, but not for SSA, while the nuclease MUS81 is involved in both pathways [38, 42].

The mechanistic HR models discussed above have been sustained qualitatively and quantitatively by analyzing the repair of DSBs induced in the genome of model plants. Early studies in tobacco showed that in the absence of DSBs, the occurrence of HR in somatic tissue is very low [43, 44]. First experiments to study DNA repair *in planta*, using transposons and their intrinsic property of cut-and-paste transposition, showed that these DSBs are mainly repaired via NHEJ [45]. An enhancement of HR-mediated repair was observed, however, when homologous sequences were localized close to the break site [46]. Using I-*Sce*I in tobacco protoplasts allowed Puchta et al. to conclude that homologous recombination can be significantly enhanced by targeted DSB induction in vivo [20]. Since then, recombination traps harboring one or two I-*Sce*I recognition sites have been used to study DSB repair mechanisms *in planta*.

When homologies are in close vicinity to the induced DSB, the SSA-mediated repair is dependent on the flanking homologies, resulting in the loss of the sequence between the homologous repeats. Quantification of SSA frequency is possible using a stably integrated *ß-glucuronidase* (*gus*) reporter, in which the *gus* gene harbors a segmentally duplicated region with an I-*Sce*I recognition site and the functional *gus* gene can be restored via SSA repair. This way, it could be demonstrated that one out of three DSBs is processed via SSA in regions with sequence homologies [47]. In a similar approach, Orel et al. established that SSA-mediated DSBR is approximately five times more efficient than interchromosomal SDSA in the presence of flanking homologies [48]. Very recently, Vu et al. analyzed the mutation spectra at three different endogenous loci in *Arabidopsis* by deep sequencing whereby it was found that the distance between the repeats has an influence on the absolute frequency of SSA, with the efficiency of SSA being inversely correlated to the distance between homologies [49]. Thus, SSA is a very efficient way of repairing DSBs in genomic regions with tandem repeats, depending on their distance. Naturally, tandem repeats and sequence duplications are not uniformly distributed over the genome, implying that repair mechanisms can differ due to the genomic context. Moreover, HDR is dependent on the accessibility of homologous sequences serving as a template. While dissolution and DSBR occur mainly during meiosis, with the dHJ and the synaptonemal complex formation being prerequisites for correct pairing of homologous chromosomes, it was shown by Puchta that the SDSA model is best suitable to describe DSBR in somatic plant cells [50]. In contrast to SSA, the frequency of SDSA-dependent repair is difficult to measure as several types of homologous donor sequences can be a substrate for SDSA. During the G_2 and S phases of the cell cycle, the homology of the sister chromatid can be utilized for repair via SDSA. The interaction of sister chromatids is mediated by the "structural maintenance of chromosome" genes SMC6A and SMC6B, factors that are required for DNA repair and HDR in *Arabidopsis* [42, 51]. Unfortunately, the frequency of such an occurrence cannot be quantified, as the accurately restored sequence is indistinguishable from the original sequence. Apart from the sister chromatid, there are different genomic sites, which can possibly be used as a repair template during HDR (Fig. 3). Allelic sequences on the homologous chromosome can act as a template as well as homologous sequences

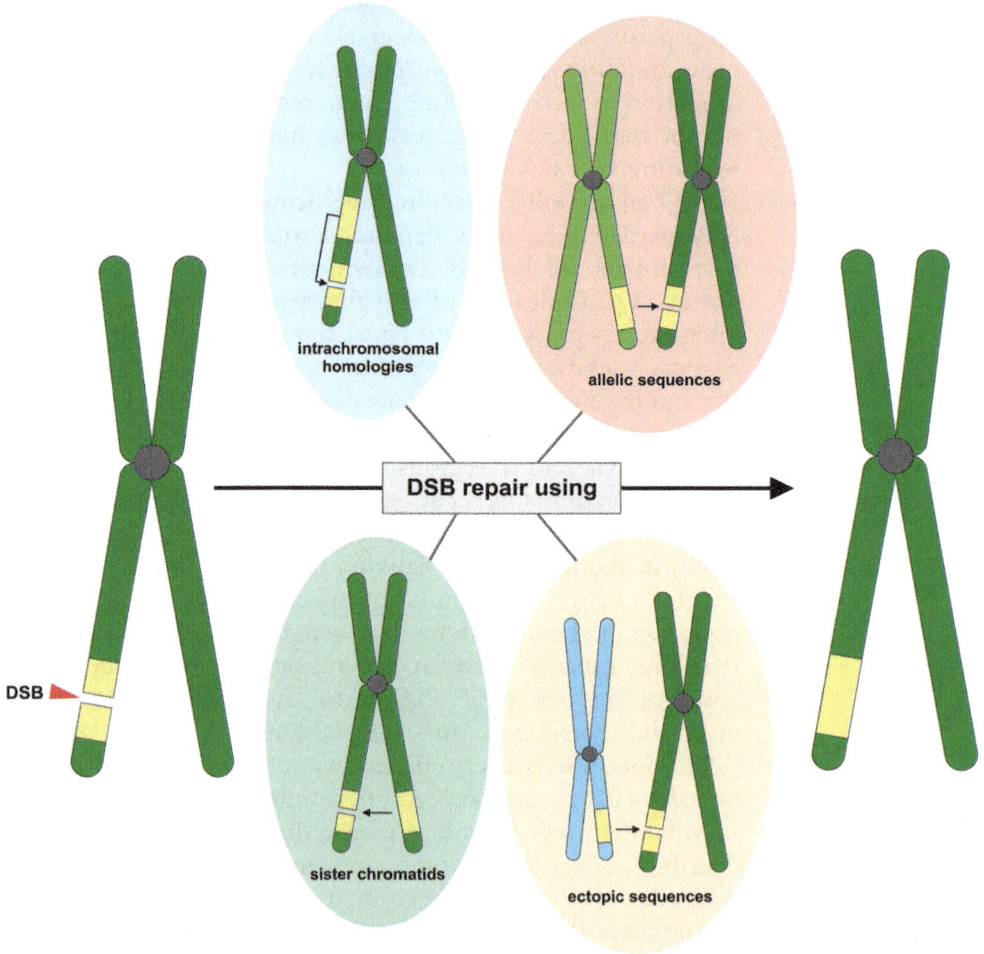

Fig. 3 Possible donor sequences for HR. Induced DSBs can be repaired using several templates for correct repair via SDSA. During G2 and S phases of the cell cycle, the sister chromatid can be used, leading to an accurate restoration of the interrupted sequence. Furthermore, intrachromosomal homologies located on the same chromosome can serve as a template for HDR, as well as allelic sequences from homologous chromosomes, primarily occurring during meiosis. Ectopic sequences located on other chromosomes or incoming ectopic DNA are also possible templates for HDR

located on the same chromosome, e.g., in the case of gene duplications or transposable elements. Additionally, ectopic sequences on different chromosomes or from extrinsic DNA are also possible templates for SDSA (for review, *see* [29]).

As discussed before, the discrimination of HDR involving the sister chromatid is not possible per se and thus relatively difficult to measure. To get an insight into how frequently the sister chromatid is used as a template during G_2 and S phases, Vu et al. displayed the sister chromatid exchange via ethynyldeoxyuridine-based staining [52]. Interestingly, it could be shown this way that most of the I-*Sce*I-induced DSBs were repaired using the sister chromatid as a template, leading to the conclusion that the frequency of this kind of repair is underestimated. Although these findings are highly interesting for fundamental research regarding DNA repair, the applications for genome engineering are negligible due to the fact that it is impossible to induce sequence changes via sister chromatid exchange.

The second most common template for DSB repair via SDSA is intrachromosomal homology. Frequencies can be measured using a nonfunctional GUS reporter with homologies on the same chromosome within close proximity, serving as a template for the restoration of the GUS marker gene. Applying this kind of reporter system, it was shown that I-*Sce*I-induced breaks can be repaired via SDSA using an intrachromosomal homology and about 1 out of 15 DSBs is repaired via this kind of HDR [48]. Besides this analysis, many studies have been published, indicating that homologous sequences on the same chromosome are an appropriate matrix for repair in plants [42, 47, 53].

During meiosis, homologies in allelic positions serve as a template for HR and it was shown that such a template can also be exploited in somatic plant cells [54]. In this study, two different transgenic lines were combined, one with an intact I-*Sce*I recognition site and the other with a disrupted I-*Sce*I site. After crossing these two lines, induction of a DSB via I-*Sce*I led to recombination events involving the allelic sequence. Besides loss-of-function mutations due to NHEJ-related repair representing the majority of events, Gisler et al. detected allelic HR frequencies of 10^{-4} [54]. Unfortunately, the transgenic lines were lacking homologies directly at the I-*Sce*I break site in this setup. Taken together, this led to the conclusion that intrachromosomal somatic HR is 100–1000-fold more efficient than gene conversion with donor sequences in ectopic or allelic positions. Very recently, a study in tomato revealed that Cas9-induced targeted DSBs are more often repaired via somatic HR, with the homologous chromosome acting as a template. In this study, an allele-specific assay was set up in *S. lycopersicum* with DSB induction occurring in a region that allows discrimination between repair footprints, originating from the homologous allele or from nonhomologous repair. In this experimental setup, 14% of all detectable DSB repair events were allele dependent and three cases could be identified in which HR-dependent repair events were transmitted through the germline [55]. In contrast to the findings of Gisler et al., the homologies were directly present at the break site, with the marker for allele-specific recombination being only a single-nucleotide polymorphism [54]. These results are promising for genome engineering approaches relying on HDR, such as the integration of an extra-chromosomal homologous template into a break site, using homologous chromosome arms.

The integration of an ectopic sequence into a break site is another possibility of repair by the use of homologous genomic sequences, especially relevant for the large plant genome with its many repetitive and widely spread genetic elements. Efficiencies for ectopic DSB-induced HR were tested in tobacco using an acceptor and a donor locus on different chromosomes. The endonuclease I-*Sce*I was transiently expressed via *Agrobacterium* in those plants, inducing a DSB in the acceptor locus. Following successful recombination involving the donor locus on a different chromosome, a kanamycin resistance cassette could be restored. Puchta established that ectopic sequences are used extremely rarely as a HR template, with frequencies in the range of 10^{-5} [56]. In a similar approach by Shalev and Levy, frequencies of 10^{-3}–10^{-5} were reported using transposons for DSB induction in *Arabidopsis* [57]. The low occurrence of such events is thought to be a prerequisite for stabilizing genetic information in somatic tissue [56] and might be the major challenge for the integration of DNA during GT. Taken together, the chromosomal site of the HR template and the occurrence of homologies close to the break site strongly influence the repair pathway choice. Furthermore, recent studies found that the genome size of barley and *Arabidopsis* correlates with the abundance of different repair outcomes, leading to the conclusion that frequencies of the entirety of repair pathways differ between plant species [52, 58]. These species-specific differences were described earlier for *Arabidopsis* and tobacco, demonstrating that there is a positive correlation between genome size and DSB processing [59, 60]. It is evident from the studies discussed here that DSB repair pathway choice and the natural prevalence of NHEJ-mediated repair in somatic plant cells pose bottlenecks for biotechnological approaches, focused on targeted genome modifications.

3.2 Gene Targeting in Plants

The GT technique is based on HDR and can be applied to integrate new genes at a specific locus or to modify the sequence of a respective gene (reviewed in [61]). GT was initially tested in tobacco protoplasts without the induction of a site-specific DSB. Targeted events could be selected by GT-mediated restoration of a nonfunctional kanamycin resistance gene with plasmids [62], as well as with incoming T-DNA or chimeric RNA/DNA

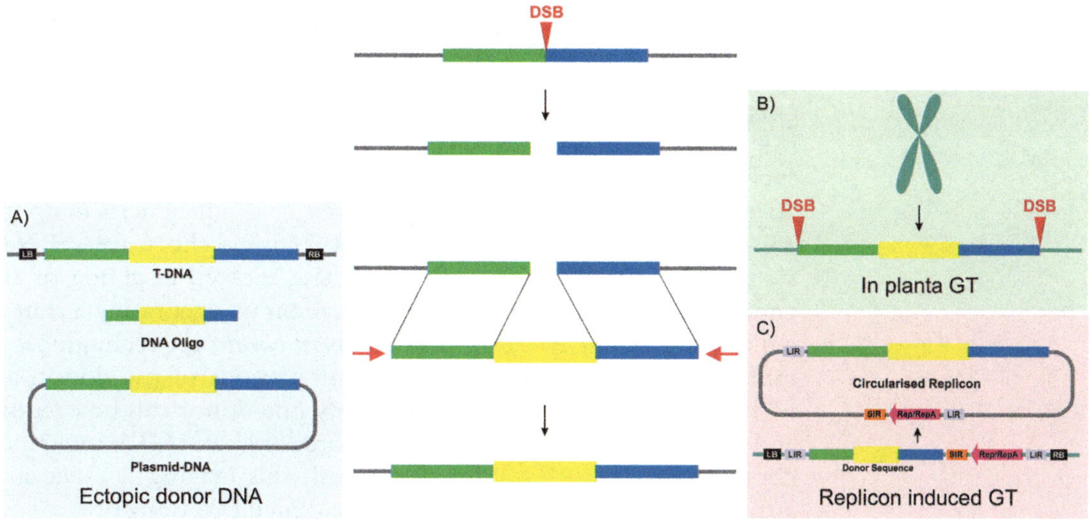

Fig. 4 Possible donor molecules for GT in plants. Following DSB induction, different donor sequences can be used for HDR. Homologies (green/blue) to the area flanking the break site need to be present for correct incorporation of the donor sequence (yellow). (**a**) Possible ectopic templates for GT are T-DNAs transferred via Agrobacterium, as well as plasmid DNA and short DNA oligos. (**b**) If the donor sequence is stably integrated into the genome, three simultaneous breaks are required to release the target sequence and simultaneously activate the donor sequence. (**c**) Enhancement of GT frequencies can be achieved by the use of geminivirus-based replicons: the transferred T-DNA consists of the donor sequence combined with a site-specific nuclease, a donor, and minimal parts necessary for geminivirus rolling-circle replication, leading to the formation of circularized replicons

oligonucleotides [63, 64] (Fig. 4a). I-*Sce*I was the first nuclease to enhance GT frequencies in tobacco by targeted DSB induction in an artificial target site integrated within the genome. The transiently expressed I-*Sce*I cassette led to the induction of a DSB and the break being repaired using a T-DNA with flanking homology stretches, leading to an increase of GT frequency by up to two orders of magnitude [65].

Besides I-*Sce*I, different synthetic nucleases have been applied to increase HR frequencies in plants, thus becoming independent of one particular recognition sequence. ZFNs were used to restore a disrupted marker gene via GT in tobacco protoplasts, reaching GT frequencies up to 10% in selected lines [66], with applications in maize, *Arabidopsis*, and tobacco cells in the following years [67–69]. Subsequently, GT experiments were performed using TALENs in tobacco protoplasts, altering the ALS gene with frequencies up to 4% [70]. A similar gene alteration frequency was successfully obtained using the Cas9-sgRNA system, combined with different donor DNAs (vector DNA, single-stranded oligos), targeting the ALS2 gene in maize protoplasts [71]. Moreover, Cas9-mediated GT approaches were also successfully applied in *Linum usitatissimum* and soybean [72, 73].

Another GT approach, referred to as "*in planta* GT," is independent of transfected donor DNA, as the donor sequence is stably integrated into the genome. For this purpose, three simultaneous cuts are required, one within the target site and two cuts to release the linear donor sequence from the genome (Fig. 4b). Applying such an *in planta* GT technique based on I-*Sce*I-mediated DSBs in *Arabidopsis*, Fauser et al. were able to reach GT efficiencies of up to 1% [74]. More recently, *in planta* GT using the Cas9-sgRNA system with an adaptable recognition site allowed targeting of an endogenous locus, without the requirement of integrating a transgene upfront [75]. An advantage of this *in planta* GT technique is a minimization of unwanted random integrations of the donor, as with the case of a hemizygous line only one donor can be excised and integrated per genome. Furthermore, *in planta* GT is independent of transformation efficiencies, with this making it especially attractive for crop plants that are more difficult to transform.

To increase the number of donor molecules, *in planta* GT can also be combined with geminivirus-based replicons (Fig. 4c). For this application, the transformed T-DNA consists of a site-specific nuclease, donor template, and the minimal parts necessary for geminivirus replication. Recently, the successful application of this technique was demonstrated using ZFNs, TALENs and Cas9, respectively [76, 77].

As HDR is a minor pathway in higher organisms, GT frequencies are relatively low. Therefore, various attempts were tested to enhance HDR by overexpression of HDR-related genes or by repression of NHEJ-related genes. It could be demonstrated in *Arabidopsis* that overexpression, as well as egg cell-specific expression of RAD54 from *S. cerevisiae*, can enhance GT efficiency [78, 79]. Furthermore, it was shown that a knockout of the SMC6B gene leads to impaired sister chromatid exchange and therefore GT efficiencies could be increased in *Arabidopsis* [80]. Besides knockout or overexpression of HDR-related proteins, different NHEJ factors are also able to influence GT. Endo et al. successfully obtained enhanced GT efficiencies by using *lig4*-deficient rice calli cells, lacking an important factor for NHEJ [81]. This indicates that besides HDR pathways, DSB repair via NHEJ plays an important role for genome engineering and deserves our attention.

3.3 NHEJ

DSBs in somatic plant cells are most frequently repaired via NHEJ, either by rejoining the broken DNA ends without or with small sequence changes. In contrast to HDR pathways, there is no need for homologous sequences during the repair process. The mechanism ensures that in the absence of homologies, a DSB can be repaired easily without significant sequence loss. The NHEJ pathway described initially is characterized by the KU70/KU80 heterodimer and is commonly referred to as canonical-NHEJ (c-NHEJ).

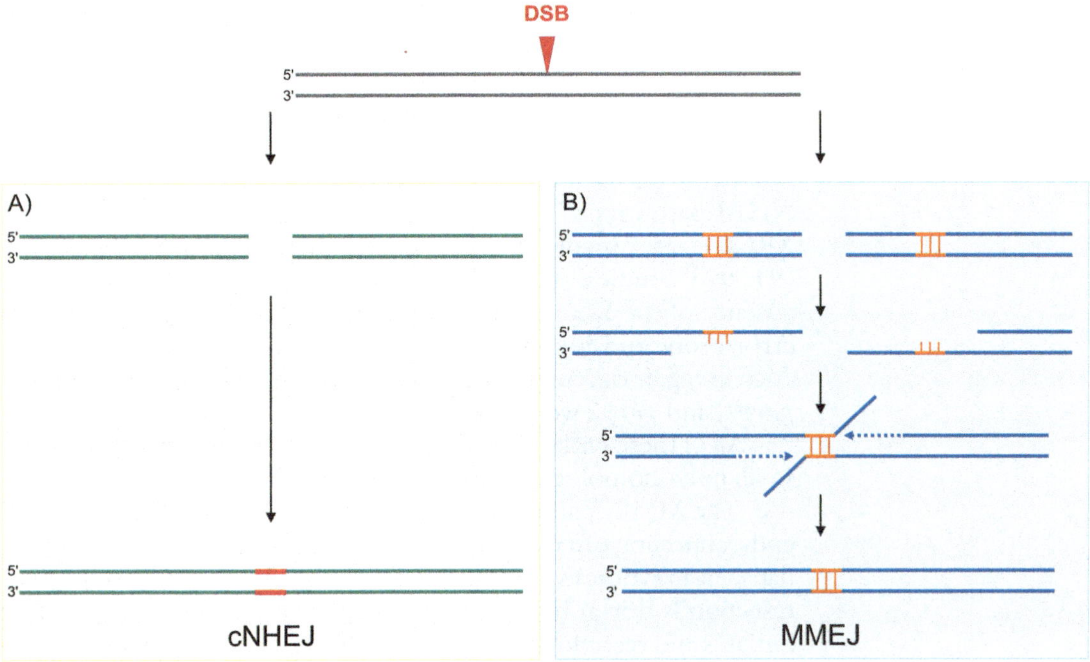

Fig. 5 DSB repair via NHEJ mechanisms. Repair of a DSB can occur via two major NHEJ pathways. (**a**) If the KU heterodimer is present, the most frequently used pathway is referred to as cNHEJ. During cNHEJ, a direct ligation of the broken ends can be performed, leading to INDELS (red) or accurate repair of the break site. (**b**) In the absence of the KU heterodimer, MMEJ takes over for DSB repair. The MMEJ pathway is dependent on microhomologies (orange) close to the break site. The first step of the MMEJ pathway is end resection, leading to 3′ overhangs. Subsequently, annealing of the microhomology region is performed, leading to 3′ flaps. The removal of heterologous flaps is followed by fill-in synthesis and finally the occurrence of a ligation reaction. MMEJ is highly error-prone as it mainly leads to deletions between microhomologies

cNHEJ results either in small insertions and/or deletions (InDels) or in an error-free ligation of the break site (Fig. 5a). The KU70 and KU80 proteins protect the free DNA ends from a significant loss of nucleotides during end resection [82]. Following binding of KU70/KU80, the cNHEJ pathway in vertebrates involves the recruitment of DNA-PKcs and Artemis, as well as X-family polymerases, followed by XLF, XRCC4, and LIG4 [83]. In plants, homologs for KU70, KU80, XRCC4, LIG4 have been described [84–86], however, no homologs for DNA-PKcs, ARTEMIS, and XLF were found. During cNHEJ, only little or no genetic information is lost; however, if only one or two nucleotides are deleted or inserted, the mutation can result in a frameshift and cause a complete knockout of gene function. The KU70/KU80 heterodimer is the key factor in cNHEJ and is supposed to repress HR in plants [87], which may explain shortening of telomeric sequences in *Arabidopsis* [88] and barley [89].

Loss-of-function mutations in *A. thaliana Ku70/80, LIG4.* and *XRCC4* genes do not cause any growth defects in the respective plants [90]. Additionally, in the absence of the KU-heterodimer, various alternatively named NHEJ pathways known as backup-NHEJ (b-NHEJ), alternative-NHEJ (a-NHEJ), and microhomology-mediated end joining (MMEJ) can take over. However, it remains unclear as to if all these designations are synonyms for each other, or if there are indeed several alternative NHEJ pathways. There are indications that PARP1/PARP2 and XRCC1 are involved in one of these alternative pathways [91, 92]. Studies in *Arabidopsis ku80* mutants suggest that in the absence of the KU heterodimer, somatic plant cells shift to a more error-prone end-joining mechanism, although no significant difference in repair outcomes was observed when triple mutants of *ku80, parp1*, and *parp2* were tested [93].

All of these backup pathways are characterized by the presence of small microhomologies, with a length of 1–20 bp, close to the break site. The MMEJ pathway consists of five steps: resection of the DSB ends, annealing of a microhomology region, removal of heterologous flaps, fill-in synthesis, and ligation [94] (Fig. 5b). The first step of end resection is driven by the MRN complex, the same complex which initiates end resection during HDR [95]. The exact annealing mechanism for sequences with microhomologies and protein factors involved in this process are not yet fully understood, but it is assumed that in mammals DNA polymerase Θ (POLΘ) plays a major role in this step [96, 97]. Subsequently, the overhanging 3′-tails are removed by a nuclease. In mammals, there are indications that MRE11 and XPF-ERCC1 are candidates for this nuclease activity [98]; however, until now there are no reports regarding the involvement of their homologs RAD1/RAD10 in plants. The following fill-in synthesis is performed by PolΘ, which seems to be the key player of MMEJ. There are several studies in higher eukaryotes supporting the prominent role of POLΘ in MMEJ [99, 100], as well as involvement in the appearance of chromosomal translocations in mice [101]. Additionally, POLΘ exhibits a RAD51-binding motif and this binding capacity may help to direct DSB repair to MMEJ and to repress the HDR pathways [102]. The plant homolog of POLΘ, TEBICHI (TEB), is involved in DNA replication, recombination, and gene expression, as well as in DNA repair [103]. *Arabidopsis teb* mutants are sensitive to MMC and MMS, indicating that TEB is involved in the repair of inter- and intra-strand cross-links [104]. Furthermore, the *Arabidopsis teb* mutant is not accessible for T-DNA integration; thus TEB is a prerequisite for T-DNA integration into the plant genome [105].

It is possible that a knockout of cNHEJ factors, like KU70 or LIG4 in plants, can lead to a shift of the repair pattern to MMEJ and therefore result in more and larger deletions. The repair patterns of different NHEJ repair mutants were analyzed in *Drosophila* and *Arabidopsis*, indicating that knockouts of KU70, LIG4, or POLΘ, for example, can influence the occurrence of deletions and insertions following DSB induction [93, 106].

3.4 Formation of Insertions

The entirety of described repair mechanisms discussed so far all result in INDELS of various sizes or an accurate restoration of the damaged DNA region. However, an explanation for larger insertions at the break site is not evident. Gorbunova and Levy obtained relatively large insertions after DSB induction in tobacco cells, with these inserted sequences originating from either internal regions of transfected plasmids or tobacco genomic DNA [107]. In 1998, Salomon and Puchta were able to demonstrate that DNA insertions can occur during DSB repair in higher eukaryotes, while repair was associated with a broad spectrum of unique and repetitive genomic sequences [108].

Based on their experiments, both groups concluded that an annealing and priming reaction seems to be involved in the formation of genomic insertions. Furthermore, it was assumed that the DSB is processed via an illegitimate recombination pathway, based on the SDSA model of recombination. During this SDSA-like mechanism, the 3′ single-stranded overhang is able to invade into a double-stranded donor and alignment takes place with only a short patch of homology. The linkage with the original break site can occur via ligation to a blunt end, with an annealing step mediated by microhomologies (Fig. 6).

NHEJ-mediated gene insertion has been already achieved with Cas9. Geisinger et al. [109] showed in human and mouse cells that precise sequence insertion is possible after DSB induction. Furthermore, Li et al. reported site-specific insertions via NHEJ after DSB induction, using the Cas9-sgRNA system in rice [110]. Targeted gene insertions were obtained at a frequency of 2.2% in this study. This recently developed approach opens up new possibilities that are promising for the field of genome engineering.

Furthermore, in 2010, Yu et al. investigated a model explaining the occurrence of direct or inverted repeats on the basis of the MMEJ pathway [106]. This synthesis-dependent MMEJ (SD-MMEJ) model assumed that POLΘ is involved in the generation of template insertions, when multiple microhomologies flank the DSB. The conclusion was drawn that direct repeats are produced by a mechanism called loop-out SD-MMEJ (Fig. 7a) and that in contrast inverted repeats can be explained by snap-back SD-MMEJ (Fig. 7b). Both processes result in the formation of a repeated motif, arising from flanking sequences on both sites. Substantial numbers of complex insertions, composed of multiple overlapping copies of nearby sequences, were also obtained in this experimental setup, leading to the assumption that multiple rounds of synthesis and dissociation can occur.

Recent findings revealed that besides DSBs, adjacent SSBs can also lead to the formation of tandem sequence duplications in plants. Schiml et al. induced SSBs on the same and opposite DNA strand, in different intervals and in different chromatic regions [111]. The data indicate that opposing SSBs are highly mutagenic

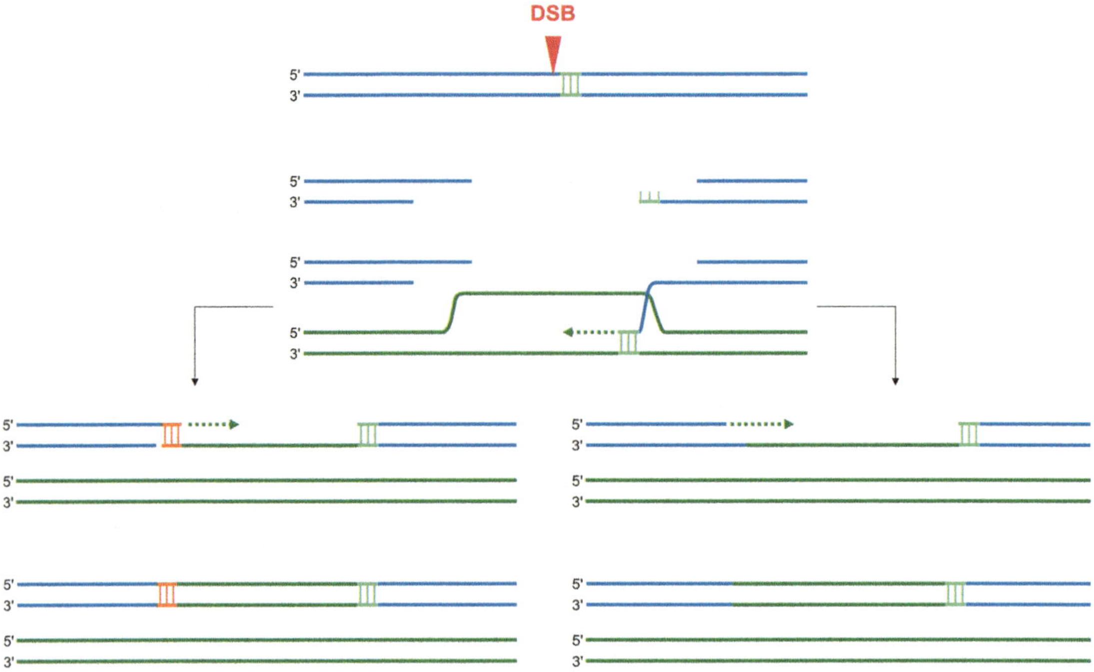

Fig. 6 SDSA-like mechanism leading to insertions. After end resection at the break site, 3′ overhangs are produced and the single strand is able to invade into the donor molecule (green). Microhomologies (light green) are a prerequisite for alignment with the donor sequence and a D-loop is formed, analogous to the SDSA mechanism. Subsequently, strand displacement can occur via a second microhomology (orange) or via blunt-end ligation. This SDSA-like repair mechanism is leading to insertions at the break site, with any genomic sequence or ectopic DNA being used as a template

and are especially suitable for the production of mutant plants. Furthermore, the tested intervals between the break sites prove that a distance of 50 bps, leading to the formation of 5′ overhangs, is suitable to generate heritable mutations. In this approach, both deletions and insertions were obtained with frequencies up to 77% and it was shown that deletion sizes correlate with SSB distance. Besides repair outcomes related to cNHEJ, deletions with flanking microhomologies were obtained, indicating that these mutations were formed via MMEJ-mediated repair.

The observed insertions originated directly from adjacent sequences or from areas in close proximity, and might be explained by a patch-mediated model for tandem duplications, based on simultaneous base excision repair [112]. During this kind of repair, single-stranded 5′ ends can hybridize via microhomologies and the internal gaps are repaired by fill-in synthesis, while 5′ flaps are removed (Fig. 8a). Schiml et al. assumed that in the absence of microhomologies at the break site, synthesis from the 3′ end and simultaneous degradation of the 5′ end, followed by ligation via cNHEJ, can lead to duplicated sequences (Fig. 8b) [111].

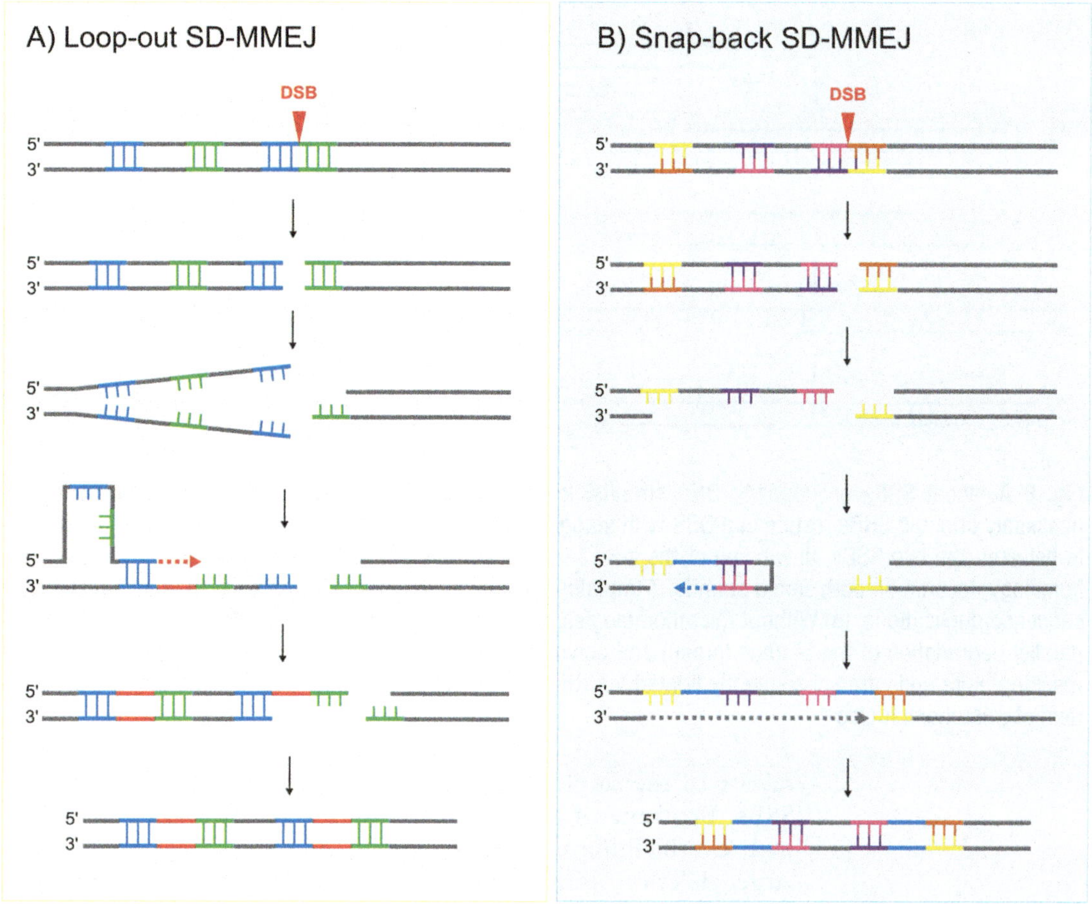

Fig. 7 Mechanisms leading to insertions via SD-MMEJ. SD MMEJ is leading to direct or inverted repeats at the site of a DSB. (**a**) For an insertion via loop-out SD-MMEJ, two pairs of short direct repeats are required, with the first pair being located on one break side (blue) and the second pair spanning the DSB (green). Initially, the two complementary strands are separated and the blue homology close to the break site aligns with the second one distal to the DSB, forming a loop-out structure. Next, a template-dependent DNA polymerase is required for elongation and the newly synthesized part forms the insertion. The loop-out structure is then dissolved and the replicated microhomology (green) aligns with the second end of the DSB. (**b**) If an insertion is created by snap-back SD-MMEJ, the DSB should be flanked by two pairs of short inverted repeats. The first pair of microhomologies spans the DSB (yellow/orange), while the second pair is oriented in parallel (pink/purple). The strand separation occurs via end resection or unwinding of the DNA, and hereafter the second pair of homologies form a hairpin. Analogous to loop-out SD-MMEJ DNA, elongation is performed using a template-dependent DNA polymerase and the newly synthesized part forms the insertion. The replicated microhomology (orange) realigns with the other break site, resulting in an inverted repeat

In contrast, the induction of two paired SSBs on one DNA strand led to relatively low mutation frequencies below 1%. Nevertheless, deletions were obtained correlating with the SSB position and additionally insertions originating directly from adjacent sequences or from areas in close proximity (20–50 bp). The formation of deletions and inversions can possibly be associated with the

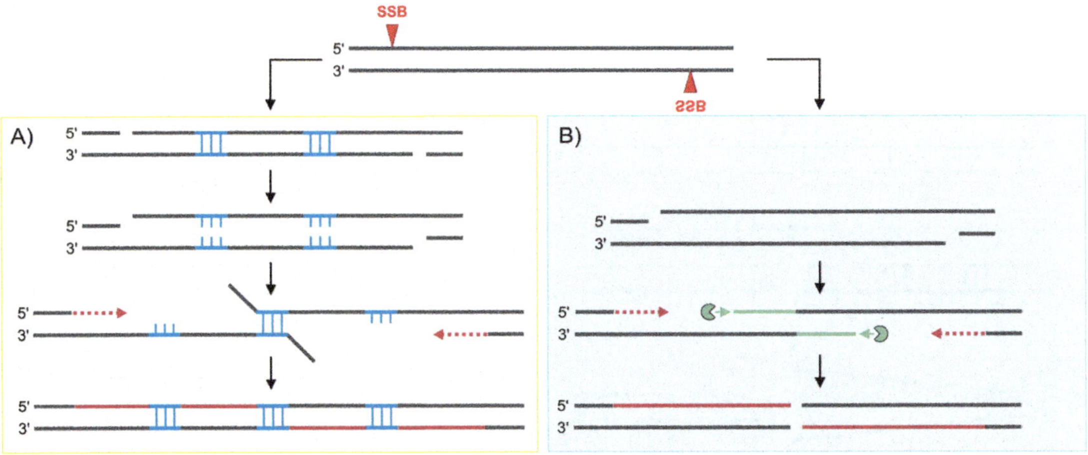

Fig. 8 Adjacent SSBs on opposite DNA strands, leading to sequence duplications. The induction of two oppositely oriented SSBs results in a DSB with staggered ends. (**a**) When microhomologies (blue) are present in between the two SSBs, unwinding of the complementary strands is followed by realignment of the distal homology, located on both single strands. Template-dependent elongation (red) fills up the gaps, resulting in sequence duplications. (**b**) Without microhomologies, unwinding of the single-stranded break ends allows for parallel degradation of the 5′ ends (green) and polymerase-dependent synthesis from the 3′ ends (red). The resulting blunt ends are subsequently ligated via NHEJ, resulting in sequence duplications that are independent of microhomologies

release of the single-stranded DNA fragment in between the two SSBs. Breakage of this single-stranded gap can end up in a DSB, with overhanging ends that need to be processed, resulting in a larger deletion. Furthermore, the formation of insertions may be supported by replication slippage. It can be assumed that adjacent SSBs on opposing strands are highly mutagenic, while adjacent SSBs on the same strand are only slightly mutagenic, and both might be relevant for plant genome evolution, when occurring naturally.

Plant genome sizes vary between different species, but the related molecular mechanisms of genome size evolution are still under discussion. It is obvious that polyploidization and transposable element proliferation can lead to increased genome size [113, 114]. Genome shrinkage can be due to a dysploid reduction of chromosome number, a loss of entire chromosomes, as well as transposon-mediated excision, replication slippage, and illegitimate recombination [115, 116]. Besides these factors, recent data suggest that inaccurate DSB repair may have an underestimated role in genome size evolution. Initially, it was shown that expansion through retrotransposon proliferation can be compensated by an increase in DNA removal during DSB repair, in small-sized genomes [117]. Thereafter, species-specific differences between *Arabidopsis thaliana* and *Nicotiana tabacum* were observed for deletions occurring during the classical NHEJ pathway, and Kirik et al. demonstrated a positive correlation between genome size and DSB processing via NHEJ [59]. In the much larger tobacco genome, deletions were about 30% smaller than in *Arabidopsis* and were accompanied, in 50% of the cases, by the insertion of filler sequences. The larger deletions in the latter species rarely coincided with insertions of only a few base pairs. Such differences are believed to originate from different exonucleolytic degradation rates of free DSB ends before downstream processing, thus shaping plant genomes during evolutionary timescales [60, 118]. In addition, the data of Vu et al. showed significant differences in DSB repair outcomes between *Arabidopsis* and barley [58]. Larger deletions were obtained on average for *Arabidopsis* and the proportion of mutated reads, containing a large deletion, was significantly higher in *Arabidopsis* than in barley. Furthermore, Vu et al. found strong distinctions between the proportion of insertions [58]. In *Arabidopsis*, only 5.3% of imprecisely repaired sequences showed insertions, as compared to 22.9% insertions in barley. Therefore, inaccurate DSB repair and repair of adjacent SSBs can be a significant source of genome variation and is involved in eukaryotic genome shaping.

4 Genomic Modifications

Combining the knowledge about DNA repair and the specific induction of DSBs using programmable site-specific nucleases provides a powerful tool for successful genome engineering. Synthetic nucleases are able to induce one or more DSBs at defined loci, predominantly repaired via NHEJ. As NHEJ is mostly error-prone, this allows for performing highly specific gene knockouts. Unfortunately, the outcome of NHEJ mediated repair can hardly be predicted. A study in human cell culture analyzing DNA repair profiles of hundreds of target sites of Cas9 indicates that repair outcomes are rather determined by the protospacer sequence than the genomic context [119]. In contrast, Vu et al. analyzed the mutation spectra after break induction via Cas9, at three selected endogenous loci, and concluded that the sequence context in close vicinity to the break has an impact on the used repair pathway and thus on repair outcomes [49]. Therefore, it remains to be elucidated as to how DSBs are repaired in a specific sequence context around the break site and how these different repair outcomes can be explained.

To bring matters to a head in the field of genome engineering, synthetic nucleases, and especially the Cas9/Cpf1 endonucleases, allow for the induction of more than one break, simultaneously. Therefore, a combination of the respective Cas9/Cpf1 nucleases with multiple spacer sequences and their suitable (s)gRNA sequences is needed for multiplexing [120–124]. Usually, the repair of more than one DSB at a time is catalyzed by mechanisms that keep broken ends in close proximity, to reduce the risk of chromosomal aberrations [125]. As these mechanisms are not always accurate, it is possible that two originally not-associated DNA strands align, resulting in deletions, inversions, or translocations. Whole-genome alignments of the two *Arabidopsis* ecotypes, *Columbia* (Col-0) and *Landsberg erecta* (Ler-1), showed that there are several inversions and inter- and intrachromosomal transpositions, which occurred naturally in those *Brassicaceae* genomes [126]. For this reason, it is possible that a low number of break sites results in the alignment of unequal strands when several breaks are induced within the genome. When two DSBs are induced within close proximity, in plants, the part between the two breaks can be deleted from the genome [47, 127–129] (Fig. 9a). In particular, deletions smaller than 100 bp can be induced easily; however it was also shown that deletions up to 245 kb can occur at a low frequency in *Arabidopsis* and rice [130, 131]. This allows for the deletion of whole genes as opposed to mutating only single bases or functional domains. To achieve a controlled deletion, different repair reactions need to be considered. The direct way is to join the broken ends via cNHEJ, leading to a more or less precise elimination of the part in between. Alternatively, when (micro-) homologies are present, the alignment of the break sites is mediated by MMEJ or SSA repair factors; thus it is possible to obtain additional deletions at the junction site. Besides the targeted deletion, two break sites can provide a basis for a flip of the flanked fragment and the integration in an inverted orientation (Fig. 9b). In *Arabidopsis*, there is a 2.1 Mbp heterochromatic Knob (hk4S) region and

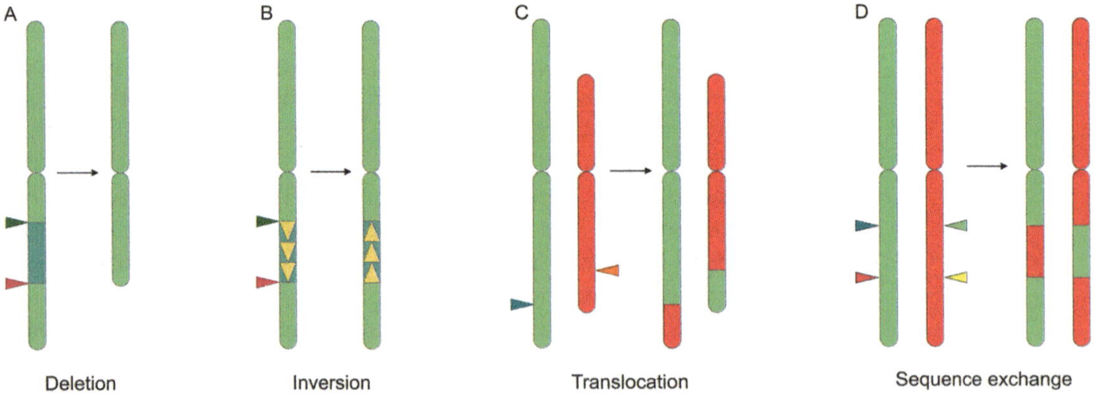

Fig. 9 Induction of genomic rearrangements using synthetic nucleases. Applications of genome engineering by the induction of several breaks at a time can lead to the formation of chromosomal rearrangements. When two breaks are induced in relative close proximity on the same chromosome arm, a deletion (**a**) or inversion (**b**) of the internal part can occur. (**c**) Furthermore, break induction can take place on different chromosomes, leading to an exchange of chromosome parts. (**d**) In the case of a simultaneous induction of four DSBs (two on each chromosome), an exchange of sequences between different chromosomes is feasible

it is postulated that this originates from an inversion event that moved DNA sequences from the pericentromere to a more distal position in the euchromatin [132]. Targeted inversions induced by ZFNs and Cas9 were shown in mammalian cells, as well as in *Arabidopsis* [133–135]. Furthermore, besides break induction on the same chromosome arm, it is possible to induce several breaks on different chromosome arms, to exchange sequences within a plant genome (Fig. 9c). Formation of a DSB-induced translocation has been demonstrated in tobacco, using the homing endonuclease I-*Sce*I, in which two previously unlinked broken ends were rejoined via NHEJ or SSA [136]. Interestingly, there is a study in mammalian cell lines showing an increase in translocation frequency when cells are lacking KU70, and being even higher if KU70 and POLΘ are mutated [96]. The induction of at least four breaks allows for the exchange of chromosomal segments. For this purpose, in each case, two DSBs should be induced on both ends containing the target sequence to be exchanged (Fig. 9d). Gene replacements, involving an incoming T-DNA, were performed using ZFNs in plant cells [137]. There are several studies investigating gene replacements with the aid of homology arms [138–140]; however, this technology is even more attractive if this can occur by NHEJ and thus without the use of existing homologies as shown for rice [110].

Taken together, the current understanding of DSB repair in plants, combined with the tremendous possibilities provided by the increasingly available site-specific and adaptable nucleases, allows for unprecedented progress in plant genome engineering and thus provides the tool for a new green revolution, with significant progress in plant breeding technologies. The following book chapters will provide a more in-depth insight into the molecular tools to be applied in state-of-the-art genome engineering approaches.

References

1. Watanabe K, Yamada N, Takeuchi Y (2006) Oxidative DNA damage in cucumber cotyledons irradiated with ultraviolet light. J Plant Res 119(3):239–246. https://doi.org/10.1007/s10265-006-0266-2
2. Sharma P, Jha AB, Dubey RS et al (2012) Reactive oxygen species, oxidative damage, and antioxidative defense mechanism in plants under stressful conditions. J Bot 2012 (1):1–26. https://doi.org/10.1155/2012/217037
3. Sikora P, Chawade A, Larsson M et al (2011) Mutagenesis as a tool in plant genetics, functional genomics, and breeding. Int J Plant Genomics 2011:314829. https://doi.org/10.1155/2011/314829
4. Chen L, Hao L, Parry MAJ et al (2014) Progress in TILLING as a tool for functional genomics and improvement of crops. J Integr Plant Biol 56(5):425–443. https://doi.org/10.1111/jipb.12192
5. Till BJ, Reynolds SH, Greene EA et al (2003) Large-scale discovery of induced point mutations with high-throughput TILLING. Genome Res 13(3):524–530. https://doi.org/10.1101/gr.977903
6. Stingele J, Jentsch S (2015) DNA-protein crosslink repair. Nat Rev Mol Cell Biol 16 (8):455–460. https://doi.org/10.1038/nrm4015
7. Schärer OD (2003) Chemistry and biology of DNA repair. Angew Chem Int Ed Engl 42

(26):2946–2974. https://doi.org/10.1002/anie.200200523

8. Stadler LJ (1928) Genetic effects of X-rays in maize. Proc Natl Acad Sci U S A 14(1):69–75
9. Muller HJ (1927) Artificial transmutation of the gene. Science 66(1699):84–87. https://doi.org/10.1126/science.66.1699.84
10. Nakagawa H (2009) Induced mutations in plant breeding and biological researches in Japan. Rome: 51
11. West CE, Waterworth WM, Story GW et al (2002) Disruption of the Arabidopsis AtKu80 gene demonstrates an essential role for AtKu80 protein in efficient repair of DNA double-strand breaks in vivo. Plant J 31 (4):517–528
12. Manova V, Gecheff K, Stoilov L (2006) Efficient repair of bleomycin-induced double-strand breaks in barley ribosomal genes. Mutat Res 601(1–2):179–190. https://doi.org/10.1016/j.mrfmmm.2006.07.004
13. Mannuss A, Trapp O, Puchta H (2012) Gene regulation in response to DNA damage. Biochim Biophys Acta 1819(2):154–165. https://doi.org/10.1016/j.bbagrm.2011.08.003
14. Alonso JM, Stepanova AN, Leisse TJ et al (2003) Genome-wide insertional mutagenesis of Arabidopsis thaliana. Science 301 (5633):653–657. https://doi.org/10.1126/science.1086391
15. Schiml S, Puchta H (2016) Revolutionizing plant biology: multiple ways of genome engineering by CRISPR/Cas. Plant Methods 12:8. https://doi.org/10.1186/s13007-016-0103-0
16. Jacquier A, Dujon B (1985) An intron-encoded protein is active in a gene conversion process that spreads an intron into a mitochondrial gene. Cell 41(2):383–394
17. Perrin A, Buckle M, Dujon B (1993) Asymmetrical recognition and activity of the I-SceI endonuclease on its site and on intron-exon junctions. EMBO J 12(7):2939–2947
18. Monteilhet C, Perrin A, Thierry A et al (1990) Purification and characterization of the in vitro activity of I-Sce I, a novel and highly specific endonuclease encoded by a group I intron. Nucleic Acids Res 18 (6):1407–1413
19. Moure CM, Gimble FS, Quiocho FA (2003) The crystal structure of the gene targeting homing endonuclease I-SceI reveals the origins of its target site specificity. J Mol Biol 334 (4):685–695
20. Puchta H, Dujon B, Hohn B (1993) Homologous recombination in plant cells is enhanced by in vivo induction of double strand breaks into DNA by a site-specific endonuclease. Nucleic Acids Res 21 (22):5034–5040
21. Kim YG, Cha J, Chandrasegaran S (1996) Hybrid restriction enzymes: zinc finger fusions to Fok I cleavage domain. Proc Natl Acad Sci U S A 93(3):1156–1160
22. Voytas DF (2013) Plant genome engineering with sequence-specific nucleases. Annu Rev Plant Biol 64:327–350. https://doi.org/10.1146/annurev-arplant-042811-105552
23. Wiedenheft B, Sternberg SH, Doudna JA (2012) RNA-guided genetic silencing systems in bacteria and archaea. Nature 482 (7385):331–338. https://doi.org/10.1038/nature10886
24. Jinek M, Chylinski K, Fonfara I et al (2012) A programmable dual-RNA-guided DNA endonuclease in adaptive bacterial immunity. Science 337(6096):816–821. https://doi.org/10.1126/science.1225829
25. Pacher M, Puchta H (2017) From classical mutagenesis to nuclease-based breeding - directing natural DNA repair for a natural end-product. Plant J 90(4):819–833. https://doi.org/10.1111/tpj.13469
26. Zhang K, Raboanatahiry N, Zhu B et al (2017) Progress in genome editing technology and its application in plants. Front Plant Sci 8:177. https://doi.org/10.3389/fpls.2017.00177
27. Puchta H (2017) Applying CRISPR/Cas for genome engineering in plants: the best is yet to come. Curr Opin Plant Biol 36:1–8. https://doi.org/10.1016/j.pbi.2016.11.011
28. Gorbunova V, Levy AA (1999) How plants make ends meet: DNA double-strand break repair. Trends Plant Sci 4(7):263–269
29. Puchta H (2005) The repair of double-strand breaks in plants: mechanisms and consequences for genome evolution. J Exp Bot 56 (409):1–14. https://doi.org/10.1093/jxb/eri025
30. Szostak JW, Orr-Weaver TL, Rothstein RJ et al (1983) The double-strand-break repair model for recombination. Cell 33(1):25–35
31. Osman K, Higgins JD, Sanchez-Moran E et al (2011) Pathways to meiotic recombination in Arabidopsis thaliana. New Phytol 190 (3):523–544. https://doi.org/10.1111/j.1469-8137.2011.03665.x
32. Eschbach V, Kobbe D (2014) Different replication protein A complexes of Arabidopsis thaliana have different DNA-binding properties as a function of heterotrimer composition.

Plant Cell Physiol 55(8):1460–1472. https://doi.org/10.1093/pcp/pcu076

33. van Dyck E, Stasiak AZ, Stasiak A et al (2001) Visualization of recombination intermediates produced by RAD52-mediated single-strand annealing. EMBO Rep 2(10):905–909. https://doi.org/10.1093/embo-reports/kve201
34. Samach A, Melamed-Bessudo C, Avivi-Ragolski N et al (2011) Identification of plant RAD52 homologs and characterization of the Arabidopsis thaliana RAD52-like genes. Plant Cell 23(12):4266–4279. https://doi.org/10.1105/tpc.111.091744
35. Serra H, Da Ines O, Degroote F et al (2013) Roles of XRCC2, RAD51B and RAD51D in RAD51-independent SSA recombination. PLoS Genet 9(11):e1003971. https://doi.org/10.1371/journal.pgen.1003971
36. Sugawara N, Ira G, Haber JE (2000) DNA length dependence of the single-strand annealing pathway and the role of Saccharomyces cerevisiae RAD59 in double-strand break repair. Mol Cell Biol 20(14):5300–5309
37. Dubest S, Gallego ME, White CI (2002) Role of the AtRad1p endonuclease in homologous recombination in plants. EMBO Rep 3(11):1049–1054. https://doi.org/10.1093/embo-reports/kvf211
38. Mannuss A, Dukowic-Schulze S, Suer S et al (2010) RAD5A, RECQ4A, and MUS81 have specific functions in homologous recombination and define different pathways of DNA repair in Arabidopsis thaliana. Plant Cell 22(10):3318–3330. https://doi.org/10.1105/tpc.110.078568
39. Bleuyard J-Y, Gallego ME, Savigny F et al (2005) Differing requirements for the Arabidopsis Rad51 paralogs in meiosis and DNA repair. Plant J 41(4):533–545. https://doi.org/10.1111/j.1365-313X.2004.02318.x
40. Charlot F, Chelysheva L, Kamisugi Y et al (2014) RAD51B plays an essential role during somatic and meiotic recombination in Physcomitrella. Nucleic Acids Res 42(19):11965–11978. https://doi.org/10.1093/nar/gku890
41. Osakabe K, Abe K, Yoshioka T et al (2006) Isolation and characterization of the RAD54 gene from Arabidopsis thaliana. Plant J 48(6):827–842. https://doi.org/10.1111/j.1365-313X.2006.02927.x
42. Roth N, Klimesch J, Dukowic-Schulze S et al (2012) The requirement for recombination factors differs considerably between different pathways of homologous double-strand break repair in somatic plant cells. Plant J 72(5):781–790. https://doi.org/10.1111/j.1365-313X.2012.05119.x
43. Dulieu HL (1975) Somatic variations on a yellow mutant in Nicotianatabacum L. (a1+/a1a2+/a2) II. Reciprocal genetic events occurring in leaf cells. Mutat Res Fundam Mol Mech Mutagen 28(1):69–77. https://doi.org/10.1016/0027-5107(75)90316-4
44. Carlson PS (1974) Mitotic crossing-over in a higher plant. Genet Res 24(1):109–112. https://doi.org/10.1017/S0016672300015123
45. Rinehart TA, Dean C, Weil CF (1997) Comparative analysis of non-random DNA repair following Ac transposon excision in maize and Arabidopsis. Plant J 12(6):1419–1427
46. Athma P, Peterson T (1991) Ac induces homologous recombination at the maize P locus. Genetics 128(1):163–173
47. Siebert R, Puchta H (2002) Efficient repair of genomic double-strand breaks by homologous recombination between directly repeated sequences in the plant genome. Plant Cell 14(5):1121–1131
48. Orel N, Kyryk A, Puchta H (2003) Different pathways of homologous recombination are used for the repair of double-strand breaks within tandemly arranged sequences in the plant genome. Plant J 35(5):604–612
49. Vu GTH, Cao HX, Fauser F et al (2017) Endogenous sequence patterns predispose the repair modes of CRISPR/Cas9-induced DNA double strand breaks in Arabidopsis thaliana. Plant J 92(1):57–67. https://doi.org/10.1111/tpj.13634
50. Puchta H (1998) Repair of genomic double-strand breaks in somatic plant cells by one-sided invasion of homologous sequences. Plant J 13(3):331–339. https://doi.org/10.1046/j.1365-313X.1998.00035.x
51. Watanabe K, Pacher M, Dukowic S et al (2009) The structural maintenance of chromosomes 5/6 complex promotes sister chromatid alignment and homologous recombination after DNA damage in Arabidopsis thaliana. Plant Cell 21(9):2688–2699. https://doi.org/10.1105/tpc.108.060525
52. Vu GTH, Cao HX, Watanabe K et al (2014) Repair of site-specific DNA double-strand breaks in barley occurs via diverse pathways primarily involving the sister chromatid. Plant Cell 26(5):2156–2167. https://doi.org/10.1105/tpc.114.126607
53. Chiurazzi M, Ray A, Viret JF et al (1996) Enhancement of somatic intrachromosomal homologous recombination in Arabidopsis

by the HO endonuclease. Plant Cell 8 (11):2057–2066. https://doi.org/10.1105/tpc.8.11.2057

54. Gisler B, Salomon S, Puchta H (2002) The role of double-strand break-induced allelic homologous recombination in somatic plant cells. Plant J 32(3):277–284
55. Filler Hayut S, Melamed Bessudo C, Levy AA (2017) Targeted recombination between homologous chromosomes for precise breeding in tomato. Nat Commun 8:15605. https://doi.org/10.1038/ncomms15605
56. Puchta H (1999) DSB-induced recombination between ectopic homologous sequences in somatic plant cells. Genetics 152:1173–1181
57. Shalev G, Levy AA (1997) The maize transposable element Ac induces recombination between the donor site and an homologous ectopic sequence. Genetics 146 (3):1143–1151
58. Vu GTH, Cao HX, Reiss B et al (2017) Deletion-bias in DNA double-strand break repair differentially contributes to plant genome shrinkage. New Phytol 214 (4):1712–1721. https://doi.org/10.1111/nph.14490
59. Kirik A, Salomon S, Puchta H (2000) Species-specific double-strand break repair and genome evolution in plants. EMBO J 19 (20):5562–5566. https://doi.org/10.1093/emboj/19.20.5562
60. Orel N, Puchta H (2003) Differences in the processing of DNA ends in Arabidopsis thaliana and tobacco: possible implications for genome evolution. Plant Mol Biol 51 (4):523–531
61. Steinert J, Schiml S, Puchta H (2016) Homology-based double-strand break-induced genome engineering in plants. Plant Cell Rep 35(7):1429–1438. https://doi.org/10.1007/s00299-016-1981-3
62. Paszkowski J, Baur M, Bogucki A et al (1988) Gene targeting in plants. EMBO J 7 (13):4021–4026
63. Offringa R, de Groot MJ, Haagsman HJ et al (1990) Extrachromosomal homologous recombination and gene targeting in plant cells after Agrobacterium mediated transformation. EMBO J 9(10):3077–3084
64. Beetham PR, Kipp PB, Sawycky XL et al (1999) A tool for functional plant genomics: chimeric RNA/DNA oligonucleotides cause in vivo gene-specific mutations. Proc Natl Acad Sci U S A 96(15):8774–8778
65. Puchta H, Dujon B, Hohn B (1996) Two different but related mechanisms are used in plants for the repair of genomic double-strand breaks by homologous recombination. Proc Natl Acad Sci U S A 93(10):5055–5060
66. Wright DA, Townsend JA, Winfrey RJ et al (2005) High-frequency homologous recombination in plants mediated by zinc-finger nucleases. Plant J 44(4):693–705. https://doi.org/10.1111/j.1365-313X.2005.02551.x
67. Shukla VK, Doyon Y, Miller JC et al (2009) Precise genome modification in the crop species Zea mays using zinc-finger nucleases. Nature 459(7245):437–441. https://doi.org/10.1038/nature07992
68. Townsend JA, Wright DA, Winfrey RJ et al (2009) High-frequency modification of plant genes using engineered zinc-finger nucleases. Nature 459(7245):442–445. https://doi.org/10.1038/nature07845
69. de Pater S, Pinas JE, Hooykaas PJJ et al (2013) ZFN-mediated gene targeting of the Arabidopsis protoporphyrinogen oxidase gene through Agrobacterium-mediated floral dip transformation. Plant Biotechnol J 11 (4):510–515. https://doi.org/10.1111/pbi.12040
70. Zhang Y, Zhang F, Li X et al (2013) Transcription activator-like effector nucleases enable efficient plant genome engineering. Plant Physiol 161(1):20–27. https://doi.org/10.1104/pp.112.205179
71. Svitashev S, Young JK, Schwartz C et al (2015) Targeted mutagenesis, precise gene editing, and site-specific gene insertion in maize using Cas9 and guide RNA. Plant Physiol 169(2):931–945. https://doi.org/10.1104/pp.15.00793
72. Sauer NJ, Narváez-Vásquez J, Mozoruk J et al (2016) Oligonucleotide-mediated genome editing provides precision and function to engineered nucleases and antibiotics in plants. Plant Physiol 170(4):1917–1928. https://doi.org/10.1104/pp.15.01696
73. Li Z, Liu Z-B, Xing A et al (2015) Cas9-guide RNA directed genome editing in soybean. Plant Physiol 169(2):960–970. https://doi.org/10.1104/pp.15.00783
74. Fauser F, Roth N, Pacher M et al (2012) In planta gene targeting. Proc Natl Acad Sci U S A 109(19):7535–7540. https://doi.org/10.1073/pnas.1202191109
75. Schiml S, Fauser F, Puchta H (2014) The CRISPR/Cas system can be used as nuclease for in planta gene targeting and as paired nickases for directed mutagenesis in Arabidopsis resulting in heritable progeny. Plant J

80(6):1139–1150. https://doi.org/10.1111/tpj.12704

76. Baltes NJ, Gil-Humanes J, Cermak T et al (2014) DNA replicons for plant genome engineering. Plant Cell 26(1):151–163. https://doi.org/10.1105/tpc.113.119792
77. Čermák T, Baltes NJ, Čegan R et al (2015) High-frequency, precise modification of the tomato genome. Genome Biol 16:232. https://doi.org/10.1186/s13059-015-0796-9
78. Even-Faitelson L, Samach A, Melamed-Bessudo C et al (2011) Localized egg-cell expression of effector proteins for targeted modification of the Arabidopsis genome. Plant J 68(5):929–937. https://doi.org/10.1111/j.1365-313X.2011.04741.x
79. Shaked H, Melamed-Bessudo C, Levy AA (2005) High-frequency gene targeting in Arabidopsis plants expressing the yeast RAD54 gene. Proc Natl Acad Sci U S A 102 (34):12265–12269. https://doi.org/10.1073/pnas.0502601102
80. Qi Y, Zhang Y, Zhang F et al (2013) Increasing frequencies of site-specific mutagenesis and gene targeting in Arabidopsis by manipulating DNA repair pathways. Genome Res 23(3):547–554. https://doi.org/10.1101/gr.145557.112
81. Endo M, Mikami M, Toki S (2016) Biallelic gene targeting in rice. Plant Physiol 170 (2):667–677. https://doi.org/10.1104/pp.15.01663
82. Mimitou EP, Symington LS (2010) Ku prevents Exo1 and Sgs1-dependent resection of DNA ends in the absence of a functional MRX complex or Sae2. EMBO J 29 (19):3358–3369. https://doi.org/10.1038/emboj.2010.193
83. Lieber MR (2010) The mechanism of double-strand DNA break repair by the nonhomologous DNA end-joining pathway. Annu Rev Biochem 79:181–211. https://doi.org/10.1146/annurev.biochem.052308.093131
84. Tamura K, Adachi Y, Chiba K et al (2002) Identification of Ku70 and Ku80 homologues in Arabidopsis thaliana: evidence for a role in the repair of DNA double-strand breaks. Plant J 29(6):771–781
85. West CE, Waterworth WM, Jiang Q et al (2000) Arabidopsis DNA ligase IV is induced by gamma-irradiation and interacts with an Arabidopsis homologue of the double strand break repair protein XRCC4. Plant J 24 (1):67–78
86. Friesner J, Britt AB (2003) Ku80- and DNA ligase IV-deficient plants are sensitive to ionizing radiation and defective in T-DNA integration. Plant J 34(4):427–440
87. Nishizawa-Yokoi A, Nonaka S, Saika H et al (2012) Suppression of Ku70/80 or Lig4 leads to decreased stable transformation and enhanced homologous recombination in rice. New Phytol 196(4):1048–1059. https://doi.org/10.1111/j.1469-8137.2012.04350.x
88. Gallego ME, Jalut N, White CI (2003) Telomerase dependence of telomere lengthening in Ku80 mutant Arabidopsis. Plant Cell 15 (3):782–789
89. Stolarek M, Gruszka D, Braszewska-Zalewska A et al (2015) Functional analysis of the new barley gene HvKu80 indicates that it plays a key role in double-strand DNA break repair and telomere length regulation. Mutagenesis 30(6):785–797. https://doi.org/10.1093/mutage/gev033
90. Bleuyard J-Y, Gallego ME, White CI (2006) Recent advances in understanding of the DNA double-strand break repair machinery of plants. DNA Repair (Amst) 5(1):1–12. https://doi.org/10.1016/j.dnarep.2005.08.017
91. Charbonnel C, Gallego ME, White CI (2010) Xrcc1-dependent and Ku-dependent DNA double-strand break repair kinetics in Arabidopsis plants. Plant J 64(2):280–290. https://doi.org/10.1111/j.1365-313X.2010.04331.x
92. Jia Q, den Dulk-Ras A, Shen H et al (2013) Poly(ADP-ribose)polymerases are involved in microhomology mediated back-up non-homologous end joining in Arabidopsis thaliana. Plant Mol Biol 82(4–5):339–351. https://doi.org/10.1007/s11103-013-0065-9
93. Shen H, Strunks GD, Klemann BJPM et al (2017) CRISPR/Cas9-induced double-strand break repair in arabidopsis nonhomologous end-joining mutants. G3 (Bethesda) 7 (1):193–202. https://doi.org/10.1534/g3.116.035204
94. Wang H, Xu X (2017) Microhomology-mediated end joining: new players join the team. Cell Biosci 7:6. https://doi.org/10.1186/s13578-017-0136-8
95. Truong LN, Li Y, Shi LZ et al (2013) Microhomology-mediated end joining and homologous recombination share the initial end resection step to repair DNA double-strand breaks in mammalian cells. Proc Natl Acad Sci U S A 110(19):7720–7725. https://doi.org/10.1073/pnas.1213431110
96. Wyatt DW, Feng W, Conlin MP et al (2016) Essential roles for polymerase θ-mediated end joining in the repair of chromosome breaks.

Mol Cell 63(4):662–673. https://doi.org/10.1016/j.molcel.2016.06.020

97. Zahn KE, Averill AM, Aller P et al (2015) Human DNA polymerase θ grasps the primer terminus to mediate DNA repair. Nat Struct Mol Biol 22(4):304–311. https://doi.org/10.1038/nsmb.2993
98. Bennardo N, Cheng A, Huang N et al (2008) Alternative-NHEJ is a mechanistically distinct pathway of mammalian chromosome break repair. PLoS Genet 4(6):e1000110. https://doi.org/10.1371/journal.pgen.1000110
99. Chan SH, Yu AM, McVey M (2010) Dual roles for DNA polymerase theta in alternative end-joining repair of double-strand breaks in Drosophila. PLoS Genet 6(7):e1001005. https://doi.org/10.1371/journal.pgen.1001005
100. Kent T, Chandramouly G, McDevitt SM et al (2015) Mechanism of microhomology-mediated end-joining promoted by human DNA polymerase θ. Nat Struct Mol Biol 22(3):230–237. https://doi.org/10.1038/nsmb.2961
101. Mateos-Gomez PA, Gong F, Nair N et al (2015) Mammalian polymerase [thgr] promotes alternative NHEJ and suppresses recombination. Nature 518(7538):254–257
102. Ceccaldi R, Liu JC, Amunugama R et al (2015) Homologous-recombination-deficient tumours are dependent on Polθ-mediated repair. Nature 518(7538):258–262. https://doi.org/10.1038/nature14184
103. Inagaki S, Nakamura K, Morikami A (2009) A link among DNA replication, recombination, and gene expression revealed by genetic and genomic analysis of TEBICHI gene of Arabidopsis thaliana. PLoS Genet 5(8):e1000613. https://doi.org/10.1371/journal.pgen.1000613
104. Klemm T, Mannuß A, Kobbe D et al (2017) The DNA translocase RAD5A acts independently of the other main DNA repair pathways and requires both its ATPase and RING domain for activity in Arabidopsis thaliana. Plant J 91(4):725–740. https://doi.org/10.1111/tpj.13602
105. van Kregten M, de Pater S, Romeijn R et al (2016) T-DNA integration in plants results from polymerase-θ-mediated DNA repair. Nat Plants 2(11):16164. https://doi.org/10.1038/nplants.2016.164
106. Yu AM, McVey M (2010) Synthesis-dependent microhomology-mediated end joining accounts for multiple types of repair junctions. Nucleic Acids Res 38(17):5706–5717. https://doi.org/10.1093/nar/gkq379
107. Gorbunova V, Levy AA (1997) Non-homologous DNA end joining in plant cells is associated with deletions and filler DNA insertions. Nucleic Acids Res 25(22):4650–4657
108. Salomon S, Puchta H (1998) Capture of genomic and T-DNA sequences during double-strand break repair in somatic plant cells. EMBO J 17(20):6086–6095. https://doi.org/10.1093/emboj/17.20.6086
109. Geisinger JM, Turan S, Hernandez S et al (2016) In vivo blunt-end cloning through CRISPR/Cas9-facilitated non-homologous end-joining. Nucleic Acids Res 44(8):e76. https://doi.org/10.1093/nar/gkv1542
110. Li J, Meng X, Zong Y et al (2016) Gene replacements and insertions in rice by intron targeting using CRISPR-Cas9. Nat Plants 2:16139. https://doi.org/10.1038/nplants.2016.139
111. Schiml S, Fauser F, Puchta H (2016) Repair of adjacent single-strand breaks is often accompanied by the formation of tandem sequence duplications in plant genomes. Proc Natl Acad Sci U S A 113(26):7266–7271. https://doi.org/10.1073/pnas.1603823113
112. Vaughn JN, Bennetzen JL (2014) Natural insertions in rice commonly form tandem duplications indicative of patch-mediated double-strand break induction and repair. Proc Natl Acad Sci U S A 111(18):6684–6689. https://doi.org/10.1073/pnas.1321854111
113. Adams KL, Wendel JF (2005) Polyploidy and genome evolution in plants. Curr Opin Plant Biol 8(2):135–141. https://doi.org/10.1016/j.pbi.2005.01.001
114. Bennetzen JL, Wang H (2014) The contributions of transposable elements to the structure, function, and evolution of plant genomes. Annu Rev Plant Biol 65:505–530. https://doi.org/10.1146/annurev-arplant-050213-035811
115. Devos KM, Brown JKM, Bennetzen JL (2002) Genome size reduction through illegitimate recombination counteracts genome expansion in Arabidopsis. Genome Res 12(7):1075–1079. https://doi.org/10.1101/gr.132102
116. Schubert I, Vu GTH (2016) Genome Stability and Evolution: Attempting a Holistic View. Trends Plant Sci 21(9):749–757. https://doi.org/10.1016/j.tplants.2016.06.003

117. Hawkins JS, Proulx SR, Rapp RA et al (2009) Rapid DNA loss as a counterbalance to genome expansion through retrotransposon proliferation in plants. Proc Natl Acad Sci U S A 106(42):17811–17816. https://doi.org/10.1073/pnas.0904339106
118. Cavalier-Smith T (2005) Economy, speed and size matter: evolutionary forces driving nuclear genome miniaturization and expansion. Ann Bot 95(1):147–175. https://doi.org/10.1093/aob/mci010
119. van Overbeek M, Capurso D, Carter MM et al (2016) DNA repair profiling reveals nonrandom outcomes at Cas9-mediated breaks. Mol Cell 63(4):633–646. https://doi.org/10.1016/j.molcel.2016.06.037
120. Cong L, Ran FA, Cox D et al (2013) Multiplex genome engineering using CRISPR/Cas systems. Science 339(6121):819–823. https://doi.org/10.1126/science.1231143
121. Li J-F, Norville JE, Aach J et al (2013) Multiplex and homologous recombination-mediated genome editing in Arabidopsis and Nicotiana benthamiana using guide RNA and Cas9. Nat Biotechnol 31(8):688–691. https://doi.org/10.1038/nbt.2654
122. Wang H, Yang H, Shivalila CS et al (2013) One-step generation of mice carrying mutations in multiple genes by CRISPR/Cas-mediated genome engineering. Cell 153 (4):910–918. https://doi.org/10.1016/j.cell.2013.04.025
123. Wang M, Mao Y, Lu Y et al (2017) Multiplex gene editing in rice using the CRISPR-Cpf1 system. Mol Plant 10(7):1011–1013. https://doi.org/10.1016/j.molp.2017.03.001
124. Zetsche B, Heidenreich M, Mohanraju P et al (2017) Multiplex gene editing by CRISPR-Cpf1 using a single crRNA array. Nat Biotechnol 35(1):31–34. https://doi.org/10.1038/nbt.3737
125. Williams GJ, Lees-Miller SP, Tainer JA (2010) Mre11-Rad50-Nbs1 conformations and the control of sensing, signaling, and effector responses at DNA double-strand breaks. DNA Repair (Amst) 9(12):1299–1306. https://doi.org/10.1016/j.dnarep.2010.10.001
126. Zapata L, Ding J, Willing E-M et al (2016) Chromosome-level assembly of Arabidopsis thaliana Ler reveals the extent of translocation and inversion polymorphisms. Proc Natl Acad Sci U S A 113(28):60. https://doi.org/10.1073/pnas.1607532113
127. Petolino JF, Worden A, Curlee K et al (2010) Zinc finger nuclease-mediated transgene deletion. Plant Mol Biol 73(6):617–628. https://doi.org/10.1007/s11103-010-9641-4
128. Kapusi E, Corcuera-Gómez M, Melnik S et al (2017) Heritable genomic fragment deletions and small indels in the putative ENGase gene induced by CRISPR/Cas9 in barley. Front Plant Sci 8:540. https://doi.org/10.3389/fpls.2017.00540
129. Qi Y, Li X, Zhang Y et al (2013) Targeted deletion and inversion of tandemly arrayed genes in Arabidopsis thaliana using zinc finger nucleases. G3 (Bethesda) 3(10):1707–1715. https://doi.org/10.1534/g3.113.006270
130. Ordon J, Gantner J, Kemna J et al (2017) Generation of chromosomal deletions in dicotyledonous plants employing a user-friendly genome editing toolkit. Plant J 89 (1):155–168. https://doi.org/10.1111/tpj.13319
131. Zhou H, Liu B, Weeks DP et al (2014) Large chromosomal deletions and heritable small genetic changes induced by CRISPR/Cas9 in rice. Nucleic Acids Res 42 (17):10903–10914. https://doi.org/10.1093/nar/gku806
132. Fransz PF, Armstrong S, de Jong JH et al (2000) Integrated cytogenetic map of chromosome arm 4S of A. thaliana: structural organization of heterochromatic knob and centromere region. Cell 100(3):367–376
133. Lee HJ, Kweon J, Kim E et al (2012) Targeted chromosomal duplications and inversions in the human genome using zinc finger nucleases. Genome Res 22(3):539–548. https://doi.org/10.1101/gr.129635.111
134. Li Y, Park AI, Mou H et al (2015) A versatile reporter system for CRISPR-mediated chromosomal rearrangements. Genome Biol 16:111. https://doi.org/10.1186/s13059-015-0680-7
135. Zhang C, Liu C, Weng J et al (2017) Creation of targeted inversion mutations in plants using an RNA-guided endonuclease. Crop J 5(1):83–88. https://doi.org/10.1016/j.cj.2016.08.001
136. Pacher M, Schmidt-Puchta W, Puchta H (2007) Two unlinked double-strand breaks can induce reciprocal exchanges in plant genomes via homologous recombination and nonhomologous end joining. Genetics 175 (1):21–29. https://doi.org/10.1534/genetics.106.065185
137. Weinthal DM, Taylor RA, Tzfira T (2013) Nonhomologous end joining-mediated gene replacement in plant cells. Plant Physiol 162

(1):390–400. https://doi.org/10.1104/pp.112.212910

138. Zheng Q, Cai X, Tan MH et al (2014) Precise gene deletion and replacement using the CRISPR/Cas9 system in human cells. BioTechniques 57(3):115–124. https://doi.org/10.2144/000114196

139. Byrne SM, Ortiz L, Mali P et al (2015) Multi-kilobase homozygous targeted gene replacement in human induced pluripotent stem cells. Nucleic Acids Res 43(3):e21. https://doi.org/10.1093/nar/gku1246

140. Arazoe T, Miyoshi K, Yamato T et al (2015) Tailor-made CRISPR/Cas system for highly efficient targeted gene replacement in the rice blast fungus. Biotechnol Bioeng 112(12):2543–2549. https://doi.org/10.1002/bit.25662

Chapter 18

Gene Stacking in Plants Through the Application of Site-Specific Recombination and Nuclease Activity

Vibha Srivastava

Abstract

Biotechnology methods for inserting genes one by one or as a block of fragment into plant genomes are needed to introduce valuable traits into crop varieties. Insertion of multiple genes into a single site, called as molecular stacking, is important to allow co-inheritance of the genes into the progeny. Generally, two approaches are available for creating gene stacks: nuclease-induced targeted gene integration into native sites and recombinase-mediated gene integration into the engineered sites. The recombinase application is attractive as several recombinases show high efficiency and precision in plant genomes. This chapter describes a gene stacking method based on the use of Cre-*lox* site-specific recombination system to integrate genes into the engineered sites and nucleases to delete selection genes leading to stacking of traits into a single genomic site. High efficiency and precision, and undetectable off-target effects of Cre-*lox* in a number of plant species, make it an attractive tool for complex applications such as gene stacking.

Key words Site-specific recombination, Marker-excision, Cre-*lox*, ZFN, I-*Sce*I, Nucleases, Gene editing

1 Introduction

Practical methods of stacking multiple genes into desired chromosomal sites are lacking in plant biotechnology. Targeted integration by gene editing reagents is yet to achieve the efficiencies needed for routine applications, whereas site-specific recombination (SSR) systems such as Cre-*lox* or FLP-*FRT* show high efficiency in precise integration of the DNA fragments into the engineered target sites [1]. The Cre-*lox* system is a remarkable genomic tool for complex genome engineering in both vertebrates and plants. FLP-*FRT* system is also a good choice for site-specific integrations in plant genomes [2, 3]. The gene stacking method described in this chapter uses Cre-*lox* system comprising Cre recombinase that carries out a conservative recombination between *loxP* sites. Depending on the relative location of the *loxP* in the DNA fragment (*cis* or *trans*) and the orientation (direct or inverse), co-integration, excision, or

Sandeep Kumar et al. (eds.), *Transgenic Plants: Methods and Protocols*, Methods in Molecular Biology, vol. 1864,
https://doi.org/10.1007/978-1-4939-8778-8_18,

inversion of the associated DNA could occur [4–7]. While the recombination is freely reversible, the selection pressure in the tissue culture phase ensures recovery of stable site-specific integration (SSI) clones that lack the integrated copy of the *cre* gene. Therefore, transient activity of Cre drives site-specific integration. As selection is important in the recovery of SSI clones, it is necessary to include mechanisms to eliminate the selection gene (SG). Of all the approaches available for eliminating SGs, excision by nucleases is most suitable for the gene stacking application. Gene editing reagents such as ZFN, TALEN, or CRISPRs or the rare cutting nucleases such as I-*Sce*I could be used to excise SGs. The chromosomal breaks created by the nuclease activity will most efficiently be repaired by micro-homology-based pairing and ligation of the two broken ends, resulting in the smearing of the nuclease target site [8]. This feature of DSB repair is ideal for the repeated use of the nuclease allowing recycling of the SG. This chapter describes a gene stacking method based on the application of Cre-*lox* for site-specific gene integration and the alternating use of I-*Sce*I [9, 10] and *CCR5*-ZFN [11, 12] for deleting SGs. This method is suitable for integrating multiple genes sequentially (iterative gene stacking) into the selected target sites. The resulting stacked locus is likely to follow the principles of single-gene inheritance, allowing rapid introgression into diverse cultivars.

The strategy of iterative gene stacking involves one SSR system, e.g., Cre-*lox*, and a pair of nucleases (I-*Sce*I and *CCR5*-ZFN) to direct transgene integration and SG deletion, respectively (Fig. 1; Ref [13]). However, availability of plant lines containing the target site for gene stacking is a prerequisite (Fig. 2). The single-copy target lines are selected for gene stacking with pDonor1 (Fig. 1a, c). The target site contains a *loxP*, coding region of neomycin phosphotransferase (*npt*), and the I-*Sce*I site (Fig. 1b). The first gene of interest (GOI-1) could be added to the target construct (Fig. 1a). Co-delivery of pDonor1 with a Cre expression vector results in SSI locus in which the SG is flanked by I-*Sce*I sites (Fig. 1c, d). Co-integration of *cre* gene is however counter-selected as SSI structure is unstable in the Cre background. Hence, the recovered SSI clones are expected to be Cre negative. The SSI clones are selectable as *npt* gene is turned on by a strong promoter (*ZmUbi1*). Cre-mediated gene integration results in the incorporation of two *cis*-positioned *loxP* sites. However, SG deletion removes one of the two *cis*-positioned *loxP* sites, leaving a single *loxP* located upstream of the coding region of hygromycin phosphotransferase (*hpt*) gene in the resulting SSI locus (ΔSSI). The ΔSSI contains target site for gene stacking with pDonor2 (Fig. 1e, f). The delivery of pDonor2 with Cre expression vector in the ΔSSI cells results in the generation of SSI that is selectable on hygromycin due to *ZmUbi1* promoter fusion with *hpt**. Further, the placement of two *CCR5* sites around the activated *hpt* gene (Fig. 1g) renders it

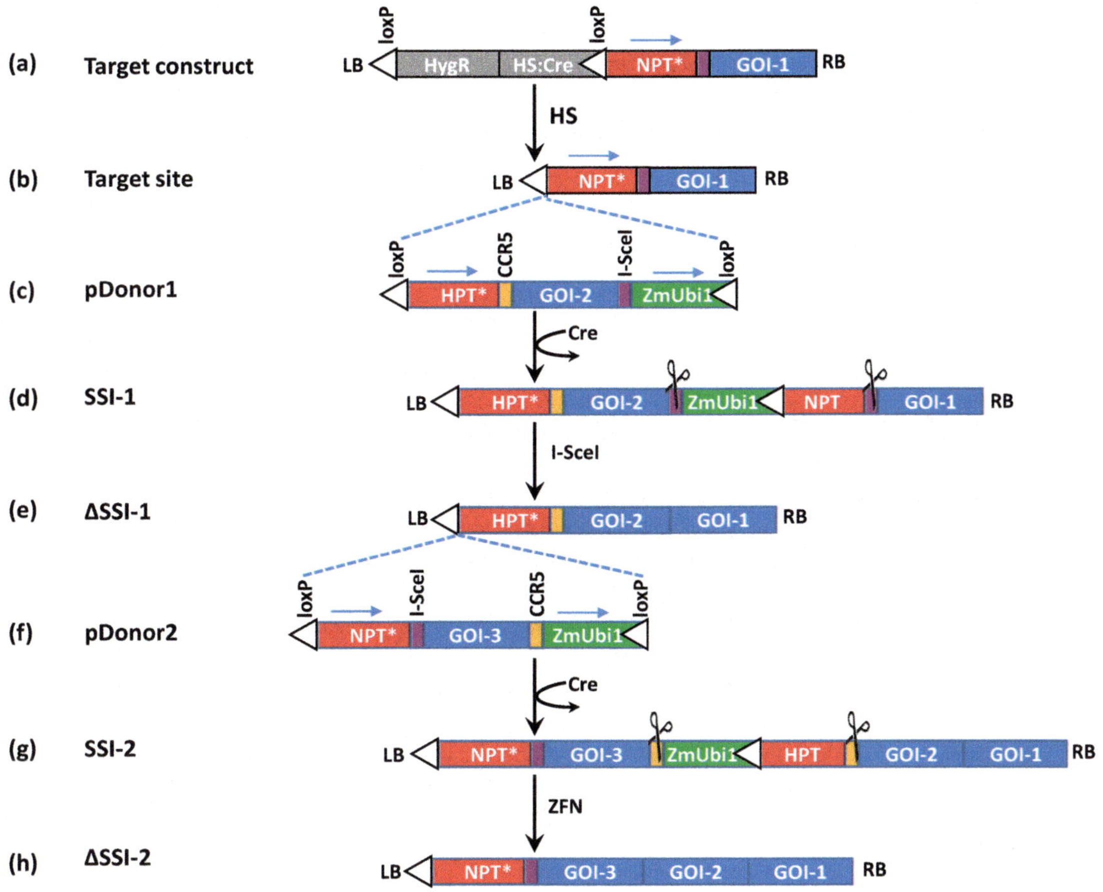

Fig. 1 Gene stacking strategy based on Cre-*lox* recombination for site-specific gene integration and nuclease activity for selection gene (SG) deletion. **(a)** Target construct in a binary vector contains *loxP*-flanked hygromycin resistance gene and the heat-inducible Cre gene (heat-shock Cre) followed by the coding region of a neomycin phosphotransferase (*NPT*) gene and the 18 bp I-*Sce*I site. A gene of interest (GOI) can be incorporated as shown. **(b)** Heat-shock treatment of cells containing the target construct leads to self-excision of selection and Cre genes, leaving the target site behind. **(c)** Structure of pDonor1 construct that contains *loxP*-flanked fragment consisting of the coding region of hygromycin phosphotransferase (*HPT*) gene, 33 bp *CCR5* sequence, GOIs, 18 bp I-*Sce*I site, and the maize ubiquitin-1 promoter (*ZmUbi1*) positioned to drive *NPT* transcription in the target site upon site-specific integration. **(d)** Delivery of pDonor1 with *cre* vector results in site-specific integration (SSI) of pDonor1 construct. The resulting SSI locus is selectable due to *NPT* expression and contains I-SceI (scissors)-flanked SG. Note that stable SSI lacks an integrated copy of *cre* gene; therefore, transient Cre expression drives recombination. **(e)** Introduction of I-*Sce*I activity deletes the SG from the SSI locus (ΔSSI). **(f)** Structure of pDonor2 construct is similar to pDonor1 except that it contains the coding region of *NPT* gene (*NPT**) and the positions of the two nuclease sites are swapped. **(g)** Delivery of pDonor2 with *cre* geneH results in the formation of SSI that is selectable on hygromycin, and contains a stack of GOIs, and *CCR5* (scissors)-flanked SG. **(h)** Introduction of ZFN leads to the deletion of the SG creating ΔSSI. The whole process can be repeated potentially unlimited times to stack more genes through additional rounds of transformation

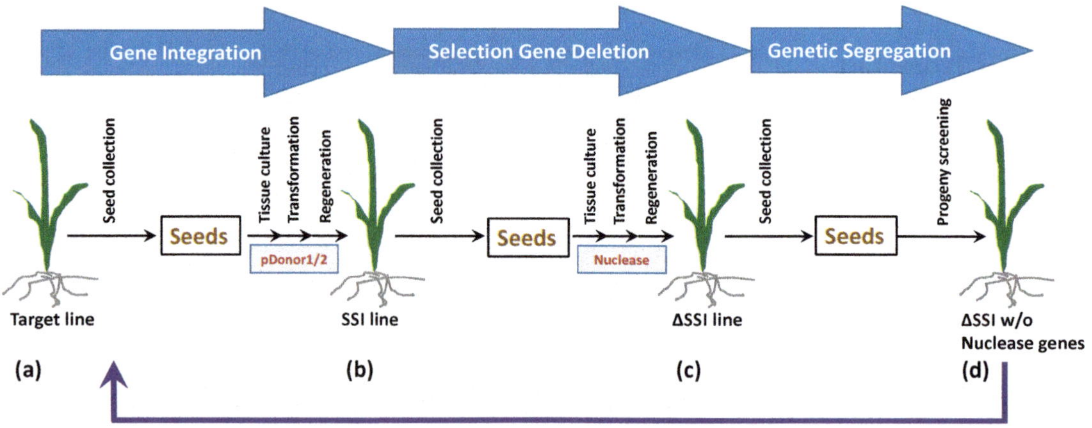

Fig. 2 Gene stacking pipeline consists of three steps: recombinase-mediated gene integration, nuclease-mediated selection gene (SG) deletion, and genetic segregation for removing nuclease genes from the stacked locus. **(a, b)** Taking the example of rice in which mature seeds serve as the starting material, the target line is subjected to tissue culture, transformation, and regeneration to obtain site-specific integration (SSI) lines using pDonor1/2 constructs. The choice of construct is based on the structure of the target site (*see* Fig. 1b, e). **(c)** The resulting SSI lines are advanced to the SG deletion step in which a specific nuclease activity is introduced by the retransformation process consisting of tissue culture from the mature seeds of SSI lines, biolistic transformation, and regeneration of transformants to obtain ΔSSI lines. The resulting ΔSSI lines could contain the input DNA vectors (nuclease gene) at random genomic sites. Therefore, the progeny of ΔSSI lines will show genetic segregation of nuclease genes and the ΔSSI locus. **(d)** Screening of the ΔSSI progeny by PCR identifies lines that contain ΔSSI and lacks nuclease genes. These lines are advanced for seed collection and the next round of gene stacking. The whole process is iterative allowing stacking of multiple genes at the selected target site

excisable by *CCR5*-ZFN activity. The delivery of ZFN activity therefore generates SG-deleted SSI locus (ΔSSI; Fig. 1h). The pDonor1 and 2 constructs contain specific choice and arrangement of SG (*npt* or *hpt*) and the nuclease target sites to allow alternating use of these vectors in the gene stacking pipeline. While Cre-*lox* is used for each gene integration step, two different nucleases (I-*Sce*I and *CCR5*-ZFN), turn by turn, are used for deleting SGs. The process could be repeated to achieve additional rounds of gene stacking into the target site (Fig. 2).

2 Materials

2.1 Vectors

2.1.1 Target Vector

The target vector was constructed in the binary vector, pPZP200 backbone (14; *see* **Note 1**). It contains following DNA fragments between the left and right borders (LB and RB):

1. A 34 bp *loxP* sequence (note the orientation; *see* **Note 2**).
2. Hygromycin resistance gene (*HygR*) consisting of cauliflower mosaic virus *35S* promoter [14], coding region of hygromycin phosphotransferase (*hpt*) gene (GenBank: ABI64010.1), and

transcription termination sequence of nopaline synthase (*nos*) gene of *Agrobacterium* C58 strain (GenBank: CS811254.1).

3. Heat-shock *cre* gene (HS:Cre) consisting of the promoter of soybean heat-shock *HSP17.5E* gene (GenBank: M28070.1), coding sequence of *cre* gene (GenBank: X03453), and *nos* transcription terminator sequence.
4. A 34 bp *loxP* sequence (note the orientation: same as the *loxP* at LB end; *see* **Note 2**).
5. Neomycin phosphotransferase II gene (*npt*) coding sequence (GenBank: AIW62898.1), and *nos* transcription termination sequence.
6. An 18 bp target site of yeast-homing endonuclease, I-*Sce*I (*see* **Note 3**).

2.1.2 Donor Vectors

pDonor1 and pDonor2 were constructed in pBluescript SK+ backbone (GenBank: X52325.1; Fig. 1c, f; *see* **Note 4**). pDonor1 contains the following fragments in the order given (left to right), while pDonor2 contains a similar design except that *npt* coding sequence is replaced by that of *hpt*, and the position of I-*Sce*I and *CCR5* sites are swapped (Fig. 1c, f):

1. A 34 bp *loxP* sequence (note the orientations indicated by white arrowheads; *see* **Note 2**).
2. Coding sequence of hygromycin phosphotransferase gene (*hpt*).
3. Transcription termination sequence of *nos* gene.
4. The 33 bp *CCR5*-ZFN target site (*see* **Note 5**).
5. Cloning sites (*Eco*RI, *Hind*III, and *Eco*RV) for inserting gene (s) of interest (GOI-2).
6. An 18 bp target site of I-*Sce*I.
7. Promoter of maize polyubiquitin gene: 2 kb *Pst*I fragment (GenBank: S94464).
8. A 34 bp *loxP* sequence (note the orientation: same as the first one and matching with that of the target construct).

2.1.3 Other Vectors

Additional vectors used in this protocol express Cre recombinase (GenBank: X03453.1), codon-optimized I-*Sce*I [9, 10], and *CCR5*-ZFN [11, 12] expression vectors. Each of these genes is expressed by strong constitutive promoter derived from maize ubiquitin-1 gene (GenBank: S94464). The Cre vector (pUbiCre) is used in co-transformation with pDonor vectors. The I-*Sce*I vector, pUbiISceI, is introduced with hygromycin resistance gene (35S:HPT:nos) vector, pHPT, and the ZFN vector, pUbiZFN, with the geneticin selection gene (35S:NPT:nos) vector, pNPT. All of these vectors are available from author's laboratory upon request (*see* **Note 6**).

2.2 Target Lines

Mature seeds of the target lines that contain a single copy of target site as shown in Fig. 1a, b serve as the donor material for generating tissue culture for the gene stacking protocol.

2.3 Equipment and Chemicals

1. Gene gun: PDS1000/He system (Bio-Rad, Inc.).
2. pDonor1 or pDonor2 (as appropriate) and pUbiCre DNA (1 μg/μL).
3. Gold particles (1 μm): Suspend 60 mg of particles in 95% ethanol, vortex vigorously, centrifuge, remove ethanol, rinse with 1 mL sterile water, and add 1 mL of water.
4. Bombardment accessories (Bio-Rad Inc.): Macrocarriers, rupture disks (1100 psi), screens.
5. Spermidine: 0.1 M in water stored at −20 °C.
6. Calcium chloride ($CaCl_2 \cdot 2H_2O$): 2.5 M in water stored at −20 °C.
7. Absolute alcohol (200 proof).
8. Geneticin 50 mg/mL.
9. Hygromycin B 50 mg/mL.
10. Callus induction medium (N6D): 3.98 g of Chu (N6) basal salt, 0.1 g of MS vitamin powder, 2 mg of 2,4-D, 0.5 g casamino acids, 2.5 g of proline, 30 g of sucrose, and 2 g of Gelrite. Adjust pH to 5.7, add water to 1 L, and autoclave.
11. Bombardment media: Callus induction media with the addition of 63 g of sorbitol per liter.
12. Selection media (N6D-S medium): Callus induction media containing geneticin (100 mg/L). Add geneticin after autoclaving.
13. Regeneration medium: 4.4 g of MS basal salt, 0.1 g of MS vitamins, 1 g of casamino acids, 2 mg of kinetin, 0.1 mg of α-naphthaleneacetic acid, 30 g of sucrose, 30 g of sorbitol, 3 g of Gelrite. Adjust pH to 5.7; add water to 1 L, and autoclave.
14. Rooting medium: 4.4 g of MS basal salt, 0.1 g of MS vitamin, 30 g of sucrose, and 2 g of Gelrite. Adjust pH to 5.7, add water to 1 L, and autoclave. Add 100 mg/L geneticin after autoclave.

3 Methods

Gene stacking pipeline involves several standard methods of molecular biology, plant tissue culture, and transformation. The pipeline is shown in Fig. 2, and the key methods, for rice as a model crop, are described below.

3.1 Generation of Target Lines by Agrobacterium-Mediated Transformation

1. Transform target vector into *A. tumefaciens* strain AGL1 using electroporator (Bio-Rad micropulser) according to the manufacturer's protocol.
2. After electroporation, add 1 mL of YEP medium to the *Agrobacterium* cells and incubate for 3 h at 28°C.
3. Plate about 100 μL of the cells on a YEP solid medium containing 100 μg/mL spectinomycin and 60 μg/mL rifampicin. Incubate the plates at 28°C for 2 days.
4. Pick a single colony and verify the presence of pGS1 by isolating plasmid and performing restriction digestion: *Eco*RV digestion will generate 2.6 and 8.9 kb fragments. Store the verified *Agrobacterium* suspension in a glycerol solution at −70 °C for future use.
5. Use *Agrobacterium* strain harboring the target vector, and follow rice transformation protocol described earlier [15]. Use 50 mg/L of hygromycin B for selection of the transformants and regeneration of transgenic plants. Determine transgene copy number by Southern blot or PCR techniques, and select single-copy plants for heat-shock treatment.
6. For heat treatment, incubate boxes or tubes containing young plants in 42 °C incubator for 3 h, followed by recovery at ambient condition for 1 day. Repeat the heat treatment, and allow the plants to grow normally for the seed set. Analyze the progeny seedlings for the excision of *loxP*-flanked fragment consisting of Cre and SG and presence of the target site [16, 17].
7. Analyze 20–50 progeny plants per parent for verifying stable inheritance of the target site. Select lines that inherit deletion locus (target site), and collect their seeds for tissue culture (*see* **Note 7**).

3.2 Generation of Site-Specific Integration (SSI) Lines

The following method is based on earlier reports of Cre-mediated site-specific gene integration in rice [18–20].

3.2.1 Preparation of Callus

1. Dehusk about 100 mature seeds of the appropriate line, e.g., target line (Fig. 1b) or ΔSSI lines (Fig. 1e, h). Surface sterilize with 70% ethanol for 1 min and then with 30% Clorox® containing 0.1% SDS solution for 30 min with vigorous shaking.
2. Remove Clorox® and rinse seeds in sterile water 4–5 times. Dry them on autoclaved filter paper.
3. Using sterile forceps place seeds on callus induction media (N6D) in Petri dishes.
4. Seal the plates, and incubate them in light at 28 °C.

5. After ~3 weeks, collect the scutellar embryogenic callus for transformation.
6. Select embryogenic callus from 10 to 15 different seeds and place them as a cluster in the middle of bombardment media on a 60 × 15 mm Petri plate.

3.2.2 Preparation of Gold Particles

1. In a 1.5 mL microcentrifuge tube, add 50 μL suspension of 60 mg/mL gold particles. Then add 5 μg each of pDonor and pUbi:Cre DNA. Mix by tapping the tube.
2. Lay the tube on its side, and place a 50 μL drop of 2.5 M $CaCl_2 \cdot 2H_2O$ solution on the wall of the tube, followed by a 20 μL drop of 0.1 M spermidine. The two drops should remain separate.
3. Carefully close the cap of the tube, quickly lift it, and hold it on vortex for vigorous mixing for 3 min.
4. Centrifuge at full speed (~17,000 × *g*), remove all liquids, and rinse with 1 mL of absolute ethanol.
5. Remove ethanol, and add 100 μL of absolute ethanol.

3.2.3 Particle Bombardment of Callus Plates

1. Rub the tube (containing DNA-coated gold particles) on an empty box of pipette tip several times to break the gold particle pellet and develop even suspension.
2. Quickly pipette 5–10 μL of gold particles on macrocarriers.
3. Let dry, and bombard on callus plates using 1100 psi rupture disks.
4. Leave bombarded callus on the bombardment media overnight at 28 °C.
5. Transfer clusters of callus-to-callus induction media (N6D) next day, and incubate plates in light at 28 °C.

3.2.4 Tissue Culture

1. After 1 week of incubation on N6D media, transfer callus to selection media (N6D containing 100 mg/L geneticin).
2. Pick geneticin-resistant clones and transfer to a fresh selection plate.
3. Transfer freshly grown geneticin-resistant callus to regeneration media, and incubate in light at 28 °C.
4. Transfer regenerated shoots to rooting media containing 100 mg/L geneticin.
5. After 2–3 weeks, rice plantlets can be transplanted to pots in the greenhouse.

3.2.5 Molecular Analysis

1. Use PCR technique with multiple sets of primers to determine the unique structure of each SSI determined by Cre-*lox* recombination (Fig. 1d, g).

2. Perform Southern blot analysis on the genomic DNA of SSI plants using DNA probes against *cre* gene and GOI located in the SSI locus.
3. Select SSI lines that contain predicted Southern hybridization pattern indicative of full-length SSI emerging from Cre-*lox* recombination, and absence of the additional copies of the donor vector or *cre* gene (random integrations).

3.3 Deletion of Selection Gene (SG)

1. Select mature seeds of the verified single-copy SSI lines for tissue culture and transformation as described above.
2. Based on whether the SSI line is derived from pDonorl or pDonor2, prepare gold particles using 5 μg each of pUbiISceI and pHPT or pUbiZFN and pNPT as described above.
3. Carry out particle bombardment, selection, and regeneration as described above (*see* **Note 8**).
4. Perform PCR with primers located around SG to determine SG deletion and presence of ΔSSI (Fig. 1e or h).
5. Select verified ΔSSI lines.

3.4 Genetic Segregation Analysis and Isolation of the "Stacked" Lines

1. Germinate 25–50 progeny seeds of the ΔSSI lines on half-strength MS media plates.
2. Extract DNA and perform PCR to determine the presence of the nuclease genes (I-*Sce*I or ZFN) and SGs (35S:HPT or 35S: NPT) (*see* **Note 9**).
3. Select progeny that lack nuclease genes and SGs (*see* **Note 10**).
4. Confirm the presence of ΔSSI by PCR; grow and collect seeds from the selected progeny plants.

4 Notes

1. The target vector developed in the author's laboratory is named pNS42 that contains all elements, as shown in Fig. 1a, except GOI. pNS42 was generated by assembling different gene fragments in pPZP200 backbone for *Agrobacterium*-mediated transformation [21].
2. *loxP* sequence (→): *5'*-ATAACTTCGTATAATGTATGCTA-TACGAAGTTAT-*3'*.
3. I-*Sce*I target sequence: *5'*-TAGGGATAA^CAGGGTATT-*3'* (^cut site).
4. pDonorl and pDonor2 are available in the author's laboratory with GFP in pDonorl, and GUS in pDonor2 as the GOIs.

5. *CCR5*-ZFN target sequence: *5′*-TGGTCATCCTCATCCTGATAAACTGCAAAAGGC-*3′* (GenBank: JB172016.1; cut site underlined).
6. pUbiZFN, the *CCR5*-ZFN expression vector for monocots, is subject to Research License and Material Transfer Agreement with Dow Agro Sciences Inc.
7. Since target line seeds may contain heterozygous target site, use fresh tissue culture, one callus cluster per seed, for maximizing the presence of target site in the experiment, and do not mix the callus lineages to prevent proliferation of callus emerging from wild-type segregants.
8. Retransformation is the best option for deleting SGs with I-*Sce*I and *CCR5*-ZFN in rice. These nucleases are effective in SG deletion through transient activity [12]. No transgenic rice with strong constitutive expression of I-*Sce*I or *CCR5*-ZFN could be developed, indicating that these nucleases at high concentrations are toxic to rice cells. However, cross-pollination could be used to introduce strong I-*Sce*I and *CCR5*-ZFN activity in other crops, where stable transgenic lines expressing these nucleases have been developed [11].
9. I-*Sce*I and ZFN gene primers: I-SceI-forward: *5′*-GCTGTCTCCTCCTCACAAG-*3′*; I-SceI-reverse: *5′*-GGGTCAGGTAGTTCTCCACC-*3′*; ZFN-F: *5′*- TGAATGGTGGAAGGTGTATCC-*3′*; ZFN-R: *5′*-AAGCTGTGCTTTGTAGTTACCCTTA-*3′*.
10. Strong constitutive expression of I-*Sce*I and *CCR5*-ZFN appears to be toxic to rice cells. Therefore, ΔSSI lines are unlikely to contain an integrated copy of the nuclease genes.

References

1. Srivastava V, Thomson J (2016) Gene Stacking by recombinases. Plant Biotechnol J 14:471–482
2. Li Z, Xing A, Moon BP, McCardell RP, Mills K, Falco SC (2009) Site-specific integration of transgenes in soybean via recombinase-mediated DNA cassette exchange. Plant Physiol 151:1087–1095
3. Nandy S, Srivastava V (2011) Site-specific gene integration in rice genome mediated by the FLP-FRT recombination system. Plant Biotechnol J 9:713–721
4. Ga JT, Sirk SJ, Barbas CF 3rd (2014) Expanding the scope of site-specific recombinases for genetic and metabolic engineering. Biotechnol Bioeng 111:1–15
5. Grindley ND, Whiteson KL, Rice PA (2006) Mechanisms of site specific recombination. Annu Rev Biochem 75:567–605
6. Ow DW (2002) Recombinase-directed plant transformation for the post-genomic era. Plant Mol Biol 48:183–200
7. Sauer B (2002) Cre/*lox*: one more step in the taming of the genome. Endocrine 19:221–228
8. Salomon S, Puchta H (1998) Capture of genomic and T-DNA sequences during double-strand break repair in somatic plant cells. EMBO J 17:6086–6089
9. Hlubek A, Biesgen C, Höffken H-W (2011) Chimeric endonucleases and uses therefore. Patent WO 2011/064750
10. Fauser F, Roth N, Pacher M, Ilg G, Sánchez-Fernández R, Biesgen C, Puchta H (2012) In

planta gene targeting. Proc Natl Acad Sci U S A 109:7535–7540

11. Perez EE, Wang J, Miller JC, Jouvenot Y, Kim KA, Liu O et al (2008) Establishment of HIV-1 resistance in CD4+ T cells by genome editing using zinc-finger nucleases. Nat Biotechnol 26:808–816
12. Petolino JF, Worden A, Curlee K, Connell J, Strange-Moynahan TL, Larsen C, Russell S (2010) Zinc finger nuclease-mediated transgene deletion. Plant Mol Biol 73:617–628
13. Nandy S, Zhao S, Pathak BP, Manoharan M, Srivastava V (2015) Gene stacking in plant cell using recombinases for gene integration and nucleases for marker gene deletion. BMC Biotechnol 15:93
14. Odell JT, Nagy F, Chua NH (1985) Identification of DNA sequences required for activity of the cauliflower mosaic virus 35S promoter. Nature 313:810–812
15. Nishimura A, Aichi I, Matsuoka M (2006) A protocol for *Agrobacterium* mediated transformation in rice. Nat Protoc 1:2796–2802
16. Srivastava V, Ow DW (2004) Marker-free site-specific gene integration in plants. Trends Biotechnol 22:627–629
17. Nandy S, Srivastava V (2012) Marker-free site-specific gene integration in rice based on the use of two recombination systems. Plant Biotechnol J 10:904–912
18. Srivastava V, Ow DW (2002) Biolistic-mediated site-specific integration in rice. Mol Breed 8:345–350
19. Srivastava V, Ariza-Nieto M, Wilson AJ (2004) Cre-mediated site-specific gene integration for consistent transgene expression in rice. Plant Biotechnol J 2:169–179
20. Srivastava V (2013) Site-specific gene integration in rice. Methods Mol Biol 956:83–93
21. Hajdukiewicz P, Svab Z, Maliga P (1994) The small, versatile pPZP family of Agrobacterium binary vectors for plant transformation. Plant Mol Biol 25:989–994

Chapter 19

CRISPR/Cas9 for Mutagenesis in Rice

Si Nian Char, Riqing Li, and Bing Yang

Abstract

CRISPR/Cas9 (clustered regularly interspaced short palindromic repeat/CRISPR-associated protein 9) provides a workhorse for genome editing biotechnology. CRISPR/Cas9 tailored for enabling genome editing has been extensively interrogated and widely utilized for precise genomic alterations in eukaryotic organisms including in plant species. The technology holds the great promise to better understand gene functions, elucidate networks, and improve the performance of crop plants such as increasing grain yields, improving nutritional content, and better combating the biotic and abiotic stresses. Various methods or protocols specific for different plant species have been established. Here, we present a CRISPR/Cas9-mediated genome editing protocol in rice, including detailed information about single-guide RNA design, vector construction, plant transformation, and mutant screening processes.

Key words CRISPR/Cas9, Genome editing, Rice, *Agrobacterium*-mediated rice transformation, Targeted mutagenesis

1 Introduction

CRISPR/Cas (clustered regularly interspaced short palindromic repeat/CRISPR-associated proteins) systems originally function in a broad range of archaea and bacteria as an adaptive immunity against invading phage and plasmids [1]. Modified genome enabling CRISPR/Cas9 system consists of the nuclease Cas9 and the single chimeric guide RNA (gRNA) derived from the trans-activating crRNA (tracrRNA) and CRISPR RNA (crRNA) (referred to as Cas9/gRNA) [2]. Cas9 protein from *Streptococcus pyogenes* is one of the most well-studied and commonly used nucleases for genome editing [3]. Cas9 protein has two nuclease domains including NHN nuclease domain that cleaves the complementary strand and RuvC domain that cleaves the noncomplementary strand of the gRNA spacer region [4]. Protein Cas9 that is guided by gRNA cleaves the target sequence 3 nucleotides upstream of the protospacer-adjacent motif (PAM, the most efficient site for SpCas9 is 5′-NGG-3′), causing site-specific DNA double-strand breaks (DSBs) in vitro or in vivo [2]. The affected

Sandeep Kumar et al. (eds.), *Transgenic Plants: Methods and Protocols*, Methods in Molecular Biology, vol. 1864,
https://doi.org/10.1007/978-1-4939-8778-8_19, © Springer Science+Business Media, LLC, part of Springer Nature 2019

cells will then initiate either nonhomologous end joining (NHEJ) or homologous recombination (HR) pathway to repair the DSBs. NHEJ is an error-prone DNA repair process which leads to insertions or deletions (indels) at the repaired sites. The indels, occurring at the site of DNA cleavage, are highly variable ranging from small indels or large deletions, some of which cause open reading frame shift of the target gene, resulting in targeted DNA mutagenesis and gene disruption. On the other hand, homologous recombination (HR) repair process can be exploitable for gene replacement or precise gene editing using donor DNA as template that contains the desired genetic change flanked by the DNA arms homologous to the target region of the genomic locus of interest [5, 6]. With the established Cas9/gRNA system, various constructs can be made just through subcloning the spacer sequence at the multiple cloning site (MCS) preceding the scaffold of the single-guide RNA (gRNA) [7]. Therefore, Cas9/gRNA system is an efficient and amenable tool for targeted mutagenesis in plants.

Cas9/gRNA-mediated mutagenesis in rice requires (1) design and construction of Cas9/gRNA plasmid based on the sequence of genomic locus of interest, (2) transfer of the plasmids into the rice immature embryo-derived callus cells through either *Agrobacterium*-mediated or biolistic particle-mediated DNA delivery method, (3) genotyping the transgenic callus lines and rice plantlets to identify the intended modifications, and [4] analysis of inheritance of the gene editing events and removal of the CRISPR transfer DNA (T-DNA). Here, we present a step-by-step protocol for each stage of the targeted mutagenesis in rice (Fig. 1). This protocol has been successfully applied to the *japonica* rice cultivar Kitaake for mutagenesis of many genes in our laboratory (Table 1).

2 Materials

2.1 Molecular Cloning Reagents

1. Restriction enzymes (BamHI, HindIII, BtgZI, BsaI, AatII).
2. Alkaline phosphatase, calf intestinal (CIAP).
3. RNase A.
4. DNA Ladder (e.g., 1 kb plus).
5. Bacterial competent cells (e.g., XL1-Blue, DH5α).
6. ExoSAP-IT (Thermo Fisher Scientific).
7. Luria-Bertani (LB) broth medium with appropriate antibiotics, kanamycin (50 mg/L), rifampicin (30 mg/L).
8. T4 DNA ligase.
9. LR recombinant clonase.
10. GENECLEAN III Kit (MP Biomedical) for DNA purification.
11. 30 and 37 °C incubators.

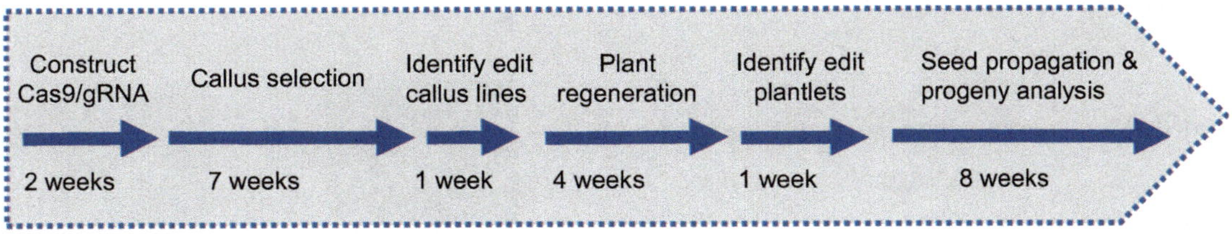

Fig. 1 Flowchart of Cas9/gRNA-mediated targeted mutagenesis in rice. This chart shows the major steps of targeted mutagenesis in rice. The timeline for design and construction of Cas9/gRNA plasmid can be accomplished within 2 weeks. The whole callus transformation and selection process take about 7 weeks. Screening for positive transgenic calli requires 1 week. Plant regeneration step needs at least 4 weeks. Genotyping of positive plants takes 1 week. Finally, it takes about 2 months for rice plants to produce seeds for further progeny analysis

Table 1
Primer and sequences used in the protocol

Primers	Sequences
OsCas9-F	5′-GGGTAATGAACTCGCTCTGC-3′
OsCas9-R	5′-TGGCGTCAAGAACTTCCTTTG-3′
U6P-F1b	5′-CGTTGAGGGGAGACAGGTTTAG-3′
pENTR4-R	5′-TGGGTCTAGATATCTCGAGTG-3′
U6T-R	5′-CTGCAGAATTGCCCTTCGAAG-3′

12. 30 and 37 °C shakers.
13. Gel electrophoresis apparatus.
14. T7 Endonuclease I.

2.2 Rice Tissue Culture and Transformation

2.2.1 Plant Expression Vectors Agrobacterium tumefaciens Strain and Rice Cultivar

1. pGW-rCas9 (available upon request) (Fig. 2).
2. pgRNA1 (available upon request) (Fig. 2).
3. *Agrobacterium tumefaciens* strain EHA105.
4. Immature embryos of the *japonica* rice (*Oryza sativa* L.) cultivar Kitaake.

2.2.2 Tissue Culture and Transformation

1. MSD medium: 4.4 g/L Murashige and Skoog (MS) basal medium with Gamborg's vitamins, 30 g/L sucrose, 2 mg/L 2,4-dichlorophenoxyacetic acid, and 8 g/L agar for solid medium, pH 5.8.
2. TY medium: 5.0 g/L Tryptone, 3.0 g/L yeast extract, 50 mg/L kanamycin, 30 mg/L rifampicin, or 15 g/L agar for solid medium, pH 5.5.

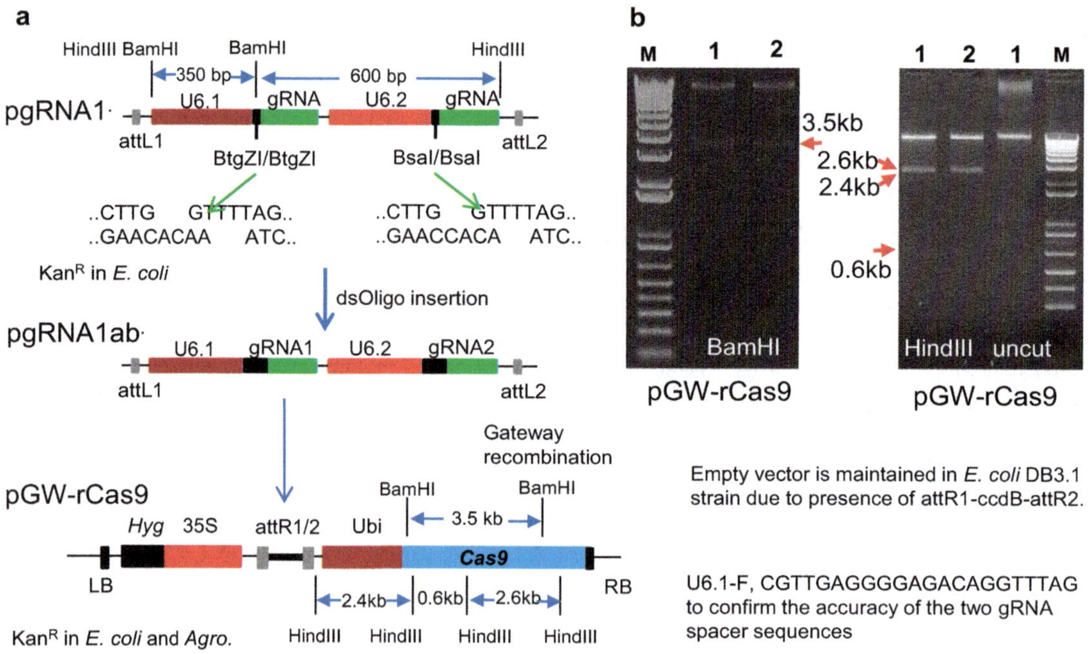

Fig. 2 Schematic diagram of rice Cas9/gRNA system. The intermediate vector pgRNA1 is sequentially digested with BtgZI and BsaI for two insertions of double-stranded oligonucleotides. The subcloning results in two gRNA expression cassettes. The insertions can be confirmed through sequencing with the primer U6P-F1b (Table 1). The cassettes flanked by the Gateway recombination sequences attL1 and attL2 are mobilized to the binary vector pGW-rCas9 through the Gateway recombination, resulting in a single plasmid of Cas9/gRNA for *Agrobacterium*-mediated gene transfer into rice callus cells

3. MSD + AS medium: MSD medium containing 0.2 mM acetosyringone (AS), or 8 g/L agar for solid medium, pH 5.8.
4. Callus selection medium: MSD medium containing 400 mg/L Timentin, 50 mg/L hygromycin B (Sigma-Aldrich), and 8 g/L agar, pH 5.8.
5. Shoot regeneration medium: MS medium containing 4.4 g/L MS basal medium with Gamborg's vitamins, 30 g/L sucrose, 50 g/L sorbitol, 3 mg/L 6-benzylaminopurine (BAP), 0.5 mg/L 1-naphthaleneacetic acid (NAA), 25–50 mg/L hygromycin B (Sigma-Aldrich), and 12 g/L agar, pH 5.8.
6. Root induction medium: ½ MS medium containing 2.2 g/LMS basal medium with Gamborg's vitamins, 10 g/L sucrose, 25 mg/L hygromycin B (Sigma-Aldrich), 6 g/L Agargellan, pH 5.8.

2.3 Analysis of Transgenic Rice Callus Lines or Plants

2.3.1 Plant Genomic DNA Extraction

1. Pestle.
2. DNA extraction buffer: 20 mM Ethylenediaminetetraacetic acid (EDTA), 100 mM Tris–HCl, 1.5 M NaCl, 2% cetyltrimethyl ammonium bromide (CTAB), 1% β-mercaptoethanol.
3. Liquid nitrogen.
4. 65 °C Water bath.
5. Chloroform:isoamyl alcohol (24:1).
6. Isopropanol.
7. 70% ethanol.
8. RNase A.
9. TE buffer: 10 mM Tris–HCl, 1 mM EDTA, pH = 8.0.

2.3.2 Genotyping by PCR

1. T7 endonuclease I (T7EI).
2. Restriction enzymes.
3. DNA polymerase.
4. 10 mM deoxynucleotide mixture.
5. PCR primers.
6. Dimethylsulfoxide (DMSO).
7. PCR thermocycler.

3 Methods

3.1 Design and Choose Candidate Cas9/gRNA Target Site in Rice Genome

Select the guide RNA spacer sequences against the rice Nipponbare reference genome sequence or sequences of interest by using the CRISPR Genome Analysis Tool (CGAT) software (*see* **Note 1**). This publicly available tool is hosted at the Iowa State University Crop Bioengineering Consortium website (http://cbc.gdcb.iastate.edu/cgat/). The online tool was previously described [8]. The CGAT bioinformatics tool is user friendly for identifying the potential target site in the genome of interest. The potential off-target sites will also be shown by using CGAT bioinformatics tool.

3.2 Construct Cas9/gRNA Expression Plasmid

3.2.1 Design and Synthesize gRNA-Specific Oligonucleotides

1. To design oligos for gRNA gene construct, synthesize two complementary oligonucleotides (22–25 nt), if both strands annealed forming a double-stranded fragment with the sense strand consisting of the 4-nt overhang of TGTT at the 5′ end and the antisense strand consisting of an overhang of AAAC at the 5′ end for BtgZI restriction site.
2. Similarly, design another two complementary oligonucleotides with sense strand containing a 4-nt overhang of GTGT at the 5′

end and antisense strand containing AAAC at the 5′ end for BsaI restriction site.

3. To construct the gRNA gene targeting the genomic locus of interest, first anneal two complementary oligonucleotides, and then ligate into the BtgZI-digested pgRNA1 vector.

3.2.2 Anneal and Clone the Synthesized Oligos into BtgZI-Digested Vector pgRNA1

1. Digest 1 μg of pgRNA1 vector with BtgZI for 3 h at 60 °C in incubator or PCR thermocycler.

X μL (1 μg)	pgRNA1
1 μL	BtgZI
0.2 μL	RNAase A
3 μL	10× Buffer
X μL	H_2O
30 μL	Digestion reaction

2. Gel-separate and purify BtgZI-digested pgRNA1 DNA using GENECLEAN III Kit (*see* **Note 2**).
3. Anneal each pair of oligos.

2 μL	Oligo 1 (10 μM)
2 μL	Oligo 2 (10 μM)
X μL	H_2O
20 μL	Total

4. Heat a beaker of water to boiling point, put the mixed oligos into boiling water, and let the water cool down to room temperature before using the reaction for next step ligation.

3.2.3 Clone the Oligo-Derived Double-Strand Fragment (dsOligo) into BtgZI-Digested gRNA Scaffold Vector

1. Set up ligation reaction.

X μL	H_2O
X μL	BtgZI-digested pgRNA1 (50 ng)
5 μL	Annealed dsOligo from **step 4** of Subheading 3.2.2
2 μL	T4 ligase
1 μL	10× Ligation buffer
10 μL	Total

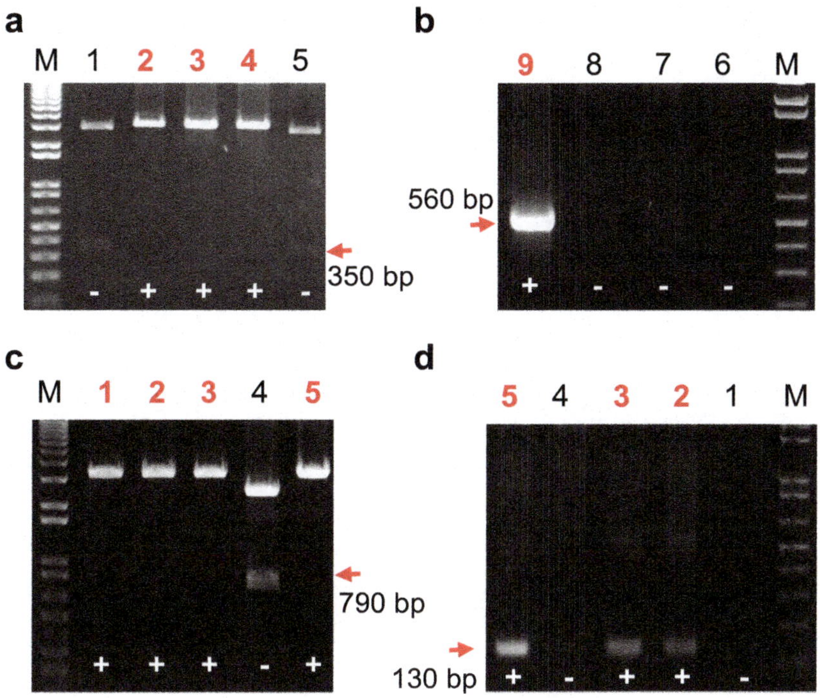

Fig. 3 Restriction and PCR analysis of clones from the ligations of two dsOligo fragments into BtgZI- and BsaI-digested pgRNA1, respectively. The individual clones are digested with BamHI (**a**) and PCR with specific primers (**b**) to confirm the right insertions of double-stranded oligonucleotides at BtgZI site, digested with BamHI and BsaI (**c**) and PCR with specific primers (**d**) to validate the successful insertions of dsOligo fragment at BsaI site

2. Incubate the ligation reaction at room temperature for 3 h or at 4 °C overnight and transfer half of the reaction into the *E. coli* XL1-Blue competent cells.
3. To identify the clones with insert, digest the plasmid DNA with BamHI. If the oligos are successfully cloned into the pgRNA1 vector, the BamHI site at the MCS will disappear, and the clones with the BamHI digestion pattern lacking a 350 bp band are potential insertion-positive clones (Fig. 3a).
4. Colony-PCR-based assay by using gRNA-F (oligo as forward primer for the spacer sequence of gRNA) and U6T-R (reverse primer complementary to the region downstream of gRNA gene in the backbone of pgRNA1) (Table 1) can also be used to identify the positive clones (Fig. 3b).
5. After the cloning step, sequence the positive candidate clone by using forward primer (U6P-F1b) or reverse primer (pENTR4-R) (Table 1) to confirm the accuracy of spacer sequence in the gRNA scaffold. The resultant plasmid is named pgRNA1a.
6. After the first dsOligo is inserted into BtgZI restriction site, ligate the second dsOligo sequentially at the BsaI restriction site.

3.2.4 Clone a Second gRNA-Specific Sequence into the BsaI-Digested gRNA Scaffold Vector

1. Digest 1 μg of pgRNA1a plasmid with BsaI for 3 h at 37 °C.

X μL (1 μg)	pgRNA1a
1 μL	BsaI
0.2 μL	RNase A
3 μL	10× Buffer
X μL	H_2O
30 μL	Total

2. Ligate the annealed double-stranded oligos into the BsaI-digested pgRNA1a plasmid that contains the first gRNA gene Set up ligation reaction.

X μL	H_2O
X μL	BsaI-digested pgRNA1a (50 ng)
5 μL	Annealed oligos
2 μL	T4 ligase
1 μL	10× Ligation buffer
10 μL	Total

3. Incubate the ligation reaction at room temperature for 3 h or at 4 °C overnight and transfer half of the reaction into the *E. coli* XL1-Blue competent cells.
4. To identify the insertion-positive clones, digest plasmid DNA from individual colonies with BamHI and BsaI. If the dsOligo is successfully cloned into the vector, there is no BsaI restriction site within the MCS and the clones with a digestion pattern lacking the 790 bp band compared to the original vector are positive (Fig. 3c).
5. Colony-PCR-based assay by using gRNA-F (oligo as forward primer for the spacer sequence of gRNA) and U6T-R (reverse primer complementary to the region downstream of gRNA gene in the backbone of pgRNA1) (Table 1) can also be used to identify the positive clones (Fig. 3d).
6. After the cloning step, sequence the positive candidate clone by using the forward primer (U6P-F1b) or reverse primer (pENTR4-R) (Table 1) to confirm the accuracy of the first and second spacer sequences that have been cloned into the two gRNA scaffolds.

3.2.5 Construct gRNA Cassettes of Four Guide RNA Genes for Multiplex Gene Targeting

Our intermediate vector can be used to construct up to four gRNAs for multiplex gene editing in rice.

1. To construct gRNA cassettes that contain four guide RNA genes, two gRNA constructs need to be made first by using pgRNA1and pgRNA2 individually. pgRNA2 is almost identical to pgRNA1 except that it does not contain the second HindIII restriction site at the 3′ end of the second guide RNA scaffold. Two dsOligo fragments can be cloned into pgRNA2 similarly to the cloning of two dsOligo fragments into pgRNA1as described above, resulting in pgRNA2a.
2. Release the whole cassette two gRNA genes from pgRNA1a with HindIII digestion. Recover the 1 kb fragment of gRNA cassette from the gel separation and purification.
3. Subclone the recovered fragment (1 kb) into HindIII-digested and subsequently CIAP-treated pgRNA2a that already contains two guide RNA genes. Digest plasmid DNA from the cloning step with HindIII and BamHI to confirm the insertion of gRNA cassette, resulting in plasmid pgRNA1ab.
4. Recombine the guide RNA cassettes of up to four gRNA genes into the destination vector pGW-rCas9 by using the Gateway LR clonase kit (*see* **Notes 3** and **4**).

Entry clone	50–150 ng
Destination vector	150 ng/μL
TE buffer	X μL
Total	5 μL

5. To confirm the gateway reaction is successful, digest the plasmid DNA from individual clones with BamHI or HindIII. A digestion pattern of 12 kb, 2.6 kb, 2.4 kb, 1 kb, 650 bp, and 125 bp from HindIII digestion or a pattern of 12 kb, 3.5 kb, and 2.8 kb due to BamHI digestion indicates a successful mobilization of gRNA cassettes into the destination Cas9 plasmid (Fig. 4).
6. Mobilize the resultant plasmid into *Agrobacterium* strain EHA105 for transformation of rice immature embryo-derived callus cells.

3.3 Generate Cas9/gRNA Transgenic Rice

Rice transformation involves method of DNA delivery into callus cells through biolistic particle bombardment or *Agrobacterium* infection. Here, we describe the *Agrobacterium*-delivered CRISPR system for mutagenesis in rice by using calli induced from the scutella of immature seeds. The transformation protocol is modified from the method described by Hiei et al. [9].

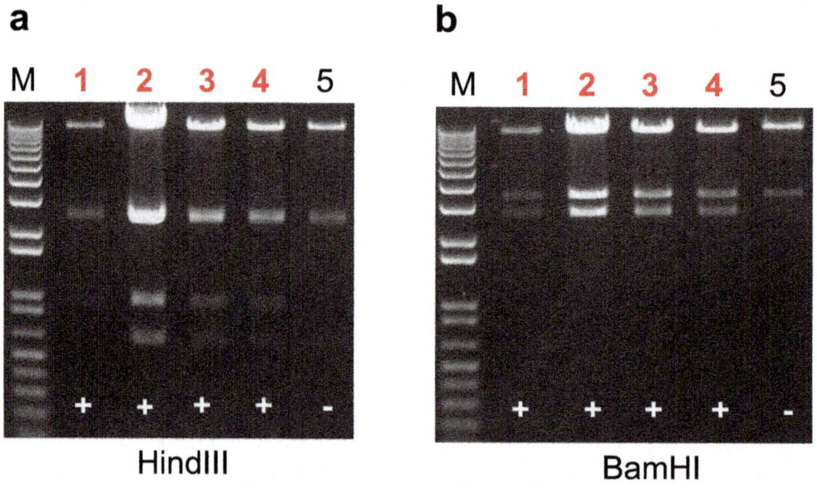

Fig. 4 Gel image of the Gateway LR recombination clones digested with HindIII and BamHI. The plasmid DNA is digested with restriction enzyme BamHI (**a**) or HindIII (**b**). The clones with the digested pattern shown in the gel picture are the positive ones that are ready for *Agrobacterium* transformation

3.3.1 Transfer Cas9/gRNA Construct into Rice Callus Cells for Gene Editing

1. Transfer the Cas9/gRNA plasmid into the *Agrobacterium* strain EHA105 through electroporation (*see* **Note 5**).
2. Callus initiation: Sterilize the dehusked immature seeds with 50% bleach (3% sodium hypochlorite) for 20 min, and rinse three times with sterilized water. Then place the seeds on MSD medium and incubate at 30 °C with continuous light for 14 days to initiate actively growing calli.
3. *Agrobacterium* infection: Grow Cas9/gRNA-containing *Agrobacterium* culture to 1.0–2.0 of OD600, mix 2 mL of *Agrobacterium* cells with 20 mL of liquid MSD medium (with 0.2 mM acetosyringone, pH 5.2), and immerse the rice calli in the suspension for 30 min. Blot dry the calli with filter paper, transfer the calli to MSD + AS plate, and keep the calli in the dark at room temperature for 2–3 days.
4. Transformed callus selection: Transfer the infected calli to callus selection medium at an interval of 2 weeks for selection of transformed callus cells. The tissue culture and regeneration at this and subsequent stages are performed in growth chamber at 30 °C and under continuous light.
5. Plantlet regeneration: Transfer the individual hygromycin-resistant callus lines to shoot regeneration medium to regenerate transgenic plants.
6. Root induction: Transfer the regenerated plants into ½ MS medium for root induction.

7. Transfer the plantlets into soil and grow them in the growth chamber set at 28 °C, 75% relative humidity, and 12-h photoperiod till plants are mature and seeds are harvested.

3.4 Screening and Identifying CRISPR/Cas9-Induced Mutations

3.4.1 Extract Genomic DNA from Transformed Calli or Plants

1. To determine the genotype of edited locus in rice callus lines or plants that contain the Cas9/gRNA transgenes, extract genomic DNA from rice callus cells or pooled leaves of individual transgenic plants by using the cetyltrimethylammonium bromide (CTAB) methods [10].
2. Measure the concentration of the isolated genomic DNA by using the Nanodrop ND1000 spectrophotometer.
3. Use the genomic DNA as the template for PCR amplification of target regions with site-specific primers designed to flank the target sites.

3.4.2 Screen for Putative Edit Lines Through Digestion with T7EI

1. Test whether the PCR amplicons are for the targeted mutations by using T7 endonuclease I (T7EI) assay.
2. Mix the PCR products originated from the transgenic tissues with respective amplicon derived from the wild-type plant, denature (95 °C for 5 min), reanneal (ramp down to 25 °C at 5 °C/min), and then subject to T7EI digestion and 1–2% agarose gel electrophoresis (Fig. 5a).

3.4.3 Sequence PCR Amplicons to Confirm the Mutant Genotypes

1. Treat the amplicons derived from the T7EI-positive callus lines or plants with ExoSAP-IT to remove the excessive primers and single nucleotides before Sanger sequencing.
2. Carefully examine the sequencing chromatograms for exact peak pattern that might contain mono-allelic or di-allelic mutations (Fig. 5b).

3.4.4 Screen for Modified Sequences Through Digestion with Restriction Enzymes

1. If a restriction site overlaps with the cleavage site (three nucleotides upstream the PAM site) of Cas9/gRNA, use the restriction digestion of the PCR products to detect the occurrence of edit events.
2. Precipitate the PCR product with 0.1× of 3 M sodium acetate and 4× of 100% ethanol, then wash the pellet with 80% ethanol, and finally dissolve the DNA pellet in H_2O.
3. Use 5 μL of the precipitated DNA for specific restriction enzyme digestion for 3 h.
4. Perform gel electrophoresis to determine the digestion pattern after the enzyme treatment.
5. If most of the PCR product is cut, it indicates that the target region would be wild-type sequence. If the intensity of the uncut DNA band is strong, it indicates that the callus or plants may have mutation with the targeted regions.

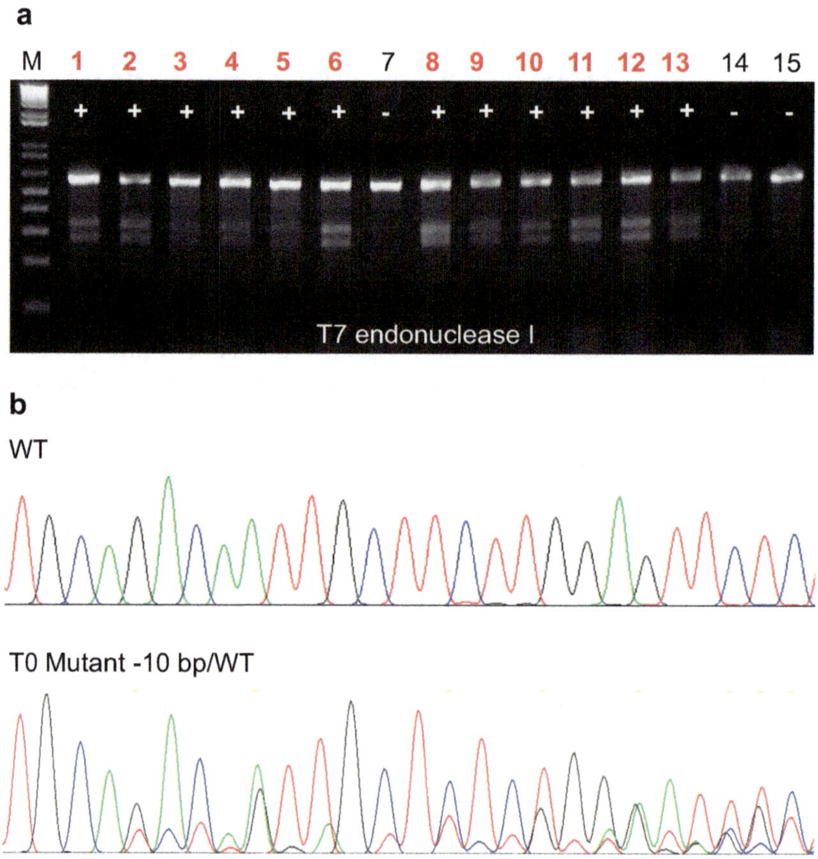

Fig. 5 Identification and confirmation of mutations induced by Cas9/gRNA in rice plants. **a**. T7EI digestion pattern of PCR amplicons from the transgenic rice plants. The digestion pattern varied according to the PCR product size and specific primers designed to flank the target sites. **b**. DNA sequencing chromatograms represent the wild-type (top) and edited mutation (bottom). A 10 bp mono-allelic deletion from the mutant sample caused the peak shift from one strand overlapping with the wild-type strand

3.4.5 Identify the Heritable Mutations from Progeny

1. Self-fertilize T_0 plants from the transgenic lines.
2. After fertilization, individually collect seeds produced from the T_0 plants, put in the 37 °C incubator for 1–2 weeks to dry, and break the seed dormancy.
3. Germinate the dried seeds in the ½ MS medium.
4. To test the heritability of the mutation from T_0 to T_1 and T_2 generation plants, extract genomic DNA and examine PCR products from the targeted region by using the T7EI assay to determine the genotypes of individual progeny plants.
5. For T7EI assay, detect heterozygous plants by T7EI digestion after denaturing and reannealing of PCR product. Homozygous mutant or wild-type plant can be determined after mixing

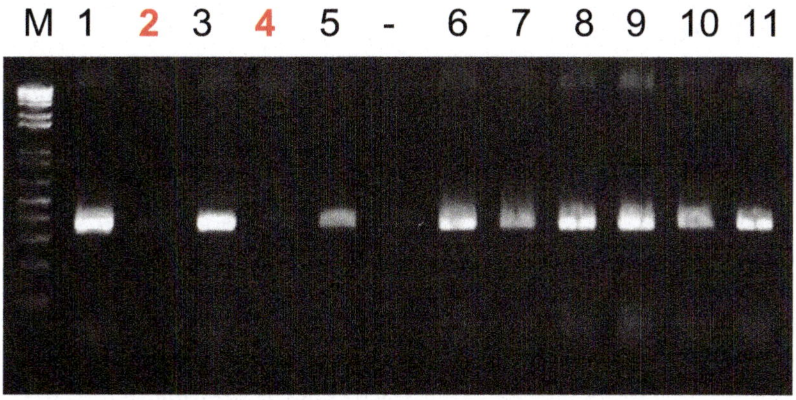

Fig. 6 Genetic segregation of the Cas9/gRNA T-DNA in T_1 progeny. Images of agarose gels for electrophoresis of PCR amplicons derived from the primers OsCas9-F and OsCas9-R. Lanes 1–10 are T_1 progeny plants, T_0 callus, or plant served as positive control and (lane 11) the wild-type Kitaake DNA served as negative control (–)

its PCR amplicon with the wild-type amplicon. The heteroduplex of DNA fragments formed from the homozygous and wild-type PCR mixture can be cut by T7 endonuclease I, indicating a homozygous plant.

6. Subject a selection of PCR products to Sanger sequencing.
7. Homozygous plants can be used for further gene functional analysis.

3.4.6 Segregate Out the T-DNA from Progeny

1. To make sure that the Cas9/gRNA T-DNA is absent in the T_1 plants, perform PCR-based assay for detection of different regions of Cas9/gRNA.
2. Design PCR primers to detect Cas9, hygromycin resistance gene (*hptII*), and gRNA genes, and for genotyping of individual plants.
3. Include positive control (T_0 generation callus line or plant) and negative control (wild-type Kitaake plant) in the assay. Plants that do not have detectable PCR products of T-DNA will be considered as T-DNA-free plants (Fig. 6).
4. Southern blot or whole-genome sequencing can also be used to further confirm the null segregants if needed.
5. The homozygous seeds from the T_1 generation of the T-DNA free lines can be used for further characterization.

4 Notes

1. A few requirements are needed for choosing the unique target sequence.
   (a) PAM sequence 5′NGG or 5′NAG is required following the target sequence of the guide RNA (PAM should not be included in the spacer sequence of gRNA).
   (b) Off-target mutagenesis could be avoided by BLAST search of the rice genome database.
   (c) Appropriate restriction enzyme site present in the target sequence is optional.
   (d) The 5′-G is required for the stability of gRNA.
2. The BtgZI completely digested vector can be used directly for gene cleaning following the kit manual without gel separation.
3. The intermediate vector is linearized with AatII restriction enzyme before used for the LR recombination. Linearization of entry plasmid can increase the efficiency of Gateway LR recombination.
4. The bacterial colonies are selected on the plate supplemented with two antibiotics (kanamycin and hygromycin B). Pick up the small colonies for further confirmation. We usually observed that right colonies are small probably due to the negative effect of Cas9 on bacterial growth.
5. The binary vector pGW-rCas9 is kanamycin resistant, so it is not suitable to use *Agrobacterium tumefaciens* strains (e.g., EHA101) that are kanamycin resistant.

Closing Remarks: *Cas9/gRNA system is a novel breakthrough technology, revolutionizing research in the area of functional genomics, enabling us to uncover the obscure gene function in basic research and cultivating crop plants with improved agronomic trait improvement. For instance, this genetic tool will be helpful to boost crop yield, increase the nutritional value, and better resist stresses of the globally staple crops.*

Acknowledgments

The authors acknowledge the funding support from National Science Foundation (IOS-1238189 to BY), the National Institute of Food and Agriculture of the US Department of Agriculture (2014-67013-21720 to BY).

References

1. Barrangou R (2013) CRISPR-Cas systems and RNA-guided interference. Wiley Interdiscip Rev RNA 4:267–278
2. Jinek M, Chylinski K, Fonfara I et al (2012) A programmable dual-RNA-guided DNA endonuclease in adaptive bacterial immunity. Science 337:816–821
3. Tsai S, Joung J (2016) Defining and improving the genome-wide specificities of CRISPR-Cas9 nucleases. Nat Rev Genet 17:300–312
4. Nishimasu H, Ran FA, Hsu PD, Konermann S et al (2014) Crystal structure of Cas9 in complex with guide RNA and target DNA. Cell 156:935–949
5. Baltes NJ, Voytas DF (2015) Enabling plant synthetic biology through genome engineering. Trends Biotechnol 33:120–131
6. Hsu P, Lander E, Zhang F (2014) Development and applications of CRISPR-Cas9 for genome engineering. Cell 157:1262–1278
7. Zhou H, Liu B, Weeks DP, Spalding MH, Yang B (2014) Large chromosomal deletions and heritable small genetic changes induced by CRISPR/Cas9 in rice. Nucleic Acids Res 42:10903–10914
8. Brazelton VA, Zarecor S, Wright DA et al (2015) A quick guide to CRISPR sgRNA design tools. GM Crops Food 6:266–276
9. Hiei Y, Ohta S, Komari T et al (1994) Efficient transformation of rice (*Oryza sativa* L.) mediated by *Agrobacterium* and sequence analysis of the boundaries of the T-DNA. Plant J 6:271–282
10. Murray MG, Thompson WF (1980) Rapid isolation of high molecular weight plant DNA. Nucleic Acids Res 8:4321–4325

Chapter 20

Plant Biotechnology Applications of Zinc Finger Technology

Stephen Novak

Abstract

With ever-increasing genomic information combined with modern tools for genome modification, we are entering a new era of plant biotechnology. One major tool used for genome modification is the zinc finger nuclease (ZFN). Here, we discuss how ZFNs have proven useful in many genome modification applications. In order to remove the function of a gene or genes, targeted mutagenesis using ZFNs has been readily demonstrated creating numerous gene knockouts, and gene deletion has been demonstrated with removal of gene segments both native and transgenic up to 9 Mb. Applications for gain of function have also been demonstrated. Precision gene editing using ZFNs has resulted in the development of herbicide tolerance, and numerous forms of targeted gene addition have been exhibited. In addition to genome modification, this chapter also describes the use of zinc finger protein transcription factors (ZFP-TFs) for gene regulation in order to provide useful modification of gene expression resulting in altered phenotypes.

Key words Zinc finger nuclease (ZFN), Zinc finger protein (ZFP), Zinc finger protein transcription factors (ZFP-TFs), Targeted mutagenesis, Gene editing, Gene targeting, Gene deletion, Gene regulation, Double-strand breaks (DSBs), Homologous recombination (HR), Nonhomologous end joining (NHEJ)

1 Introduction

The field of plant biotechnology has made remarkable progress in the last few decades. With the advent of robust plant transformation technologies, greater understanding of plant genomes, and the discovery and creation of tools for genetic manipulation, progress has only accelerated. One such genetic manipulation tool, which is the subject of this chapter, is the recombinant zinc finger domain containing protein. Zinc finger proteins have been used in plant biotechnology in two forms, as zinc finger nucleases (ZFNs) and zinc finger protein transcription factors (ZFP-TFs). The modular nature of zinc finger proteins (ZFPs) offers an attractive framework for designing ZFNs with desired sequence specificities to create targeted double-strand breaks (DSBs) to enable genome engineering. Genome engineering includes targeted mutagenesis, editing

Sandeep Kumar et al. (eds.), *Transgenic Plants: Methods and Protocols*, Methods in Molecular Biology, vol. 1864,
https://doi.org/10.1007/978-1-4939-8778-8_20,

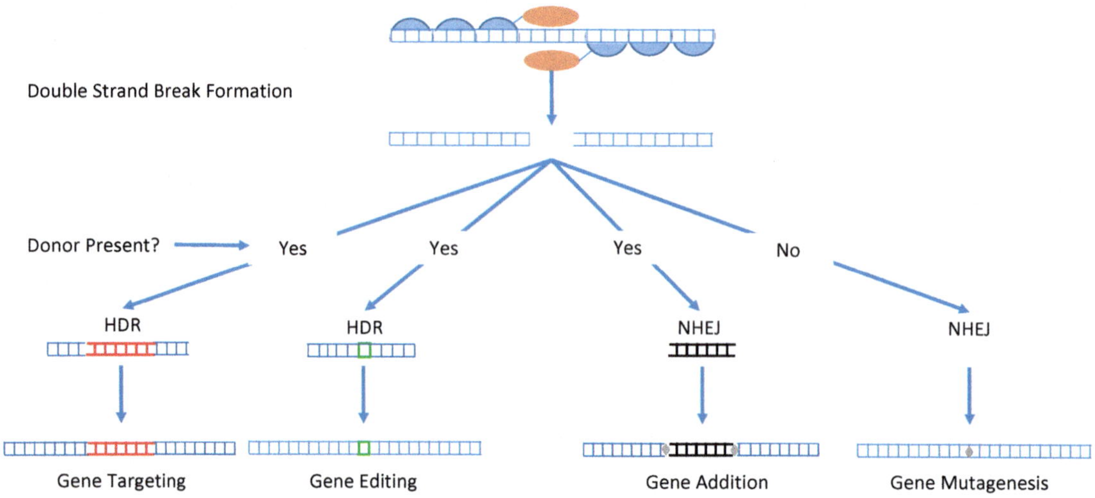

Fig. 1 Examples of different types of DNA modifications using zinc finger nucleases. Upon induction of a DSB by a ZFN, four potential outcomes are possible dependent upon the presence and type of donor molecule present. Gene targeting occurs when a donor is present which contains flanking homology (both 5′and 3′) to the site of the DSB (thus utilizing HDR). In this instance, genetic material situated between the homology flanks is inserted into the genome. Gene editing also utilized a donor which contains flanking homology (both 5′and 3′) to the site of the DSB (thus utilizing HDR). However, in this instance, genetic changes situated between the homology flanks are edited into the genome with no additional DNA inserted. NHEJ-based gene addition occurs when the donor molecule does not contain flanking homology, and instead the donor DNA is inserted into the site of the DSB as a result of imperfect NHEJ repair. When a DSB occurs without a template for DNA repair, imperfect NHEJ repair of the DSB takes place resulting in gene mutagenesis

(precisely replacing a small segment of DNA), gene deletion, and gene addition through precise insertion (Fig. 1).

There are also alternative nucleases which can also be deployed to achieve genome engineering including meganucleases, transcription activator-like effector nucleases (TALENs), and clustered regularly interspaced short palindromic repeats (CRISPR)/CRISPR-associated protein 9 (Cas9) [application for plant biotechnology reviewed in [1–3]].

Zinc finger proteins (ZFPs) also have another novel application for plant biotechnology which is their ability to act as transcriptional regulators. This chapter provides an overview of zinc finger technology development and application to plant biotechnology.

2 Zinc Finger Nuclease Development

The first publication on the development of ZFNs, which were originally called chimeric restriction enzymes, appeared in 1996 [4]. In this pioneering work, an engineered ZF chimeric restriction enzyme was created by linking the nonspecific cleavage domain of the FokI endonuclease with two different ZFP DNA-binding

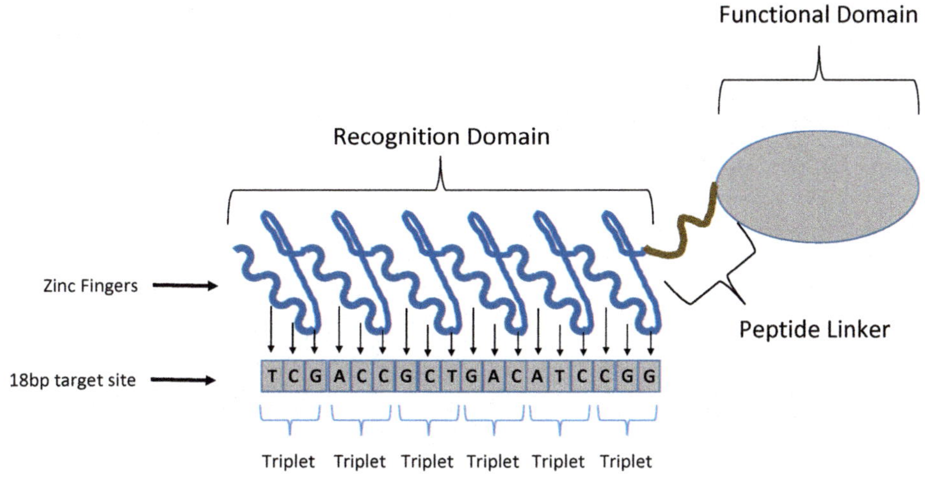

Fig. 2 Anatomy of a zinc finger protein. The chimeric structure of a ZFP consists of an N-terminal DNA recognition domain derived from individual zinc finger proteins in an array. Each zinc finger protein recognizes a DNA triplet. The N-terminal recognition domain is then connected via a peptide linker to a C-terminal functional domain. The characteristics of the functional domain determine the type of activity seen by the ZFP. Examples of functional domains include nucleases, transcriptional activators, transcriptional repressors, and domains which stimulate DNA methylation

domains. Demonstration of the activity of this fusion of ZFP and FokI catalytic domain opened up a new possibility for the creation of custom nucleases which would cut DNA at chosen sites. While much of the original work was focused upon creation of new custom enzymes, it also provided a gateway into the ever-expanding world of biotechnology for targeted genome modification. The chimeric structure of ZFNs consists of an N-terminal DNA-binding domain derived from ZFPs. These are then connected via a peptide linker to a C-terminal FokI nuclease domain (Fig. 2). The initial in vitro work demonstrated ZFN cleavage as a monomer [5]; however, further studies established that ZFN cleavage is more efficient when used as a dimer [6, 7]. Bibikova et al. demonstrated in living *Xenopus laevis* oocytes the ability of ZFNs to find and cleave targeted DNA sites [7]. Authors were able to measure both DNA cleavage and homologous recombination (HR) in these living cells. They also noted that when ZFNs were deployed without a linker between the binding and cleavage domain, binding sites must be 6 bp apart for efficient cleavage. Taking these observations together, it was concluded that in order to create an efficient functional ZFN, two different nucleases oriented in an inverted orientation allow each arm of the ZFN to make contact with a different strand of the DNA (Fig. 3). The sequence in the space between the two different zinc finger binding regions seems to be unimportant outside its 6 bp length.

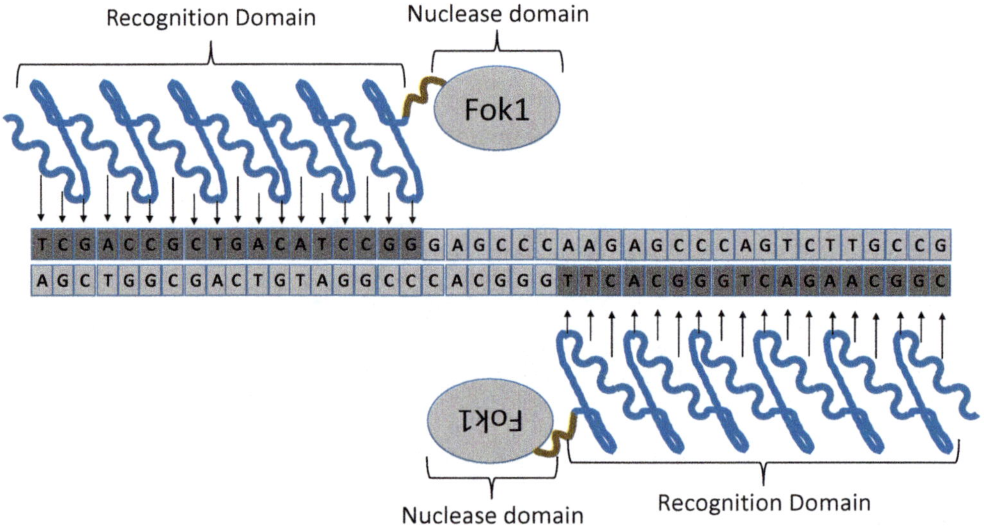

Fig. 3 Zinc finger nuclease architecture. ZFN are effectively deployed when two different ZFNs are oriented in an inverted orientation which allows for each arm of the ZFN to make contact with a different strand of the DNA. Each ZFN specifically recognizes and binds to a segment of DNA (dark gray). The overlapping Fok1 nuclease domains nonspecifically cleave the DNA in the space between the two different zinc finger binding regions

An important step for biotechnology applications was the realization that the creation of double-strand breaks (DSBs) in vivo is known to be an active substrate for both homologous recombination [8] and nonhomologous end joining (NHEJ) [9]. Previous studies had shown the ability to utilize ZF domains with known recognition sites to bind and cut DNA; the next step in ZFN development was to create chimeric fingers which could bind endogenous target genes [10–12].

In order to improve the ability to bind a specific DNA sequence, multiple ZF peptides are typically assembled together in an array. While each individual finger recognizes a 3 bp sequence (Fig. 2), design challenges can arise as assembled ZF arrays exhibit context-related effects [13]. By constructing ZF arrays consisting of three two-finger domains instead of two three-finger units, greater specificity for the target sequence has been achieved. This architecture also enables the capability to span short gaps of unbound DNA [14]. Using the existing design of ZFNs, cleavages by either homodimers (where the same ZFN monomer forms a pair to induce a DSB) or heterodimers (where two different ZFN monomers form a pair to induce a DSB) were both possible. It is also possible that homodimeric ZFNs in a few instances could induce off-target DSBs at unintended genomic sites. To combat this, a novel ZFN architecture using two variants of the FokI cleavage domains improved ZFN fidelity [15]. In one study, it was observed that ZFNs containing these new high-fidelity cleavage domains demonstrated *a* > 40-fold reduction in homodimer formation and a much lower level of off-target cleavage without any reduction in efficiency of target modification.

One of the first examples of gene addition through targeting using ZFNs was accomplished in the fruit fly *Drosophila melanogaster* [16]. In this work, a designed pair of ZFNs was used to target a unique site in the *Drosophila* yellow (*y*) gene. This work demonstrated that ZFN-induced DSBs stimulated targeted gene addition using homologous recombination (HR) [17, 18]. These initial studies in animal systems indicated that ZFNs were a useful tool that enabled targeted modifications in model organisms and paved the way toward application of this tool in plant biotechnology for genome engineering.

3 Targeted Mutagenesis

Various techniques for random mutagenesis have been used in the past (radiation, EMS mutagenesis, T-DNA insertion, to name a few). However, the advent of ZFNs provided the ability to create targeted knockouts by designing sequence-specific ZFNs to create DSBs at pre-defined genomic locations. Creating new alleles through targeted mutations unlocks a whole new area of potential gain in plant biotechnology. The first application of ZFNs in a plant involved targeted mutagenesis of a transgenic target locus in *Arabidopsis* [19]. Constructs containing heat shock-inducible ZFN as well as a target containing a ZFN recognition sequence were introduced into *Arabidopsis*. During seedling development ZFN expression was induced to create DSBs and targeted mutations. The mutations occurred at rates as high as 0.2 per target with 10% of mutants transmitting the mutation to the next generation. Additionally, frameshift mutations comprised 77% of the mutations obtained. Using an OPEN ZFN design scheme to create ZFNs to induce mutations in native plant genes [20], a ZFN pair was used to introduce cleavage of a native gene in tobacco, with one plant showing biallelic cleavage. Targeted mutagenesis has also been shown using a non-transgenic system for ZFN delivery and expression utilizing the tobacco rattle virus (TRV)-based expression system [21].

Targeted modification of an endogenous gene resulting in a demonstrable phenotype is an important application in biotechnology. The targeted inactivation of the endogenous *Arabidopsis* gene *ABA-INSENSITIVE4* (*ABI4*) was accomplished using ZFNs [22]. Resulting lines homozygous for the ABI4 mutation exhibited the expected phenotypes of ABA and glucose insensitivity. Additionally, somatic mutations of endogenous genes in *Arabidopsis* (*ALCOHOL DEHYDROGENASE1* (*ADH1*) and *TRANSPARENT TESTA4* (*TT4*) genes) have also been shown [23]. This was accomplished with the fidelity of the ZFNs also being confirmed, as upon examination of genomic sequences most similar to the *ADH1* and *TT4* ZFN target sites there was no evidence of off-target cleavage.

The first example of ZFN-induced mutation of an endogenous gene in a crop plant [24] was demonstrated in maize (*Zea mays*) cells by expressing an engineered ZFN targeted to exon 2 of the maize *inositol-1,3,4,5,6-pentakisphosphate 2-kinase* (*IPK1*) gene. Targeted mutations of endogenous genes in the cells of economically important plant species such as soybean (*Glycine max*)[25], canola (*Brassica napus*) [26], rice (*Oryza sativa*), tomato [27], fruit trees (apple and fig) [28], and hybrid poplar [29] have also been generated.

As our knowledge and understanding of plant genomes expands, greater opportunities exist for the application of biotechnology to unravel the complex interaction of gene and phenotypes. Targeted gene mutations are very useful to enable new attributes/phenotypes in the altered plants. As described above, the use of ZFNs to create mutations in genomes is certainly well documented. ZFNs have been designed and applied to alter endogenous genes, multigene families, and transgenic loci in model plants, crops, and trees. Taking advantage of the plant's endogenous DNA repair machinery, DSB-induced frameshift mutations occur at a reasonably high rate, and ZFN-induced knockouts have been used to acquire expected phenotypes. ZFN technology has proven quite robust in producing gene mutations, with multiple expression methods (constitutive or induced) as well as multiple methods of DNA delivery (both transient and stable). The mutations created using ZFNs have been shown to be heritable, and the fidelity of the ZFNs has been quite high, with individual genes from multigene families being targeted with little evidence of mutations at nontarget sites. Taken together, these studies describe the application of ZFNs to induce DNA mutations is a useful tool for plant biotechnology.

4 Gene Editing

While creating gene knockouts using ZF technology is a useful application for creating loss of function mutants, gene editing via precise alteration (changing one or more bases in a gene sequence) offers endless possibilities in plant biotechnology. Gene editing could allow alteration of gene activity to create a desired phenotype. As you will soon read, tolerance to several herbicides has been achieved through editing of endogenous genes. However, if applied on a mass scale, gene editing could be used for generating allelic diversity.

Gene editing occurs when, in addition to supplying a ZFN for targeted DSB formation, a donor molecule containing the desired modification (edit) is also provided. When the donor molecule is used as a repair template for the DSB, precise modification of the target site can occur (Fig. 1). One such example of gene editing is the modification of the endogenous tobacco *acetolactate synthase* genes (*ALS SuRA* and *SuRB*) which is the target enzyme for the sulfonylurea and imidazolinone herbicides [30]. Through the use of a ZFN and a donor molecule, specific codon mutations were induced into the *SuR* locus, thus generating *acetolactate synthase* (*ALS*) mutants which were herbicide resistant. Similarly, mutagenesis, using ZFNs designed to cleave the natural *protoporphyrinogen oxidase* (*PPO*) gene in *Arabidopsis*, resulted in another demonstration of gene editing [31]. The edited *PPO* gene contained two separate mutations allowing the PPO enzyme to be insensitive to the butafenacil herbicide. The plant application of ZFNs in gene editing has so far been limited to the target loci that provide herbicide selection for the gene edits. The efficiencies of the process need to be improved for large-scale editing of alleles and creating genetic diversity.

5 Targeted Gene Addition

While creation of gene edits provides minor modification, targeted gene addition using ZFNs allows a much greater range of potential modifications and opens up entirely new possibilities. Targeted gene addition (gene targeting) could allow, for instance, the stacking of traits in a single locus. This form of trait stacking is favorable in the agricultural biotechnology industry as it would streamline the deployment of multiple traits when introduced across germplasms. Moreover, the targeting of a trait to a specific favorable location in the genome could increase the probability of success of a new trait.

Similar to gene editing, gene targeting also utilizes a ZFN and a donor molecule; however, unlike editing, gene targeting results in substantial addition of the donor at the target site (instead of merely changing bases, bases are added). In a tobacco test system, HR-based gene targeting was obtained using a donor and ZFN delivered into protoplasts of transgenic tobacco target lines containing nonfunctional reporter genes with an internal ZFN recognition site [32]. Working with a tobacco cell culture system which was pre-engineered to contain a reporter construct having dual nonfunctional reporter genes, ZFNs were used to test several types of recombination [33]. Firstly, intrachromosomal recombination leading to reconstituted functional reporter gene was accomplished following DSBs induced by a ZFN. Secondly, induced DSBs upstream of previously integrated target locus containing intron and 3'fragment of the selectable marker were repaired by a donor DNA containing the 5′ fragment of the selectable marker and an intron as homology to the target region. Homology-directed repair (HDR) was used to produce gene targeting leading to reconstitution of selectable marker. Lastly, gene targeting into the endogenous tobacco *endochitinase* gene was demonstrated using a construct containing both a ZFN and a selectable marker-containing donor introduced into tobacco cells. Resistant isolates obtained on selection were PCR confirmed indicating HDR gene targeting to the native *endochitinase* gene.

ZFNs designed to induce DSBs at the endogenous maize *inositol-1,3,4,5,6-pentakisphosphate 2-kinase* (*IPK1*) gene were used to facilitate gene targeting [24]. Disruption of the IPK1 gene by gene addition was accomplished through the simultaneous delivery of both a ZFN and donor molecule via WHISKERS mediated transformation of maize cells. Two different donor constructs were utilized, each containing short homology arms: one carried an autonomous selectable marker gene expression cassette, whereas the second carried a nonautonomous donor that relied on precise trapping of the endogenous *IPK1* promoter for expression of the selectable marker. Heritability of the gene targeting to the next generation was demonstrated, and the *IPK1* disruptions caused by the gene targeting resulted in herbicide tolerance as well as alteration in the seed inositol phosphate profile. This work demonstrates that precise modification of an endogenous gene in an agriculturally important crop plant is indeed possible by creating ZFN-mediated DSB formation and homology-directed repair of DSB by exogenously supplied donor DNA.

Creation of stacked transgenic traits using ZFNs is another important application in plant biotechnology. The current gene stacking process relies on breeding stacks of randomly inserted events that need to be sorted independently, characterized, and introgressed into elite germplasm, which is a time-consuming and expensive process. Creation of transgenic trait stacks using ZFNs to precisely integrate a new transgene into a previously integrated trait landing pad (TLP) has been demonstrated (Fig. 4a) [34]. In this work, herbicide resistance traits were stacked using precision gene targeting. The modification in the resulting targeted stacked events was inherited in the subsequent generation, with the traits co-segregating as linked transgenes in a Mendelian fashion. After stacking, protein expression and herbicide resistance for both genes were confirmed.

Further advances in the efficiency of transgenic trait stacking were made using a system that allows selection for ZFN-induced gene targeting via nuclease-mediated cassette exchange (NMCE) (Fig. 4b) [35, 36]. ZFN-induced DSB formation enabled gene targeting at a high frequency in an existing transgene using the homology and selection for precise recombination. To overcome

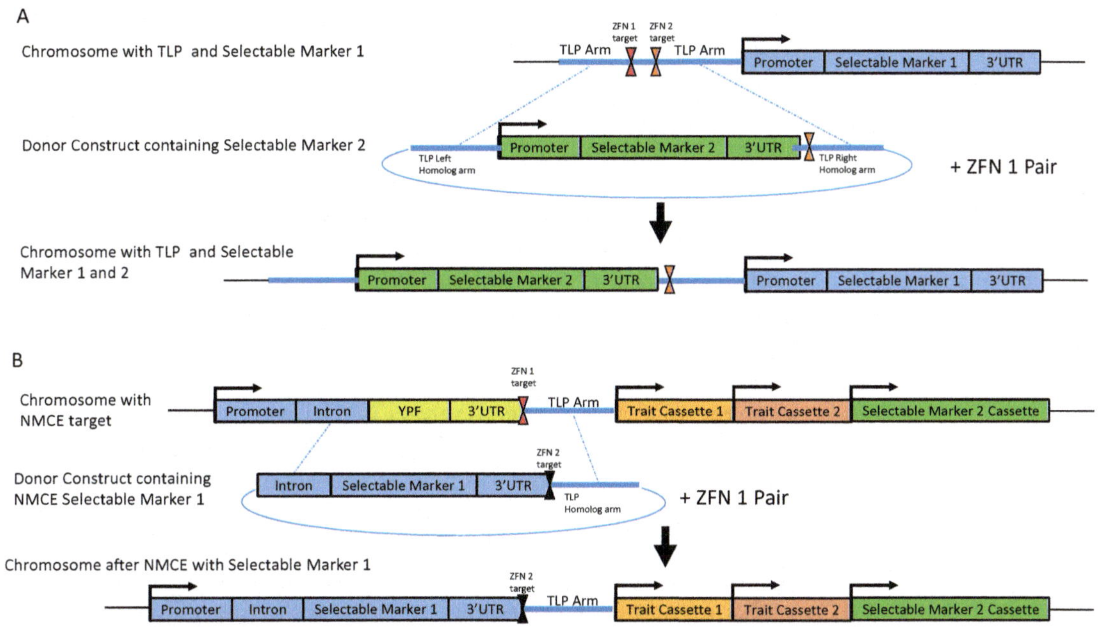

Fig. 4 Gene targeting into a transgenic event. Two different methods for gene targeting into a transgenic event are illustrated. (**a**) (top) An integrated trait landing pad (TLP) construct containing a TLP (bold blue line) with two ZFN target sites and selectable marker 1 cassette (blue boxes) is integrated into the chromosome of a plant (thin black line). The event is subsequently transformed with both a ZFN pair and a donor DNA construct containing the selectable marker 2 cassette (green boxes). Results show predicted targeting via HDR-directed donor integration into the TLP utilizing the TLP homology arms. (**b**) (bottom) A nuclease-mediated cassette exchange (NMCE) construct containing a promoter with an intron (blue boxes), a YFP cassette and 3′UTR (yellow boxes) with a ZFN recognition site (red) at the 3′end, a TLP arm (bold blue line), along with two trait cassettes (two orange boxes) and selectable marker 2 cassette (green boxes), is integrated into the chromosome of a plant (thin black line). The event is subsequently transformed with both a ZFN pair and a donor DNA construct containing an intron homologous to the aforementioned selectable marker 1 and 3′UTR (blue boxes) with a ZFN recognition site (black) at the 3′end. Results show predicted precise cassette exchange utilizing HDR-directed donor integration into the transgene driven by homology of the intron and TLP arm

another bottleneck for gene targeting in plants due to low efficiencies of plant transformation, a more efficient system for deployment of ZFNs for gene targeting was developed. To this end, a system to convert two unlinked transgenic loci into a targeted single genetic stack using ZFNs was created [37]. In this work, stably integrated target and donor transgenes were stacked via intra-genomic homologous recombination (IGHR) by crossing to a transgenic line expressing a corresponding ZFN. Using this approach, 1% of treated embryos generated plants with the donor sequence precisely integrated into the target locus. This work demonstrates improvement of the gene targeting process through the use of nuclease-mediated cassette exchange (NMCE) and stably integrated ZFN and donor to more effectively deliver gene targeting components to cells and overcome the inefficiencies of plant transformation. The use of a geminivirus replicon as an alternative means for delivery of reagents for ZFN-induced gene targeting was demonstrated in potato [38]. It was found that there was a twofold increase in targeting activity when reagents were delivered using a geminivirus replicon over conventional reagent delivery methods.

In addition to gene targeting using homologous recombination, NHEJ-mediated gene replacement has also been achieved in both tobacco and *Arabidopsis* [39]. A previously integrated transgene harboring flanking ZFN recognition sites between the coding region and the promoter/3'UTR of reporter gene was created as a target locus. Evidence of ZFN-mediated mutagenesis, and reporter gene removal and replacement via NHEJ were observed following transformation of the target line with a ZFN expressing transgene and a donor with a promoterless selectable marker gene flanked by ZFN recognition sequences. The insertion of the selectable marker gene was found to be linked to the target DNA construct, which allowed the plants to survive selection. The NHEJ-mediated gene exchange was stably transmitted to the following generation.

Randomly inserted transgenic events need to be sorted and characterized in order to create biotech plant products [40]. When transgenes are inserted randomly in the plant genome, the majority of events insert into locations which are unfavorable for a desired gene expression needed for a robust and durable product. For example, insertion into repetitive regions of the genome could cause issues with both insertion detection and long-term stability of expression. It is also undesirable for insertions to take place inside an endogenous gene, as this may have an effect on plant health. Other properties of the locus, such as recombination frequency as well as regions which allow for high expression, are also important factors. One potential application of precision targeting technology is to insert transgenes into previously identified safe harbor locations in the genome [41].

Targeted gene addition has long been sought in plant biotechnology and has now been deployed in many forms. Gene targeting using ZFNs has been demonstrated in transgenic and endogenous loci of both model and crop plants alike. Two different types of DNA repair (NHEJ and HDR) have been utilized for targeted insertion of transgene. As new advances are made in the field, inefficiencies in the process are being addressed. One of the main applications of this technology is be the stacking of biotech traits, which has been exemplified but not yet commercialized. The potential value in agricultural biotechnology is enormous, especially considering the size of the effort dedicated to breeding and introgression of biotech traits in commercial agriculture.

6 Gene Deletion

Targeted mutagenesis is not the only means of removing the function of a gene. In fact, a more thorough and elegant means to remove a gene's function would be to delete it entirely. Gene deletion can be accomplished by creating a ZFN-mediated DSB on both sides of the target sequence and thus removing the intervening segment. In addition, ZFN-induced DSB made adjacent to genomic regions of high homology can trigger HDR resulting in the deletion of the intervening segment of DNA.

ZFNs have been used to induce the deletion of a 2.8 kb stretch of DNA [33] via intrachromosomal recombination of homologous sequences, which allowed reconstitution of a reporter gene in tobacco cells. ZFNs were also used for targeted gene deletions in transgenic tobacco plants [42]. Stably transformed tobacco plants containing a target transgene harboring a reporter gene flanked by ZFN binding sites were created. The target plants were crossed using transgenic plants containing a corresponding ZFN expression cassette and the resulting progeny analyzed. The frequency of reporter gene removal ranged from 0% up to 35% depending upon the cross with an overall average of 6.42%. Deletion of the 4.3 kb reporter cassette was confirmed by Southern blot, sequencing and through phenotype observation. The deletion was also transmitted to the F2 progeny.

The use of ZFNs to create deletions in tandemly arrayed genes (TAG) has been demonstrated in *Arabidopsis* [43]. Using ZFNs, seven target genes for three different TAGs were successfully targeted for deletion. The size of deletions ranged from 4.5 kb to 55 kb with a frequency of ~1% in vegetative tissue, but germline transmission of these deletions was not observed. Deletions as large as ~9 Mb have been obtained but with a much lower frequency (0.046%). In addition to deletions, ZFN-induced inversions and duplications of these gene clusters have also been obtained, thus showing the ability to create novel genomic regions through targeted DSBs creating chromosome rearrangements and/or chimeric genes.

Whether deployed in native or transgenic loci, ZFNs have removed segments of DNA (with sizes as large as ~9 Mb). Depending upon the concept, this may have great utility in creating genetic diversity. This possibility for genetic alteration is profound considering that deletions, inversion, and duplications have all been demonstrated as a result of ZFN-induced DSBs.

7 Gene Regulation

The capability to regulate gene expression offers intriguing potential for basic biology and plant biotechnology applications. As we gain greater understanding how gene expression impacts plant phenotypes, the potential benefits for gene regulation grow. By combining understanding of the interface between the ZF and DNA [44] with knowledge about transcription regulators [45] and engineered DNA-binding domains [10], all the necessary ingredients are available to create engineered zinc finger protein transcription factors (ZFP-TFs) to regulate gene expression.

In an early example, ZFP-TFs were designed to the mapped DNase I accessible regions within the *vascular endothelial growth factor A* (*VEGF-A*) gene [46]. Fused to the acidic activation domain VP16 which is derived from the herpes simplex virus, or the activation domain from the p65 subunit of the human NF-κB transcription factor p65 subunit, these ZFP-TFs induced activation of *VEGF-A* expression in HEK293 cells.

The application of this technology in plants initiated with a study on the *Arabidopsis APETALA3* (*AP3*) gene. ZFP-TFs were designed to target ~50 bp upstream of the TATA box [47]. When introduced into leaf protoplasts, ZFPs fused to the VP16 activation domain resulted in both increased expression of a transgenic reporter and floral phenotypes consistent with expectation in stable plants. When the ZFP was fused to the mammalian repression domain (mSIN3), dramatic repression was also observed. These findings indicate that engineered ZFP-TFs were capable of gene regulation in plant cells. A further demonstration of ZFP-TFs has shown an astounding 450-fold increase in reporter gene expression in stable tobacco plants. In addition, tissue-specific activation has also been shown in vascular tissue using a ZFP-TF expression driven by phloem-specific promoter [48]. Additionally, ZFP-TF-induced activation was shown to respond in a dose-dependent manner when induced chemically [49]. Moreover, the specificity afforded by the ZFP-TF for gene activation has been confirmed as a single bp mutation in the ZFP-TF recognition site led to a 33-fold decrease in expression levels. The location of the ZFP-TF binding site has a great influence on the level of activation, as activation reduced significantly when the binding site was moved further away from the TATA box [50]. Furthermore, transcription activation domains derived from plants were created, which allowed higher activation of expression compared to the VP16 domain used previously [51].

The utility of ZFP-TFs has been exemplified in the model crop *Arabidopsis*, where ZFP-TFs completely derived from plant sequences exhibited the capability to upregulate expression of an endogenous target gene in a whole plant. Also, seed-specific expression of ZFP-TFs resulted in elevated seed α-tocopherol for several lines [52]. The application of engineered ZFP-TFs to up- and downregulate expression of an endogenous plant gene was demonstrated. Transgenic ZFP-TF lines with an activation domain showed increased lignin content, while transgenic ZFP-TF lines with a repression domain showed decreased lignin content. [53].

Engineered ZFP-TFs have also been used to increase the expression of an endogenous gene in the crop plant canola (*Brassica napus*) [54]. Using an engineered ZFP-TF with the VP16 activation domain designed to bind the 5'UTR of two different *β-ketoacyl-ACP synthase II* (*KASII*) genes, transgenic plants with as much as 3.5-fold increase in *KASII* expression were produced resulting in leaves and seeds exhibiting statistically significant decreases in palmitic acid content, reduced total saturated fatty acid content, and increased total C18 content. Further alteration of expression and fatty acid content was achieved by both activation and repression using ZFP-TF designed for the canola (*Brassica napus*) *fatty acid thioesterase B4* (*FATB4*) and *fatty acid thioesterase B5* (*FATB5*) genes [55].

Another exciting application of ZFPs for gene regulation is the modulation of gene expression by targeted DNA methylation. Recruitment of RNA polymerase V (Pol V) was accomplished using a ZFP tethered to the nonredundant SET domain protein SUVH2 [56]. As a result, DNA methylation surrounding the ZFP-binding sites was observed, with subsequent gene silencing leading to an observable phenotype. Of note, the DNA methylation and gene silencing were maintained even after the ZFP had been segregated away, indicating that targeted DNA methylation and gene silencing were heritable.

Engineered ZFPs have tremendous utility in the regulation of plant genes. Regulation of endogenous genes as well as transgenes has been demonstrated in both model and crop plants. In many instances this expression regulation has led to modification of phenotypes. The use of ZFPs enables expression modulation leading to a range of phenotypes, which may not be possible by knockout created via targeted mutagenesis. Many possibilities exist in the application of this technology. Through the use of tissue-specific promoters, one can envision tissue-specific gene regulation which would provide for an enhanced realm of possible phenotypes. This method would be particularly useful for regulation of genes which have severe phenotypes when altered constitutively. The use of ZFPs to produce heritable epigenetic modification also creates many new possibilities.

8 Conclusions

It is quite clear that the future holds many possibilities for the applications of ZF technology to plant biotechnology. As we discover and understand more about the biology of plants and the complexities of the genome, the opportunity for the application of biotechnology based upon this knowledge expands as well. Zinc finger proteins provide a diverse set of tools to take advantage of this information and leverage it for the improvement of plants through the discovery and creation of new traits. Zinc fingers have proven to be capable tools for the precise modification of gene expression and for precise modification of genomes. Based upon the numerous examples given, whether the application is a new trait created through expression modulation, precision mutagenesis, gene editing, creation of new alleles/germplasm through gene deletion, or faster deployment of new transgenic trait stacks, zinc fingers have demonstrable utility as a versatile tool for plant biotechnology.

References

1. Voytas DF (2013) Plant genome engineering with sequence-specific nucleases. In: Merchant SS (ed) Annual review of plant biology, vol 64. pp 327–350
2. Petolino JF (2015) Genome editing in plants via designed zinc finger nucleases. In Vitro Cell Dev Biol-Plant 51(1):1–8
3. Cardi T, Stewart CN (2016) Progress of targeted genome modification approaches in higher plants. Plant Cell Rep 35 (7):1401–1416
4. Kim YG, Cha J, Chandrasegaran S (1996) Hybrid restriction enzymes: Zinc finger fusions to Fok I cleavage domain. Proc Natl Acad Sci U S A 93(3):1156–1160
5. Smith J, Berg JM, Chandrasegaran S (1999) A detailed study of the substrate specificity of a chimeric restriction enzyme. Nucleic Acids Res 27(2):674–681
6. Smith J, Bibikova M, Whitby FG, Reddy AR, Chandrasegaran S, Carroll D (2000) Requirements for double-strand cleavage by chimeric restriction enzymes with zinc finger DNA-recognition domains. Nucleic Acids Res 28(17):3361–3369
7. Bibikova M, Carroll D, Segal DJ, Trautman JK, Smith J, Kim YG, Chandrasegaran S (2001) Stimulation of homologous recombination through targeted cleavage by chimeric nucleases. Mol Cell Biol 21(1):289–297
8. Choulika A, Perrin A, Dujon B, Nicolas JF (1995) Induction of homologous recombination in mammalian chromosomes by using the I-SceI system of saccharomyces-cerevisiae. Mol Cell Biol 15(4):1968–1973
9. Lieber MR (2010) The mechanism of double-strand DNA break repair by the nonhomologous DNA end-joining pathway. Annu Rev Biochem 79:181–211
10. Pabo CO, Peisach E, Grant RA (2001) Design and selection of novel Cys(2)His(2) zinc finger proteins. Annu Rev Biochem 70:313–340
11. Isalan M, Choo Y (2001) Engineering nucleic acid-binding proteins by phage display. DNA-Protein Interactions: Principles and Protocols:417–429
12. Beerli RR, Barbas CF (2002) Engineering polydactyl zinc-finger transcription factors. Nat Biotechnol 20(2):135–141
13. Urnov FD, Rebar EJ, Holmes MC, Zhang HS, Gregory PD (2010) Genome editing with engineered zinc finger nucleases. Nat Rev Genet 11(9):636–646
14. Moore M, Klug A, Choo Y (2001) Improved DNA binding specificity from polyzinc finger peptides by using strings of two-finger units. Proc Natl Acad Sci 98(4):1437–1441
15. Miller JC, Holmes MC, Wang J, Guschin DY, Lee Y-L, Rupniewski I, Beausejour CM, Waite AJ, Wang NS, Kim KA et al (2007) An improved zinc-finger nuclease architecture for

highly specific genome editing. Nat Biotechnol 25(7):778–785
16. Bibikova M, Golic M, Golic KG, Carroll D (2002) Targeted chromosomal cleavage and mutagenesis in Drosophila using zinc-finger nucleases. Genetics 161(3):1169–1175
17. Bibikova M, Beumer K, Trautman JK, Carroll D (2003) Enhancing gene targeting with designed zinc finger nucleases. Science 300 (5620):764–764
18. Porteus MH, Carroll D (2005) Gene targeting using zinc finger nucleases. Nat Biotechnol 23 (8):967–973
19. Lloyd A, Plaisier CL, Carroll D, Drews GN (2005) Targeted mutagenesis using zinc-finger nucleases in Arabidopsis. Proc Natl Acad Sci U S A 102(6):2232–2237
20. Maeder ML, Thibodeau-Beganny S, Osiak A, Wright DA, Anthony RM, Eichtinger M, Jiang T, Foley JE, Winfrey RJ, Townsend JA et al (2008) Rapid “open-source” engineering of customized zinc-finger nucleases for highly efficient gene modification. Mol Cell 31 (2):294–301
21. Marton I, Zuker A, Shklarman E, Zeevi V, Tovkach A, Roffe S, Ovadis M, Tzfira T, Vainstein A (2010) Nontransgenic genome modification in plant cells. Plant Physiol 154 (3):1079–1087
22. Osakabe K, Osakabe Y, Toki S (2010) Site-directed mutagenesis in Arabidopsis using custom-designed zinc finger nucleases. Proc Natl Acad Sci U S A 107(26):12034–12039
23. Zhang F, Maeder ML, Unger-Wallace E, Hoshaw JP, Reyon D, Christian M, Li X, Pierick CJ, Dobbs D, Peterson T et al (2010) High frequency targeted mutagenesis in Arabidopsis thaliana using zinc finger nucleases. Proc Natl Acad Sci U S A 107(26):12028–12033
24. Shukla VK, Doyon Y, Miller JC, DeKelver RC, Moehle EA, Worden SE, Mitchell JC, Arnold NL, Gopalan S, Meng X et al (2009) Precise genome modification in the crop species Zea mays using zinc-finger nucleases. Nature 459 (7245):437–U156
25. Curtin SJ, Zhang F, Sander JD, Haun WJ, Starker C, Baltes NJ, Reyon D, Dahlborg EJ, Goodwin MJ, Coffman AP et al (2011) Targeted mutagenesis of duplicated genes in soybean with zinc-finger nucleases. Plant Physiol 156(2):466–473
26. Gupta MP, Palta AM, Stephen N, Fyodor U, Sunita G (2013) Engineered zinc finger proteins targeting 5-enolpyruvyl shikimate-3-phosphate synthase genes. In: USPTO (ed) USPTO. Dow Agrosciences, LLC, Sangamo Bio Sciences, Inc., United States of America
27. Shukla V, Gupta M, Urnov F, Guschin D, De Both MJ, Bundock P, Sastry-Dent L (2016) Targeted modification of malate dehydrogenase. In: USPTO (ed) USPTO. Dow Agrosciences LLC, Sangamo Biosciences, Inc., United States of America
28. Peer R, Rivlin G, Golobovitch S, Lapidot M, Gal-On A, Vainstein A, Tzfira T, Flaishman MA (2015) Targeted mutagenesis using zinc-finger nucleases in perennial fruit trees. Planta 241 (4):941–951
29. Lu HW, Klocko AL, Dow M, Ma C, Amarasinghe V, Strauss SH (2016) Low frequency of zinc-finger nuclease-induced mutagenesis in Populus. Mol Breed 36(9):121
30. Townsend JA, Wright DA, Winfrey RJ, Fu FL, Maeder ML, Joung JK, Voytas DF (2009) High-frequency modification of plant genes using engineered zinc-finger nucleases. Nature 459(7245):442–U161
31. de Pater S, Pinas JE, Hooykaas PJJ, van der Zaal BJ (2013) ZFN-mediated gene targeting of the Arabidopsis protoporphyrinogen oxidase gene through Agrobacterium-mediated floral dip transformation. Plant Biotechnol J 11(4):510–515
32. Wright DA, Townsend JA, Winfrey RJ Jr, Irwin PA, Rajagopal J, Lonosky PM, Hall BD, Jondle MD, Voytas DF (2005) High-frequency homologous recombination in plants mediated by zinc-finger nucleases. Plant J 44 (4):693–705
33. Cai CQ, Doyon Y, Ainley WM, Miller JC, DeKelver RC, Moehle EA, Rock JM, Lee YL, Garrison R, Schulenberg L et al (2009) Targeted transgene integration in plant cells using designed zinc finger nucleases. Plant Mol Biol 69(6):699–709
34. Ainley WM, Sastry-Dent L, Welter ME, Murray MG, Zeitler B, Amora R, Corbin DR, Miles RR, Arnold NL, Strange TL et al (2013) Trait stacking via targeted genome editing. Plant Biotechnol J 11(9):1126–1134
35. Kumar S, AlAbed D, Worden A, Novak S, Wu H, Ausmus C, Beck M, Robinson H, Minnicks T, Hemingway D et al (2015) A modular gene targeting system for sequential transgene stacking in plants. J Biotechnol 207:12–20
36. Schneider K, Schiermeyer A, Dolls A, Koch N, Herwartz D, Kirchhoff J, Fischer R, Russell SM, Cao ZH, Corbin DR et al (2016) Targeted gene exchange in plant cells mediated by a zinc finger nuclease double cut. Plant Biotechnol J 14(4):1151–1160

37. Kumar S, Worden A, Novak S, Lee R, Petolino JF (2016) A trait stacking system via intra-genomic homologous recombination. Planta 244(5):1157–1166
38. Butler NM, Baltes NJ, Voytas DF, Douches DS (2016) Geminivirus-mediated genome editing in potato (Solanum tuberosum L.) using sequence-specific nucleases. Front Plant Sci 7:1045
39. Weinthal DM, Taylor RA, Tzfira T (2013) Nonhomologous end joining-mediated gene replacement in plant cells(1 C). Plant Physiol 162(1):390–400
40. Mumm RH, Walters DS (2001) Quality control in the development of transgenic crop seed products. Crop Sci 41(5):1381–1389
41. Cantos C, Francisco P, Trijatmiko KR, Slamet-Loedin I, Chadha-Mohanty PK (2014) Identification of "safe harbor" loci in indica rice genome by harnessing the property of zinc-finger nucleases to induce DNA damage and repair. Front Plant Sci 5:302
42. Petolino JF, Worden A, Curlee K, Connell J, Moynahan TLS, Larsen C, Russell S (2010) Zinc finger nuclease-mediated transgene deletion. Plant Mol Biol 73(6):617–628
43. Qi Y, Li X, Zhang Y, Starker CG, Baltes NJ, Zhang F, Sander JD, Reyon D, Joung JK, Voytas DF (2013) Targeted deletion and inversion of tandemly arrayed genes in Arabidopsis thaliana using zinc finger nucleases. G3 (Bethesda) 3(10):1707–1715
44. Pavletich NP, Pabo CO (1991) Zinc finger DNA recognition – crystal-structure of a Zif268-DNA complex at 2.1-A. Science 252 (5007):809–817
45. Sainz MB, Goff SA, Chandler VL (1997) Extensive mutagenesis of a transcriptional activation domain identifies single hydrophobic and acidic amino acids important for activation in vivo. Mol Cell Biol 17(1):115–122
46. Liu PQ, Rebar EJ, Zhang L, Liu Q, Jamieson AC, Liang YX, Qi H, Li PX, Chen BL, Mendel MC et al (2001) Regulation of an endogenous locus using a panel of designed zinc finger proteins targeted to accessible chromatin regions - Activation of vascular endothelial growth factor A. J Biol Chem 276 (14):11323–11334
47. Guan X, Stege J, Kim M, Dahmani Z, Fan N, Heifetz P, Barbas CF, Briggs SP (2002) Heritable endogenous gene regulation in plants with designed polydactyl zinc finger transcription factors. Proc Natl Acad Sci U S A 99(20):13296–13301
48. Ordiz MI, Barbas CF, Beachy RN (2002) Regulation of transgene expression in plants with polydactyl zinc finger transcription factors. Proc Natl Acad Sci U S A 99 (20):13290–13295
49. Sanchez JP, Ullman C, Moore M, Choo Y, Chua NH (2002) Regulation of gene expression in Arabidopsis thaliana by artificial zinc finger chimeras. Plant Cell Physiol 43 (12):1465–1472
50. Stege JT, Guan X, Ho T, Beachy RN, Barbas CF (2002) Controlling gene expression in plants using synthetic zinc finger transcription factors. Plant J 32(6):1077–1086
51. Li JQ, Blue R, Zeitler B, Strange TL, Pearl JR, Huizinga DH, Evans S, Gregory PD, Urnov FD, Petolino JF (2013) Activation domains for controlling plant gene expression using designed transcription factors. Plant Biotechnol J 11(6):671–680
52. Van Eenennaam AL, Li GF, Venkatramesh M, Levering C, Gong XS, Jamieson AC, Rebar EJ, Shewmaker CK, Case CC (2004) Elevation of seed alpha-tocopherol levels using plant-based transcription factors targeted to an endogenous locus. Metab Eng 6(2):101–108
53. Sanchez JP, Ullman C, Moore M, Choo Y, Chua NH (2006) Regulation of Arabidopsis thaliana 4-coumarate:coenzyme-A ligase-1 expression by artificial zinc finger chimeras. Plant Biotechnol J 4(1):103–114
54. Gupta M, DeKelver RC, Palta A, Clifford C, Gopalan S, Miller JC, Novak S, Desloover D, Gachotte D, Connell J (2012) Transcriptional activation of Brassica napus β-ketoacyl-ACP synthase II with an engineered zinc finger protein transcription factor. Plant Biotechnol J 10 (7):783–791
55. DeKelver RG, Manju G, Miller Jeffrey C, Stephen N, Petolino Joseph F (2013) Engineered zinc finger proteins targeting plant genes involved in fatty acid biosynthesis. In: USPTO (ed) USPTO. Dow Agrosciences LLC, Sangamo Biosciences, Inc., Unites States of America
56. Johnson LM, Du J, Hale CJ, Bischof S, Feng S, Chodavarapu RK, Zhong X, Marson G, Pellegrini M, Segal DJ et al (2014) SRA- and SET-domain-containing proteins link RNA polymerase V occupancy to DNA methylation. Nature 507(7490):124–128

Part III

Utility of Transgenic Traits

Chapter 21

Overview of Biotechnology-Derived Herbicide Tolerance and Insect Resistance Traits in Plant Agriculture

Tejinder Mall, Manju Gupta, Tarlochan Singh Dhadialla, and Sarria Rodrigo

Abstract

Biotechnology has been central for the acceleration of crop improvement over the last two decades. Since 1994, when the first commercial biotechnology-derived tomato crop was commercialized, the cultivated area for genetically modified crops has reached 185.1 million hactares worldwide. Both the number of crops and the number of traits developed using biotechnology have accounted for this increase. Among the most impactful biotechnology-derived traits are insect resistance and herbicide tolerance, which have greatly contributed to the worldwide increase in agricultural productivity and stabilization of food security. In this chapter, we provide an overview of the history of the biotechnology-derived input traits, the existing genetically engineered commercial crop products carrying insect resistance and herbicide tolerance traits, as well as a perspective on how new technologies could further impact the development of new traits in crops. With the projection of the world population to increase to 9.8 billion by the year 2050 and reduction in available farmland, one of the biggest challenges will be to provide sustainable nourishment to the projected population. Biotechnology will continue to be the key enabler for development of insect resistance and herbicide tolerance traits to overcome that imminent challenge.

Key words Biotechnology, Insect resistance, Herbicide tolerance, Molecular characterization

1 Introduction

Over the last two decades, genetically engineered (GE) crops have been developed in multiple plant species (maize, soybean, cotton, canola, eggplant, etc.) to safeguard against insect pests, disease, weed, and abiotic stress damage to obtain higher yields. Up until 2016, these GE crops have been planted in 30 countries covering about 457 million acres and adopted by ~18 million farmers [1]. The numbers represent a very fast adoption of GE crops since 1996 when the crop acreages increased by more than 100-fold from only 4.2 million acres.

Currently, the major countries growing GE crops are the USA, Brazil, and Argentina. These crops have also been adopted in other

Sandeep Kumar et al. (eds.), *Transgenic Plants: Methods and Protocols*, Methods in Molecular Biology, vol. 1864, https://doi.org/10.1007/978-1-4939-8778-8_21, © Springer Science+Business Media, LLC, part of Springer Nature 2019

countries including other parts of North America, Latin America, China, India, Australia, the Philippines, Myanmar, South Africa, and Spain, and their adoption is expected to increase in additional countries in the near future [2]. So far the major focus for the development of GE crops has been in soybean, maize, canola, cotton, and sugar beet. However, GE traits have been produced and cultivated in other crops such as alfalfa, eggplant, papaya, potato, pineapple, squash, apple, plum, eucalyptus, and poplar [3].

It is projected that by 2050, the world population will increase to 9.8 billion [4] and this population growth is projected to outpace the current rate of crop yield increase (1.3% per year) due to declining land resources for agriculture and global warming adding to stressed environmental conditions for optimal crop productivity [5]. Hence, there is an urgent need to develop crops for stressed environments, e.g., abiotic, insect pests, weeds, and crop disease, that would also give higher crop yield in order to feed the evergrowing population. The grain yield for corn increased by 0.8 bushels/acre/year from 1937 to 1955, due to adoption of hybrid corn production. Subsequently, a second phase of yield increase of 1.9 bushels/acre/year occurred from 1955 until now due to continuous improvements in corn genetics, biotechnology-derived traits, and better agronomic production practices [6]. A potential third phase of yield increase is yet to come through the integration of new technologies beyond those used in currently commercialized GE crops.

In this chapter, we provide an overview of insect resistance (IR) and herbicide tolerance (HT) GE crops, including historical perspectives, commercialized crop products, and future perspectives. For more in-depth knowledge of the topics covered in this chapter, the reader is referred to excellent reviews by other authors [7–16].

2 Historical Perspective on Input Trait Biotechnology

2.1 Herbicide Tolerance

An important breakthrough came around the year 1983 when deliberate efforts were made to transform genes into crop plants [17–19]. It was not until over a decade later in 1996 that the first commercial transgenic crop—"Flavr Savr" fresh tomatoes—was marketed by Calgene (now Monsanto). These tomatoes were the first example of the commercial value and impact biotechnology could have in a wide variety of traits in plant agriculture. Biotechnology techniques were also used to introduce genes for herbicide tolerance (HT) into soybean, cotton, corn, and canola crops [20]. HT crops help farmers in reducing production costs and environmental impact of using herbicides for weed control. Genetically engineered HT crops also eliminate the need for manual removal of weeds that compete with the crop of interest (COI)

[21] and help to preserve topsoil microflora by encouraging the use of non-tillage agricultural practices [22]. A variety of HT traits have been developed for selective control of weeds using glyphosate, glufosinate, 2,4-D, and other herbicides maintaining other agronomic attributes of GE crops. A premise for the development of HT traits is the availability and understanding of the herbicide chemistry, their environmental impact, ease of use, cost-effectiveness, and their mode of action. Furthermore, it is also important to consider the type of trait stewardship program needed after the trait is commercialized to minimize environmental impact, achieve trait longevity, and avoid development of resistant weeds in specific geographies where the technology is used. Without doubt, the HT traits presently available provide farmers great convenience, increase crop productivity and management efficiency, and also provide environmental benefits that impact soil health and water quality in their farms.

2.2 Insect Resistance

To date most of the bioinsecticides and insecticidal proteins have come from the soil bacterium *Bacillus thuringiensis* (*Bt*). The first *Bacillus* bacterium (*Bacillus sotto*) was discovered and isolated from silkworms in 1901, and subsequently *Bt* was isolated from Mediterranean flour moth caterpillars after the dead caterpillars were found to be loaded with *Bt* spores and crystals. This observation led to the conclusion that *Bt* spores and crystals had insecticidal properties, which led Mattes [23] to isolate the strain for subsequent use in field trials against the European corn borer, *Ostrinia nubilalis*, with promising results. This work eventually led to the development of Sporeine, a commercial *Bt* insecticide for lepidopteran pest control, used initially in 1938 in France, and registered for use in the USA in 1961. Since then, through strain improvement and improved fermentation processes, a number of *Bacillus* bioinsecticides have been commercialized for control of lepidopteran (e.g., Biobit, Dipel, Thuricide, and Javelin), dipteran (e.g., Tekar, Bactimos, Vectolex GC, and Acrobe), and coleopteran (Novodor and Trident) pests [8]. The next stage in the evolution of *Bacillus* bioinsecticides was the production of hybrid *Bacillus* strains with increased toxicity and insecticidal specificity. This was achieved by either conjugation of two different strains with different insecticidal protein-encoding genes or direct transformation of one strain with insecticidal genes from another *Bacillus* strain [24, 25]. This practice would be equivalent to today's "pyramiding" of insecticidal genes for purposes of insect resistance management and broader target insect pest specificity. An example of a trans-conjugant commercial product included 'Foil', which was based on a strain EG2424 created at Ecogen for activity against lepidopteran and coleopteran insects and carried the *cry1Ac* gene from *Bt* var. *kurstaki* (active against *O. nubilalis*) and the *cry3A* gene from *Bt* var. *tenebrionis* (active against Colorado potato beetle, *Leptinotarsa decemlineata*) [26]. An example of a recombinant strain EG2348 commercialized under the name of Raven by Ecogen for control of lepidopteran larvae expressed a combination of endogenous (*cry1Ac*) and transgenic (*cry3A* and *cry3Bb*) insecticidal proteins from different *Bacillus* strains.

Use of *Bacillus* strains or their spores directly as sprays to control certain insects was attractive in being specifically toxic to host pests, nontoxic to humans and mammals, and biodegradable, but had disadvantages too. Some of the major disadvantages were that *Bt* strains or spores were susceptible to rapid inactivation with UV light, heat, and extreme pH rendering them ineffective. Additionally, the *Bt* sprays were susceptible to protease inactivation by host plant leaf exudates and were also easily washed away by wind and rain. This meant that for effective control of target pests, *Bt* sprays had to be applied and reapplied frequently, which resulted in increased cost to control target insect pests. Additionally, the *Bt* sprays were effective against specific chewing insects (Lepidoptera or beetle larvae on plants) but not effective against sap-sucking insect pest stages of hemipteran and homopteran orders like aphids, whiteflies, and others [27].

A solution to circumvent the challenges for effective use of insecticidal *Bt* strains or spores was to isolate genes encoding the insecticidal proteins and express them directly within the COI through genetic engineering techniques. Using this approach, the insecticidal proteins would be expressed in the COI and be readily available for intoxication of target insect pests for immediate control. However, technological hurdles were discovered in expressing bacterial genes in plants. *Bt* or other insecticidal toxin-encoding genes eventually required codon optimization and use of appropriate promoters for effective *in planta* expression [28, 29] to deliver the desired insect resistance. It was also discovered that the full-length gene sequence of the insecticidal protein may not be required for plant transformation as only the domains responsible for insecticidal activity and protein expression stability may be enough for *in planta* pest control efficacy for certain insecticidal proteins. Hence, Hoffmann [30] genetically engineered tobacco plants using a truncated gene encoding the N-terminal portion of a *cry1A* gene from *Bt* var. *kurstaki* HD-73 under the control of cauliflower mosaic virus (CaMV) 35S promoter. The engineered tobacco plants were protected from damage with infested *Helicoverpa zea* larvae. This demonstration of effective control of an insect pest using transgenic tobacco, though never meant to be developed for commercialization, paved the way to develop other approaches to create GE insect resistance traits in other COI. One of the first crops targeted for commercial development and marketed was the potato NewLeaf variety carrying the *cry3A* from *Bt var. tenebrios* developed by Monsanto in 1995 for the control of the Colorado potato beetle. This potato variety received the first registration from the Environmental Protection Agency (EPA) for a crop resistant to an insect pest. Subsequently a number of other transgenic crops providing tolerance to specific insect pests have been developed (Table 1).

3 Important Technologies in Input Trait Development

3.1 Genetic Engineering of Crops

Plant breeding had long been a method of crop improvement to develop new varieties with improved characteristics. Use of traditional plant breeding approaches has led to tremendous success in crop improvement. Major plant breeding achievements such as introduction of hybrid corn and improved wheat and rice varieties were among the most important advancements that resulted in dramatic increases in crop productivity [31].

However, traditional plant breeding methods have limitations in that genes can be transferred only within the same or closely related species, and often can have significant yield drag from other genes linked to a given trait. Moreover, as indicated above, in the case of IR and HT traits, most of the more impactful genes have originated from the non-plant sources, requiring the use of gene transfer technologies to move genes across noncompatible species. Therefore, the discovery, development, and use of *Agrobacterium*- and gene gun-mediated gene transfer methods delivered additional tools and approaches for transfer of specific and desired segments of DNA from any organism into a wide range of crop species. Later, other methods of transformation were also discovered and overall these methods can be grouped under two broad categories: (1) indirect transformation methods (i.e., bacterium mediated) and (2) direct transformation methods (i.e., nonorganismic mediated by chemical and mechanical introduction of DNA into cells). An overview of these methods is presented in another chapter of this book (Chap. 1).

3.2 Molecular Characterization of Transgenic Events

Molecular characterization of transgenic events (where an event is a plant that contains the transgene(s) and expresses the inserted gene phenotype) is foundational to the advancement of high-quality events in the commercial trait development process. Hundreds of events are created in the beginning of the trait development process using any of the plant transformation methods described above, and eventually one event is selected for commercialization [32]. Molecular characterization encompasses multiple parameters for evaluation of transgenic events including identification of inserted gene copy number, number of insertion sites, integrity and location of inserts in the target genome, identification of "junction" sequences at both termini of a transgene(s) cassette, unintended DNA presence including transformation vector backbone, and stability of the insertions across generations [33]. All this

Table 1
List of insect resistance and herbicide-tolerance transgenic crops, product names, transgenes used, phenotype expressed, and company that produced the product

Crop	Product	IR/HT/IR_HT	Gene	Phenotype	Company
Soybean	Enlist Soybean	HT	aad-12, pat	Glufosinate tolerance	Dow AgroSciences
Corn	Herculex I Insect Protection × Roundup Ready Corn 2	IR/HT	cry34Ab1, cry35Ab1, pat, cp4 epsps	Coleoptera resistance, glufosinate and glyphosate tolerance	DAS, DuPont Pioneer
Corn	Optimum Leptra	IR	cry1F, pat, cry1Ab, Vip3Aa19, cp4 epsps	Lepidoptera resistance, glyphosate tolerance	DuPont Pioneer
Corn	Herculex I Insect Protection	IR	cry1F, pat	Lepidoptera resistance, glufosinate tolerance	DuPont Pioneer
Corn	Optimum Intrasect	IR	cry1f, pat, cry1Ab, cp4 epsps		DuPont Pioneer
Corn	Optimum Intrasect Xtra	IR	cry1F, pat, cry1Ab, cry34Ab1, cry35Ab1 cp4 epsps	Lepidoptera and coleoptera resistance, glufosinate and glyphosate tolerance	DuPont Pioneer
Corn	Optimum TRIsect	IR	cry1Fa2, pat, mcry3A, cp4 epsps	Lepidoptera and coleoptera resistance, glufosinate and glyphosate tolerance	DuPont Pioneer
Corn	Qrome	IR	cry1F, cry34Ab1, cry35Ab1, pat, cry1Ab, mcry3A, cp4 epsps	Lepidoptera resistance, rootworm resistance, glyphosate and glufosinate tolerance	DuPont Pioneer
Corn	Herculex I × Roundup Ready Corn 2	IR/HT	cry1F, pat, cp4 epsps	Lepidoptera resistance, glufosinate and glyphosate resistance	Dow AgroSciences, DP
Corn	PowerCore	IR/HT	cry1A.105, cry2Ab1, cry1F, pat	Lepidoptera resistance, glufosinate resistance	Dow AgroSciences, Monsanto
Corn	Herculex I	IR	cry1F, pat	Lepidoptera resistance, glufosinate resistance	Dow AgroSciences

Corn	SmartStax	IR	cry1A.105, cry2Ab1, cry1F, cry3Bb1, cry34Ab1, cry35Ab1, cp4 epsps, pat, aad-1	Lepidoptera resistance, rootworm resistance, glyphosate/glufosinate tolerance	Dow AgroSciences, Monsanto
Corn	SmartStax® Enlist™ Corn	IR/HT	cry1A.105, cry2Ab1, cry1F, cry3Bb1, cry34Ab1, cry35Ab1, cp4 epsps, pat, aad-1	Lepidoptera resistance, rootworm resistance, glyphosate/glufosinate/2,4-D tolerance	Dow AgroSciences
Corn	PowerCore Enlist	IR/HT	cry1A.105, cry2Ab1, cry1F, pat, aad-1	Lepidoptera resistance, glyphosate/glufosinate/2,4-D tolerance	Dow AgroSciences
Corn	Herculex XTRA	IR	cry1F, pat, cry34Ab1, cry35Ab1	Lepidoptera and coleopteran resistance, glufosinate tolerance	Dow AgroSciences, DP
Cotton	WideStrike	IR	cry1F, cry1Ac,	Lepidoptera resistance	Dow AgroSciences
Cotton	WideStrike × FLEX	IR/HT	cry1F, cry1Ac, vip3A	Lepidoptera resistance	Dow AgroSciences
Cotton	WideStrike × RR1445	IR/HT	cry1F, cry1Ac, cp4 epsps, bar	Lepidoptera resistance, glyphosate and glufosinate tolerance	Dow AgroSciences
Cotton	WideStrike 3 × RR Flex × Enlist Cotton	IR/HT	cry1F, cry1Ac, vip3A, cp4 epsps, pat, aad-12	Lepidoptera resistance, glyphosate and glufosinate tolerance, 2,4-D	Dow AgroSciences
Cotton	WideStrike 3 × RR Flex	IR/HT	cry1F, cry1Ac, vip3A, cp4 epsps, pat	Lepidoptera resistance, glyphosate and glufosinate tolerance	Dow AgroSciences
Cotton	Enlist Cotton	HT	aad-12, pat	2,4-D tolerance	Dow AgroSciences
Cotton	WideStrike 3 × Enlist Cotton	HT	cry1F, cry1Ac, vip3A, aad-12	Lepidopteran resistance, 2,4-D tolerance	Dow AgroSciences
Cotton	WideStrike × RR Flex × Vipcot	IR/HT	cry1Ac, vip3A(a), cry1F, cp4 epsps, pat	Lepidoptera and coleoptera resistance, glufosinate and glyphosate tolerance	Dow AgroSciences
Corn	Herculex RW	IR/HT	pat, cry34Ab1, cry35Ab1	Coleoptera resistance, glufosinate tolerance	Dow AgroSciences
Soybean	Roundup Ready Soybeans	HT	cp4 epsps	Glyphosate tolerance	Monsanto

(continued)

Table 1 (continued)

Crop	Product	IR/HT/IR_HT	Gene	Phenotype	Company
Soybean	Roundup Ready 2 Xtend Soybeans	HT	cp4 epsps	Glyphosate tolerance	Monsanto
Soybean	Genuity Roundup Ready 2 Yield Soybeans	HT	cp4 epsps, dmo	Glyphosate, dicamba tolerance	Monsanto
Soybean	Intacta RR2 PRO Soybeans	IR/HT	cp4 epsps, cry1Ac	Glyphosate tolerance, lepidopteran resistance	Monsanto
Soybean	Vistive Gold	HT	cp4 epsps	Glyphosate tolerance, high oleic acid	Monsanto
Corn	Roundup Ready Corn 2	HT	cp4 epsps	Glyphosate tolerance	Monsanto
Corn	Roundup Ready Corn 2/LibertyLink Maize	HT	cp 4 epsps, pat	Glyphosate/glufosinate tolerance	Monsanto
Corn	SmartStax	IR/HT	cry1A.105, cry2Ab1, cry1F, cry3Bb1, cry34Ab1, cry35Ab1, cp4 epsps, pat	Lepidoptera resistance, rootworm resistance, glyphosate/glufosinate tolerance	Monsanto/DAS
Corn	YieldGard Corn Borer Corn	IR	cry1Ab	Lepidoptera resistance	Monsanto
Corn	YieldGard Corn Borer with Roundup Ready Corn 2	IR/HT	cry1Ab, cp4 epsps	Lepidoptera resistance, glyphosate tolerance	Monsanto
Corn	YieldGard Plus	IR	cry1Ab, cry3Bb1	Lepidoptera resistance, rootworm resistance	Monsanto
Corn	YieldGard Plus with Roundup Ready Corn 2	IR/HT	cry1Ab, cry3Bb1, cp4 epsps	Lepidoptera resistance, rootworm resistance, glyphosate tolerance	Monsanto

Corn	YieldGard Rootworm Corn	IR	cry3Bb1	Rootworm resistance	Monsanto
Corn	YieldGard Rootworm with Roundup Ready Corn 2	IR/HT	cry3Bb1, cp4 epsps	Rootworm resistance, glyphosate tolerance	Monsanto
Corn	YieldGard VT PRO	IR	cry1A.105, cry2Ab1,	Lepidoptera resistance	Monsanto
Corn	YieldGard VT Rootworm	IR	cry3Bb1, cp4 epsps	Rootworm resistance, glyphosate tolerance	Monsanto
Corn	YieldGard VT Triple	IR/HT	cry1Ab, cry3Bb1, cp4 epsps	Lepidoptera and rootworm resistance, glyphosate tolerance	Monsanto
	YieldGard VT triple PRO	IR/HT	cry1A.105, cry2Ab1, cry3Bb1, cp4 epsps	Lepidoptera and rootworm resistance, glyphosate tolerance	Monsanto
Cotton	Bollgard 3 Roundup Ready Flex	HT	cry1Ac. cry2Ab, vip3A, cp4 epsps	Lepidoptera resistance and herbicide tolerance	Monsanto
Cotton	Bollgard 3 XtendFlex cotton	IR/HT	cry1Ac. cry2Ab, vip3A, cp4 epsps, pat, DMO*	Lepidoptera resistance and herbicide tolerance each	Monsanto
Cotton	Bollgard Cotton	IR	cry1Ac	Lepidoptera resistance	Monsanto
Cotton	Bollgard II XtendFlex	IR	cry1Ac, cry2Ab, cp4 epsps, pat, DMO	Lepidoptera resistance and herbicide tolerance	Monsanto
Cotton	Genuity Bollgard II Cotton	IR	cry1Ac, cry2Ab	Lepidopteran resistance	Monsanto
Cotton	Genuity Bollgard II with Roundup Ready Flex Cotton	IR/HT	cry1Ac, cry2Ab, cp4 epsps	Lepidoptera resistance, glyphosate tolerance	Monsanto
Cotton	Genuity Roundup Ready Flex Cotton	HT	cp4 epsps	Glyphosate tolerance	Monsanto
Cotton	Roundup Ready Cotton	HT	cp4 epsps	Glyphosate tolerance	Monsanto

(continued)

Table 1
(continued)

Crop	Product	IR/HT/IR_HT	Gene	Phenotype	Company
Cotton	Roundup Ready with Bollgard Cotton	IR/HT	cp4 epsps, cry1Ac	Lepidoptera resistance, glyphosate tolerance	Monsanto
Cotton	XtendFlex Cotton	HT	epsps, pat, DMO	Herbicide tolerance	Monsanto
Canola	Genuity Roundup Ready Canola	HT	cp4 epsps	Glyphosate tolerance	Monsanto
Alfalfa	Roundup Ready Alfalfa	HT	cp4 epsps	Glyphosate tolerance	Monsanto
Alfalfa	HarvXtra with Roundup Ready	HT	ccomt(inverted repeat), cp4 epsps	Glyphosate tolerance	Monsanto
Sugarbeet	Roundup Ready Sugarbeets	HT	cp4 epsps	Glyphosate tolerance	Monsanto
Corn	Agrisure 3000GT	IR/HT	cry1Ab, mcry3A, pat, mepsps	Lepidoptera and coleoptera resistance and glyphosate tolerance	Syngenta
Corn	Agrisure 3120	HT	cry1Ab, cry1F, pat, mepsps	Lepidoptera resistance, herbicide tolerance	Syngenta
Corn	Agrisure 3122	iR/HT	cry1Ab, cry1Fa2, pat, mepsps, mcry3A, cry34Ab1, cry35Ab1	Lepidoptera and coleoptera resistance, glufosinate and glyphosate resistance	Syngenta
Corn	Agrisure CB/LL	iR/HT	cry1Ab, pat	Lepidoptera resistance, glufosinate tolerance	Syngenta
Corn	Agrisure CB/LL/RW	iR/HT	cry1Ab, pat, mcry3A	Lepidoptera and coleoptera resistance, glufosinate tolerance	Syngenta
Corn	Agrisure Duracade E-Z Refuge 5122	iR	ecry3.1Ab (synthetic/chimeric)	Lepidoptera resistance	Syngenta

Corn	Agrisure Duracade E-Z Refuge 5222	iR/HT	ecry3.1Ab, mcry3A (synthetic/chimeric), cry1Ab, pat, cry1Fa2, mepsps, vip3A20	Lepidoptera resistance, glufosinate and glyphosate tolerance	Syngenta
Corn	Agrisure GT	HT	mepsps	Glyphosate tolerance	Syngenta
Corn	Agrisure GT/CB/LL	IR/HT	cry1Ab, pat, mepsps	Lepidoptera resistance, glufosinate and glyphosate tolerance	Syngenta
Corn	Agrisure GT/RW	IR/HT	mcry3A, mepsps	Coleoptera resistance, glyphosate tolerance	Syngenta
Corn	Agrisure RW	IR	mcry3A	Coleoptera resistance	Syngenta
Corn	Agrisure Viptera	IR	vip3A20	Lepidoptera resistance	Syngenta
Corn	Agrisure Viptera 3110	IR/HT	cry1Ab, vip3Aa20, pat, mepsps	Lepidoptera resistance, glufosinate and glyphosate tolerance	Syngenta
Corn	Agrisure Viptera 3111	IR/HT	cry1Ab, vip3Aa20, mcry3A, pat, mepsps	Lepidoptera and coleoptera resistance, glufosinate and glyphosate tolerance	Syngenta
Corn	Agrisure Viptera 3220	IR/HT	cry1Ab, cry1Fa2, vip3Aa20, pat, mepsps	Lepidoptera resistance, glufosinate and glyphosate tolerance	Syngenta
Corn	Enogen/Agrisure 3000GT	OT/IR/HT	amy797E, cry1Ab, pat, mepsps, mcry3A	Amylase, lepidoptera, and coleoptera resistance, glufosinate and glyphosate resistance	Syngenta
Corn	Knockout Insect resistance corn	IR/HT	cry1Ab, bar	Lepidoptera resistance and glufosinate tolerance	Syngenta
Cotton	Bollgard III	IR	vip3A(a), cry1Ac, cry2Ab2	Lepidoptera resistance	Monsanto
Cotton	Vipcot Roundup Ready Flex	IR/HT	vip3A(a), cry1Ab, cp4 epsps	Lepidoptera resistance and glyphosate tolerance	Syngenta
Cotton	Vipcot Roundup Ready Flex	IR	vip3A(a), cry1Ab, cp4 epsps	Lepidoptera resistance	Syngenta
Soybean	LibertyLink Soybeans	HT	pat	Glufosinate tolerance	Bayer Crop Sciences

(continued)

Table 1 (continued)

Crop	Product	IR/HT/IR_HT	Gene	Phenotype	Company
Corn	LibertyLink Maize	HT	pat	Glufosinate tolerance	Bayer Crop Sciences
Cotton	LibertyLink Cotton	HT	bar	Glufosinate tolerance	Bayer Crop Sciences
Cotton	Liberty Link × Bollgard II Cotton	IR/HT	cry1Ac, cry2Ab2, bar	Lepidoptera resistance, glufosinate tolerance	Bayer Crop Sciences
Cotton	GlyTol Cotton	HT	2mepsps	Lepidoptera resistance, glufosinate and glyphosate tolerance	Bayer Crop Sciences
Cotton	GlyTol × Liberty Link	HT	2mepsps, bar	Herbicide tolerance	Bayer Crop Sciences
Canola	LibertyLink Canola	HT	bar	Herbicide tolerance	Bayer Crop Sciences
Canola	SeedLink Canola	HT	bar	Herbicide tolerance	Bayer Crop Sciences

information is used in determining proper trait functionality and stability, and for safety assessment of the commercial event(s) for regulatory approval [34, 35] leading to event deregulation. The technologies used for the molecular characterization of transgenes are discussed in another chapter in this book (Chap. 24).

3.3 Protein Detection and Quantitation

Gene expression analysis is a crucial component of trait product development for selection of transgenic events that express genes at desired levels, in specific tissues and temporally in the COI. The first step of gene expression analysis can be performed at the messenger RNA (mRNA) level. However, quantitative protein expression analysis is of utmost importance due to posttranscriptional and posttranslational modifications that can affect the correlation of mRNA levels with the final protein quantity and quality and the resulting phenotypic expression. This is a key step in the evaluation of gene expression for input traits. Immunoassays, e.g., western blots and enzyme-linked immunosorbent assays (ELISA), have been the primary analytical approaches for detection and quantitation of proteins. However, these assays require the development of antibodies that can take several months, may exhibit cross-reactivity to other protein(s), and have high development cost for routine use. More recently, significant progress has been made in the development of liquid chromatography in conjunction with tandem mass spectrometry (LC-MS/MS) as the next-generation protein detection technology toward solving the challenges posed by immunoassays [36]. The basis of the LC-MS/MS technology is the detection of specific signature peptides as surrogates representing the target proteins. It has significant advantage over immunoassays in that assay development requires only a few days as compared to multiple months needed for immunoassays. These technologies are also reviewed in detail in another chapter in this book (Chap. 25).

4 Currently Available Insect Resistance and Herbicide Tolerance Genes in GE Crops

4.1 Insect Resistance Traits: Gene Sources and Approaches

4.1.1 *Bacillus thuringiensis*-Derived Traits

Although it was realized in the early 1900s that *Bt* strains have insecticidal spores and crystals, it was not until 1986 when the first *Bt* gene encoding a lepidopteran active insecticidal protein was isolated, cloned, and used to protect tobacco plants from damage by *Helicoverpa zea* [30]. Since then a number of *Bt* genes, encoding insecticidal proteins, have been isolated, cloned, and screened for insecticidal properties against a range of pests. In addition to the insecticidal proteins in *Bt* crystals produced from *cry* (crystal) genes, *Bt* also produce cytolytic (*cyt*) insecticidal proteins and soluble insecticidal proteins, which are produced during the vegetative growth phases. Genes from *Bt's* vegetative growth phase called *vips* (*v*egetative insecticidal proteins), either alone or in combination with *cry* or *cyt* genes, have been isolated and used to develop transgenic crops with protection against insect damage.

Despite a wealth of knowledge of *Bt* insecticidal proteins and their genes, reviewed by Crickmore et al. and Narva et al. [37, 38], only a handful of the genes (*cry1A*, *cry2A*, *cry1F*, *cry3A*, *cry3B*, *cry34A*, *cry35A*, and *vip3A* and their variants) have been used to genetically engineer commercial crops for resistance to lepidopteran and coleopteran pests. Initially, GE crops were commercialized using a single insect resistance (IR) gene products, although in several instances IR traits were developed in combination with the HT traits. Another approach has been to stack insecticidal proteins with different modes of action targeting a specific order of insect pests to delay or prevent the development of resistance in the insects. Agricultural companies have mostly discovered homologs of *Bt cry* genes, and in very few cases produced synthetic variants of those *Bt* genes to create novel gene versions, like in the case of the mutated *mcry3A* gene for resistance to coleopteran pests and the synthetic chimeric gene, *ecry3.1Ab*, that also confers resistance to coleopteran pests (Table 1). Transgenic insect control traits have been widely beneficial and their commercial success is demonstrated by their widespread use by farmers in many industrialized and developing countries [39].

4.1.2 Non-Bacillus thuringiensis-Derived Traits

While *Bacillus thuringiensis* strains continue to provide a rich source of insecticidal proteins, there is a need to look for alternative sources or approaches for developing crops for IR traits to broaden the spectrum of modes of action available given the diverse pests that threaten commercial crops. The need for this is not only to increase the spectrum of activity to control primary and secondary insect pests using multiple modes of action of the insecticidal proteins, but also to delay the development of resistance by susceptible insects. However, after 20 years of *Bt* crop products in the market, there have been very few confirmed cases of insect pests developing resistance to the *Bt* gene-carrying crops [40]. This is mainly due to the implementation of strict insect resistance management strategies like inclusion of certain percentage of non-transgenic seed (refuge plants) along with the trait-containing transgenic seed provided. Also more recently, the industry has moved to stacked insecticidal products with more than one mode of action against key insect pests to help slow the development of resistance. However, it is expected that insects will eventually become resistant to the existing transgenic traits and in anticipation of this there have been numerous reports of alternative approaches being developed and insecticidal proteins being evaluated that are not derived from *Bt* sources.

4.1.2.1. Non-Bacillus thuringiensis Bacterial Gene Sources

Certain other bacteria have been the focus of extensive research as potential sources of insecticidal proteins. Such examples are the insect pathogenic bacteria, *Photorhabdus* and *Xenorhabdus* species, both of which are symbionts of the insect pathogenic nematode, *Heterorhabditis* [41]. These bacteria are nonpathogenic to the nematode itself. However, in insects, once the bacteria enter the insect host blood system or hemocoel, it becomes very pathogenic to the insect and rapidly overcomes the insect immune system resulting in insect mortality. Extensive research has been carried out to biochemically and molecularly characterize a complex (multiple proteins needed for insecticidal activity) of insecticidal *Photorhabdus* and *Xenorhabdus* toxin proteins, and attempts made to produce transgenic crops using these insecticidal proteins. Although much is known of the structure and function of the toxin complex proteins [42], little progress has been made in developing transgenic crops using genes encoding these proteins, possibly due to host plant phytotoxicity issues. It is possible that future research will allow the engineering of toxin complex genes with high insect control efficacy. Alternatively, other bacterial sources can be used to identify proteins with insecticidal activity.

4.1.2.2. Other Sources for Enhancing Insect Resistance in Plants

Production of proteases in the insect gut or in saliva is important for efficient protein digestion that sustains amino acid and metabolite pools essential for insect growth and development. Herbivorous insects feed either by chewing the plant tissues, as in the case of Lepidoptera and Coleoptera, or by piercing or rasping and sucking intra- or intercellular fluids from plants, as in the case of hemipteran and thysanopteran insects, respectively. Although not always effective, plants use proteinase inhibitors (PIs) as a common defense mechanism to defend from insect attack. When plant-produced PIs are ingested in artificial diets or overexpressed in plants, increased mortality and reduced growth and development have been reported for lepidopteran, coleopteran, or orthopteran immature stages [43–47].

Another approach explored for insect control is the expression of proteins that inhibit function and stability of α-amylases. Insects and other animals use α-amylases for digestion of starch ingested in their diet. Proteins that inhibit α-amylases are found widely in the plant kingdom. Such inhibitory proteins are directed specifically for the inhibition of α-amylase activity from insects and microorganisms without affecting the plant enzymes. Multiple examples of the use of α-amylase inhibitors have been reported with mixed results. The α-amylase inhibitor 1 gene from common bean (*Phaseolus vulgaris*) was expressed in pea (*Pisum sativum*) plants and it provided complete protection against the coleopteran pest, *Bruchus pisorum* L., commonly known as pea weevil. However, the α-amylase inhibitor 2 when expressed in pea plants only resulted in delayed maturation of pea weevil larvae [48]. Some bifunctional PIs have been identified that simultaneously inhibited both proteinases and α-amylases [49]. These bifunctional PIs offer the benefit of controlling pests via two different modes of action, perhaps enhancing their effectiveness against insect pests.

Other types of plant proteins have also been identified with potential but with varying levels of efficacy for insect control. Lectins are a class of multivalent carbohydrate-binding proteins found in plants that have been tested against insects for growth inhibition and mortality. In a study by Shukle and Murdock [50], lectins from soybean seed were found to inhibit growth of tobacco hornworm larvae. Similar results have been obtained using wheat germ agglutinin and snowdrop lectin when tested against European corn borer [51] and tomato pinworm, *Lacanobia oleracea*, respectively [52]. Other non-plant proteins, chitinases, produced by insects, have also been tested for insect control and results demonstrated impact on insect growth inhibition by disruption of the peritrophic membrane lining in the gut of *Heliothis virescens* [53].

Although enhanced protection of plants against lepidopteran pests has been reported by expressing PIs, lectins, α-amylases, and chitinases [9, 54], there has yet to be a commercial product using PI-encoding genes in crops. Some of the challenges that may have prevented this success could be related to either the potency of the PI proteins expressed *in planta*, including their expression levels, or the insect's ability to use endogenous proteases that circumvent the plant-expressed PIs. This highlights the need to understand the physiology and temporal expression of proteases in the insect gut and salivary glands in response to different PIs. Other approaches may also be needed like the use of multiple and different PI genes in transgenic plants or even chimeric genes encoding bifunctional gene products, such as α-amylase/trypsin inhibitors [9], trypsin/carboxypeptidase A inhibitors [55], and cystatin/serine PI [56].

Among other strategies that have been used to identify new insecticidal toxins, the screening of small peptide toxins produced in venoms of predatory spiders was investigated [57]. The venoms of these predatory spiders contain a number of disulfide-rich small (3–4 kDa in size) peptides that are selectively toxic to insects upon injection into the insect hemocoel. However, one of the functional challenges that need to be circumvented first is the fact that these peptides are not very active when ingested orally. Promising results have been obtained in several studies where expression of venom proteins from the Australian funnel-web spider, *Atrax robustus*, in tobacco [58, 59] and cotton [60] showed significantly increased protection against *Helicoverpa armigera*. Results from these experiments offer the possibility to engineer improvements of the peptides for effective crop protection

4.1.3 RNA-Mediated Technologies for Insect Control: dsRNA (RNAi)

Double-stranded RNA (dsRNA) to interfere with specific gene expression, also referred to as RNA *interference* (RNAi), has been used extensively to suppress or knock out specific genes in eukaryotic cells [61–63]. This approach takes advantage of a control mechanism that occurs naturally in eukaryotic cells whereby short stretches of diced mRNA or microRNA produced within a cell direct the degradation of complementary endogenous mRNA to suppress the expression of specific genes when required in that cellular environment. Similar effects can be achieved by design involving delivery and expression of an exogenous dsRNA molecule against a target transcript in cells of an organism. Timmons and Fire [64] and Timmons et al. [65] first demonstrated that ingestion of a dsRNA by the nematode, *Caenorhabditis elegance*, suppressed expression of a target gene. This work and many other investigations with insects [66–69] offered that a similar approach could be used for insect pest control, provided that (1) target genes critical for the survival or the growth and development of insects could be identified and (2) an effective method for delivery of dsRNA could be developed.

A number of insect target genes have been identified as critical for either insect growth and development or survival. An example of such genes is the vacuolar ATPase A subunit [62]. When a dsRNA against the vacuolar ATPase A was either fed mixed in artificial diet or injected into the hemocoel of coleopteran insect pests like corn rootworm and others, it affected larval development [70]. Baum et al. [67] also reported that western corn rootworm (WCR) larvae which fed on corn transformed with a dsRNA hairpin against the vacuolar ATPase A subunit gene, exhibited severe stunting of growth, and inhibited feeding damage to roots. Similarly, Mao et al. [71] demonstrated that *Arabidopsis* plants transformed with dsRNA hairpin that targets a cytochrome P450 monooxygenase gene in cotton bollworm led to decreased bollworm tolerance to the cotton sesquiterpene aldehyde, namely gossypol, thus decreasing insect damage.

While there has been significant progress either to identify target genes for dsRNA or in the delivery methods (e.g., via injection, in artificial diet, or through transgenic crop tissues), different levels of insect susceptibility to these approaches were observed. For example, the coleopteran insects, corn rootworm, Colorado potato beetle, and the canola flea beetle are sensitive to ingested dsRNA (e.g., V-ATPase subunit) with LC_{50} values ranging from 1 to 10 ppb in both larval and adult stages [67, 72, 73]. Insects outside of the order Coleoptera (like Diptera, Lepidoptera, and Hemiptera) that have been tested with the V-ATPase subunit dsRNA are almost a 1000-fold less sensitive ($LC_{50} \geq 10$ ppm). Even within the order Coleoptera, not all coleopterans, like red flour beetle, are as sensitive to dsRNA as corn rootworm or Colorado potato beetle [67, 74]. Of the 290 gene targets identified by

Baum et al. [67], dsRNA was effective for only about 40% of the targets using a surface overlay diet bioassay with about 0.1 ppm for each target dsRNA, causing rootworm mortality or growth stunting. At the same time, when *H. armigera* and *Spodoptera exigua* larvae were fed transgenic crops expressing *H. armigera* ecdysone receptor dsRNA, the phenotypic effects (e.g., elevated mortality, growth inhibition, and suppression of ecdysone receptor transcripts over controls) on the larvae were similar [75].

Despite the tremendous gain in the understanding of gene targets that are are essential for development and survival of insects, it has taken long to commercialize transgenic plants using the RNAi approach. In insects that are much less susceptible to sub-ppm dsRNA concentrations, the biggest barriers may be their gut anatomy that could hinder the effective delivery of dsRNA to target tissues and cells outside midgut. Therefore, challenges may remain in developing methods for improved delivery of dsRNA for effective adsorption in the insect gut of certain insects and delivery to target tissues.

The durability of RNAi-based approaches remains uncertain, mainly due to the lack of understanding of the mechanisms of resistance, i.e., whether there is a common mechanism of inactivation of dsRNA or multiple mechanisms. A single mechanism of resistance among target insect pests would severely reduce the durability of RNAi-based crop products.

Finally, the reader is referred to the volume in Advances in Insect Physiology [57], which covers many of the aspects discussed above in greater depth. Some other suggested readings covering the above topic extensively are [70, 76–81].

4.2 Herbicide Tolerance Traits and Weed Control

Weed control practices used by early farmers included hand or mechanical weed control, cultural practices like crop rotation, and even mechanical means through plowing and cultivation. These methods were often tedious and slow, and came with a high cost. Therefore, better, more efficient, and easier weed control methods were needed. The major breakthrough that came in the weed management practices was the discovery of synthetic auxin herbicides in 1940s which spawned an era of innovation yielding other selective and some nonselective herbicides in the intervening decades. Some important nonselective herbicides were glyphosate and glufosinate and selective herbicides including acetolactate synthase (ALS), acetyl coenzyme A carboxylase (ACCase), hydroxyphenylpyruvate dioxygenase (HPPD), phytoene desaturase (PDS), and protoporphyrinogen oxidase (PPO). Today, more than 200 herbicide-active ingredients are available that are classified in 29 different modes of action categories of which six modes of action cover more than 80% of the herbicide market [82].

Farmers that controlled weeds using selective herbicides needed to have a good understanding of weed identification in order to choose appropriate herbicide applications and often multiple herbicides were required to control different categories of weeds. In addition to crop-selective herbicide use, another breakthrough in the field of weed control was the introduction of herbicide-tolerant crops, with the first set of crops developed tolerant to glyphosate and glufosinate (Table 1). Prior to the availability of HT crops, glyphosate was used to control weeds in non-crop areas and its use against weeds closely associated with crops was not possible due to its broad-spectrum activity which could damage the crops. In 1996, glyphosate-tolerant crops were introduced making weed control in crops much easier for the farmers, since a single application of glyphosate killed all weeds in the field while having no effect on the transgenic glyphosate-tolerant crops [83].

The glyphosate mode of action is based on its binding affinity to the 5-enolpyruvylshikimate-3-phosphate synthase (EPSPS) enzyme, which is required for synthesis of essential aromatic amino acids in plants, and blocking EPSPS activity results in plant death [84]. In order to generate transgenic glyphosate-tolerant crops, three strategies have been attempted: (1) overexpression of the endogenous COI EPSPS enzyme to avoid glyphosate toxicity; (2) introduction of genes that metabolize and detoxify glyphosate in the COI; and (3) introduction of a glyphosate-insensitive form of the EPSPS enzyme t [85]. The first strategy using overexpression of the native EPSPS enzyme resulted in glyphosate tolerance of plant cells in culture and transgenic *Petunia* plants [86]. However, overall the strategy has not been very successful since application of glyphosate still affected the growth of the plants. Until now, no transgenic plants overexpressing a native EPSPS enzyme have been commercialized [85]. The second strategy for metabolic degradation of glyphosate has been in which glyphosate-metabolizing enzymes, like GOX (glyphosate oxidoreductase) and GAT (glyphosate N-acetyl transferase), have been engineered to impart glyphosate tolerance to plants [87]. The third strategy of using a glyphosate-insensitive form of the target enzyme and its variants has been the most effective and successfully used for the production of transgenic crops [82, 88]. Mutated forms of this enzyme, which are resistant to the action of glyphosate, have been introduced into crops resulting in commercial levels of tolerance in transgenic plants [89].

Another weed control system that became widely used was the production of genetically modified crops tolerant to glufosinate. Glufosinate inhibits the glutamine synthetase enzyme leading to accumulation of ammonia in the cells, which affects photosynthesis and eventually leads to plant death. Glufosinate is another nonselective herbicide but is faster acting than glyphosate. Glufosinate has gained greater adoption to help manage weeds that have developed glyphosate resistance [90]. In contrast to glyphosate, which is a systemic herbicide, glufosinate is a very effective contact herbicide. It needs to be applied when plants are smaller to have a good leaf coverage. The gene that encodes the enzyme phosphinothricin acetyltransferase (PAT) detoxifies glufosinate by acetylation of the amino group and hence makes plants tolerant to glufosinate [91]. Using this technology, transgenic canola, corn, cotton, and soybean herbicide-tolerant crops were introduced into the market starting 1997 [92].

Glyphosate-tolerant soybean and canola were introduced and deregulated for commercialization in 1996, followed by glyphosate-tolerant cotton and maize in 1997 and 1998, respectively [83]. Within 10 years of the introduction, the adoption rate for these crops reached >90% in the USA [93]. Glyphosate- and glufosinate-resistant crops were initially widely adopted crops by farmers. First, the glyphosate- and glufosinate-resistant crops were introduced as single-trait crops. However, later with the discovery of insect-resistance traits, the herbicide-tolerance traits were stacked with insect-resistant traits. Moreover, multiple modes of action for insect resistance as well as genes imparting resistance to different herbicides were stacked. In 2016, globally, 455 million acres were planted with crops that include herbicide tolerance, insect resistance, and stacked gene products. This includes 227.6 million acres under herbicide-tolerant crops only [2]. This broad adoption of HT crops has resulted in increased value to farmers due to reduced weed control costs and increased productivity, cumulatively worth around $63 billion, over a period of 1996–2014 [94].

Another herbicide target is the enzyme acetohydroxyacid synthase (AHAS), also called acetolactate synthase (ALS). This enzyme is required for the synthesis of branched-chain amino acids and inhibition of its activity results in plant death due to the depletion of essential amino acids [95]. More than 50 ALS inhibitors have been discovered that fall in five chemical classes: sulfonylurea [SU], imidazolinones [IMI], triazolopyrimidines [TP], pyrimidinylthiobenzoates [PTB], and sulfonylamino-carbonyl-triazolinones [96]. Different mutations in the ALS gene are known to impart tolerance to some or all of the above herbicides. These mutated forms of the enzyme have been used to develop crops tolerant to specific herbicide chemical classes. Such crops have been commercialized under the name Clearfield from 1992 onwards [97]. Largely, the Clearfield crops were developed using non-GE plant breeding methods and were nontransgenic in nature. However, recently BASF launched genetically modified soybean with the trade name Cultivance in collaboration with Embrapa in Brazil [98]. Following the introduction of Clearfield corn in 1992 by BASF, a large number of crops including rice, wheat, brassicas, corn, soybean, lettuce, sunflower, etc. have been developed for tolerance to ALS-inhibitory herbicides [97].

Because glyphosate herbicide-tolerant crops have been the most widely adopted weed control system, the extensive use of glyphosate among the farmers and lack of integrated weed management practices have led to the development of glyphosate resistance in weeds. By mid-2017, around 37 glyphosate-resistant weed species had been reported [93]. Therefore, new strategies have been under development to control emerging glyphosate-resistant weeds. This has stimulated the discovery of new trait genes encoding new modes of action. The most prominent next-generation weed control systems include Enlist and Roundup Ready Xtend in corn, cotton, and soybean [99]. The Enlist technology includes the use of 2,4-dichlorophenoxyacetic acid (2,4-D), which is a broad-spectrum herbicide and controls more than 90 weed species including glyphosate-resistant weeds such as horseweed and water hemp [100]. The trait is encoded by the bacterial aryloxyalkanoate dioxygenase (AAD) gene product, which inactivates 2,4-D and provides excellent tolerance to this chemical in plants [101]. Dow AgroSciences, LLC, has used stacked HT-tolerant gene strategy to transform soybean, corn, and cotton for tolerance to 2,4-D and other herbicides, for effective weed control [102] and to allow implementation of this weed management system. Monsanto has also developed a Roundup Ready Xtend Crop system that stacks dicamba and glyphosate tolerance for weed control [102]. Dicamba tolerance is provided by the bacterial dicamba monooxygenase gene (DMO) that inactivates dicamba when sprayed on crop plants [99]. Another category of herbicides is the 4-hydroxyphenylpyruvate dioxygenase (HPPD) inhibitors. HPPD is a key enzyme required for the production of pigments and for the development of chloroplasts. Inhibition of this enzyme affects chloroplast biosynthesis, which leads to bleaching followed by plant death. The overexpression of the resistant forms of HPPD, that is less sensitive to HPPD inhibitors, is the strategy used to impart tolerance to HPPD-inhibiting herbicides such as Balance Flexx, Callisto, and Lumax [103–105]. Recently Bayer Crop Sciences has developed transgenic soybean, Balance GT, that carries a gene stack of both the glyphosate-tolerance trait combined with tolerance to isoxaflutole-based HPPD-inhibiting herbicide [106]. In general, the stacking of multiple modes of action for herbicide tolerance will be more effective for weed control and is expected to delay the development of resistance in weeds, thus increasing the durability of the HT traits.

Other high-potential herbicides are PPO (protoporphyrinogen oxidase) inhibitors and ACCase (acetyl coenzyme A carboxylase) inhibitors. PPO is a gene product that acts in the chlorophyll/heme biosynthetic pathway. Chlorophyll is an essential light-harvesting pigment and heme is required as a cofactor in many enzymes. Inhibiting the synthesis of this enzyme causes light-dependent membrane damage resulting in burning, desiccation, and growth inhibition of the plants [107]. Overexpression of naturally resistant bacterial PPO genes results in resistance to PPO-inhibiting herbicides [108, 109]. Monsanto and Sumitomo have partnered to develop the next generation of crops tolerant to PPO-inhibiting herbicides [110]. Similarly, ACCase carries the first step in fatty acid biosynthesis. Inhibiting the activity of ACCase results in cessation of fatty acid biosynthesis that eventually leads to plant death. Point mutations in this gene provide resistance to the application of ACCase-inhibitory chemicals, like aryloxyphenoxypropionate class of herbicides abbreviated as FOP herbicides [111]. Development of herbicide-tolerant crops using these different modes of action will diversify the weed control tool box and provide more effective weed controls for a wide variety of weeds while reducing the probability of developing herbicide resistance. Although multiple and diverse HT modes of action are available for stacking, and despite the challenges with glyphosate weed resistance, glyphosate is still widely used and is a component of new HT stacks. Glyphosate is expected to continue to be used in the near future in commercial HT stack releases [82, 110].

Though multiple modes of action for herbicide tolerance are available today, these are based on the existing herbicide chemistries. The discovery of new herbicides is not being as actively pursued as for insecticides and fungicides where the search for novel chemistries and modes of action are areas of focus. The 1970s was the era when herbicides with new modes of action were discovered every 2–3 years [93]. Today, the availability of a good number of herbicide chemistries bearing different modes of action has reduced the need to further discover and develop new ones [82].

5 Next Generation of Insect Resistance and Herbicide-Tolerant Traits

Site-specific nucleases (SSNs) have emerged as a powerful technology for genome engineering. Of the SSNs, the CRISPR-Cas system is most easy to design, affordable, and versatile. Using SSNs, it is possible to add, delete, and edit genes or modulate their expression [112, 113]. The add and delete functions permit transgene manipulations including molecular stacking at preselected ("preferred") genomic locations in plants [114, 115]. As a result, the trait introgression process is simplified, as multiple transgenes (traits) can be managed or tracked as a single locus to ensure physical and expression stability of the transgenes and surrounding host genome genes. SSN-based development of molecular stacks is an elegant example of genome engineering that impacts superior organization of transgenes relative to the native genes in the genomes of the major crops.

The "edit" function of SSNs is aimed at introducing gene modifications to create gene knockouts and expression modulations, develop new alleles, or engineer genes that express new protein versions. Therefore, using the edit function, new alleles can be added to the repertoire of the existing alleles to increase allelic diversity. Learnings from the mutation technologies practiced in the past few decades, e.g., T-DNA insertion [116], transposon tagging [117], ethyl methane sulfonate (EMS)-induced mutations screened with TILLING approach [118, 119], and fast neutron (FN)-induced deletions [120], have illustrated the development of a large number of new crop varieties worldwide by essentially modifying genes or alleles to create new versions with new or additional functions. At least 3270 new varieties were developed from the mutation research alone [121].

6 Trait Reengineering

Another important application of the SSN technology is to modify/reengineer the existing transgene(s) already present in commercial GE crops. This could be conceptually explained as the generation of new allelic variants from the existing transgenes. Many of the insect resistance genes currently present in commercial products, as discussed earlier in this chapter, are continuously challenged by insect pests trying to overcome their susceptibility to IR traits. With the advances in protein engineering, reengineered versions of the existing IR genes may become readily available in the future, which when replacing the older version could quickly reestablish protection in the COI, thus avoiding financial losses to the farmers and the industry in general. This approach would bypass the need for recreating a completely new stack, which is a process that could take more than a decade before commercial release. The same concept could apply to HT genes with the purpose to generate new transgene alleles that provide protection to active ingredient analogs for broader or more effective weed control spectrum. Alternatively, and as mentioned for IR genes, HT trait genes can be removed from commercial events and replaced with new/improved genes in the stack.

There are immense opportunities that CRISPR-CAS, a simple and precise genetic scissor, has to offer in crop improvement and its potential impact on trait development and agriculture in general.

7 Future Perspectives

GE traits have been very impactful in crop production. These traits have made possible the industrialization of several crops, contributing to food security worldwide. Traits for IR and HT are among the most impactful and widely used in agriculture, thus improving land management, reducing operational costs, and decreasing environmental impact by minimizing the number of chemical applications of insecticides and herbicides, which in turn reduces the use of fossil fuels. GE traits have also reduced the exposure of farmers to chemicals by reducing the application of herbicides by 50%, thus improving the quality of their lives [39]. However, despite all those benefits and the great rate of adoption of these technologies, greater adoption of GE crops is needed worldwide to increase productivity in other parts of the world where these technologies have not yet been adopted.

The development of GE traits is not possible without the advancement and development of enabling technologies that improve the ability to discover and develop new and better input traits. Technologies like crop transformation and genome sequencing (like with next-generation sequencing (NGS)) play a key role in the discovery, testing, and development of IR and HT GE crops. Similarly, analytical technologies for molecular and biochemical characterization make possible the analyses of transgenic events and also allow for the generation of key information to assess their safety and support deregulation of these traits. Last but not least, further development of phenotyping technologies will continue to enhance the ability to make clearer determinations of the efficacy of traits and also help assess agronomic performance.

To date GE IR traits commercially available have been mainly derived from *Bacillus thuringiensis* genes. The development of IR traits is very complex and has to meet the criteria not only for trait efficacy, or insect control, but also for minimal to no impact on nontarget organisms and agronomic performance of COI. Due to these technical challenges, the current toolbox has relatively few insecticidal genes with different modes of action available. To increase the availability of genes encoding for insecticidal activity and with a diversity of modes of action for managing insect resistance and prolonged product life, gene discovery efforts have been focused on non-*Bt* gene sources.

In the case of GE HT, stacked modes of action for HT are becoming available to widen the spectrum of weeds that can be controlled, improve weed control efficacy, and also extend trait durability. In recent years, greater emphasis has been given to the implementation of integrated weed management programs whereby herbicides with different modes of action should be applied in rotation or in combination to slow the development of herbicide-resistant weeds.

To date IR and HT GE crops have been produced using random gene insertions into the plant genome using primarily *Agrobacterium*-based or biolistic-mediated transformation methods. However, technologies using SSNs are now available, which allow delivery of transgenes to precise locations in the plant genome. Furthermore, SSNs can be used to edit native genes and create new and trait-specific allelic variants in the host genome. This technology could also be used to reengineer traits developed by

GE, or to create completely new genes with novel modes of action or functions. Such approaches can play a key role in accelerating the development and availability of IR or HT traits in the near future.

Further increases in crop yields will continue to be delivered through greater adoption of GE crops, discovery and development of new IR and HT trait genes, better integrated pest management systems, improved cultural practices, and use of new and novel technologies to enhance trait performance without negative agronomic effects in the elite varieties of target crops. IR and HT crops will continue to be, without a doubt, among the most impactful traits to meet the challenge of feeding the world's growing population.

Acknowledgments

We are grateful to Lauren Clark for providing information on analytical technologies for transgene analysis and to Laura Tagliani, Phil Poirier, Robert Lampe, Stephen Novak, and Terry Wright for critical review of the manuscript.

References

1. James C (2016) Global status of commercialized biotech / GM crops: 2015. ISAAA Brief No. 51. ISAAA, Ithaca, NY
2. McDougall P (2017) Seed industry overview. Phillips McDougall, Midlothian, UK
3. Mall TK, Han L, Tagliani L, Christensen C (2018) Transgenic crops: status, potential and challenges. In the book - Biotechnologies of Crop Improvement, Volume 2, Springer, Chapter 16
4. United Nations Department of Economic and Social Affairs Population Division. http://www.un.org/en/development/desa/population/theme/trends/index.shtml. Accessed 30 Oct 2017
5. Li L, Zhang Q, Huang D (2014) A Review of Imaging Techniques for Plant Phenotyping. Sensors 14:20078–20111. https://doi.org/10.3390/s141120078
6. Nielsen RL (2017) Historical corn grain yields for the US. Purdue communication. https://www.agry.purdue.edu/ext/corn/news/timeless/yieldtrends.html
7. Schwember AR (2008) An update on genetically modified crops. Cien Inv Agr 35:231–250
8. Sanahuja G, Banakar R, Twyman RM, Capell T, Christou P et al (2011) Bacillus thuringiensis: a century of research, development and commercial applications. Plant Biotechnol J 9:293–300
9. Stevens J, Dunse K, Fox J, Evans S, Anderson M (2012) Biotechnological approaches for the control of insect pests in crop plants. INTECH. http://creativecommons.org/licenses/by/3.0
10. Hilscher J, Burstmayr H, Stoger E (2017) Targetted modification of plant genomes from precision crop breeding. Biotechnol J 12:1600173–1600187
11. Sovova T, Kerins G, Demnerova K, Ovesna J et al (2017) Genome editing with engineered nucleases in economically important animals and plants: State of the art in the research pipeline. Curr Issues Mol Biol 21:41–62
12. Yau Y-Y, Stewart CN Jr (2013) Less is more: strategies to remove marker genes from transgenic plants. BMC Biotechnol 13:36
13. Bauer EK, Singh DK, Popescu SC (2014) Next-generation plant science: putting big data to work. Genome Biol 15:301
14. Nocker SV, Gardiner SE (2014) Breeding better cultivars, faster: applications of new technologies for the rapid deployment of superior horticultural crops. Hortic Res 1:14022. https://doi.org/10.1038/hortres.2014.22

15. Varshney RK, Singh VK, Hickey JM, Xun X, Marshall DF, Wang J, Edwards D, Ribaut J-M (2016) Analytical and decision support tools for genomics-assisted breeding. Trends Plant Sci 21:354–366. https://doi.org/10.1016/j.tplants.2015.10.018
16. Bates SL, Zhao JZ, Roush RT, Shelton AM et al (2005) Insect resistance management in GM crops: past, present and future. Nat Biotechnol 23:57–62
17. Murai N, Kemp JD, Sutton DW, Murray MG, Slightom JL, Merlo DJ, Reichert NA, Sengupta-Gopalan C, Stock CA, Barker RF, Hall TC (1983) Phaseolin gene from bean is expressed after transfer to sunflower via tumor-inducing plasmid vectors. Science 222 (4623):476–482
18. Fraley RT, Rogers SG, Horsch RB, Sanders PR, Flick JS, Adams SP, Bittner ML, Brand LA, Fink CL, Fry JS, Galluppi GR, Goldberg SB, Hoffmann NL, Woo SC (1983) Expression of bacterial genes in plant cells. Proc Natl Acad Sci 80(15):4803–4807
19. Herrera-Estrella L, Depicker A, Van Montagu M, Schell J (1983) Expression of chimaeric genes transferred into plant cells using a Ti-plasmid-derived vector. Nature 303:209–213. https://doi.org/10.1038/303209a0
20. ISAAA Briefs (2016) Global status of commercialized biotech/GM crops 2016. http://www.isaaa.org/resources/publications/briefs/52/download/isaaa-brief-52-2016.pdf
21. Gianessi LP (2005) Economic and herbicide use impacts of glyphosate-resistant crops. Pest Manag Sci 61:241–245
22. Fuchs RL, Mackey MA (2003) Genetically modified foods. In: Encyclopedia of food sciences and nutrition, 2nd edn. Academic Press, Cambridge, pp 2876–2882
23. Mattes O (1927) Parasitare krankheiten der mehlmottenlarven und versuche uber ihre vermendbarkeit als biologisches bekiampfungusmittel. Sitzber Ges Beforder Ges Naturw Marburg 62:381–417
24. Kronstad JW, Schnepf HE, Whiteley HR (1983) Diversity of locations for Bacillus thuringiensis crystal protein genes. J Bacteriol 154:419–428
25. Carlton BC, Gonzales JN (1985) Plasmids and delta-endotoxin production in different subspecies of Bacillus thuringensis. In: Hoch JA, Setlow P (eds) Molecular biology of microbial differentiation. American Society for Microbiology, Washington DC
26. Carlton BC, Gawron-Burke G (1993) Genetic improvement of Bacillus thuringiensis for bioinsecticide development. In: Kim L (ed) Advanced engineered pesticides. Marcel Dekker, Inc, New York
27. Kaur S (2000) Molecular approaches towards development of novel *Bacillus thuringiensis* biopesticides. World J Microbiol Biotechnol 16:781–793
28. Perlak FJ, Fuchs RL, Dean DA, McPherson SL, Fischoff DA (1991) Modificatioon of the coding sequence enhances plant expression of insect control genes. Proc Natl Acad Sci U S A 88:3324–3328
29. Koziel MG, Beland GL, Bowman C, Carozzi NB et al (1993) Field performance of elite transgenic maize plants expressing an insecticidal protein derived from *Bacillus thuringiensis*. Nat Biotechnol 11:194–200
30. Hoffmann MP, Zalom FG, Wilson LT et al (1992) Field evaluation of transgenic tobacco containing genes encoding Bacillus thuringiensis delta-endotoxin or cowpea trypsin inhibitor: efficacy against Helicoverpa zea (Lepidoptera: Noctuidae). J Econ Entomol 85:2516–2522
31. Moose SP, Mumm RH (2008) Molecular plant breeding as the foundation for 21st century crop improvement. Plant Physiol 147 (3):969–977
32. Gupta M, Ram R (2004) Development of genetically modified agronomic crops. In: Parekh SR (ed) The GMO handbook. Humana Press, New Jersey
33. Guttikonda SK, Marri P, Mammadov J, Ye L, Soe K, Richey K, Cruse J, Zhuang M, Gao Z, Evans C, Rounsley S, Kumpatla SP (2016) Molecular characterization of transgenic events using next generation sequencing approach. PLoS One 11(2):e0149515. https://doi.org/10.1371/journal.pone.0149515
34. Codex A (2003) Guideline for the conduct of food safety assessment of foods derived from recombinant-DNA plants. CAC/GL 45:1–18
35. Privalle LS, Chen J, Clapper G, Hunst P, Spiegelhalter F, Zhong CX (2012) Development of an agricultural biotechnology crop product: testing from discovery to commercialization. J Agric Food Chem 60:10179–10187. https://doi.org/10.1021/jf302706e
36. Liebler DC, Zimmerman LJ (2013) Targeted quantitation of proteins by mass spectrometry. Biochemistry 52:3797–3806. https://doi.org/10.1021/bi400110b

37. Crickmore N, Zeigler DR, Feitelson J et al (1998) Revision of the nomenclature from the Bacillus thuringiensis pesticidal crystal proteins. Microbiol Mol Biol Rev 62:807–813
38. Narva KE, Storer NP, Meade T (2014) Disciovery and development of insect-resistant crops using genes from Bacillus thuringiensis. In: Dhadialla TS, Gill SS (eds) Advances in insect physiology: Insect midgut and insecticidal proteins, vol 47. Elsevier, San Francisco
39. James C (2015) 20th anniversary (1996 to 2015) of the global commercialization of biotech crops and biotech crop highlights in 2015. ISAAA Brief No. 51. ISAAA, Ithaca, NY
40. Bohn T, Lovei GL (2017) Complex outcomes from insect and weed control with transgenic plants: ecological surprises. Front Environ Sci 5(60):1–8. https://doi.org/10.3389/fenvs.2017.00060
41. French-Constant RH, Dowling AJ (2014) Photorhabdus toxins. In: Dhadialla TS, Gill SS (eds) Advances in insect physiology: insect midgut and insecticidal proteins, vol 47. Elsevier, San Francisco
42. Sheets JJ, Hey TD, Fencil KJ et al (2011) Insecticidal toxin complex proteins from Xenorhabdus nematophilus: structure and pore formation. J Biol Chem 286:22742–22749
43. Swati M, Mishra PK, Lokya V, Swaroop V, Mallikarjuna N, Gupta AD, Padmasree K (2016) Purufication and partial characterization of trypsin specific proteinase inhibitors from pigeonpea wild relative cajanus platycarpus L. (Fabaceae) active against gut proteases of lepidopteran pest Helicoverpa armigera. Front Physiol 7:1–13
44. Tanpure RS, Barbole RS, Dawkar VV, Waichal YA, Joshi RS, Giri AP, Gupta VS (2017) Inproved tolerance against Helicoverpa armigera in transgenic tomato over-expressing multi domain proteinase inhibitor gene from capsicum annuum. Physiol Mol Biol Plants 23 (3):597–604
45. Rao DE, Divya K, Prathyusha IVSN, Krishna CR, Chaitnya KV (2017) Insect resistant plants. In: Dubey SK, Pandey A, Sangwan RS (eds) Current developments in biotechnology and bioengineering. Elsevier, Amsterdam
46. Ravindran BM (2016) Transgenic pest resistance. Devagiri J Sci 2(1):01–31
47. Stevens J, Dunse K, Fox J, Evans S, Anderson M (2012) Biotechnological approaches for the control of insects pests in crop plants. In: Soundararajan RP (ed) Pesticides - advances in chemical and botanical pesticides. InTech, Rijeka, Croatia, pp 269–308
48. Morton RL, Schroeder HE, Bateman KS, Chrispeels MJ, Armstrong E, Higgins TJV (2000) Bean alpha-amylase inhibitor 1 in transgenic peas (Pisum sativum) provides complete protection from pea weevil (Bruchus pisorum) under field conditions. Proc Natl Acad Sci USA 97:3820–3825
49. Haq SK, Atif SM, Khan RH (2004) Protein proteinase inhibitor genes in combat against insects, pests, and pathogens: natural and engineered phytoprotection. Arch Biochem Biophys 431:145–159
50. Shukle RH, Murdock LL (1983) Lipoxygenase, trypsin inhibitor and lectin from soybeans effects on larval growth of Manduca sexta (Lepidoptera: Sphingidae). Environ Entomol 12:787–1128
51. Czapla TH, Lang BA (1990) Effects of plant lectins on the larval development of European corn borer (Lepidoptera: Pyralidae) and southern corn rootworm (Coleoptera: Chrysomelidae). J Econ Entomol 83:2480–2485
52. Fitches E, Gatehouse JA (1998) A comparison of the short and long term effects of insecticidal lectins on the activities of soluble and brush border enzymes of tomato moth larvae (Lacanobia oleracia). J Insect Physiol 44:1213–1224
53. Ding XF, Gopalakrishnan B, Johnson LB, White FF, Wang XR, Morgan TD, Kramer KJ, Muthukrishnan S (1988) Insect resistance of transgenic tobacco expressing an insect chitinase gene. Transgenic Res 7:77–84
54. Ussuf KK, Laxmi NH, Mitra R (2001) Proteinase inhibitors: Plant –derived genes of insecticidal protein for developing insect-resistant transgenic plants. Curr Sci 80 (7):847–852
55. Chiche L, Heitz A, Padilla A, Lenguyen D, Castro B (1993) Solution conformation of a synthetic Bis-headed inhibitor of trypsin and carboxypeptidase A: New structural alignment between the squash inhibitors and the potato carboxypeptidase inhibitor. Protein Eng 6:675–682
56. Urwin PE, McPherson MJ, Atkinson HJ (1998) Enhanced transgenic plant resistance t nematodes by dual proteinase inhibitor constructs. Plants 204:472–479
57. Herzig V, Bende NJ, Alam MS et al (2014) Methods for deployment of spider venom peptides as bioinsecticides. In: Dhadialla TS, Gill SS (eds) Advances in insect physiology:

insect midgut and insecticidal proteins, vol 47. Elsevier, San Francisco

58. Jiang H, Zhu YX, Che ZL (1996) Insect resistance of transformed tobacco plants with the gene of the spider insecticidal peptide. J Integr Plant Biol Acta Bot Sin 38:95–99
59. Khan SA, Zafar Y, Briddon RW et al (2006) Spider venom toxin ptotects plants from insect attack. Transgenic Res 15:349–357
60. Omar A, Chatha KA (2012) National Institute for Biotechnology and Genetic Engineering (NIBGE): genetically modified spider cotton. Asian J Manag Cases 9:33–58
61. Hannon GJ (2002) RNA interference. Nature 418:244–251
62. Lee JT (2012) Epigenetic regulation by long noncoding RNAs. Science 338:1435–1439
63. Djebali S, Davis CA, Merkel A, Dobin A, Lassman T et al (2012) Landscape of transcription in human cells. Nature 489:101–108. https://doi.org/10.1038/nature11233
64. Timmons L, Fire A (1998) Specific interference by ingested dsRNA. Nature 395:854
65. Timmons L, Court DL, Fire A (2001) Ingestion of bacterially expressed dsRNA can produce specific and potent genetic interference in Caenorhabditis elegans. Gene 263:103–112
66. Fu KY, Li Q, Zhou LT, Meng QW, Lu FG, Guo WC, Li GQ (2016) Knockdown of juvenile hormone acid methyl transferase severely affects the performance of Leptinotarsa decemlineata (Say) larvae and adults. Pest Manag Sci 72:1231–1241
67. Baum JA, Bogaert T, Clinton W, Heck GR, Feldman P, Ilagan O, Johnson S, Plaetinck G, Munyikwa T, Pleau M, Vaughn T, Roberts J (2007) Control of coleopteran insect pests through RNA interference. Nat Biotechnol 25:1322–1326
68. Manakuchi C, Namiki T, Yoshiyama M, Shinoda T (2008) RNAi-mediated knockdown of juvenile hormone acid O-methyltransferase gene causes precocious metamorphosis in the red flour beetle Tribolium castaneum. FEBS J 275:2919–2931
69. Tian G, Cheng L, Qi X et al (2015) Transgenic cotton plants expressing double-stranded RNAs target HMG-CoaA reductase (HMGR) gene inhibits the growth development and survival of cotton bollworms. Int J Biol Sci 11:1296–1305
70. Baum JA, Roberts JK (2014) Progress towards RNAi-mediated insect pest management. In: Dhadialla TS, Gill SS (eds) Advances in insect physiology: insect midgut and insecticidal proteins, vol 47. Elsevier, San Francisco
71. Mao YB, Cai WJ, Wang JW, Hong GJ, Tao XY, Wang LJ, Huang YP, Chen XY (2007) Silencing a cotton bollworm P450 monooxygenase gene by plant-mediated RNAi impaires larval tolerance of gossypol. Nat Biotechnol 25:1307–1313
72. Bolognesi R, Ramaseshadri P, Anderson J, Bachman P, Clinton W, Flannagan R, Ilgan O, Lawrence C, Levine S, Moar W, Mueller G (2012) Characterizing the mechanism of action of double-stranded RNA activity against western corn rootworm Diabrotica virgifera virgifera LeConte. PLoS One 7(10): e47534
73. Zhao YY, Yang G, Wang-Pruski G, You MS (2008) Phyllotreta striolata (Coleoptera: Chrysomelidae): Arginine kinase cloning and RNAi-based pest control. Eur J Entomol 105:815–822
74. Whyard S, Singh AD, Wong S (2009) Double-stranded RNAs can act as species-specific insecticide. Insect Biochem Mol Bio 39(11):824–832
75. Zhu JQ, Liu S, Ma Y, Zhang JQ, Qi HS, Wei ZJ, Yao Q, Zhang WQ, Li S (2012) Improvement of pest resistance in transgenic tobacco plants expressing dsRNA of an insect-associated gene *EcR*. PLoS One 7(6):e38572
76. Belles X (2010) Beyond Drosophila, RNAi in vivo and functional genomics in insects. Annu Rev Entomol 55:111–128
77. Burand JP, Hunter WB (2013) RNAi, future in insect management. J Invertebr Pathol 112 (Suppl. 1):S68–S74
78. Gu L, Knipple DC (2013) Recent advances in RNA interference research in insects: Implications for future insect pest management strategies. Crop Prot 45:36–40
79. Huvenne H, Smagghe G (2010) Mechanism of dsRNA uptake in insects and potential of RNAi for pest control: a review. J Insect Physiol 56(3):227–235
80. Zhang X, Zhang J, Zhu K (2012) Advances and prospects of RNAi technologies in insect pest management. In: Liu T, Kang L (eds) Recent advances in entomological research. Springer, Berlin Heidelberg
81. Zhang M, Zhou Y, Wang H et al (2013) Identifying potential RNAi targets in grain aphid Sitobian avenae F. based on transcriptome profiling of its alimenrtry canal after feeding on wheat plants. BMC Genomics 14:560

82. Green JM (2014) Current state of herbicides in herbicide resistant crops. Pest Manag Sci 70:1351–1357
83. Duke SO, Powles SB (2008) Glyphosate: a once-in-a-century herbicide. Pest Manag Sci 64(4):319–325
84. Baylis AD (2000) Why glyphosate is a global herbicide: Strengths, weaknesses and prospects. Pest Manag Sci 56:299–308
85. Dill GM (2005) Glyphosate resistant crops: history, status and future. Pest Manag Sci 61:219–224
86. Shah DM, Horsch RB, Klee HJ, Kishore GM, Winter JA, Tumer NE, Hironaka CM, Sanders PR, Gasser CS, Aykent S, Siegel NR, Rogers SG, Fraley RT (1986) Engineering herbicide tolerance in transgenic plants. Science 133(4762):478–481
87. Pline-Srnic W (2006) Physiological mechanisms of glyphosate resistance. Weed Technol 20(2):290–300
88. Funke T, Han H, Healy-Fried M, Fischer M, Schonbrunn E (2006) Molecular basis for the herbicide resistance of roundup ready crops. PNAS 103(35):13010–13015
89. Eschenburg S, Healy ML, Priestman MA, Lushington GH, Schonbrunn E (2002) How the mutation glycine96 to alanine confers glyphosate insensitivity to 5-enolpyruvyl shikinate-3-phosphate synthase from Escherichia coli. Planta 216:129–135
90. Green JM (2011) Herbicide resistant crops: utilities and limitations for herbicide resistant weed management. J Agric Food Chem 59:5819–5829
91. Shakula S, Braverman MP, Linscombe SD (1997) Response of BAR transformed rice (Oryza sativa) and red rice (Oryza sativa) to glufosinate application timing. Weed Technol 11:303–307
92. Duke SO (1997) Weed management: implications of herbicide resistant crops. Workshop on ecological effect of pest resistance genes in management of ecosystems p. 8. January 31 to February 3, 1999, Bethesda, MD, USA
93. Duke SO (2017) The history and current status of glyphosate. Pest Manag Sci 74 (5):1027–1034. https://doi.org/10.1002/ps.4652
94. ISAAA (International Service for the Acquisition of Agri-Biotech Applications) (2016) Pocket K No. 10: Herbicide tolerance technology: glyphosate and glufosinate. http://isaaa.org/resources/publications/pocketk/10/default.asp. Last updated August 2016
95. Yu Q, Powles SB (2014) Resistance to AHAS inhibitor herbicide: current understanding. Pest Manag Sci 70(9):1340–1350
96. Green JM (2007) Review of glyphosate and ALS inhibiting herbicide crop resistance and resistant weed management. Weed Technol 21:547–558
97. Tan S, Evans RR, Dahmer ML, Singh BK, Shaner DL (2005) Imidazolinone tolerant crops: history, current status and future. Pest Manag Sci 61:246–257
98. BASF Communication (2015) https://agriculture.basf.com/en/Crop-Protection/News-Events/Press-releases/BASF-and-Embrapa-launch-Cultivance.contact.html
99. Chahal PS, Aulakh JS, Jugulam M, Jhala AJ (2015) Herbicide resistant palmer amaranth (Amaranthus palmeri S. Wats.) in the United States - Mechanisms of resistance, impact and management. INTECH Chapter 1:1–30
100. Craigmyle BD, Ellis JM, Bradley KW (2013) influence of herbicide programs on weed management in soybean with resistance to glufosinate and 2,4-D. Weed Technol 27:78–84
101. Wright TR, Shan G, Walsh TA, Lira JM, Cui C, Song P, Zhuang M, Arnold N, Lin G, Yau K, Russell SM, Cicchillo RM, Peterson MA, Simpson DM, Zhou N, Ponsamuel J, Zhang Z (2010) Robust crop resistance to broadleaf and grass herbicide provided by aryloxyalkanoate dioxygenase transgenes. PNAS 107(47):20240–20245
102. Ricroch AE, Henard-Damave MC (2016) Next biotech plants: new traits, crops, developers and technologies for addressing global challenges. Crit Rev Biotechnol 36 (4):675–690. https://doi.org/10.3109/07388551.2015.1004521
103. Poree F, Heinrichs V, Lange G, Laber B, Peters C and Schouten L (2014) HPPD variants and methods of use. United States patent application # US20150267180A1
104. Jain R, Miller B R, Vail G D and Ulmer B J (2016) Improved weed control methods. United states patent # US20160058014 A1
105. Owen MDK (2013) 2013 herbicide guide for iowa corn and soybean production. Agriculture and Environment Extension Publications
106. Gullickson G (2017) Balance GT soybean performance system full scale launch likely for 2018 growing season. Successful farming at agriculture.com
107. Li X, Nicholl D (2005) Development of PPO inhibitor resistant cultures and crops. Pest Manag Sci 61(3):277–285

108. Choi KW, Han O, Lee HJ, Yun YC, Moon YH, Kim MK, Kuk YI, Han SU, Guh JO (1998) Generation of resistance to the diphenyl ether herbicide, oxyfluorfen, via expression of the Bacillus subtilis protoporphyrinogen oxidase gene in transgenic tobacco plants. Biosci Biotechnol Biochem 62:558–560

109. Lee HJ, Lee SB, Chung JS, Han SU, Han O, Guh JO, Jeon JS, An G, Back K (2000) Transgenic rice plants expressing a Bacillus subtilis protoporphyrinogen oxidase gene are resistant to diphenyl ether herbicide oxyfluorfen. Plant Cell Physiol 41:743–749

110. Green JM (2018) The rise and future of glyphosate and glyphosate resistant crops. Pest Manag Sci 74(5):1035–1039. https://doi.org/10.1002/ps.4462

111. Hinga M, Griffin S, Moon M S, Rasmussen R D and Cuevas F (2017) Methods and composition to produce rice resistant to ACCase inhibitors. United states patent # US20170002375 A1

112. Gilbert LA, Larson MH, Morsut L, Liu Z, Brar GA, Torres SE, Stern-Ginossar N, Brandman O, Whitehead EH, Doudna JA, Lim WA, Weissman JS, Qi LS (2013) CRISPR-mediated modular RNA-guided regulation of transcription in eukaryotes. Cell 154:442–451. https://doi.org/10.1016/j.cell.2013.06.044

113. Gupta M, DeKelver RC, Palta A et al (2012) Transcriptional activation of Brassica napus b-ketoacyl-ACP synthase II with an engineered zinc finger protein transcription factor. Plant Biotechnol J 10(7):783–791. https://doi.org/10.1111/j.1467-7652.2012.00695.x

114. Ainley WM, Sastry-Dent L, Welter ME et al (2013) Trait stacking via targeted genome editing. Plant Biotechnol J 11:1126–1134. https://doi.org/10.1111/pbi.12107

115. Kumar S, AlAbed D, Worden A et al (2015) A modular gene targeting system for sequential transgene stacking in plants. J Biotechnol 207:12–20. https://doi.org/10.1016/j.jbiotec.2015.04.006

116. Feldman KS (1991) T-DNA insertion mutagenesis in Arabidopsis: mutational spectrum. Plant J 1:71–83

117. Martienssen RA (1998) Functional genomics: Probing plant gene function and expression with transposons. Proc Natl Acad Sci U S A 95:2021–2026

118. CM MC, Comai L, Greene EA, Henikoff S (2000) Targeting induced local lesions IN genomes (TILLING) for plant functional genomics. Plant Physiol 123:439–442. https://doi.org/10.1104/pp.123.2.439

119. McCallum CM, Comai L, Greene EA, Henikoff S (2000b) Nat Biotechnol 18:455–457

120. Koornneef N, Dellaert LWM, van der Veen JH (1982) EMS- and radiaton-induced mutagenesis at individual loci in Arabidopsis thaliana (L.) Heynh. Mutat Res 93:109–123

121. IAEA (2010) Welcome to the mutant variety database. http://mvgs.iaea.org/AboutMutantVarieties.aspx

Chapter 22

Developing Transgenic Agronomic Traits for Crops: Targets, Methods, and Challenges

John P. Davies and Cory A. Christensen

Abstract

The last two decades have witnessed a surge of investment by the agricultural biotechnology industry in the development of transgenic agronomic traits. These are traits that improve yield performance by modifying endogenous physiological processes such as energy capture, nutrient utilization, and stress tolerance. In this chapter we provide a foundation for understanding these fundamental processes and then outline approaches that have been taken to use this knowledge for yield improvement. We characterize the current status of product development pipelines in the industry and illustrate the trait discovery process with three important examples—bacterial cold-shock proteins, alanine aminotransferase, and auxin-regulated genes. The challenges with developing and commercializing an agronomic trait product are discussed.

Key words Transgenic traits, Agronomic traits, Yield, Nitrogen-use efficiency, Stress tolerance

1 Introduction

The first generation of crops with transgenic traits were introduced in the 1990s and provided herbicide tolerance (HT) and insect resistance (IR). These traits benefited growers by reducing costs, simplifying pest management, increasing yield, and enhancing growers' profits [1]. In subsequent years, as companies have developed the second generation of HT and IR traits that complement or replace the first-generation traits, they have also been developing new classes of so-called agronomic traits that target increased yield potential by manipulating metabolic processes such as photosynthesis or carbon and nitrogen metabolism or that target yield stability by enhancing tolerance to abiotic stresses such as drought [2, 3]. These traits hold the promise of helping to feed the earth's growing population by increasing farm yields and farmer profits. In this chapter we review progress and challenges in developing these agronomic traits. Our goal is not to provide an exhaustive review of all the approaches to improve agronomic traits, but rather to

Sandeep Kumar et al. (eds.), *Transgenic Plants: Methods and Protocols*, Methods in Molecular Biology, vol. 1864,
https://doi.org/10.1007/978-1-4939-8778-8_22,

highlight examples of the strategies employed and to discuss a few traits commercialized or in development that illustrate these strategies.

Transgenic trait development is a time-consuming and expensive process requiring approximately 13 years and costing an estimated $136 million dollars for each deregulated event [4]. The release of transgenic crops is regulated by numerous agencies in the jurisdictions where the crops are grown and consumed. Because of this expense, transgenic crop development has been mostly limited to high-acreage row crops such as maize, soybean, canola, and cotton [5]. The benefits of the herbicide tolerance and insect resistance traits, which simplify herbicide treatment and kill yield-reducing insect pests, are readily recognized by farmers and explain their rapid adoption [1]. The value of a successful agronomic trait is its ability to enhance yield. Agronomic traits differ from IR and HT traits in that they are not typically measured in a binary fashion (e.g., resistant, sensitive), but quantitatively in terms of improved agronomic performance. Performance may be measured in physiological terms such as growth rate, nutrient uptake, or drought tolerance, but ultimately economic yield is the measure that matters most to the market.

2 Physiological Foundations for Crop Improvement Through Agronomic Traits

2.1 Candidates in Primary Metabolism and Plant Physiology

Crop yields are a function of the crop's ability to capture and store energy, and then to convert it into harvestable, consumable biomass [6]. The balance of these processes is understood within the framework of "source-sink" interactions [7]. Within a plant, "source" activity is exhibited wherever resources (molecules and energy) are primarily exported, while sink activity occurs at the site of consumption and/or storage. Genes that can influence either the "source" or the "sink" activities may be candidates for agronomic trait development.

Photosynthetic efficiency, which comprises the capture of light energy and its conversion to chemical energy, is an attractive target for improving source processes because it was not improved much during the domestication of crop plants and it falls short of the predicted biological limit [6, 8]. Photosynthesis is composed of two main processes; the light reactions, which absorb light energy, use it to elevate the redox state of electrons split from water and store the energy in the form of NADPH and ATP, and the dark reactions which use the NADPH and ATP produced by the light reactions to reduce CO_2 and form sugar. These processes work efficiently under low-light, non-stressed conditions, but their efficiency declines as light levels and various stresses increase. Light reaction targets include changes to the light-harvesting pigments for absorption of a broader spectrum of light, thereby increasing the amount of energy captured per unit area of the plant [9, 10] and enhancements to the relaxation of non-photochemical quenching (NPQ), which can limit photosynthetic efficiency as plants pass from high-light conditions to low-light conditions [11, 12].

The dark reactions of photosynthesis have inefficiencies that are also targets for improvements. The efficiency of the dark reactions is considered to be co-limited by the carboxylation of ribulose bisphosphate (RuBP) and the capacity to regenerate it [13, 14]. The enzyme ribulose bisphosphate carboxylase/oxygenase (RUBISCO) catalyzes the carboxylation of RuBP, but it can also catalyze the oxygenation of RuBP; carboxylation of RuBP leads to the production of sugars in an energy-efficient process while oxygenation of RuBP produces 2-phosphoglycolate which is then metabolized to 3-phospho-glycerate in a process called photorespiration. Photorespiration is an energy-costly process that results in the loss of one molecule of fixed CO_2 as well as the energy required to fix it [15]. The balance of carboxylation/oxygenation reactions carried out by RUBISCO depends on the concentration of CO_2 around RUBISCO, with higher CO_2 concentrations favoring the carboxylation reaction.

Some photosynthetic organisms such as C4 plants, green algae, and cyanobacteria have developed mechanisms to limit photorespiration by increasing the concentration of CO_2 near RUBISCO. In C4 plants, carbonic anhydrases of the mesophyll cells convert CO_2 to bicarbonate, which is added to phosphoenolpyruvate (PEP) by PEP carboxylase to form a four-carbon (C4) dicarboxylic acid. The C4 acid is transported to the bundle sheath cells where it is decarboxylated, thereby increasing the partial pressure of CO_2 in the proximity of RUBISCO for efficient RuBP carboxylation. This process effectively eliminates photorespiration in C4 plants and greatly enhances photosynthetic efficiency. C3 plants, which include important crops such as soybean, wheat, canola, and rice, do not have this carbon-concentrating system and have lower photosynthetic efficiencies. Thus, there is interest in enhancing the carboxylation of RuBP in C3 crops while limiting its oxygenation.

Strategies have been developed to engineer C3 plants with the C4 carbon-concentrating mechanism [16, 17] but the implementation of these strategies faces significant challenges since both metabolic and anatomical changes must be made in the plant. Cyanobacteria and green algae use a different strategy to increase the concentration of CO_2 near RUBISCO. These organisms actively take up bicarbonate which is then moved to specialized organelles where carbonic anhydrase and RUBISCO are co-localized. The carbonic anhydrase decarboxylates bicarbonate to release CO_2 and effectively increases the CO_2 concentration around RUBISCO. This type of carbon-concentrating system may be easier to engineer than C4 photosynthesis while still enhancing the efficiency of photosynthesis in C3 plants [18].

Another approach is to make the photorespiration reactions less energetically costly [19–21] and several examples of using enzymes from various bacteria to engineer photorespiration to do this have been reported [19, 22, 23]. Crop models also indicate that improving the regeneration of RuBP by increasing the production of the dark reaction intermediates sedoheptulose-1,7-bisphosphatase and/or fructose-1,6-bisphosphate aldolase, which limit CO_2 fixation under some conditions [8], may increase the rate of RuBP regeneration and remove a factor-limiting photosynthesis.

Leaves are the primary site of photosynthesis and their premature senescence can cause a decrease in productivity [24]. Senescence can be triggered by environmental cues such as drought and nutrient stress, so controlling environmentally induced senescence can lead to increases in yield [25]. Several plant hormones are known to be associated with senescence; ethylene, jasmonic acid, salicylic acid, auxins, and brassinosteroids are inducers of senescence while cytokinins can repress it. Since cytokinins reduce premature senescence, researchers have pursued a strategy of increasing the biosynthesis of cytokinins under stressful conditions. This has been done by expressing isopentyl transferase, the rate-limiting enzyme in the cytokinin biosynthetic pathway, from stress-induced or developmentally regulated promoters [25–28]. Transcription factors that show senescence-specific patterns of transcript accumulation such as the NAC family of transcription factors have also been used in transgenic plants to overcome premature senescence [25].

Nutrient acquisition and utilization are also important determinants of plant health and yield. Nitrogen is an essential nutrient and its availability often limits yield [29]. Therefore it is important that crop plants acquire and use this essential nutrient efficiently. Nitrogen-use efficiency (NUE) is defined as the amount of yield that a crop produces per unit of nitrogen applied and is the product of nitrogen uptake efficiency (NUpE) and nitrogen utilization efficiency (NUtE) [30]. NUpE has been addressed by targeting nitrate and ammonium transporters and root architecture [31] while NUtE has been addressed by targeting storage and remobilization of nitrogen-containing compounds [31–33] and by expressing amino acid transporters [34] as well as various nitrogen metabolic enzymes such as alanine aminotransferase [35, 36], glutamine synthetase [37–39], GOGAT [40], and glutamate dehydrogenase [41–43]. Recently *DEP1*, a signal transduction gene that regulates various aspects of nitrogen uptake and assimilation has become a target for enhancing NUE [44, 45].

In addition to enhancing the "source" part of the source-sink relationship, researchers have attempted to increase yield by targeting processes and pathways involved in creating "sinks" for photosynthetically derived energy. One of the processes that have been targeted is increasing starch biosynthesis in tissues such as seeds and tubers. This has been approached by increasing the activity of ADP-glucose pyrophosphorylase, which catalyzes the first committed step in starch biosynthesis [46, 47], or by increasing expression of transporters that move glucose-6-phosphate and nucleotide triphosphate into the plastid [48]. The latter approach was more effective when starch biosynthesis in source tissues (i.e., leaves) was inhibited by downregulating ADP-glucose pyrophosphorylase or a soluble pyrophosphatase [49]. In addition to manipulating the enzymes involved in sugar transport and starch biosynthesis, reprogramming signal transduction pathways to increase the size of "sink" tissues or organs has been reported to increase yields [50–52]. The plant brassinosteroid and auxin hormones are important regulators of organ size and many of the genes in the signal transduction pathways controlling responses to these hormones are being investigated for their ability to increase organ size, biomass, and yield in plants [52].

2.2 Candidates in Abiotic Stress Tolerance Mechanisms

Crop losses due to abiotic stresses such as flooding, drought, and salinity are costly both to producers and to consumers who rely on crop production. These stress-inducing events have increased in frequency and are predicted to become even more common [53]. Enhancing crop tolerance to abiotic stresses is an essential component of the strategy to improve crop productivity. Plants exhibit both adaptation and avoidance strategies to stressful conditions. Upon the recognition of the stress condition, regulatory pathways are induced that enable the plant to acclimate or avoid the condition. Numerous attempts have been made to enhance crop tolerance to these stresses by understanding the recognition, signaling, and adaptation or avoidance strategies and using them to build resilience into the crop.

Annual losses to crop production due to flooding are in the billions of dollars [54]. Flooding is already an important constraint for rice production in flood-prone, lowland regions of South and Southeast Asia and West Africa, and the frequency of flooding in these regions is expected to increase as a result of global climate change. One crop engineering success story is the adaptive tolerance to flooding stress in rice conferred by the *SUB1* gene. When rice is submerged, this gene enhances expression of two transcription factors that limit gibberellic acid-induced growth, thereby limiting consumption of energy reserves. *SUB1* plants also accumulate less damage from reactive oxygen species and can recover faster [55]. Avoidance strategies to flooding include those employed by the "deep-water" landraces of rice that grow elongated porous stems that are connected to root tissues and allow gas exchange between the submerged and non-submerged tissues [53].

Drought can also devastate crop production and annually costs growers in the United States $6–8 billion [56]. Numerous attempts, using a variety of approaches, have been made to make plants more resilient to this stress. Drought tolerance or drought avoidance may occur through a number of mechanisms, including developmental/physiological (e.g., root branching/depth, stomatal number/responsiveness), biochemical (e.g., compatible solutes like betaine, proline), and signaling (e.g., hormonal like ABA, transcription factor networks, signaling cascades like kinases/phosphatases). Since water is absorbed through the roots of plants, root architecture has been an obvious target for enhancing tolerance to water-limited environments. Various ideotypes have been proposed for root systems to increase tolerance to water-limited environments [57]. One that has received support is the "steep, cheap, and deep" ideotype that targets long and thick primary roots with few lateral roots so that the root system can penetrate the soil deeper where water is more likely to be available during times of drought while the seminal roots are shallow, thin, and highly branched so they can access water (and nutrients) efficiently while seedlings are getting established [58, 59]. Because transpiration occurs through the leaves, leaf characteristics are also important for acclimating to drought conditions. Phenotypes such as leaf rolling, epicuticular wax, stomatal density, and stomatal aperture have been associated with enhanced tolerance to drought, as have physiological processes such as osmotic adjustment and biochemical processes such as membrane stability [60]. The mechanisms that control these characteristics are targets for transgenic approaches to improve drought tolerance. Although elucidating and exploiting these control mechanisms is a challenge, there have been some notable successes. For example, the *DRO1* gene from rice influences root growth angle and morphology of the root system and has been demonstrated to improve drought tolerance [61]. Additionally, a large number of genes influencing the biosynthesis and deposition of epicuticular wax have been identified [62].

Many processes that enable plants to acclimate to drought conditions are mediated by plant hormones; particularly important are ABA [63] and ethylene [64]. Both hormone receptors [65, 66] and transcription factors such as ethylene response factors (ERFs) and drought-responsive element-binding proteins (DREBs) [67] controlling these responses have also been used to enhance crop responses to drought. In addition to manipulating a plant's normal response to drought, some researchers have transferred stress adaptation processes found in other organisms to plants. One example of this is the stress-induced mRNA-stabilizing process from *Bacillus subtilis* that is encoded by the *cspB* gene which has been used to confer drought tolerance in crop plants.

Approximately 20% of agricultural lands have salinity problems that have been caused by both natural events and human activities. Salinity imparts both ionic and osmotic stresses and can affect crop performance by inhibiting cell expansion and photosynthesis and promoting premature senescence caused by elevated levels of Na^+

and Cl^- in the leaves [53]. Salinity issues can be minimized by mechanisms that control Na^+ uptake and transport into cells and vacuoles in roots and shoots and the resulting oxidative damage. Genes that encode proteins that participate in and/or regulate these processes may be candidates for improving salt tolerance in crops [68].

Enhancing crop tolerance to abiotic stress is critical for making plants more resilient to environmental changes, improving yield stability and providing food security. Transgenic agronomic traits are one of the important tools to achieving this.

3 Discovery of Genes to Improve Agronomic Traits

Transgenic approaches to improving agronomic traits in crops are based on the idea that expressing genes from other species or expressing endogenous genes at elevated levels or in different tissues and at different times can produce useful phenotypes not available within crop germplasm. These approaches, therefore, diversify the genetics of the target crop. Researchers have used many avenues to identify genes to improve agronomic traits in crops. Some have built on the understanding of the biochemical and physiological properties that have been studied for many years and taken a targeted approach, while others have pursued less targeted approaches by testing random genes or many genes of certain classes (e.g., transcription factors). Trait discovery systems have often been developed in experimentally facile model plant systems, such as *Arabidopsis*, and then applied to crop plants.

The examples highlighted in the section above demonstrate how biochemical and physiological dissection of an important biological process led to the identification of potential candidate genes for improving agronomic traits. Photosynthesis has been at the center of study in plant biology for many years and understanding of the light and dark reactions has been foundational. This understanding has led researchers to ask whether these highly evolved systems that are adapted to reproductive success in natural ecosystems are the most efficient for producing high yields in agricultural systems. To look for ways to improve photosynthesis researchers have compared photosynthesis in different vascular plants as well as in nonvascular plant systems and performed modeling studies to identify processes that are not operating as efficiently as possible [8, 9]. This has led to several proposals for improving photosynthesis such as using light-harvesting pigments from other organisms in crop plants to broaden the spectrum of light that they absorb and use for photosynthesis [10], transferring carbon-concentrating systems to C3 plants, and reengineering photorespiratory pathways [16].

The understanding and comparative analyses of biological processes and biochemical pathways in different species can provide candidate genes from other (e.g., non-plant) species for improving agronomic traits. These genes may have different functions and different biochemical properties or may be regulated in a manner different from the endogenous plant genes. One prominent example is the *ipt* gene from *Agrobacterium tumefaciens* which encodes an isopentenyl transferase that performs a function analogous to the endogenous *ipt* genes of plants in catalyzing the rate-limiting step in cytokinin biosynthesis. The *Agrobacterium* enzyme has substrate affinities and kinetic properties different from most plant-encoded IPT enzymes [69]. Transgenic studies in several crop species show that expressing the Agrobacterium *ipt* from various senescence-associated gene promoters can improve stress tolerance and yield [28, 70–74].

Arabidopsis has been a powerful model system to discover gene function [75] and its usefulness was enhanced after its genome was published [76]. Genomics-based experiments with *Arabidopsis* have identified numerous genes that may improve agronomic traits in crop plants. Among these techniques are gene expression profiling and reverse genetics which identified *AtERF* that conferred broad-spectrum disease resistance in tomato [77], activation tagging which identified the *AtHRD2* gene that improves drought tolerance in rice [78], and systematic overexpression of the set transcription factors from *Arabidopsis* which identified the *NF-YB2* gene that led to the identification of a maize ortholog which confers drought tolerance in maize [79]. The agricultural industry recognized the power of *Arabidopsis* genetics and several biotechnology start-up companies including Paradigm Genetics [80, 81], Ceres [82, 83], Mendel Biotechnology [84], and Exelixis Plant Sciences [85, 86] used *Arabidopsis*-based technology in their trait discovery efforts.

4 Assessing the Performance of Transgenic Traits

Whether taking an approach that targets the genes of a particular pathway or one that is not targeted, the process of developing a transgenic trait requires testing many candidate constructs (one or more plant transcriptional units comprising promoter-gene-terminator elements in a delivery vector, which is frequently T-DNA) in order to identify ones that are effective in delivering the desired trait characteristics. In non-targeted approaches, hypotheses about the mode of action for a candidate gene are typically developed after there is an indication of efficacy, whereas approaches that target known pathways (such as those described previously) are derived from hypotheses about ways the pathway can be modulated to improve plant performance. Even in this case, adequately testing a hypothesis can require many constructs to determine whether the gene or genes can improve the physiology or biochemistry of the organism. Factors that impact the success of a construct include the ability of the construct as designed to express the genes at the correct level in the desired tissues at the desired time, the presence or absence of mechanisms within the plant to compensate for target gene expression, the accuracy of the hypothesized gene function in the pathway, and consequences stemming from construct insertion into the host genome.

In this early stage, researchers are focused on demonstrating that the hypothesis has merit and the constructs have the potential to increase yield in the target crops. Because many constructs are tested, a high-throughput testing system such as *Arabidopsis*, *Brachypodium*, *Setaria*, rice, or cell culture systems is often used. The assay system to evaluate the efficacy of the constructs in these systems must also be of high throughput so that a large number of constructs may be evaluated to identify those that demonstrate the desired phenotype. These systems are often comprised of high-throughput greenhouse or growth chamber screens using sophisticated imaging processes that can measure growth and development often under different conditions [80, 87, 88]. Additional screening platforms may include analysis of biochemical components that are indicative of the trait.

Once the hypothesis has been shown to be effective in improving the targeted biochemical or physiological pathway in a model system and reached proof of concept, constructs are prepared for transformation of the target crop. Prior to transformation of constructs into the crop of interest, the protein-coding sequences in the construct are examined for similarity to known allergens and toxins. This process follows internationally accepted standards and guidance from regulatory agencies to ensure the safety of the construct [89, 90]. Each construct is carefully designed so that the proteins are expressed when and where they are required and the proteins accumulate in the desired subcellular location. To hedge against the inability to translate effects from the model system or undesired effects in the crop of interest, multiple constructs are made and transformed into the crop, and multiple events (primary transformants) from each transformation experiment are generated. Multiple events from each construct are tested in high-throughput screens to get a quick evaluation of the construct. Multiple rounds of transformation and testing may be required to demonstrate efficacy and mitigate undesired effects. The goal of this phase of the development process is to identify constructs that have the desired characteristics in the crop of interest. The assays used in this stage may leverage learnings from the proof-of-concept stage.

Constructs that pass the initial screens are advanced to field screens to evaluate whether they perform in the crop of interest in environments similar to where they will be grown commercially. If they pass these criteria, they are often stacked with other constructs to test whether the concept will work with other transgenes that will be in commercial trait packages. The initial tests of these stacked traits are performed in a few select genetic backgrounds, and events are evaluated on whether all the genes are expressed and whether they provide a yield benefit. Additional data may be collected to demonstrate that they have the expected effect on the biochemistry and/or physiology of the crop. This phase of testing further reduces the number of constructs and events being tested.

Once a construct has reproducibly demonstrated the desired effect on yield without unintended effects, a new set of transformations may be performed to identify events that may be commercialized. Events are molecularly characterized to identify those that have a single copy of the complete set of transgenes integrated in a site in the genome and that contain no backbone sequences from the transformation vector. The sequences flanking the insertion site are identified to determine if any other genes or regulatory sequences are in the region that may be affected and possibly hurt plant performance or cause safety concerns. The insertion site sequence also provides the basis of an event-specific assay that can be used to test for the presence of the event in breeding populations [5].

Events that have been molecularly characterized are then thoroughly tested in field studies in multiple environments over several years to ensure that the event is stable and imparts the beneficial trait package. These events are also tested in different genetic backgrounds to demonstrate that the construct is effective across different genetics. This requires introgressing the events into these alternative genetic backgrounds. The resulting lines are tested in multiple environments for several seasons.

Once events are identified that meet the requirements of the product concept including the phenotypic performance, molecular characterization, and commercial requirements, they are subjected to additional experiments for regulatory approval. Regulatory submissions are made on specific events and the data packages are extensive [91]. For these reasons only one or two events are included in these analyses. The data package for regulatory approval is to assess the safety of the transgenic product. Internationally recognized approaches and guidelines are followed to assess the toxicological and safety risks for food and feed [89, 92]. These data are collected in dossiers and submitted to the relevant regulatory agencies throughout the world where the product is likely to be grown or consumed.

5 Agronomic Trait R&D Pipelines

Over the past couple of decades, work has been done by universities, institutes, small biotechnology companies, and major multinational corporations to discover and develop agronomic traits. However, because of the costs involved, only large or well-funded organizations have the wherewithal to bring a trait to market. Recently, Parisi et al. (2016) undertook a study to characterize transgenic trait product development pipelines. They conducted a search of regulatory filings, company publications, and public genetically modified crop databases to assemble a dataset, which was then vetted at an international conference in 2014 by participating companies, institutes, and government agencies. The resulting dataset captured information about companies, trait descriptions, crops, and stage of R&D. Table 1 summarizes the 46 entries comprising agronomic trait descriptions. Not surprisingly, maize, rice, and soybean have received the most attention among the crop species, and intrinsic yield, NUE, and abiotic stress tolerance (including drought) received the most focus as trait targets. This data summary illustrates the broad scope of agronomic trait discovery efforts to date in terms of species and types of trait. Attrition as potential trait products advance from early stages to commercial launch is expected. However, the sharp decline in numbers of traits at more advanced R&D stages and the fact that only one agronomic trait gene has been commercialized to date are testaments to the difficulty in developing commercially viable products.

The genes in these company product development pipelines represent a small portion of the many hundreds of agronomic trait candidate genes that have been identified in the past couple of decades. A large number of these have been tested as transgenes for their ability to confer beneficial phenotypes in discovery programs that are too preliminary to register even as "Early R&D Stage" in Table 1. In these programs, the agricultural biotechnology industry has collectively spent hundreds of millions of dollars (USD), yet very few traits have advanced to late stages of R&D (Table 1). The primary reasons for this limited market impact to date are the difficulty of modifying a plant system to sufficiently improve economic yield and the translation of those improvements across species, germplasms within a species, and diverse environments.

6 Examples of Agronomic Traits in Commercial Product Development

In this section, we highlight three examples of transgenic agronomic traits in commercial crops. We describe the function and

Table 1
Number of distinct agronomic traits in R&D pipelines based on Parisi et al., 2016, and summarized by crop, trait description, or R&D stage

Crop		Trait description		R&D stage	
Maize	8	Abiotic stress tolerance	13	Early	23
Rice	7	Abiotic stress tolerance (drought)	9	Advanced	16
Soybean	5	Increased yield (nitrogen)	8	Regulatory	5
Sugar beet	3	Increased yield	5	Pre-commercial	1
Sugarcane	3	Drought and saline stress tolerance	2	Commercial cultivation	1
Barley	2	Increased yield and lignin composition	1		
Eucalyptus	2	Dwarfing	1		
Mustard	2	Higher cellulose content	1		
Cotton	2	Early flowering and color modification	1		
Alfalfa	2	Increased yield (heterosis)	1		
Wheat	1	Cold stress tolerance	1		
Oilseed rape	1	Heat stress tolerance	1		
Potato	1	Iron stress tolerance	1		
Chickpea	1	Frost resistance	1		
Sugarcane	1				
Rubber	1				
Switchgrass	1				
Sorghum	1				
Groundnut	1				
Pineapple	1				

potential of these genes and use publicly available information to highlight the twists and turns of their development. Our goal is not to give a comprehensive overview of all advanced trait concepts, but to detail a few examples in order to illustrate some of the challenges in developing transgenic traits for crop plants.

6.1 CSPB: Cold-Shock Protein B

After all of the effort directed at the discovery and development of agronomic trait transgenes, there is only one that has been launched as a product in a major crop. It was developed by Monsanto and is marketed as DroughtGard. The trait is delivered by the MON 87460 event expressing cold-shock protein B (CSPB) from *Bacillus subtilis* and is an example of moving a stress tolerance mechanism from another organism into a crop plant to develop an agronomic trait. Bacterial cold-shock proteins (CSP) were first reported in the literature in 1991 and since that time have been the subject of numerous articles and reviews. At the cellular level, bacterial CSPs function as RNA chaperones to prevent secondary structure formation or to facilitate the degradation of structured RNA that would otherwise inhibit translation under cold stress conditions [93, 94]. Plant genomes also contain *CSP* orthologs that play a role in posttranscriptional RNA metabolism and regulate plant growth, development, and stress responses [95].

Based on the stress tolerance observed in bacteria, researchers at Monsanto carried out studies demonstrating that expression of *cspA* from *E. coli* or *cspB* from *B. subtilis* conferred improved growth under cold conditions in *Arabidopsis*. They went on to demonstrate that the expression of these genes also conferred tolerance to multiple abiotic stresses (cold, heat, and water deficit) in rice (*Oryza sativa*) as measured by improved growth relative to controls. Following these studies, they advanced the *cspB* gene into further characterization in greenhouse and field-grown maize. *cspB* expression was found to improve vegetative growth in maize under controlled water-deficit conditions [96]. Monsanto researchers further characterized the nature of water-deficit tolerance in a whole-plant physiology study conducted in the field. This study showed that *cspB* expression had little to no effect under well-watered conditions, but under water-limiting conditions resulted in reduced leaf area and increased grain yield. The study authors propose a model wherein *cspB* expression caused a transient reduction in leaf growth under water-deficit stress. These smaller leaves reduced overall water usage, thereby increasing water-use efficiency and minimizing stress exposure during silking. This resulted in increased ear partitioning and growth. The authors conclude, "the larger ears were able to attract more assimilates during grain fill, resulting in increased kernel set, increased harvest index and increased grain yield" [97]. At this time, however, the mechanism linking *cspB* to reduced leaf area expansion under water deficit remains unknown. In MON 87640, *cspB* is expressed under the control of the constitutive rice actin1 promoter [97] and no difference in protein levels was observed between well-watered and water-deficient treated plants [98]. Therefore, there is no support for differential expression of *cspB* to explain the conditional response.

It has been shown that *cspB* binds RNA *in planta* and localizes to rapidly growing tissues and reproductive organs that are important for yield [98] and the linkage between RNA-binding activity of this protein and drought tolerance was established by mutating the RNA-binding domain of *cspB* [96]. However, how these biochemical processes produce the effects observed at the level of whole-plant physiology remain unclear. In a systematic study for pleiotropic effects beyond the observed tolerance to water deficit, Monsanto researchers did not find any significant effects of *cspB* expression in maize other than the reduced leaf area observed under drought stress [99]. The linkage of the water-deficit tolerance to reduced leaf area leaves open the possibility elaborated extensively in a review by Lawlor (2013), that the beneficial effect of *cspB* expression is merely an inadvertent result of a negative impact on growth [100]. Indeed the drought tolerance literature is replete with studies that neglect fundamental physiological principles in their design resulting in many cases where reports of drought tolerance can be explained simply by stunted plants using less water and therefore not experiencing the effects of water deficit to the same extent as control plants [100]. However, the effects observed in MON 87460 appear to be specific to water-deficit conditions [99] unlike many of these other reported genes, where reduced size is seen under control and water-deficit conditions. Thus, the trait conferred by *cspB* does not appear to result simply from an inadvertent reduction in plant size. Understanding the chain of events linking the RNA-binding activity to the conditional reduction in leaf growth will require further investigation.

DroughtGard has enjoyed some success in the marketplace. Maize area being planted with it has increased since introduction in 2012 up to 810,000 hectares in 2015 and Monsanto has entered an agreement to make the trait available to farmers in Africa through a public-private partnership [101].

6.2 AlaAT: Alanine Aminotransferase

A large number of candidate genes affecting NUE in crop plants and associated with nitrogen metabolism have been identified. Many of these have been tested for their ability to confer an NUE trait when expressed as a transgene, but the results have been inconsistent [102]. One that has shown more promise and is perhaps the most broadly evaluated biotech agronomic trait in terms of plant species transformed and industry collaborations is the NUE trait conferred by tissue-preferential expression of alanine aminotransferase (*AlaAT*). The ability of *AlaAT* to confer NUE was first observed serendipitously when Good and colleagues were using mis-expression of *AlaAT* to investigate the potential of alanine as a compatible solute in a drought stress experiment [35, 103]. In the seminal study conducted in canola, an ortholog of the gene from barley (*HvAlaAT*) was expressed using the promoter from a *Brassica napus* turgor-responsive gene (*btg26*). The *btg26* promoter in this study is thought to be preferentially expressed in root tissue with some expression in the vascular tissue of leaves and cotyledons. Canola plants transformed with the *btg26:AlaAT* construct showed >30% increased seed yield in field trials under insufficient nitrogen conditions compared to controls and no yield penalty under sufficient nitrogen conditions [35].

Similar results were obtained in rice when a tissue-specific promoter from a rice ortholog of *btg26* (*OsAnt1*) was used to drive expression of the same barley *AlaAT* gene. The expression pattern of *OsAnt1* is comparable to *btg26* in that it was also detected most highly in root tissues with some, more weakly observed, expression in shoot vascular tissue. Under nitrogen-limiting conditions plants containing the *OsAnt1:AlaAT* construct showed increased biomass and grain yield (30–54%) relative to non-transgenic control plants. The increase was achieved primarily through an increase in the number of tillers [36]. In contrast to the results observed with the *btg26* and *OsAnt1* promoters, positive results in terms of biomass and grain yield were not obtained when the *AlaAT* gene was expressed under a constitutive promoter (*CaMV35S*) in *Arabidopsis* [104] or canola [35] indicating that the NUE trait is dependent on tissue-specific expression. Increased levels of alanine in root tissues and reduced levels of glutamine in the shoot tissue of these transgenic plants demonstrate that mis-expression of *AlaAT* alters amino acid metabolism. Increased nitrate in the shoot tissue suggests that the plants recognize the altered amino acid levels as nitrogen deficiency and respond by increasing nitrogen uptake [35].

The patented invention of the NUE trait conferred by the tissue-specific expression of *AlaAT* has been licensed to Arcadia Biosciences, Inc. whose business model addresses the early validation of candidate genes, but achieves trait commercialization through partnership with other seed companies (as described in Arcadia's SEC filing of form S-1, 2015). The licensed technology has been evaluated for commercial purposes in several crops. It has been evaluated in wheat and barley in collaboration with the Commonwealth Scientific and Industrial Research Organization (CSIRO) and the Australian Centre for Plant Functional Genomics (ACPFG) [102], in maize with DuPont Pioneer [3], in sugarcane with South African Sugarcane Research Institute [105], and further evaluated in rice for cultivation in Africa with the International Center for Tropical Agriculture [39]. In this recent rice work, the *OsAnt1:HvAlaAT* construct was evaluated in paddy and rain-fed conditions under a range of nitrogen treatments and over two seasons. Lead transgenic events demonstrated a 25–30% yield increase under nitrogen-limiting conditions in both paddy and rain-fed fields [39]. Additional work in sugarcane has also shown the broad potential application of this technology [106]. From publicly available information, it is not clear why this trait has not yet been commercialized.

6.3 ARGOS

A gene family named *ARGOS*—auxin-regulated gene involved in organ size—was identified in *Arabidopsis* based on differential expression in response to auxin treatment. When overexpressed as a transgene in *Arabidopsis*, *ARGOS* showed a prolonged period of growth and increased cell proliferation, but not elongation, ultimately resulting in larger leaves and increased seed yield per silique [107]. Additional studies have been carried out to evaluate the potential of this gene family in crop species. *AtARGOS* or orthologs have been expressed in tobacco, poplar, and wheat where similar effects on organ size were observed [108–110]. Interestingly, this effect was not observed when the rice ortholog, *OsARGOS*, was overexpressed in rice [111].

The story of *ARGOS* took an interesting turn with a recent report by DuPont Pioneer scientists [112]. They identified a family of eight *ARGOS* genes in maize and named them *ZAR* (Zm *ARGOS*). Their study focused on *ZAR1*, the closest ortholog to the *Arabidopsis* gene. As in other species, constitutive overexpression of *ZAR1* resulted in larger vegetative and reproductive organs due to increased cell number, not increased cell size. Notably, transgenic maize plants grew faster rather than longer as in *Arabidopsis*. It was observed, however, that the transgene effect depended on field location and year with some locations showing significant increases in plant size and yield and others showing significant decreases. Characterization of environmental conditions at these sites for each year revealed that the *ZAR1* transgenic plants were outperforming controls in hotter and drier environments with more cumulative solar radiation.

An independent study by Rai et al. (2015) indicated that *ARGOS* acts in the ethylene signal transduction pathway to desensitize plants to ethylene, thereby increasing the dynamic range of ethylene response. Ethylene inhibits cell proliferation and expansion, and the results of this study suggest that phenotypes observed in transgenic *ARGOS* plants are directly related to attenuation of the ethylene signal [113]. In an extensive study conducted on field-grown maize, DuPont Pioneer researchers demonstrated that attenuation of the ethylene signal through downregulation of ACC synthase, which catalyzes the first committed step in ethylene biosynthesis, resulted in improved performance under water-deficit conditions [64]. Thus, it is possible that the positive transgenic *ZAR1* effect under water-deficit stress is directly linked to its effects on ethylene signaling and modulation of hormone-mediated stress conditioning. Alternatively, the accelerated development observed by Guo et al. (2014) suggests a response resulting from altered phenology rather than a physiological adaptation to stress. Early flowering is a common drought-avoidance mechanism that would allow a maize plant to transit sensitive developmental stages before the onset of severe water deficit.

Further investigation of the linkage between *ZAR1*, ethylene signaling, and drought stress tolerance is likely forthcoming. Nevertheless, it remains to be seen whether a transgenic *ZAR1* event can confer a commercially viable agronomic trait. The positive or negative performance based on environmental conditions may limit its utility.

6.4 Summary

These examples show that the development of an agronomic trait is an iterative process that builds on the initial hypothesis and requires careful experimentation and interpretation. In the highlighted work, the initial hypothesis that led to the first experiments was ultimately modified based on the results of these experiments. For example, the hypothesized function of *cspB* was initially to confer cold tolerance, but experiments showed that it could also be used to confer drought tolerance; *AlaAT* was initially hypothesized to be a compatible solute for acclimation to drought stress, but subsequent experiments showed that it could enhance NUE and yield; and *ARGOS* was investigated initially for its ability to increase organ size, but subsequent analysis showed that it was involved in ethylene signaling, affected plant phenology, and showed positive effects when plants were grown in hotter and drier environments. The development of these agronomic trait genes has also benefited from the use of a model system; *cspB* and *ARGOS* were first demonstrated to confer beneficial phenotypes in *Arabidopsis* [96, 107], and orthologs of *AlaAT* from several species have been shown to confer similar but slightly different characteristics in *Arabidopsis* [114]. In addition, the *AlaAT* example highlights the importance of having the transgene expressed in the right tissues since expression of *AlaAT* from the *btg26*-like promoters conferred beneficial characteristics while expression from strong, constitutive promoters did not.

An understanding of environmental and genetic interactions is also critical for the development of transgenic traits. The examples show that *ARGOS* was able to confer a benefit in some environments but not others and the beneficial characteristics imparted by expression of *AlaAT* in root tissues are more apparent when plants are grown in low or moderate nitrogen levels than when grown at high nitrogen levels. This is critical information for the development of transgenic traits, as it provides clues for further refinement of the hypothesized function of the gene, information on product placement, and potential economic value. In addition to finding a transgene or set of transgenes that deliver a commercially viable yield increase in a given genetic background, traits need to provide similar value across different genotypes representing the elite varieties in the commercial product portfolio. Because these traits are modulating endogenous physiological processes, they may be more sensitive to genetic interactions than first-generation IR and HT traits.

7 Conclusions

While our knowledge of plant biology has grown exponentially in recent decades driven by advances in "omics" technologies and public and private research initiatives, we still do not have a robust capability to predict the outcome of simple system perturbations derived from the mis-expression of one or a few transgenes. Thus, the development of agronomic traits is a highly empirical exercise. The long product development timelines are well illustrated by the *AlaAT* gene, which was first associated with NUE over 23 years ago and has not as yet made it to commercial launch. With the examples we have highlighted in this chapter, we can begin to see what may be needed to discover, develop, and bring these traits to the market. Perhaps the long product development timelines will call for new funding models as outlined by Rothstein et al. (2014) where publicly funded consortia conduct more of the fundamental research equivalent to the early phases of R&D and then business completes the development by moving validated traits into commercial germplasm [103]. With respect to the quest to engineer plant systems to achieve ever higher levels of productivity using biotechnology, it is highly likely (paraphrasing Sir Winston Churchill) that we are much closer to the end of the beginning, than the beginning of the end.

Developing agronomic trait products is inherently challenging in terms of sustaining business investment and solving the technical problems. The past 20 years have seen an explosion of plant science knowledge and a wave of biotech industry investment. Currently, the industry appears to be in a denouement with respect to investment in transgenic agronomic traits while we see an increase in effort on alternative approaches—gene editing, biologicals, and precision agriculture—that represent new promises to improve economic yield. Ultimately, the demand for increasing agricultural productivity and our ever-increasing knowledge of plant systems are likely to converge again and set off a new wave of investment and enthusiasm. In the meantime, there are still some interesting candidates working their way through the development pipelines from the first wave of investment and time will tell whether any of them are able to achieve a level of success on par with the first generation of biotech traits.

References

1. Klümper W, Qaim M (2014) A meta-analysis of the impacts of genetically modified crops. PLoS One 9(11):e111629
2. Ricroch AE, Hénard-Damave M-C (2016) Next biotech plants: new traits, crops, developers and technologies for addressing global challenges. Crit Rev Biotechnol 36 (4):675–690
3. Parisi C, Tillie P, Rodríguez-Cerezo E (2016) The global pipeline of GM crops out to 2020. Nat Biotechnol 34(1):31
4. McDougall P (2011) The cost and time involved in the discovery, development and authorisation of a new plant biotechnology derived trait. Consultancy Study for Crop Life International
5. Prado JR, Segers G, Voelker T, Carson D, Dobert R, Phillips J, Cook K, Cornejo C, Monken J, Grapes L (2014) Genetically engineered crops: from idea to product. Annu Rev Plant Biol 65:769–790
6. Wu A, Song Y, Van Oosterom EJ, Hammer GL (2016) Connecting Biochemical Photosynthesis Models with Crop Models to Support Crop Improvement. Front Plant Sci 7:1518
7. White AC, Rogers A, Rees M, Osborne CP (2015) How can we make plants grow faster? A source–sink perspective on growth rate. J Exp Bot 67(1):31–45
8. Zhu X-G, de Sturler E, Long SP (2007) Optimizing the distribution of resources between

enzymes of carbon metabolism can dramatically increase photosynthetic rate: a numerical simulation using an evolutionary algorithm. Plant Physiol 145(2):513–526

9. Zhu X-G, Long SP, Ort DR (2008) What is the maximum efficiency with which photosynthesis can convert solar energy into biomass? Curr Opin Biotechnol 19(2):153–159
10. Blankenship RE, Tiede DM, Barber J, Brudvig GW, Fleming G, Ghirardi M, Gunner M, Junge W, Kramer DM, Melis A (2011) Comparing photosynthetic and photovoltaic efficiencies and recognizing the potential for improvement. Science 332(6031):805–809
11. Zhu XG, Ort DR, Whitmarsh J, Long SP (2004) The slow reversibility of photosystem II thermal energy dissipation on transfer from high to low light may cause large losses in carbon gain by crop canopies: a theoretical analysis. J Exp Bot 55(400):1167–1175
12. Kromdijk J, Głowacka K, Leonelli L, Gabilly ST, Iwai M, Niyogi KK, Long SP (2016) Improving photosynthesis and crop productivity by accelerating recovery from photoprotection. Science 354(6314):857–861
13. Long SP, Ainsworth EA, Rogers A, Ort DR (2004) Rising atmospheric carbon dioxide: plants FACE the future. Annu Rev Plant Biol 55:591–628
14. Farquhar GD, von Caemmerer S, Berry JA (2001) Models of photosynthesis. Plant Physiol 125(1):42–45
15. Farquhar G, Von Caemmerer S (1982) Modelling of photosynthetic response to environmental conditions. In: Physiological plant ecology II. Springer, Berlin Heidelberg, pp 549–587
16. Long SP, Marshall-Colon A, Zhu X-G (2015) Meeting the global food demand of the future by engineering crop photosynthesis and yield potential. Cell 161(1):56–66
17. von Caemmerer S, Quick WP, Furbank RT (2012) The development of C4 rice: current progress and future challenges. Science 336 (6089):1671–1672
18. McGrath JM, Long SP (2014) Can the cyanobacterial carbon-concentrating mechanism increase photosynthesis in crop species? A theoretical analysis. Plant Physiol 164 (4):2247–2261
19. Kebeish R, Niessen M, Thiruveedhi K, Bari R, Hirsch H-J, Rosenkranz R, Stäbler N, Schönfeld B, Kreuzaler F, Peterhänsel C (2007) Chloroplastic photorespiratory bypass increases photosynthesis and biomass production in Arabidopsis thaliana. Nat Biotechnol 25(5):593–599
20. Peterhansel C, Krause K, Braun HP, Espie G, Fernie A, Hanson D, Keech O, Maurino V, Mielewczik M, Sage R (2013) Engineering photorespiration: current state and future possibilities. Plant Biol 15(4):754–758
21. South P, Ort DR (2017) Engineering biochemical bypass to photorespiration to improve photosynthesis and crop production. FASEB J 31(1 Supplement):628.623
22. de FC Carvalho J, Madgwick PJ, Powers SJ, Keys AJ, Lea PJ, Parry MA (2011) An engineered pathway for glyoxylate metabolism in tobacco plants aimed to avoid the release of ammonia in photorespiration. BMC Biotechnol 11(1):111
23. Maier A, Fahnenstich H, Von Caemmerer S, Engqvist MK, Weber AP, Flügge U-I, Maurino VG (2012) Transgenic introduction of a glycolate oxidative cycle into A. thaliana chloroplasts leads to growth improvement. Front Plant Sci 3:38
24. Egli DB (2004) Seed-fill duration and yield of grain crops. Adv Agron 83:243–279
25. Guo Y, Gan S-S (2014) Translational researches on leaf senescence for enhancing plant productivity and quality. J Exp Bot 65 (14):3901–3913
26. Kant S, Burch D, Badenhorst P, Palanisamy R, Mason J, Spangenberg G (2015) Regulated expression of a cytokinin biosynthesis gene IPT delays leaf senescence and improves yield under rainfed and irrigated conditions in canola (Brassica napus L.). PLoS One 10 (1):e0116349
27. Peleg Z, Blumwald E (2011) Hormone balance and abiotic stress tolerance in crop plants. Curr Opin Plant Biol 14(3):290–295
28. Robson PR, Donnison IS, Wang K, Frame B, Pegg SE, Thomas A, Thomas H (2004) Leaf senescence is delayed in maize expressing the Agrobacterium IPT gene under the control of a novel maize senescence-enhanced promoter. Plant Biotechnol J 2(2):101–112. https://doi.org/10.1046/j.1467-7652.2004.00054.x
29. Kraiser T, Gras DE, Gutiérrez AG, González B, Gutiérrez RA (2011) A holistic view of nitrogen acquisition in plants. J Exp Bot 62(4):1455–1466
30. Moll R, Kamprath E, Jackson W (1987) Development of nitrogen-efficient prolific hybrids of maize. Crop Sci 27(2):181–186
31. Han M, Okamoto M, Beatty PH, Rothstein SJ, Good AG (2015) The genetics of nitrogen use efficiency in crop plants. Annu Rev Genet 49:269–289

32. Below FE, Christensen LE, Reed AJ, Hageman RH (1981) Availability of reduced N and carbohydrates for ear development of maize. Plant Physiol 68(5):1186–1190
33. Cliquet J-B, Deléens E, Mariotti A (1990) C and N mobilization from stalk and leaves during kernel filling by 13C and 15N tracing in Zea mays L. Plant Physiol 94(4):1547–1553
34. Yadav UP, Ayre BG, Bush DR (2015) Transgenic approaches to altering carbon and nitrogen partitioning in whole plants: assessing the potential to improve crop yields and nutritional quality. Front Plant Sci 6:275
35. Good AG, Johnson SJ, De Pauw M, Carroll RT, Savidov N, Vidmar J, Lu Z, Taylor G, Stroeher V (2007) Engineering nitrogen use efficiency with alanine aminotransferase. Can J Bot 85(3):252–262. https://doi.org/10.1139/b07-019
36. Shrawat AK, Carroll RT, DePauw M, Taylor GJ, Good AG (2008) Genetic engineering of improved nitrogen use efficiency in rice by the tissue-specific expression of alanine aminotransferase. Plant Biotechnol J 6(7):722–732
37. Martin A, Lee J, Kichey T, Gerentes D, Zivy M, Tatout C, Dubois F, Balliau T, Valot B, Davanture M, Tercé-Laforgue T, Quilleré I, Coque M, Gallais A, Gonzalez-Moro M-B, Bethencourt L, Habash DZ, Lea PJ, Charcosset A, Perez P, Murigneux A, Sakakibara H, Edwards KJ, Hirel B (2006) Two Cytosolic Glutamine Synthetase Isoforms of Maize Are Specifically Involved in the Control of Grain Production. Plant Cell 18(11):3252–3274. https://doi.org/10.1105/tpc.106.042689
38. Thomsen HC, Eriksson D, Møller IS, Schjoerring JK (2014) Cytosolic glutamine synthetase: a target for improvement of crop nitrogen use efficiency? Trends Plant Sci 19 (10):656–663
39. Selvaraj MG, Valencia MO, Ogawa S, Lu Y, Wu L, Downs C, Skinner W, Lu Z, Kridl JC, Ishitani M (2017) Development and field performance of nitrogen use efficient rice lines for Africa. Plant Biotechnol J 15(6):775–787
40. Tabuchi M, Abiko T, Yamaya T (2007) Assimilation of ammonium ions and reutilization of nitrogen in rice (Oryza sativa L.). J Exp Bot 58(9):2319–2327
41. Abiko T, Wakayama M, Kawakami A, Obara M, Kisaka H, Miwa T, Aoki N, Ohsugi R (2010) Changes in nitrogen assimilation, metabolism, and growth in transgenic rice plants expressing a fungal NADP(H)-dependent glutamate dehydrogenase (gdhA). Planta 232(2):299–311. https://doi.org/10.1007/s00425-010-1172-3
42. Zhou Y, Liu H, Zhou X, Yan Y, Du C, Li Y, Liu D, Zhang C, Deng X, Tang D (2014) Over-expression of a fungal NADP (H)-dependent glutamate dehydrogenase PcGDH improves nitrogen assimilation and growth quality in rice. Mol Breed 34 (2):335–349
43. Tercé-Laforgue T, Bedu M, Dargel-Grafin C, Dubois F, Gibon Y, Restivo FM, Hirel B (2013) Resolving the role of plant glutamate dehydrogenase: II. Physiological characterization of plants overexpressing the two enzyme subunits individually or simultaneously. Plant Cell Physiol 54(10):1635–1647
44. Huang X, Qian Q, Liu Z, Sun H, He S, Luo D, Xia G, Chu C, Li J, Fu X (2009) Natural variation at the DEP1 locus enhances grain yield in rice. Nat Genet 41(4):494–497
45. Sun H, Qian Q, Wu K, Luo J, Wang S, Zhang C, Ma Y, Liu Q, Huang X, Yuan Q (2014) Heterotrimeric G proteins regulate nitrogen-use efficiency in rice. Nat Genet 46 (6):652–656
46. Wang Z, Chen X, Wang J, Liu T, Liu Y, Zhao L, Wang G (2007) Increasing maize seed weight by enhancing the cytoplasmic ADP-glucose pyrophosphorylase activity in transgenic maize plants. Plant Cell Tissue Organ Cult 88(1):83–92
47. Meyer FD, Smidansky ED, Beecher B, Greene TW, Giroux MJ (2004) The maize Sh2r6hs ADP-glucose pyrophosphorylase (AGP) large subunit confers enhanced AGP properties in transgenic wheat (Triticum aestivum). Plant Sci 167(4):899–911
48. Zhang L, Häusler RE, Greiten C, Hajirezaei MR, Haferkamp I, Neuhaus HE, Flügge UI, Ludewig F (2008) Overriding the co-limiting import of carbon and energy into tuber amyloplasts increases the starch content and yield of transgenic potato plants. Plant Biotechnol J 6(5):453–464
49. Jonik C, Sonnewald U, Hajirezaei MR, Flügge UI, Ludewig F (2012) Simultaneous boosting of source and sink capacities doubles tuber starch yield of potato plants. Plant Biotechnol J 10(9):1088–1098
50. Busov VB, Brunner AM, Strauss SH (2008) Genes for control of plant stature and form. New Phytol 177(3):589–607
51. Gonzalez N, Beemster GT, Inzé D (2009) David and Goliath: what can the tiny weed Arabidopsis teach us to improve biomass production in crops? Curr Opin Plant Biol 12 (2):157–164

52. Krizek BA (2009) Making bigger plants: key regulators of final organ size. Curr Opin Plant Biol 12(1):17–22
53. Mickelbart MV, Hasegawa PM, Bailey-Serres J (2015) Genetic mechanisms of abiotic stress tolerance that translate to crop yield stability. Nat Rev Genet 16(4):237–251
54. Pucciariello C, Voesenek LA, Perata P, Sasidharan R (2014) Plant responses to flooding. Front Plant Sci 5
55. Septiningsih EM, Collard BC, Heuer S, Bailey-Serres J, Ismail AM, Mackill DJ (2013) Applying genomics tools for breeding submergence tolerance in rice. In: Translational genomics for crop breeding: abiotic stress, yield and quality, vol 2, pp 9–30
56. Agency FEM (1995) National mitigation strategy-partnerships for building safer communities
57. Meister R, Rajani M, Ruzicka D, Schachtman DP (2014) Challenges of modifying root traits in crops for agriculture. Trends Plant Sci 19(12):779–788
58. Lynch JP (2013) Steep, cheap and deep: an ideotype to optimize water and N acquisition by maize root systems. Ann Bot 112 (2):347–357
59. Gao Y, Lynch JP (2016) Reduced crown root number improves water acquisition under water deficit stress in maize (Zea mays L.). J Exp Bot 67(15):4545–4557
60. Tuberosa R, Salvi S (2006) Genomics-based approaches to improve drought tolerance of crops. Trends Plant Sci 11(8):405–412
61. Uga Y, Sugimoto K, Ogawa S, Rane J, Ishitani M, Hara N, Kitomi Y, Inukai Y, Ono K, Kanno N (2013) Control of root system architecture by DEEPER ROOTING 1 increases rice yield under drought conditions. Nat Genet 45(9):1097–1102
62. Xue D, Zhang X, Lu X, Chen G, Chen Z-H (2017) Molecular and evolutionary mechanisms of cuticular wax for plant drought tolerance. Front Plant Sci 8
63. Cutler SR, Rodriguez PL, Finkelstein RR, Abrams SR (2010) Abscisic acid: emergence of a core signaling network. Annu Rev Plant Biol 61:651–679
64. Habben JE, Bao X, Bate NJ, DeBruin JL, Dolan D, Hasegawa D, Helentjaris TG, Lafitte RH, Lovan N, Mo H, Reimann K, Schussler JR (2014) Transgenic alteration of ethylene biosynthesis increases grain yield in maize under field drought-stress conditions. Plant Biotechnol J 12(6):685–693. https://doi.org/10.1111/pbi.12172
65. Okamoto M, Peterson FC, Defries A, Park S-Y, Endo A, Nambara E, Volkman BF, Cutler SR (2013) Activation of dimeric ABA receptors elicits guard cell closure, ABA-regulated gene expression, and drought tolerance. Proc Natl Acad Sci 110(29):12132–12137
66. González-Guzmán M, Rodríguez L, Lorenzo-Orts L, Pons C, Sarrión-Perdigones A, Fernández MA, Peirats-Llobet-M, Forment J, Moreno-Alvero M, Cutler SR (2014) Tomato PYR/PYL/RCAR abscisic acid receptors show high expression in root, differential sensitivity to the abscisic acid agonist quinabactin, and the capability to enhance plant drought resistance. J Exp Bot 65 (15):4451–4464
67. Todaka D, Shinozaki K, Yamaguchi-Shinozaki K (2015) Recent advances in the dissection of drought-stress regulatory networks and strategies for development of drought-tolerant transgenic rice plants. Front Plant Sci 6
68. Agarwal PK, Shukla PS, Gupta K, Jha B (2013) Bioengineering for salinity tolerance in plants: state of the art. Mol Biotechnol 54 (1):102–123
69. Sakakibara H (2006) Cytokinins: activity, biosynthesis, and translocation. Annu Rev Plant Biol 57:431–449
70. McCabe MS, Garratt LC, Schepers F, Jordi WJRM, Stoopen GM, Davelaar E, van Rhijn JHA, Power JB, Davey MR (2001) Effects of P-SAG12-IPT gene expression on development and senescence in transgenic lettuce. Plant Physiol 127(2):505–516. https://doi.org/10.1104/Pp.127.2.505
71. Rivero RM, Kojima M, Gepstein A, Sakakibara H, Mittler R, Gepstein S, Blumwald E (2007) Delayed leaf senescence induces extreme drought tolerance in a flowering plant. Proc Natl Acad Sci USA 104 (49):19631–19636. https://doi.org/10.1073/pnas.0709453104
72. Peleg Z, Reguera M, Tumimbang E, Walia H, Blumwald E (2011) Cytokinin-mediated source/sink modifications improve drought tolerance and increase grain yield in rice under water-stress. Plant Biotechnol J 9 (7):747–758. https://doi.org/10.1111/j.1467-7652.2010.00584.x
73. Huynh LN, VanToai T, Streeter J, Banowetz G (2005) Regulation of flooding tolerance of SAG12:ipt Arabidopsis plants by cytokinin. J Exp Bot 56(415):1397–1407. https://doi.org/10.1093/jxb/eri141
74. Li Q, Robson P, Bettany A, Donnison I, Thomas H, Scott I (2004) Modification of senescence in ryegrass transformed with IPT

under the control of a monocot senescence-enhanced promoter. Plant Cell Rep 22 (11):816–821

75. Meyerowitz EM (1989) Arabidopsis, a useful weed. Cell 56(2):263–269
76. Buell CR, Last RL (2010) Twenty-first century plant biology: impacts of the arabidopsis genome on plant biology and agriculture. Plant Physiol 154(2):497–500. https://doi.org/10.1104/pp.110.159541
77. Lacombe S, Rougon-Cardoso A, Sherwood E, Peeters N, Dahlbeck D, Van Esse HP, Smoker M, Rallapalli G, Thomma BP, Staskawicz B (2010) Interfamily transfer of a plant pattern-recognition receptor confers broad-spectrum bacterial resistance. Nat Biotechnol 28(4):365–369
78. Karaba A, Dixit S, Greco R, Aharoni A, Trijatmiko KR, Marsch-Martinez N, Krishnan A, Nataraja KN, Udayakumar M, Pereira A (2007) Improvement of water use efficiency in rice by expression of HARDY, an Arabidopsis drought and salt tolerance gene. Proc Natl Acad Sci 104(39):15270–15275. https://doi.org/10.1073/pnas.0707294104
79. Nelson DE, Repetti PP, Adams TR, Creelman RA, Wu J, Warner DC, Anstrom DC, Bensen RJ, Castiglioni PP, Donnarummo MG, Hinchey BS, Kumimoto RW, Maszle DR, Canales RD, Krolikowski KA, Dotson SB, Gutterson N, Ratcliffe OJ, Heard JE (2007) Plant nuclear factor Y (NF-Y) B subunits confer drought tolerance and lead to improved corn yields on water-limited acres. Proc Natl Acad Sci 104(42):16450–16455. https://doi.org/10.1073/pnas.0707193104
80. Boyes DC, Zayed AM, Ascenzi R, McCaskill AJ, Hoffman NE, Davis KR, Görlach J (2001) Growth stage–based phenotypic analysis of Arabidopsis a model for high throughput functional genomics in plants. Plant Cell 13 (7):1499–1510
81. Kjemtrup S, Boyes DC, Christensen C, McCaskill AJ, Hylton M, Davis K (2003) Growth stage-based phenotypic profiling of plants. In: Grotewold E (ed) Plant functional genomics. Humana Press, Totowa, NJ, pp 427–441. https://doi.org/10.1385/1-59259-413-1:427
82. Christensen CA, Feldmann KA (2007) Biotechnology approaches to engineering drought tolerant crop. In: Jenks MA, Hasegawa PM, Jain SM (eds) Advances in molecular breeding toward drought and salt tolerant crops. Springer, Dordrecht, pp 333–357. https://doi.org/10.1007/978-1-4020-5578-2_14
83. Haas BJ, Volfovsky N, Town CD, Troukhan M, Alexandrov N, Feldmann KA, Flavell RB, White O, Salzberg SL (2002) Full-length messenger RNA sequences greatly improve genome annotation. Genome Biol 3 (6):research0029.0021. https://doi.org/10.1186/gb-2002-3-6-research0029
84. Riechmann JL, Heard J, Martin G, Reuber L, Jiang C-Z, Keddie J, Adam L, Pineda O, Ratcliffe O, Samaha R (2000) Arabidopsis transcription factors: genome-wide comparative analysis among eukaryotes. Science 290 (5499):2105–2110
85. Walsh TA, Neal R, Merlo AO, Honma M, Hicks GR, Wolff K, Matsumura W, Davies JP (2006) Mutations in an auxin receptor homolog AFB5 and in SGT1b confer resistance to synthetic picolinate auxins and not to 2,4-dichlorophenoxyacetic acid or indole-3-acetic acid in Arabidopsis. Plant Physiol 142 (2):542–552. https://doi.org/10.1104/pp.106.085969
86. Walsh TA, Bauer T, Neal R, Merlo AO, Schmitzer PR, Hicks GR, Honma M, Matsumura W, Wolff K, Davies JP (2007) Chemical genetic identification of glutamine phosphoribosylpyrophosphate amidotransferase as the target for a novel bleaching herbicide in Arabidopsis. Plant Physiol 144 (3):1292–1304. https://doi.org/10.1104/pp.107.099705
87. Luhua S, Hegie A, Suzuki N, Shulaev E, Luo X, Cenariu D, Ma V, Kao S, Lim J, Gunay MB (2013) Linking genes of unknown function with abiotic stress responses by high-throughput phenotype screening. Physiol Plant 148(3):322–333
88. Fahlgren N, Gehan MA, Baxter I (2015) Lights, camera, action: high-throughput plant phenotyping is ready for a close-up. Curr Opin Plant Biol 24:93–99
89. Alimentarius C (2003) Guideline for the conduct of food safety assessment of foods derived from recombinant-DNA plants. CAC/GL 45:1–18
90. Nations FaAOotU (2001) Evaluation of allergenicity of genetically modified foods. Food and Agriculture Organization of the United Nations (FAO)
91. Kuiper HA, Kleter GA, Noteborn HP, Kok EJ (2001) Assessment of the food safety issues related to genetically modified foods. Plant J 27(6):503–528
92. Organization WH (2009) Foods derived from modern biotechnology. Foods derived from modern biotechnology (Ed. 2)

93. Barria C, Malecki M, Arraiano CM (2013) Bacterial adaptation to cold. Microbiology 159:2437–2443. https://doi.org/10.1099/mic.0.052209-0
94. Horn G, Hofweber R, Kremer W, Kalbitzer HR (2007) Structure and function of bacterial cold shock proteins. Cell Mol Life Sci 64 (12):1457
95. Lee K, Kang H (2016) Emerging roles of RNA-binding proteins in plant growth, development, and stress responses. Mol Cells 39 (3):179
96. Castiglioni P, Warner D, Bensen RJ, Anstrom DC, Harrison J, Stoecker M, Abad M, Kumar G, Salvador S, D'Ordine R, Navarro S, Back S, Fernandes M, Targolli J, Dasgupta S, Bonin C, Luethy MH, Heard JE (2008) Bacterial RNA chaperones confer abiotic stress tolerance in plants and improved grain yield in maize under water-limited conditions. Plant Physiol 147(2):446–455. https://doi.org/10.1104/pp.108.118828
97. Nemali KS, Bonin C, Dohleman FG, Stephens M, Reeves WR, Nelson DE, Castiglioni P, Whitsel JE, Sammons B, Silady RA (2015) Physiological responses related to increased grain yield under drought in the first biotechnology-derived drought-tolerant maize. Plant Cell Environ 38(9):1866–1880
98. Petition for the determination of non-regulated status for MON 87460 (2009). Monsanto Company. www.aphis.usda.gov/brs/aphisdocs
99. Whitsel J, Stork LG, Reeves W, Horak M (2014) Characterization of drought-tolerant maize MON 87460 for use in environmental risk assessment. Crop Sci 54(2):719–729
100. Lawlor DW (2012) Genetic engineering to improve plant performance under drought: physiological evaluation of achievements, limitations, and possibilities. J Exp Bot 64 (1):83–108
101. James C (2015) 20th anniversary (1996 to 2015) of the global commercialization of biotech crops and biotech crop highlights in 2015. ISAAA brief 51
102. McAllister CH, Beatty PH, Good AG (2012) Engineering nitrogen use efficient crop plants: the current status. Plant Biotechnol J 10(9):1011–1025
103. Rothstein SJ, Bi Y-M, Coneva V, Han M, Good A (2014) The challenges of commercializing second-generation transgenic crop traits necessitate the development of international public sector research infrastructure. J Exp Bot 65(19):5673–5682
104. Miyashita Y, Dolferus R, Ismond KP, Good AG (2007) Alanine aminotransferase catalyses the breakdown of alanine after hypoxia in Arabidopsis thaliana. Plant J 49 (6):1108–1121
105. Snyman S, Hajari E, Watt M, Lu Y, Kridl J (2015) Improved nitrogen use efficiency in transgenic sugarcane: phenotypic assessment in a pot trial under low nitrogen conditions. Plant Cell Rep 34(5):667–669
106. Hajari E, Watt M, Snyman S (2013) Towards improved nitrogen use efficiency in sugarcane by overexpression of alanine aminotransferase. S Afr J Bot 86:174
107. Hu Y, Xie Q, Chua N-H (2003) The Arabidopsis auxin-inducible gene ARGOS controls lateral organ size. Plant Cell 15 (9):1951–1961
108. Kuluev B, Knyazev A, Iljassowa A, Chemeris A (2011) Constitutive expression of the ARGOS gene driven by dahlia mosaic virus promoter in tobacco plants. Russ J Plant Physiol 58(3):507–515
109. Kuluev B, Knyazev A, Mikhaylova E, Ermoshin A, Nikonorov Y, Chemeris A (2016) The poplar ARGOS-LIKE gene promotes leaf initiation and cell expansion, and controls organ size. Biol Plant 60 (3):513–522
110. Zhao Y, Tian X, Li Y, Zhang L, Guan P, Kou X, Wang X, Xin M, Hu Z, Yao Y (2017) Molecular and functional characterization of wheat ARGOS genes influencing plant growth and stress tolerance. Front Plant Sci 8:170
111. Wang B, Sang YL, Song H, Gao XQ, Zhang XS (2009) Expression of a rice OsARGOS gene in Arabidopsis promotes cell division and expansion and increases organ size. J Genet Genomics 36(1):31–40
112. Guo M, Rupe MA, Wei J, Winkler C, Goncalves-Butruille M, Weers BP, Cerwick SF, Dieter JA, Duncan KE, Howard RJ (2013) Maize ARGOS1 (ZAR1) transgenic alleles increase hybrid maize yield. J Exp Bot 65(1):249–260
113. Rai MI, Wang X, Thibault DM, Kim HJ, Bombyk MM, Binder BM, Shakeel SN, Schaller GE (2015) The ARGOS gene family functions in a negative feedback loop to desensitize plants to ethylene. BMC Plant Biol 15(1):157
114. McAllister CH, Good AG (2015) Alanine aminotransferase variants conferring diverse NUE phenotypes in Arabidopsis thaliana. PLoS One 10(4):e0121830

Chapter 23

Transgenic and Genome Editing Approaches for Modifying Plant Oils

Laura L. Wayne, Daniel J. Gachotte, and Terence A. Walsh

Abstract

Vegetable oils are important for human and animal nutrition and as renewable resources for chemical feedstocks. We provide an overview of transgenic and genome editing approaches for modifying plant oils, describing useful model and crop systems and different strategies for transgenic modifications. We also describe new genome editing approaches that are beginning to be applied to oilseed plants and crops. These approaches are illustrated with examples for modifying the nutritional quality of vegetable oils by altering fatty acid desaturation, producing non-native fatty acids in oilseeds, and enhancing the overall accumulation of oil in seeds and leaves.

Key words Lipid biosynthesis, Metabolic engineering, Fatty acid, Polyunsaturated fatty acids (PUFAs), Healthy oils, Industrial oils, CRISPR, Genome editing, Fatty acid desaturase, Acyl transferase

1 Introduction

Plant oils are critical components of food, feed, and fuel, as well as feedstocks for manufacturing. The demand for sustainable sources of plant oils for these applications has brought intense interest in improving plant oils, both in the quantity of desired oils and in the quality of the oils based on their specific lipid constituents. Differences in the fatty acid moieties of triacylgycerol (TAG) oils, the major storage form of lipid in plants, influence the dietary health benefits and the functional attributes of the oil. The high energy density of TAG in oilseeds is a key attribute for animal feed and aquaculture formulations and is also the basis for fuel applications such as biodiesel. The remarkable diversity of plant-derived fatty acids has been exploited for industrial applications and human consumption. For example, castor bean oil is a source of ricinoleic acid, a hydroxylated fatty acid with many manufacturing applications, and palm kernel and coconut oils are natural sources of shorter chain fatty acids used for soaps and shortening.

Sandeep Kumar et al. (eds.), *Transgenic Plants: Methods and Protocols*, Methods in Molecular Biology, vol. 1864,
https://doi.org/10.1007/978-1-4939-8778-8_23,

Modification of plant lipids has occurred throughout the historical domestication of crops and more recent intensive plant breeding to select varieties with improved oil qualities. Mutagenesis techniques have been applied to derive additional variation and have been introduced into crop-breeding systems. Modern canola (*Brassica napus*) represents a success story in modifying an oilseed crop. The progenitor oilseed rape naturally produces high levels of the long-chain fatty acid, erucic acid, which is undesirable in food oils. In the 1960s, introduction of lines with key mutations in fatty acid elongases led to the elimination of erucic acid and, combined with reductions in seed glucosinolates, led to modern canola genotypes. Further improvements in canola oil for food applications have occurred with the introduction of additional mutations that decrease the amount of unstable polyunsaturated fatty acids (PUFAs) in the oil [1]. This modification results in oils with high levels of stable monounsaturated fatty acids (MUFAs) combined with naturally low levels of saturated fatty acids (SFAs). These "high oleic" canola oils are consequently more stable in frying and cooking applications and can replace partially hydrogenated soybean oils that contain unhealthy *trans* fatty acids. The precise genetic basis for these modifications to an oilseed crop, that were originally driven by phenotypic selections, has been determined and demonstrates the potential for targeted alterations in lipid biosynthetic pathways to deliver commercial value.

With the availability of plant genomic information across an increasing number of species, tools for targeted gene modifications via reverse genetic approaches ("gene-first" versus "phenotype-first") such as TILLING populations have been developed, first in Arabidopsis [2] and now extended to oilseed crops [3]. These approaches can be powerful tools for uncovering specific mutations in polyploid species such as *Brassica* where there may be several redundant genes within a gene family that need to be targeted to result in a significant phenotype, or where the precise function of a particular gene may be unknown.

There are of course limitations to the types of genetic variation within natural or mutagenized populations of sexually compatible species that can be of utility in conventional plant-breeding programs. Transgenic technologies can surmount this barrier and the first examples of transgenic oilseed crops are now entering the marketplace [4] with more in earlier stages of development [5]. However, significant hurdles remain in routinely bringing transgenic oilseeds to the marketplace in terms of the costs of development of these sophisticated technologies, the global regulatory acceptance of the resulting crops and products, and the public acceptance of genetically modified (GM) crops, especially as sources of food.

A "third wave" of technology, genome editing, is beginning to impact oilseed engineering. Genome editing tools enable targeting of specific sites within the plant genome for precise modification, in some cases leaving little or no footprint [6]. These tools offer ways to rapidly introduce genome modifications for oilseed improvement and could reduce or circumvent some of the development costs and regulatory burdens associated with "conventional" transgenesis and GM crops.

In this review, we focus on transgenic tools and technology that can be applied to oilseeds and oil biosynthesis, as well as looking forward to the evolving application of genome editing tools to modify lipids in plants. We describe major oilseed model and crop systems and various approaches for oilseed modifications, with specific examples of modulating fatty acid desaturation, introducing non-native fatty acids into plants, and increasing oil accumulation in plants. An overview of plant fatty acid and TAG biosynthesis and the genes involved is shown in Fig. 1 and briefly described in Subheading 4.1. For more details, *see* reviews [7, 8]. Fatty acids are referred to by their common name, and by the number of carbons and number of desaturated double bonds in the acyl chain for example, oleic acid (18:1).

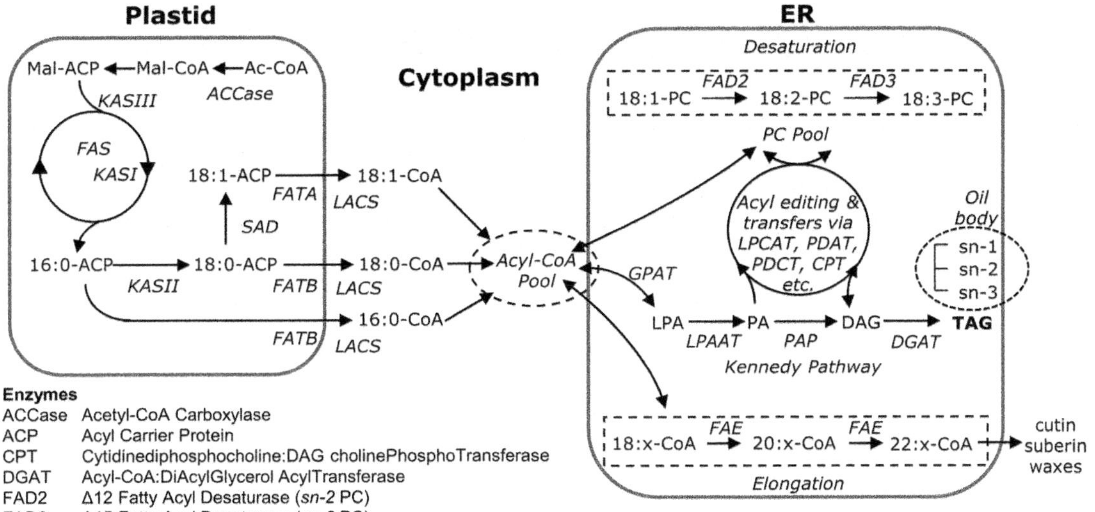

Enzymes

ACCase	Acetyl-CoA Carboxylase
ACP	Acyl Carrier Protein
CPT	Cytidinediphosphocholine:DAG cholinePhosphoTransferase
DGAT	Acyl-CoA:DiAcylGlycerol AcylTransferase
FAD2	Δ12 Fatty Acyl Desaturase (*sn-2* PC)
FAD3	Δ15 Fatty Acyl Desaturase (*sn-2* PC)
FAE	Fatty Acid Elongase complex
FAS	Fatty Acid Synthase complex
FATA	Acyl-ACP Thioesterase (18:1)
FATB	Acyl-ACP Thioesterase (16:0,18:0)
GPAT	*sn-1* Glycerol-3-Phosphate AcylTransferase
KAS	Ketoacyl-ACP Synthase
LACS	Long-chain Acyl-CoA Synthetase
LPAAT	Acyl-CoA:LysoPhosphatidic Acid AcylTransferase
LPCAT	Acyl-CoA:LysoPhosphatidylCholine AcylTransferase
PAP	Phosphatidic Acid Phosphatase
PDAT	Phosphatidylcholine:DAG AcylTransferase
PDCT	Phosphatidylcholine:DAG CholinephosphoTransferase
SAD	Δ9 Stearoyl-ACP Desaturase

Metabolites

Ac-CoA	Acetyl CoA
CoA	Coenzyme-A
DAG	DiAcylGlycerol
LP	Lysophosphatidic acid
Mal-CoA	Malonyl-CoA
PA	Phosphatidic Acid
PC	PhosphatidylCholine
TAG	TriAcylGlycerol

Fig. 1 Simplified schematic of fatty acid and TAG biosynthesis in plants

2 Plant Systems for Transgenic Oil Studies

2.1 Model Systems

The study of seed oil phenotypes requires plants to reach maturity and be at the reproductive stage, which extends time to analysis. Readily transformable model systems such as Arabidopsis (*Arabidopsis thaliana*) and *Nicotiana* species allow for rapid iteration for testing diverse sets of genes, regulatory elements (promoters and terminators), and subcellular targeting. These systems help to prioritize constructs for the optimum oil content in the desired tissues at the appropriate time of plant development. Iterative testing in model systems can also diagnose undesirable phenotypes that may result from the expressed transgenic protein, such as reduced seed yield, size, weight, or chlorosis. The best performing constructs can be selected prior to the more laborious and costly process of crop transformation.

2.1.1 Arabidopsis

In addition to advantages such as small size and a relatively nonredundant well-annotated genome, Arabidopsis is an oilseed plant in the *Brassicaceae* family and thus is an appropriate model for oil studies. The use of Arabidopsis as a transgenic model is greatly simplified by the straightforward floral dip transformation method [9]. The characterization of lipid mutants through forward screening [10–13] using EMS mutagenesis or reverse genetics using T-DNA insertion as a mutational agent (*see* reviews [14, 15]) became the basis for transgenic complementation experiments. The sequencing of the Arabidopsis genome and its annotation accelerated the establishment of a lipid gene catalog [16], available at http://aralip.plantbiology.msu.edu/pathways/pathways, and presently encompasses over 700 genes. Arabidopsis seeds contain about 35–45% oil which is sensitive to abiotic factors and can vary depending on the environmental conditions, such as light intensity [17, 18], daylight amount/duration, and temperature [19]. When grown in the cold (10 °C), accumulation of 18:3 is increased [19]. For well-designed studies, at least 11 biological replicates of each treatment and control group are recommended [18]. When analyzing transgenic events, a sufficient number of untransformed, mock-transformed, or sibling-null plants are also recommended to control for variations caused by environmental factors.

Analysis of transgenic Arabidopsis events for a seed oil phenotype can be performed on T_1 generation seed harvested from the parental T_0 plant that was dipped in *Agrobacterium*. Generally about 1–5% of the seeds are transformed using the floral dip method. T_1 seed can be visually sorted when a fluorescent reporter gene such as *GFP*, *RFP*, or *dsRed* is included within the T-DNA [20]. Fatty acid methyl ester (FAME) analysis can then be performed via gas chromatography (GC) equipped with a flame ionization detector (GC-FID) to determine the fatty acid profile of a single Arabidopsis seed [20]. This analysis provides results 10–12 weeks earlier than T_2 bulk seed FAME analysis and is helpful in decision-making and prioritizing constructs. Fluorescent markers can also be used to separate segregating T_2 seed into non-transformed and transformed groups for GC-FID analysis [21]. High-throughput analysis of bulk seed, useful for screening the segregating T_2 seed, can be achieved through the use of liquid handlers and GC-FID equipped with a 96-well microtiter plate automatic sampler.

2.1.2 Nicotiana Species

Transient expression of transgenes in *Nicotiana benthamiana* is a fast cycle time system [22], which can generate results within a week after transfection. It has been adapted to the study of oil metabolism by increasing oil synthesis and accumulation in the leaf [23]. As transfection is performed in leaf tissue, transfected lipid genes of interest have been combined with lipid biosynthesis-related transcription factors (*see* Subheading 6.2), such as *WRI1* [24] or *LEC2* [25], to enhance accumulation of oil [26]. The *N. benthamiana* transient expression system also allows for the iterative testing of gene combinations [23, 27], regulatory elements, and codon usage with a faster cycle time than stable transformation methods, helping prioritize and optimize genetic engineering strategies. Stable expression in *Nicotiana tabacum* can also be used to generate homozygous events for increasing lipid accumulation in leaves (*see* Subheading 6.2).

2.2 Crop Systems

Oilseed food crops such as canola and soy are used for production of edible vegetable oils. Industrial oils are preferably expressed in oilseeds that have limited potential for human consumption, such as Camelina (*Camelina sativa*), *B. carinata*, and *Crambe*. Early-phase development of oilseed traits in crops has highlighted differences between crop species in how fatty acids are channeled into seed TAG. Depending on the crop, seed oil content can vary from 20% in soybean (*Glycine max*) to 40% in oil palm (*Elaeis guineensis*) and can accumulate in the embryo for canola, mesocarp and kernel for oil palm, endosperm for castor bean, or kernel for coconut (*Cocos nucifera*). Oilseed crops are cultivated from tropical through temperate zones. A primary focus for transgenics has been on temperate annual row crops amenable to genetic transformation and grown on global industrial scale, such as soybean, cotton, and canola (https://www.fas.usda.gov/data/oilseeds-world-markets-and-trade). Although cottonseed oil has been modified using transgenic tools (*see* Subheading 4.4), it has become less widely used than soybean. With the advancement of genetic tools and transformation capabilities, other less widely grown crops have become of interest, especially for the production of nonfood-use industrial oils.

2.2.1 Soybean

Transformation of soybean embryonic tissue containing the oil biosynthesis pathway greatly accelerated the design of constructs and functional characterization of lipid genes expressed in soybean [28]. This system, using particle bombardment, allows for oil modifications to be tested prior to regenerating plants [29, 30]. Other *Agrobacterium*-mediated transformation methods using cotyledonary explants or hairy root systems have been optimized to increase infection rate and transformation efficiency [31, 32]. The hairy root system has also been adapted to screen constructs for precision genome editing using zinc-finger nucleases [33] or TALENs [34] for lipid modifications (*see* Subheading 3.4).

2.2.2 Oilseed Rape or Canola (Brassica napus)

Oilseed rape (in Europe and Asia) or canola (in North America) is an ideal crop for oilseed modifications as it is a major oilseed crop that is readily transformable. The cotyledonary method [35] has been used for transforming *Limnanthes douglasii LPAAT* and *B. napus FAE1* genes (Fig. 1) into *B. napus* to increase erucic acid (22:1) content [36]. The hypocotyl explant method [37] has been widely used for transforming *B. napus* with lipid biosynthesis genes, e.g., *DGAT1* [38] and RNAi knockdown of *FAD2* and *FAE1* [39]. Floral dip methods are also reported for *Brassica* species, including *B. napus* and *B. carinata* [40]. *B. napus* cultivars amenable to transformation by *Agrobacterium* are often not agronomically improved and transgenic oil traits may require introgression into elite germplasm, especially to gauge oil yield. Optimization of hypocotyl transformation has led to improved transformation efficiencies of additional *B. napus* cultivars [41].

2.2.3 Camelina (Camelina sativa)

Camelina, another member of the *Brassicaceae*, is not widely grown commercially and is relatively unimproved as an oilseed crop. Research interest in Camelina has increased because of the ease of transformation via *Agrobacterium* floral dip, circumventing the need for tissue culture [42]. Individual seeds are also of sufficient size for direct biochemical analysis (which is difficult for Arabidopsis). The short growing season, ability to grow on marginal land, and other agronomic benefits also make Camelina an attractive crop for producing industrial oils (*see* review [43]). These reasons have made Camelina a preferred crop to test construct designs and lipid metabolic engineering strategies (*see* review [44]).

2.2.4 Other Brassicaceae Species

Other *Brassica* species (*B. rapa*, *B. juncea*, and *B. carinata*) can be transformed via tissue culture of cotyledonary or hypocotyl explants (*see* review [45]). *Crambe* is a *Brassicaceae* species of interest for industrial oil production and metabolic engineering and can be transformed via cotyledonary explants [46] or floral dip transformation [40]. Crambe oil has high levels (~60%) of erucic acid (22:1). Lesquerella (*Physaria fendleri*) is of interest as it accumulates hydroxylated fatty acids, which have industrial uses (*see*

Subheading 5.2). An in vitro tissue culture method has been developed for Lesquerella [47] and was used to generate plants expressing castor bean LPAAT2 for directing hydroxylated fatty acids into the *sn*-2 position of TAG [48].

2.2.5 Other Oilseed Species

Other oilseed crops are of interest due to their yield, oil profile, or production of specialty fatty acids. Palm oil is the most widely used vegetable oil by volume and produces the highest oil yields per hectare. It has a high SFA content useful in food applications and for biodiesel and has become a preferred replacement for partially hydrogenated soybean oil as it is free of *trans* fatty acids (*see* Subheading 4.1). Oil palm is grown primarily in Southeast Asia and increasing demand has led to tropical deforestation. This demand has resulted in efforts to grow palm more sustainably and to develop replacement oils [49] (*see* Subheading 4.2). Oil palm transformation methods have been developed [50] to modify the oil profile, but as it is a perennial crop and trees take several years to mature, these methods are slow to implement.

Castor bean (*Ricinus communis*) accumulates ricinoleic acid, a hydroxylated fatty acid, which has many industrial applications. It is a tropical plant with limited agronomic improvement and seeds contain the deadly toxin ricin. Much effort has been placed on producing ricinoleic acid in crops with more suitable agronomics (*see* Subheading 5.2), although castor bean has also undergone breeding efforts to eliminate ricin content [51] and to establish in vitro tissue culture techniques for biotechnology improvements [52].

Jojoba (*Simmondsia chinensis*) accumulates very-long-chain wax esters, which may offer sustainable alternatives for high-temperature lubricants. Jojoba grows in arid tropical areas and is not improved for large-scale agriculture. Nevertheless, biotechnology applications are being developed for jojoba [53]. *Jatropha curcas* is another developing tropical oilseed crop grown primarily for biodiesel. Varieties of Jatropha accumulate from 25 to 45% oil in seeds. Understanding the genetic differences between high and low oil varieties will help in developing Jatropha as a reliable oilseed crop for biodiesel [54].

3 Transgenic Approaches for Oil Modifications

3.1 Knockout/Knockdown of Oil Biosynthesis Genes

Two transgenic strategies have been widely used to eliminate or reduce the expression of oil biosynthesis genes: gene disruption by random transfer-DNA (T-DNA) insertion with a selectable genetic element [55] or by RNA interference (RNAi) [56]. T-DNA insertions in a gene can result in complete loss of function (knockout) or reduced expression or activity (knockdown), particularly if the insertion occurs in a promoter or regulatory region, and has been used in reverse genetic approaches to identify the function of lipid genes [14]. RNAi technology has largely replaced earlier antisense RNA methods and relies on introduction of short RNA sequences to specifically alter the level of transcripts of targeted genes via posttranscriptional gene silencing. Many strategies for construct design have advanced as basic understanding of RNAi has developed [57] using various forms of delivery like hairpin RNA (hpRNA) or artificial microRNAs (amiRNAs). Examples of RNAi technology to target oil biosynthesis genes or increase carbon flux toward lipid biosynthesis are described throughout this review.

3.2 Expression of Oil Biosynthesis Transgenes

Many studies have investigated the effects of introducing transgenes from different species to investigate the phenotypic effects on oilseed fatty acid profiles, especially in Arabidopsis. Typically this approach relies on using seed-specific promoters, such as those for fatty acid biosynthesis genes like *FAE1* [58], oil body protein oleosin [59], or storage protein promoters like napin from canola or 2S albumin from Arabidopsis [60]. Storage protein promoters are typically very strong relative to lipid biosynthetic gene expression and are therefore useful to diagnose the effect of a transgene superimposed on normal lipid metabolism. However, they often express in the middle to late stages of seed development and may not coincide with other seed lipid-associated metabolic processes.

3.3 Multiple Transgene Stacking

More than one gene is frequently required to modify oils at significant levels in host plants and several genes from a pathway may be required to accumulate non-native fatty acids (*see* Subheading 5). An appropriate selection of regulatory elements (promoters and terminators) is essential, as tissue and developmental stages affect outcomes in the desired tissue. A tool kit of plasmids containing seed-specific promoter cassettes for cloning and binary plant expression vectors has been developed [61] and can be used to stack up to seven genes per binary vector. Repeated use of the same regulatory elements within a T-DNA may lead to silencing [62]. Spacers and insulators between genes and the relative orientation and position of genes within the construct can also impact gene expression [63], and thus affect accumulation of products. Assessing the phenotype of homozygous transgenic plants over several generations is important to ensure that the trait is stable [64]. Additionally, the accumulation of a non-native fatty acid may cause undesired phenotypes, such as reduced germination or lower oil yield [65, 66].

3.4 Genome Editing

Targeted editing of plant genomic sequences relies on sequence-specific targeting of an endonuclease to create a double-stranded break (DSB) and subsequent DSB repair. DSBs are repaired naturally by homology-directed recombination (HDR) or nonhomologous end joining (NHEJ) (*see* reviews [67–69]). Targeted DSBs can be introduced by protein-based endonucleases such as zinc-finger nucleases (ZFN) and transcription activator-like effector nucleases (TALEN) that rely on engineered proteins (fused to a nuclease domain such as FokI) to recognize and bind to a specific genomic DNA sequence. In contrast, RNA-dependent endonuclease CRISPR (clustered regularly interspaced short palindromic repeats)-associated CAS9 protein uses a complementary guide RNA to recognize the target DNA sequence to be cleaved. Genome editing applications for the development of plant traits have been reviewed for gene deletions [70] and recombination at transgene loci [71, 72]. Targeted gene deletion by introducing indels via NHEJ is now widely reported. However, introduction of recoded DNA sequences or "perfect" replacement of gene sequences without disruption of reading frames is more challenging and requires design and delivery of appropriate donor DNA molecules.

Of the increasing reports of genome editing in plants, relatively few have focused on oil modifications to date and most describe the creation of targeted gene deletions via DSBs that undergo NHEJ and imperfect repair. TALEN-mediated knockout of FAD2 [34] and FAD3 [73] has been demonstrated in soybean. In Camelina, the CRISPR-CAS9 system was used to target three *FAD2* genes [74], and *DGAT1* and *PDAT1* homologs [75]. A different approach was used to target a zinc-finger transcription activator fusion to the *KAS II* promoter in canola to produce a decrease in SFAs [76]. These studies have successfully recapitulated targeted modified oil phenotypes obtained via conventional random mutagenic methods (*see* Subheading 4.3).

4 Modulating Fatty Acid Desaturation in Edible Oils

The degree of desaturation of plant oils can profoundly affect their physical properties and health benefits. Coconut and palm kernel oils are very rich in SFAs (10:0–16:0) and are semisolid at room temperature, whereas olive and canola oils have high levels of MUFA (18:1) and are liquid at room temperature. Many health studies have linked high dietary saturated fat to obesity and cardiovascular disease and so oils with lower SFA content are desirable to address this health issue [77]. Oils that are naturally rich in ω-3 fatty acids such as α-linolenic acid (18:3) are recognized as heart-healthy [78]. However, oil with high levels of PUFAs, such as flax oil, can be susceptible to oxidation over time leading to off-flavors and discoloration of oils, and so low PUFA levels are desirable for food applications requiring more stable oils. The ω-6 fatty acid, linoleic acid (18:2), is an essential dietary fatty acid, but there is evidence that high ω-6:ω-3 ratios in dietary lipids may have pro-inflammatory consequences contributing to chronic diseases such as heart disease [79]. Another major health concern has been the presence of *trans* fatty acids in processed vegetable oils. *Trans* fatty acids are generated during partial hydrogenation of polyunsaturated vegetable oils (particularly soybean and cottonseed oils) via partial isomerization of residual double bonds. Hydrogenation is performed to stabilize vegetable oils and also raises the melting point to make them useful for baking, frying, and other high-heat food applications. Furthermore, because *trans* fatty acids have a strong association with heart disease, the US Food and Drug Administration does not recognize *trans* fats as safe and requires that *trans* fat content of foods be listed on nutrition labels. These reasons have led to generating low-cost vegetable oils that are high in MUFA and require no dehydrogenation and increase health benefits as well as stability for frying and baking applications. Blending of oils with different compositions has also met the need for *trans*-fat-free vegetable oils and margarines.

4.1 Pathways

The levels and types of fatty acid desaturation can be changed by modulating the expression of several different classes of lipid biosynthetic genes. The initial steps of plant lipid biosynthesis occur in the plastid via the fatty acid synthase complex, where the growing fatty acid chain is attached to an acyl carrier protein (ACP) (Fig. 1). Condensation reactions to add two carbon units to the growing acyl chain (supplied by malonyl-ACP) are performed by ketoacyl synthase (KAS) enzymes. KAS III performs the initial condensation reactions, KAS I performs intermediary elongations, and KAS II specializes in converting palmitoyl-ACP to stearoyl-ACP. Initial desaturation of stearoyl-ACP occurs via stearoyl-ACP desaturase (SAD) to form oleoyl-ACP. Synthesis reactions are terminated by cleavage of the fatty acid from ACP by fatty acyl-ACP thioesterases (FATs) that have varying specificities. In Arabidopsis, FATA has preference for palmitoyl-ACP and FATB for oleoyl-ACP [80]. The newly synthesized fatty acids are incorporated into plastidial lipids in the highly specialized thylakoid membranes or are exported to the cytosol.

After export from the plastid, fatty acids are esterified to coenzyme A (CoA) and resulting acyl-CoAs can either enter the Kennedy pathway for TAG biosynthesis, be further elongated by fatty acid elongase complexes, or supply substrates for a variety of acyl editing and exchange reactions that occur within the ER (*see* Subheadings 5 and 6 for examples). Further fatty acyl desaturations are performed with acyl chains attached to phosphatidylcholine (PC) (Fig. 1). FAD2 is a Δ12-desaturase and converts oleic acid (18:1) to linoleic acid (18:2) and FAD3 is a Δ15-desaturase and converts linoleic acid (18:2) to linolenic acid (18:3). These two enzymes are instrumental in affecting the degree of unsaturation in seed oils and have opened routes to develop oils highly enriched in oleic or linolenic acids in oilseed crops.

Through adaptations in substrate specificities and level of expression of these fatty acid biosynthesis enzymes, different species of oilseed plants have evolved to produce a remarkable array of vegetable oil compositions. These enzymes therefore offer a range of targets and sources for transgenic strategies to create modified oil profiles.

4.2 Manipulating Plastidial Fatty Acid Biosynthesis Genes

Several enzymes are available to affect desaturation by controlling which fatty acids are made and at what stage they are exported out of the plastid. The fatty acyl-ACP thioesterases and KAS II enzymes are the primary targets for modifying desaturation within the plastid and are discussed in the next subheadings. Additionally, the plastidic SAD can be downregulated to increase the amount of SFAs. Mutations in a seed-expressed soybean SAD result in elevated 18:0 [81] and antisense downregulation of *SAD* increases SFA content in *B. napus* [82].

4.2.1 Fatty Acyl-ACP Thioesterases

The acyl-ACP-thioesterase FATB has a preference for palmitic acid (16:0) as a substrate. Loss of FATB function in several plant species results in lower levels of 16:0 in seed oils [83–85]. An Arabidopsis T-DNA insertional mutant of FATB exhibited a 50% decrease in SFAs [86] but with deleterious phenotypes including reduced growth rate and decrease in seed germination. Downregulation of *FATB* by expression of an antisense construct with a constitutive viral promoter also results in reduction of 16:0 in all tissues including the seeds by 45% and without a noticeable phenotype [87]. Downregulation of *FatB* using a microRNA with a napin seed-specific promoter results in reduction of 16:0 by 50% in Arabidopsis seed oil [88]. In contrast, seed-specific overexpression of FATB1 in Arabidopsis increases the level of 16:0 up to fourfold (38 mol%) in seeds and reduces the level of 18-carbon unsaturated fatty acids and elongated 20-carbon fatty acids [87].

Solid fats rich in SFAs such as palm kernel oil are widely used in the food industry, particularly in baking shortenings and spreadable margarines. Mangosteen (*Garcinia mangostana*), a tropical tree with edible fruit, contains up to 56% 18:0 in the seed oil. In the mangosteen FAT family, FATA1 has the highest affinity for 18:0-ACP [89]. This gene was expressed in canola under the control of the napin seed-specific promoter. The resulting events accumulate up to 22% 18:0 and have increases in elongated SFAs (20:0 and 22:0) for up to 30% total SFAs in the seed oil, and concomitant decreases in 18:1 and 18:3 [89].

High yields of medium-chain fatty acid lauric acid (12:0) were achieved in canola by expressing the California bay laurel (*Umbellularia californica*) 12:0-ACP thioesterase. The oil was approved for commercialization in North America as Laurical™ canola for a potential cocoa butter alternative. Lauric acid comprises almost half of the total fatty acids in these transgenic canola seeds and can be further increased (at the *sn*-2 position of TAG) by expressing a coconut LPAAT specific for medium-chain fatty acids [90].

4.2.2 KAS II

The gene encoding plastidial KAS II (or FAB1) is an essential single-copy gene in Arabidopsis and the knockout is embryo lethal [91]. A partially deficient mutant allele *fab1-1* has a twofold increase in 16:0 in leaf and seed [92, 93]. Downregulating *FAB1* using a hpRNA driven by a seed-specific promoter results in a sevenfold increase in 16:0 (from 9 to 53%) without the embryo-lethal phenotype [91]. Seed-specific RNAi downregulation of *KAS II* in cottonseed increases 16:0 content by twofold to 53% [94]. Introduction of soybean *KAS II* transgenes decreases 16:0 in canola [95] and maize [96]. Thus, adjusting KAS II activity can modulate 16:0 content in oil. This observation was exploited by Gupta et al. who introduced a designed zinc-finger (ZF) protein fused to a transcription-enhancing domain into canola [97]. The ZF protein was designed to bind to a sequence upstream of promoters of two endogenous canola *KAS II* genes. Expression of the artificial ZF-enhancer gene driven by a constitutive promoter results in increased *KAS II* transcript levels in leaves and seeds and a decrease in total seed SFAs from 7.5 to 5.2% in one of the events. In contrast, constitutive overexpression of several plant KAS III enzymes in *N. tabacum* and seed-specific overexpression of the *Cuphea* KAS III enzymes in Arabidopsis and *B. napus* lead to an increase of 16:0 [98].

4.3 Modulation of Desaturation by Extra-Plastidial Desaturases

FAD2 and *FAD3* represent excellent targets for the development of genetic engineering strategies, given the ease of phenotypic measurements and the benefits of high-oleic oils. In soybean, conventional mutations in *FAD2* genes provide higher 18:1 and lower PUFA levels, but are associated with poor agronomic performance [99]. This phenotype has been circumvented by the use of seed-specific antisense RNA expression targeting of *FAD2-1* genes that has resulted in high-oleic soybeans (with lower SFA levels) [100], currently being commercially marketed by DuPont Pioneer under the trade name Plenish®. All three soybean *FAD3* genes have also been successfully targeted using a seed-specific RNAi construct to create lines with extremely low levels of 18:3 [101].

Several genome editing studies have now demonstrated the ability to target *FAD* genes. Using TALENs, two *FAD2* gene homologs were modified via NHEJ imperfect repair to increase 18:1 and decrease PUFA content [34]. The TALENs were designed to recognize a region of homology in the second exon and 4 of 19 events had mutations in both *FAD2* loci. PUFAs decreased in these seeds (from 63 to 8%) with a concomitant increase in 18:1 (from 20 to 80%) [34]. The TALEN transgene and selectable marker were removed by screening T_2 segregants.

This study demonstrates that plants with two modified loci can be edited in a single round of transformation. In further work, a *FAD3*-targeting TALEN was introduced into the double-FAD2 knockout [73] and 16 events were confirmed to have indels at the targeted locus. Again segregants were identified free of the TALEN transgene. These soybean events produce seeds with greater than 80% 18:1 content and lower combined PUFA levels [73].

The more flexible and rapid CRISPR-Cas9 system has been similarly used to modify Camelina seed oils. RNA-guided CRISPR-Cas9 targeting of the three *FAD2* homologs corresponding to the three Camelina sub-genomes were introduced via *Agrobacterium*-mediated transformation [74]. The T_4 seeds accumulate up to 50% oleic acid with a decrease of PUFAs from 46 to 14%. In another study targeting *FAD2* genes, a population of CRISPR-Cas9-mutated Camelina plants were generated, including a suite of alleles at all three *FAD2* loci [102]. Both studies illustrate the potential power and some of the challenges of current genome editing technology. Clearly a suite of targeted mutations in three *FAD2* homologs can be rapidly generated. However, the continued presence of the CRISPR system in each generation (if not removed by segregation) can continue to introduce additional mutations and can also introduce somatic mutations that can give rise to chimeric plants. Each sub-genome *FAD2* homolog has a different expression profile, so the cumulative effect of different suites of mutations can vary depending on the targeted gene(s), the exact targeted site within the genes, and the resulting suite of NHEJ-introduced mutations. Careful experimental design with in-depth analysis of introduced mutations and associated genotypes is needed to isolate valuable haplotypes for optimal oil profiles and crop agronomics.

These examples illustrate the power of modifying endogenous desaturase genes. However, non-native desaturase transgenes have also been shown to have utility. Introduction of a 16:0-specific Δ9-desaturase from *Caenorhabditis elegans* into Arabidopsis leads to a 66% reduction in 16:0 with a concomitant increase in palmitoleic acid (16:1,Δ9) and total SFAs being reduced by half in seeds [112]. There was no significant effect on 18:0, indicating that lowering only 16:0 by this strategy can be tolerated.

The ω-3 fatty acid stearidonic acid (SDA, 18:4) has been produced in soybeans by expressing Arabidopsis FAD3 (to increase the level of 18:3 substrate) and a Δ-6 desaturase from *Borage officinalis*, resulting in 29% SDA and total ω-3 fatty acids of 60% in seeds [103]. In 2010, Monsanto announced development of a high-SDA soybean, Soymega™ with 20% SDA. SDA has also been produced in flax via a Δ-6 desaturase from *Primula vialii* [104].

The previously described studies illustrate that modified oils can be made by targeting one or two classes of lipid biosynthesis genes.

4.4 Combining Approaches to Alter Desaturated Fatty Acid Profiles

However, combining approaches can result in more profound changes in oil profiles that can be of commercial relevance. For example, in Monsanto's Vistive Gold® soybeans expression of both plastidial *FATB* and extra-plastidial *FAD2* genes is suppressed via RNAi in a *fad3* mutant background. These alterations result in seed oil with 72% 18:1 and 6% SFAs (compared to 22% and 15%, respectively, in conventional soybeans), with 16:0 lowered from 11 to 2.5% [105].

The utility of combining approaches has also been demonstrated by modifications of cottonseed oil to improve stability and nutrition by increasing 18:0 content, thought to be a healthier SFA than 16:0. Cottonseed oil is rich in 16:0 (25%) and also has high PUFA content (58% 18:2) contributing to lower oxidative stability in high-heat applications. Two seed-specific RNAi constructs were designed to lower SAD and FAD2 activities [106, 107]. Transgenic seed expressing a *FAD2* hpRNA decreased 18:2 from 58.5 to 3.7% with a concomitant increase in 18:1 to 78.2%. The best line expressing the *SAD* hpRNA resulted in an increase in 18:0 from 2.3 to 39.8%. Crossing the best transgenic lines from each construct enabled simultaneous reduction of activities of both genes, demonstrating the additive effects of modulating both targets. To increase the nutritional value of cottonseed oil and compete with high-oleic soybean and canola oils, cottonseed oil has been modified to increase the oleic acid content by using hpRNAi to downregulate FAD2-1 and FATB, although the seeds have reduced germination [108].

In soybean, a mangosteen stearoyl-ACP FATB isoform was introduced into soybean to elevate 18:0 in oil, and then crossed with a high-oleic transgenic line in which *FAD2* and *FATB* were silenced via generation of ribozyme-terminated transcripts [109]. This strategy resulted in a stable high-oleic (>70%), high-stearate (>17%) phenotype that represents a prototype for a stable, *trans*-fat-free oil with potential for margarine applications.

Another combination was used to make a palm oil-equivalent oil in canola by increasing SFAs. This was achieved by seed-specific overexpression of a native FATB combined with amiRNA-mediated downregulation of eight endogenous *SAD* genes. This downregulation led to increases of SFAs from 7 to 45%, with a sevenfold increase in 16:0 [110]. However the high level of SFAs was correlated with poorer seedling vigor at lower temperatures, likely due to the higher melting point of the modified lipid composition.

Omega-7 fatty acids such as palmitoleic acid (16:1), found at high levels in macadamia nuts for example, are thought to have health benefits as MUFAs. They can also be useful as industrial feedstocks for fatty acid metathesis into 1-octene building blocks. Initial work in Arabidopsis demonstrated that a combination of four different mechanisms results in extraordinarily high levels of ω-7 fatty acids (up to 71%) in seeds [111]. These mechanisms were

(1) expression of COM25, a modified SAD engineered to be 16:0 specific (to convert 16:0 to 16:1 in the plastid); (2) downregulation of FATB by hpRNAi (to increase 16:0 in the plastid); (3) downregulation of KAS II by a *KAS II* hpRNA and use of a *fatb fae1* mutant background (to prevent 16:0 being elongated to 18:0); and (4) expression of two fungal extra-plastidial acyl-CoA desaturases (to convert 16:0 to 16:1 in the cytoplasm). This strategy was successfully translated into Camelina where KAS II, FAE1, and FatB were all downregulated via hpRNAi and two desaturases, COM25 plastidial 16:0-specific acyl-ACP desaturase and the *C. elegans* cytosolic 16:0-specific acyl-CoA Δ9-desaturase [112], were expressed using seed-specific promoters. This combination led to an increase in ω-7 fatty acids from 2% in conventional Camelina to 62% total, with 32% 16:1 and 30% vaccenic acid (18:1,Δ11), in the modified seeds [113].

These studies demonstrate that combining strategies for modulating expression of key genes and targeting processes in different cellular compartments can lead to additive or synergistic effects on the type and degree of desaturation in vegetable oils. Achieving desired oil profiles with specific degrees of unsaturation at robust levels will require these types of combined approaches.

5 Introducing Non-native Fatty Acids into Plant Oils

Transgenic production of non-native fatty acids in oilseed crops has been more challenging than altering endogenous fatty acid profiles and accumulation in the new host plants has frequently been low. Therefore, transgenes of interest have been expressed in different mutant backgrounds or co-expressed with additional accessory enzymes to enhance production of the oil trait. Several case studies follow, which outline the development of metabolic engineering strategies and barriers that still exist to produce sufficient levels of non-native fatty acids.

5.1 Producing ω-3 Long-Chain (LC) PUFAs in Plants

The ω-3 LC-PUFAs docosahexaenoic acid (DHA) and eicosapentaenoic acid (EPA) are key dietary fatty acids found in fish oils. DHA and EPA are primarily synthesized by marine algae and bioaccumulate in fish via the food chain. More sustainable sources of these LC-PUFAs are needed so this has led to efforts to produce DHA and EPA in plants. Two different pathways have been introduced into plants to produce DHA in seed oil (reviewed in [114, 115]). The first pathway uses a series of elongases and desaturases to convert native plant fatty acids into LC-PUFAs. A diverse array of elongase and desaturase genes from marine microalgae, fungi, and other sources have been examined [116, 117]. A significant learning was the use of acyl-CoA-dependent rather than PC-dependent desaturases to improve LC-PUFA production in seed oils by avoiding exchanges between CoA and PC pools [118]. Lipidomic analysis indicates that directing non-native DHA-CoA through phospholipids into TAG is challenging [116]. Nevertheless, up to 15% DHA in Arabidopsis and Camelina has been achieved by expressing several different gene sets containing seven or eight transgenes each [116, 117, 119, 120]. Lipidomic analysis of the distribution of DHA and EPA in polar and neutral lipids has provided insights into the complexity of how LC-PUFAs are channeled into TAG [121]. This information is important for further optimization of metabolic engineering strategies for this complex pathway.

A second pathway to produce DHA in plants is to synthesize it directly from malonyl-CoA using polyketide synthase-like PUFA synthase systems from marine microalgae [122]. DHA is currently produced commercially by microalgal fermentation [123]. The microalgal PUFA synthase system comprises three large multidomain polypeptides and requires a dedicated phosphopantetheinyl transferase (PPTase) to activate the PUFA synthase ACP domains [124]. A PUFA synthase system plus PPTase was expressed first in Arabidopsis and then in canola to yield up to 3.7% DHA and 0.7% EPA in seed oil and required the introduction of the 638 kDa enzyme system in the cytosol, encoded by a 31 kb T-DNA [125]. The PUFA synthase system does not require plant fatty acid precursors; thus DHA production simply supplements the native fatty acid profile and can be introduced into high-oleic canola backgrounds [125]. DHA accumulates in the *sn*-1,3 positions of TAG, suggesting that there are selectivities in enzymes transferring DHA into TAG in canola [125]. This result is similar to observations made with the desaturase-elongase pathway expressed in Arabidopsis and Camelina [117, 126].

5.2 Designing New Sources of Castor Oil for Industrial Feedstocks

Castor oil contains up to 90% ricinoleic acid and is used as an industrial feedstock for lubricants and as a component of plasticizers, cosmetics, paints, and nylon. The castor bean plant is relatively undomesticated and the seeds contain the deadly toxin ricin. Over 20 years of research have been dedicated to the goal of producing ricinoleic acid in an alternative oilseed crop. The castor bean fatty acid hydroxylase (FAH) converts oleic acid (18:1) to ricinoleic acid (18:1-OH). Expression of the castor bean FAH in Arabidopsis yields 17% hydroxylated fatty acids [127, 128] in the *fae1* mutant. When castor FAH is co-expressed with the castor bean diacylglycerol acyltransferase-2 (DGAT2), there is an increase in accumulation of up to 30% hydroxylated fatty acids [65, 129]. Reducing DGAT isozyme competition by crossing the Arabidopsis line co-expressing castor bean FAH and DGAT with an Arabidopsis *dgat1* mutant line further increases hydroxylated fatty acid accumulation [130]. Although achieving commercial levels without agronomic effects remains a challenge, accumulation of hydroxylated fatty acids in Camelina has demonstrated that this trait can be produced in a crop plant [42].

Other strategies to increase ricinoleic acid accumulation have been tried without marked improvements but have expanded our understanding of lipid metabolism and modification. Cleaving ricinoleic acid from the transgenic Arabidopsis membrane PC via castor bean phospholipase A2-α results in a decrease in hydroxylated fatty acids [131], presumably due to either low efficiency of TAG assembly via acyltransferases or redirection to β-oxidation. Co-expressing the castor bean electron transport system (cytochrome b_5 and cytochrome b_5 reductase) to increase reductant for FAH failed to increase the accumulation of hydroxylated fatty acids [132]. Modifications of fatty acids by desaturases and FAH occur on PC lipids and there is an acyl-editing exchange between neutral (DAG) lipids and PC lipids in the synthesis of polyunsaturated and hydroxylated fatty acids [133, 134]. Understanding the flux of fatty acids between different substrate and product pools can generate hypotheses on how to address bottlenecks. Arabidopsis lines accumulating hydroxylated fatty acids have reduced oil content, caused by inhibition of acetyl-CoA carboxylase (ACCase; the first step of fatty acid biosynthesis and a key regulator of fatty acid synthesis), and lead to a decreased rate of fatty acid synthesis [135]. Inhibition of ACCase can be rescued by overexpression of *WRI1*, albeit with detrimental effects to seed germination [21].

5.3 Producing Other Novel Oils in Plants

Industrial oils with a functional group (hydroxyl-, epoxy-, cyclopropyl-, etc.), conjugated double bonds, or acetylene bonds within the fatty acid can make them suitable for industrial applications. These oils are typically synthesized in tropical species that can produce high levels of these novel fatty acids and many of these species are not amenable to temperate agronomic production. The genes responsible for modifying the novel fatty acids have been identified and expressed in host species. Acetylenases, epoxygenases, and conjugases have similar mechanisms to FAD2 and contain a non-heme di-iron coordinated by three histidine-rich motifs which modify the substrate on phosphatidylcholine (PC) [136–138]. Transgenic seeds expressing conjugases in the *fae1/fad3* double-mutant background (for maximum 18:1 substrate availability) accumulate conjugated fatty acids in the PC, indicating that these unusual *trans*-fatty acids are not effectively removed from the ER membrane [139, 140]. Some *Asteraceae* species accumulate vernolic acid, an epoxy fatty acid. Transgenic expression of *Crepis palaestina* epoxygenase in Arabidopsis seeds leads to substrate inhibition and can be recovered by heterologous expression of the *C. palaestina FAD2* [141, 142]. The triple *fad3/fad7/fad8* (plastidial desaturases) mutant background was used for heterologous expression of *C. palaestina* epoxygenase in order to maximize the 18:2 substrate availability [143]. *Vernonia galamensis*

DGATs increase epoxy fatty acid accumulation in soybean seeds when co-expressed with the *Stokesia laevis* epoxygenase and mitigate the lower oil and protein levels of the epoxygenase expressed alone [144, 145]. The *E. coli* cyclopropane synthase expressed in Arabidopsis produces higher levels of cyclopropane fatty acids than cyclopropane synthases from plants and the co-expression of a tropical tree *Sterculia foetida* LPAAT in Arabidopsis and Camelina further increases the cyclopropane fatty acids [66, 146].

Camelina, crambe, and *B. carinata* have been transformed with jojoba fatty acyl-CoA reductase and wax synthase genes for production of wax esters, which are used as an industrial lubricant. *FAE1* was co-expressed in these species with the wax synthase genes to increase long-chain fatty acid substrate supply. Greater than 20% wax esters accumulated in the transgenic crop seed oil, highlighting these species as potential biotechnology platforms for producing industrial oils [147]. Another set of fatty acyl-CoA and wax synthase genes from a variety of species were tested in Camelina with acyl-ACP thioesterases and FAH to produce wax esters [148].

6 Increasing Oil Accumulation in Plants

For dedicated oilseed crops, much effort has been placed on increasing total oil yield per hectare, by increasing the proportion of oil in the seed without reducing seed size or number of seeds per plant. In soybean, where the major economic driver is primarily protein content and oil is a valuable coproduct, it is desirable to increase both protein and oil content while reducing carbohydrate content [149]. Additionally, the oil biosynthesis pathway present in vegetative tissues can be upregulated to enable accumulation of TAG in leaves for potentially higher oil yields per acre than in seeds. Whether accumulating oil in seeds or leaves, the carbon flux must be redirected into oil by removing bottlenecks, downregulating carbon toward starch, and stabilizing TAG storage into oil bodies.

6.1 Increasing Oil Accumulation in Seeds

Total oil can be increased by modifying enzymes directly involved in TAG synthesis. This pathway (the Kennedy pathway) involves three sequential acylations of glycerol via three acyltransferases (Fig. 1). The final acyltransferase DGAT acylates *sn-1-2*-diacylglycerol (DAG) using acyl-CoA as a substrate and is considered to be the rate-limiting step [150]. There are two classes of DGATs in plants: typical DGAT1 proteins are about 20 kDa larger than DGAT2 and are constitutively expressed [151–153], whereas DGAT2 has a less defined physiological function and is highly expressed in the seeds [154, 155]. Overexpression of DGAT1 in Arabidopsis and canola increases seed oil content [156–158], demonstrating that oil content is limited in part by this step.

Directed evolution of the *B. napus* DGAT1 has been used to identify amino acid substitutions that lead to an increase in oil accumulation via increased enzyme activity [159]. Expression of *DGAT2A* from *Umbelopsis ramanniana* (a soil fungus) in soybean leads to an increase of 1.5% in seed oil on a dry seed matter basis across multi-year field trials [160]. The *Corylus avellana DGAT1* gene was mutated via gene shuffling and screened in yeast to identify 14 mutations, which were introduced into the soy *DGAT1b* and expressed in soybean [149]. The optimized soybean DGAT1 produces an average oil increase of 3% on a dry seed basis in field-grown T_3 seeds [149]. The increase in oil content is accompanied by decreases in soluble sugars and a slight increase in protein amount. Thus, expression of the modified *DGAT1* maintains the total protein level and shifts carbon from sugars to oil.

Other approaches to increase total seed oil involve overexpression of WRI1 and LEC2 lipid biosynthesis transcription factors and altering starch metabolism to direct carbon partitioning to lipid synthesis. Expressing *B. napus WRI1* genes in Arabidopsis leads to an increase in total seed oil content and larger seeds [161]. Total seed oil content can be increased in Arabidopsis by replacing the Arabidopsis starch-branching enzymes with constitutive expression of maize orthologs [162]. These transgenic plants have greater seed set and thus an increase in yield per plant. It is proposed that the shift in carbon allocations due to the heterologous expression of the maize orthologs allows for more growth during reproductive development [162].

6.2 Increasing Oil Accumulation in Leaves

Accumulating oil in vegetative tissues such as leaves or tubers, rather than seeds, may enable increases in total vegetable oil production per hectare. Oil-rich vegetative tissue may also have improved energy density for livestock consumption. Tobacco has been a useful model system for accumulating lipids in leaves. Transgenic *N. tabacum* plants expressing the Arabidopsis WRI1 and DGAT1 and *Sesamum indicum* oleosin (to stabilize and help form oil bodies) proteins accumulate up to 15% dry weight TAG in mature leaves [163]. The use of a temporal RuBisCO small subunit promoter was used for expressing *WRI1* and oleosin genes to coincide with a sufficient supply of photosynthate during the day [163]. Transcriptomics and pulse-chase labeling identified a futile cycle of lipid turnover in *N. tabacum* expressing these three genes [164]. Two approaches were used to address this issue; RNAi silencing of *SDP1* lipase and overexpression of Arabidopsis LEC2 transcription factor (working upstream of WRI1) increase TAG accumulation to 33% leaf dry weight [164]. Expression of thioesterases in the triple-gene line further increased leaf oil by promoting export of fatty acids from the chloroplast [165]. These studies illustrate the array of metabolic engineering strategies required to shift carbon from starch to oil in non-seed photosynthetic tissues.

Gene selection, spatial and temporal expression, substrate supply, product stabilization, and terminating futile cycling all need to be considered in order to produce oil in tissues that do not normally accumulate substantial amounts of oil.

A next logical step is to accumulate non-native fatty acids in leaf oils. LC-PUFAs have been produced transiently in *N. benthamiana* leaves by co-expression of the desaturase and elongases, along with a AtDGAT1 to increase accumulation of TAG [23]. Eleostearic acid, a conjugated fatty acid from tung tree, has been produced in Arabidopsis leaves. The accumulation of eleostearic acid in Arabidopsis leaves was accomplished by constitutive expression of the tung tree conjugase and *DGAT2* genes in a mutant Arabidopsis line *pdx1* [166], which is disrupted in the peroxisome ABC-transporter 1 to reduce β-oxidation in the peroxisome and thus reduce fatty acid turnover. In this system, eleostearic acid is efficiently removed from the membrane, presumably by a phospholipase A2, and shuttled into TAG via the tung tree DGAT2 [166]. These studies provide understanding of carbon allocation for oil storage with the ultimate goal of producing specialty oils in organs other than the seeds for greater oil accumulation per unit of biomass.

7 Conclusion

Knowledge of how the diverse array of fatty acids in plant oils are synthesized, channeled, exchanged, and converted has developed tremendously in the last several years. Transgenic work in model systems and crops has revealed subtle and not-so-subtle differences in how novel fatty acids are made and accumulated in different species. In addition to applications toward generating novel oils, this knowledge can be used to minimize environmental effects on oil profiles and yield to create more uniform and improved oil compositions. Transgenic and, more recently, genome editing approaches to modifying plant oils have expanded in recent years. Remarkable changes in the fatty acid composition of model plants have been demonstrated at the investigative and "proof-of-principle" level and then translated into oilseed crops in "proof-of-concept" studies. Novel fatty acids that have been produced in crop plants include highly desaturated fatty acids such as LC-PUFAs, as well as hydroxylated, epoxygenated, conjugated, and acetylated fatty acids.

From much of this work, it is becoming clear that a combination of approaches will be needed to successfully deliver improved oils in crop plants and to establish transgenic crops with robust and valuable commercial oilseed traits. Overcoming bottlenecks such as feedback inhibition, isozyme competition, substrate depletion, substrate channeling, conversion to storage forms, and futile cycling via β-oxidation will provide focus and direction for future metabolic engineering strategies. Holistic and comprehensive approaches are essential to understanding lipid metabolism and modifying plant oils. Lipidomics tools [19, 167, 168], as well as flux analyses [134, 169], metabolic control analysis [158], and biochemical analysis experiments, will aid in generating transgenic and genome editing hypotheses to determine which bottlenecks obstruct efficient accumulation of the desired oil. Much transgenic work in oil biosynthesis has relied on overexpression of genes that are superimposed on the host's normal metabolism, or occasionally in mutant backgrounds that are deficient in a particular process, such as cytoplasmic elongation (*fae1* in Arabidopsis) to increase substrate supply. Genome editing tools now open up more sophisticated ways to modify oilseeds by *replacing* native genes with alternative transgenes derived from another plant, alga, or microbe, or editing genes in situ to introduce residue changes that have been pre-scouted in vitro to be effective, or to fine-tune levels of expression. This approach becomes similar to metabolic refactoring approaches in synthetic biology, for example, to optimize novel natural product synthesis [170].

Despite this technical progress, to date only a few transgenic traits have advanced into commercialization. Major hurdles include the time, cost, and risks associated with the development and required deregulation of transgenic crops, estimated at over US $136 million by CropLife International in 2011 [171]. New oilseed traits must provide sufficient novelty and value so that these significant development costs and risks are worthwhile for producers. For high-value components such as DHA/EPA, megatrends (such as the importance of dietary ω-3 LC-PUFAs and the need for protecting marine resources) may indicate that the transgenic journey is worthwhile. For other oilseed traits, the outlook is less clear. It is anticipated that the precision nature of genome editing may lead to reduced time to market and reduced regulatory costs. In addition to development costs, there are risks associated with consumer perception and acceptance of GM traits, especially for food applications. It remains to be seen if genome-edited crops and products will receive improved consumer acceptance.

Additional development costs may be involved as specialty-modified oilseeds require investment in separate downstream handling procedures ("identity preservation") to maintain the integrity and value of the novel oils within the production channel. Enhancements to the oil must be valuable enough to justify separation of specialized or dedicated seed crushers and steps taken to prevent cross-contamination at generic processing plants. New postharvest processing pipelines may also be required to process and store novel oils with different physical properties and functional attributes. Similarly, accumulating oil in vegetative tissues such as leaves is appealing on a yield-per-hectare basis; however procedures for harvesting, transporting, processing, and extracting oil on a commercial scale will need to be developed.

For smaller scale oilseed crops that are less agronomically developed, such as Camelina and *B. carinata*, further breeding efforts are needed to improve agronomics to match the initial investment in optimizing the oilseed modifications. However, these "boutique crops" are appealing for production of industrial oils due to the relatively low human and animal consumption.

In the coming years, we hope to see transgenic oilseed traits such as canola producing DHA, soybeans with increased oil yield, and Camelina producing specialty industrial oils come to fruition and become commercially successful. These traits and other modifications of vegetable oils have the potential to revolutionize the sustainable sourcing and value of specialty oils from plants.

References

1. Wells R et al (2014) The control of seed oil polyunsaturate content in the polyploid crop species Brassica napus. Mol Breed 33:349–362
2. Colbert T et al (2001) High-throughput screening for induced point mutations. Plant Physiol 126(2):480–484
3. Stephenson P et al (2010) A rich TILLING resource for studying gene function in Brassica rapa. BMC Plant Biol 10:62
4. Pollack A (2013) In a Bean, a Boon to Biotech, in New York Times
5. Napier JA et al (2014) Understanding and manipulating plant lipid composition: metabolic engineering leads the way. Curr Opin Plant Biol 19:68–75
6. Voytas DF (2013) Plant genome engineering with sequence-specific nucleases. Annu Rev Plant Biol 64(1):327–350
7. Bates PD, Stymne S, Ohlrogge J (2013) Biochemical pathways in seed oil synthesis. Curr Opin Plant Biol 16(3):358–364
8. Li-Beisson Y et al (2010) Acyl-lipid metabolism. Arabidopsis Book 8:e0133
9. Clough SJ, Bent AF (1998) Floral dip: a simplified method for agrobacterium-mediated transformation of Arabidopsis thaliana. Plant J 16(6):735–743
10. Browse J, McCourt P, Somerville CR (1985) A mutant of Arabidopsis lacking a chloroplast-specific lipid. Science 227(4688):763–765
11. Gachotte D, Meens R, Benveniste P (1995) An arabidopsis mutant deficient in sterol biosynthesis - heterologous complementation by ERG-3 encoding a delta(7)-sterol-C-5-desaturase from yeast. Plant J 8(3):407–416
12. Katavic V et al (1995) Alteration of seed fatty-acid composition by an ethyl methanesulfonate-induced mutation in arabidopsis-thaliana affecting diacylglycerol acyltransferase activity. Plant Physiol 108 (1):399–409
13. Kumar R et al (2006) A mutation in Arabidopsis cytochrome b5 reductase identified by high-throughput screening differentially affects hydroxylation and desaturation. Plant J 48(6):920–932
14. Miquel M, Browse J (1998) Arabidopsis lipids: a fat chance. Plant Physiol Biochem 36(1–2):187–197
15. Somerville C, Browse J (1991) Plant lipids: metabolism, mutants, and membranes. Science 252(5002):80
16. Beisson F et al (2003) Arabidopsis genes involved in acyl lipid metabolism. A 2003 census of the candidates, a study of the distribution of expressed sequence tags in organs, and a web-based database. Plant Physiol 132 (2):681–697
17. Burgos A et al (2011) Analysis of short-term changes in the Arabidopsis thaliana glycerolipidome in response to temperature and light. Plant J 66(4):656–668
18. Li Y et al (2006) Oil content of Arabidopsis seeds: the influence of seed anatomy, light and plant-to-plant variation. Phytochemistry 67 (9):904–915
19. Li Q et al (2015) Understanding the biochemical basis of temperature-induced lipid pathway adjustments in plants. Plant Cell 27 (1):86–103

20. Stuitje AR et al (2003) Seed-expressed fluorescent proteins as versatile tools for easy (co) transformation and high-throughput functional genomics in Arabidopsis. Plant Biotechnol J 1(4):301–309
21. Adhikari ND, Bates PD, Browse J (2016) WRINKLED1 rescues feedback inhibition of fatty acid synthesis in hydroxylase-expressing seeds. Plant Physiol 171(1):179–191
22. Llave C, Kasschau KD, Carrington JC (2000) Virus-encoded suppressor of posttranscriptional gene silencing targets a maintenance step in the silencing pathway. Proc Natl Acad Sci 97(24):13401–13406
23. Wood CC et al (2009) A leaf-based assay using interchangeable design principles to rapidly assemble multistep recombinant pathways. Plant Biotechnol J 7(9):914–924
24. Cernac A, Benning C (2004) WRINKLED1 encodes an AP2/EREB domain protein involved in the control of storage compound biosynthesis in Arabidopsis. Plant J 40 (4):575–585
25. Stone SL et al (2001) LEAFY COTYLEDON2 encodes a B3 domain transcription factor that induces embryo development. Proc Natl Acad Sci U S A 98 (20):11806–11811
26. Vanhercke T et al (2013) Synergistic effect of WRI1 and DGAT1 coexpression on triacylglycerol biosynthesis in plants. FEBS Lett 587(4):364–369
27. Liu F et al (2015) Enhanced seed oil content by overexpressing genes related to triacylglyceride synthesis. Gene 557(2):163–171
28. Kinney AJ (1996) Development of genetically engineered soybean oils for food applications. J Food Lipids 3(4):273–292
29. Cahoon EB et al (1999) Biosynthetic origin of conjugated double bonds: production of fatty acid components of high-value drying oils in transgenic soybean embryos. Proc Natl Acad Sci U S A 96(22):12935–12940
30. Cahoon EB et al (2000) Production of fatty acid components of meadowfoam oil in somatic soybean embryos. Plant Physiol 124 (1):243–251
31. Li S et al (2017) Optimization of agrobacterium-mediated transformation in soybean. Front Plant Sci 8:246
32. Govindarajulu M et al (2008) Evaluation of constitutive viral promoters in transgenic soybean roots and nodules. Mol Plant-Microbe Interact 21(8):1027–1035
33. Curtin SJ et al (2011) Targeted mutagenesis of duplicated genes in soybean with zinc-finger nucleases. Plant Physiol 156 (2):466–473
34. Haun W et al (2014) Improved soybean oil quality by targeted mutagenesis of the fatty acid desaturase 2 gene family. Plant Biotechnol J 12(7):934–940
35. Moloney MM, Walker JM, Sharma KK (1989) High efficiency transformation ofBrassica napus usingAgrobacterium vectors. Plant Cell Rep 8(4):238–242
36. Nath UK et al (2009) Increasing erucic acid content through combination of endogenous low polyunsaturated fatty acids alleles with Ld-LPAAT + Bn-fae1 transgenes in rapeseed (Brassica napus L.). Theor Appl Genet 118 (4):765–773
37. De Block M, De Brouwer D, Tenning P (1989) Transformation of Brassica napus and Brassica oleracea using agrobacterium tumefaciens and the expression of the bar and neo genes in the transgenic plants. Plant Physiol 91(2):694–701
38. Guo XJ et al (2017) Identification and characterization of an efficient acyl-CoA: diacylglycerol acyltransferase 1 (DGAT1) gene from the microalga Chlorella ellipsoidea. BMC Plant Biol 17:16
39. Shi JH et al (2017) Depressed expression of FAE1 and FAD2 genes modifies fatty acid profiles and storage compounds accumulation in Brassica napus seeds. Plant Sci 263:177–182
40. Verma SS, Chinnusamy V, Bansa KC (2008) A simplified floral dip method for transformation of Brassica napus and B. carinata. J Plant Biochem Biotechnol 17(2):197–200
41. Maheshwari P, Selvaraj G, Kovalchuk I (2011) Optimization of Brassica napus (canola) explant regeneration for genetic transformation. New Biotechnol 29(1):144–155
42. Lu C, Kang J (2008) Generation of transgenic plants of a potential oilseed crop Camelina sativa by agrobacterium-mediated transformation. Plant Cell Rep 27(2):273–278
43. Bansal S, Durrett TP (2016) Camelina sativa: an ideal platform for the metabolic engineering and field production of industrial lipids. Biochimie 120:9–16
44. Haslam RP et al (2016) Synthetic redesign of plant lipid metabolism. Plant J 87(1):76–86
45. Rani T et al (2013) Genetic transformation in oilseed brassicas - a review. Indian J Agr Sci 83 (4):367–373
46. Qi WC et al (2014) Regeneration and transformation of Crambe abyssinica. BMC Plant Biol 14:12

47. Chen GQ (2011) Effective reduction of chimeric tissue in transgenics for the stable genetic transformation of lesquerella fendleri. Hortscience 46(1):86–90
48. Chen GQ et al (2016) Expression of castor LPAT2 enhances ricinoleic acid content at the sn-2 position of triacylglycerols in lesquerella seed. Int J Mol Sci 17(4):507
49. Parveez GKA et al (2015) Biotechnology of oil palm: strategies towards manipulation of lipid content and composition. Plant Cell Rep 34(4):533–543
50. Izawati AM, Parveez GK, Masani MY (2012) Transformation of oil palm using agrobacterium tumefaciens. Methods Mol Biol 847:177–188
51. McKeon TA et al (2014) Toxin content of commercial castor cultivars. J Am Oil Chem Soc 91(9):1515–1519
52. Zhang JX et al (2016) In vitro establishment of a highly effective method of castor bean (Ricinus communis L.) regeneration using shoot explants. J Integr Agric 15 (6):1417–1422
53. Kumar S et al (2012) Biotechnological advances in jojoba Simmondsia chinensis (link) Schneider : recent developments and prospects for further research. Plant Biotechnol Rep 6(2):97–106
54. Sood A, Chauhan RS (2015) Regulation of FA and TAG biosynthesis pathway genes in endosperms and embryos of high and low oil content genotypes of Jatropha curcas L. Plant Physiol Biochem 94:253–267
55. Feldmann KA et al (1989) A dwarf mutant of arabidopsis generated by T-DNA insertion mutagenesis. Science 243(4896):1351–1354
56. Kamthan A et al (2015) Small RNAs in plants: recent development and application for crop improvement. Front Plant Sci 6:208
57. Saurabh S, Vidyarthi AS, Prasad D (2014) RNA interference: concept to reality in crop improvement. Planta 239(3):543–564
58. Rossak M, Smith M, Kunst L (2001) Expression of the FAE1 gene and FAE1 promoter activity in developing seeds of Arabidopsis thaliana. Plant Mol Biol 46(6):717–725
59. Plant AL et al (1994) Regulation of an arabidopsis oleosin gene promoter in transgenic brassica-napus. Plant Mol Biol 25 (2):193–205
60. Guerche P et al (1990) Differential expression of the arabidopsis 2S albumin genes and the effect of increasing gene family size. Plant Cell 2(5):469–478
61. Shockey J et al (2015) Development and analysis of a highly flexible multi-gene expression system for metabolic engineering in Arabidopsis seeds and other plant tissues. Plant Mol Biol 89(1–2):113–126
62. Butaye KMJ et al (2005) Approaches to minimize variation of transgene expression in plants. Mol Breed 16(1):79–91
63. Gudynaite-Savitch L, Johnson DA, Miki BL (2009) Strategies to mitigate transgene-promoter interactions. Plant Biotechnol J 7 (5):472–485
64. Sidorenko LV et al (2017) GC-rich coding sequences reduce transposon-like, small RNA-mediated transgene silencing. Nat Plant 3:875–884
65. van Erp H et al (2011) Castor phospholipid: diacylglycerol acyltransferase facilitates efficient metabolism of hydroxy fatty acids in transgenic arabidopsis. Plant Physiol 155 (2):683–693
66. Yu XH et al (2014) Coexpressing Escherichia coli cyclopropane synthase with sterculia foetida lysophosphatidic acid acyltransferase enhances cyclopropane fatty acid accumulation. Plant Physiol 164(1):455–465
67. Zhang K et al (2017) Progress in genome editing technology and its application in plants. Front Plant Sci 8:177
68. Rani R et al (2016) CRISPR/Cas9: a promising way to exploit genetic variation in plants. Biotechnol Lett 38(12):1991–2006
69. Petolino JF, Srivastava V, Daniell H (2016) Editing plant genomes: a new era of crop improvement. Plant Biotechnol J 14 (2):435–436
70. van de Wiel CCM et al (2017) New traits in crops produced by genome editing techniques based on deletions. Plant Biotechnol Rep 11(1):1–8
71. Webb SR et al (2013) The EXZACT (TM) precision transformation & gene stacking platform: design, development, deployment and implications on new plant product discovery and development. In Vitro Cell Dev Biol Anim 49:S14–S14
72. Ainley WM et al (2013) Trait stacking via targeted genome editing. Plant Biotechnol J 11(9):1126–1134
73. Demorest ZL et al (2016) Direct stacking of sequence-specific nuclease-induced mutations to produce high oleic and low linolenic soybean oil. BMC Plant Biol 16:8
74. Jiang WZ et al (2017) Significant enhancement of fatty acid composition in seeds of the allohexaploid, Camelina sativa, using CRISPR/Cas9 gene editing. Plant Biotechnol J 15(5):648–657

75. Aznar-Moreno JA, Durrett TP (2017) Simultaneous targeting of multiple gene homeologs to alter seed oil production in Camelina sativa. Plant Cell Physiol 58(7):1260–1267
76. Gupta M et al (2012) Transcriptional activation of Brassica napus beta-ketoacyl-ACP synthase II with an engineered zinc finger protein transcription factor. Plant Biotechnol J 10(7):783–791
77. Hammad S, Pu SH, Jones PJ (2016) Current evidence supporting the link between dietary fatty acids and cardiovascular disease. Lipids 51(5):507–517
78. Jakobsen MU et al (2009) Major types of dietary fat and risk of coronary heart disease: a pooled analysis of 11 cohort studies. Am J Clin Nutr 89(5):1425–1432
79. Simopoulos AP, DiNicolantonio JJ (2016) The importance of a balanced ω-6 to ω-3 ratio in the prevention and management of obesity. Open Heart 3(2):e000385
80. Salas JJ, Ohlrogge JB (2002) Characterization of substrate specificity of plant FatA and FatB acyl-ACP thioesterases. Arch Biochem Biophys 403(1):25–34
81. Zhang P et al (2008) Mutations in a delta(9)-stearoyl-ACP-desaturase gene are associated with enhanced stearic acid levels in soybean seeds. Crop Sci 48(6):2305–2313
82. Knutzon DS et al (1992) Modification of brassica seed oil by antisense expression of a stearoyl-acyl carrier protein desaturase gene. Proc Natl Acad Sci U S A 89(7):2624–2628
83. Cardinal AJ et al (2007) Molecular analysis of soybean lines with low palmitic acid content in the seed oil. Crop Sci 47(1):304–310
84. Li L et al (2011) An 11-bp insertion in Zea mays fatb reduces the palmitic acid content of fatty acids in maize grain. PLoS One 6(9): e24699
85. Zheng Honggang, F.C.C.O.U.S., et al. (2013) Brassica plants yielding oils with a low total saturated fatty acid content. US
86. Bonaventure G et al (2003) Disruption of the FATB gene in Arabidopsis demonstrates an essential role of saturated fatty acids in plant growth. Plant Cell 15(4):1020–1033
87. Dormann P, Voelker TA, Ohlrogge JB (2000) Accumulation of palmitate in arabidopsis mediated by the acyl-acyl carrier protein thioesterase FATB1. Plant Physiol 123 (2):637–643
88. Belide S et al (2012) Modification of seed oil composition in arabidopsis by artificial microrna-mediated gene silencing. Front Plant Sci 3:168
89. Hawkins DJ, Kridl JC (1998) Characterization of acyl-ACP thioesterases of mangosteen (Garcinia mangostana) seed and high levels of stearate production in transgenic canola. Plant J 13(6):743–752
90. Knutzon DS et al (1999) Lysophosphatidic acid acyltransferase from coconut endosperm mediates the insertion of laurate at the sn-2 position of triacylglycerols in lauric rapeseed oil and can increase total laurate levels. Plant Physiol 120(3):739–746
91. Pidkowich MS et al (2007) Modulating seed beta-ketoacyl-acyl carrier protein synthase II level converts the composition of a temperate seed oil to that of a palm-like tropical oil. Proc Natl Acad Sci U S A 104(11):4742–4747
92. James DW, Dooner HK (1991) Novel seed lipid phenotypes in combinations of mutants altered in fatty-acid biosynthesis in arabidopsis. Theor Appl Genet 82(4):409–412
93. Carlsson AS et al (2002) A KAS2 cDNA complements the phenotypes of the Arabidopsis fab1 mutant that differs in a single residue bordering the substrate binding pocket. Plant J 29(6):761–770
94. Liu Q et al (2017) Genetic enhancement of palmitic acid accumulation in cotton seed oil through RNAi down-regulation of ghKAS2 encoding beta-ketoacyl-ACP synthase II (KASII). Plant Biotechnol J 15(1):132–143
95. Kinney AJ (1996) Beta-ketoacyl-ACP synthetase II genes from plants. E.I.D.P.d. Nemours, et al. (ed). US
96. Rubin Wilson BC, Young SA, Folkerts O (2001) Nucleotide sequences of maize and soybean beta-ketoacyl-acyl carrier protein synthase II and their use in the regulation of fatty acid content of oil, I.I.N. Dow AgroSciences Llc (ed). US
97. Gupta M et al (2012) Transcriptional activation of Brassica napus ss-ketoacyl-ACP synthase II with an engineered zinc finger protein transcription factor. Plant Biotechnol J 10(7):783–791
98. Dehesh K et al (2001) Overexpression of 3-ketoacyl-acyl-carrier protein synthase IIIs in plants reduces the rate of lipid synthesis. Plant Physiol 125(2):1103–1114
99. Primomo VS et al (2002) Genotype X environment interactions, stability, and agronomic performance of soybean with altered fatty acid profiles. Crop Sci 42(1):37–44
100. Delaney B et al (2008) Subchronic feeding study of high oleic acid soybeans (event DP-3Ø5423-1) in Sprague–Dawley rats. Food Chem Toxicol 46(12):3808–3817

101. Flores T et al (2008) Silencing of GmFAD3 gene by siRNA leads to low alpha-linolenic acids (18:3) of fad3-mutant phenotype in soybean [Glycine max (Merr.)]. Transgenic Res 17(5):839–850
102. Morineau C et al (2017) Selective gene dosage by CRISPR-Cas9 genome editing in hexaploid Camelina sativa. Plant Biotechnol J 15(6):729–739
103. Eckert H et al (2006) Co-expression of the borage Delta 6 desaturase and the Arabidopsis Delta 15 desaturase results in high accumulation of stearidonic acid in the seeds of transgenic soybean. Planta 224(5):1050–1057
104. Ruiz-Lopez N et al (2009) The synthesis and accumulation of stearidonic acid in transgenic plants: a novel source of 'heart-healthy' omega-3 fatty acids. Plant Biotechnol J 7 (7):704–716
105. Ulmasov T et al (2012) High-oleic, low-saturate soybeans offer a sustainable and nutritionally enhanced solution for food applications requiring high oil stability A2 - Wilson, Richard F. In: Designing soybeans for 21st century markets. AOCS Press, Urbana, pp 277–295
106. Liu Q, Singh S, Green A (2002) High-oleic and high-stearic cottonseed oils: nutritionally improved cooking oils developed using gene silencing. J Am Coll Nutr 21(3):205S–211S
107. Liu Q, Singh S, Green A (2000) Genetic modification of cotton seed oil using inverted-repeat gene-silencing techniques. Biochem Soc Trans 28:927–929
108. Liu F et al (2017) Simultaneous silencing of GhFAD2-1 and GhFATB enhances the quality of cottonseed oil with high oleic acid. J Plant Physiol 215:132–139
109. Park H et al (2014) Stacking of a stearoyl-ACP thioesterase with a dual-silenced palmitoyl-ACP thioesterase and Δ12 fatty acid desaturase in transgenic soybean. Plant Biotechnol J 12(8):1035–1043
110. Sun J-Y et al (2014) Simultaneous over-expressing of an acyl-ACP thioesterase (FatB) and silencing of acyl-acyl carrier protein desaturase by artificial microRNAs increases saturated fatty acid levels in Brassica napus seeds. Plant Biotechnol J 12 (5):624–637
111. Nguyen HT et al (2010) Metabolic engineering of seeds can achieve levels of omega-7 fatty acids comparable with the highest levels found in natural plant sources. Plant Physiol 154(4):1897–1904
112. Fahy D et al (2013) Reducing saturated fatty acids in Arabidopsis seeds by expression of a Caenorhabditis elegans 16:0-specific desaturase. Plant Biotechnol J 11(4):480–489
113. Nguyen HT et al (2015) Redirection of metabolic flux for high levels of omega-7 monounsaturated fatty acid accumulation in camelina seeds. Plant Biotechnol J 13 (1):38–50
114. Haslam RP et al (2012) The modification of plant oil composition via metabolic engineering-better nutrition by design. Plant Biotechnol J 11(2):157–168
115. Ruiz-Lopez N et al (2012) Metabolic engineering of the omega-3 long chain polyunsaturated fatty acid biosynthetic pathway into transgenic plants. J Exp Bot 63 (7):2397–2410
116. Ruiz-Lopez N et al (2013) Reconstitution of EPA and DHA biosynthesis in Arabidopsis: iterative metabolic engineering for the synthesis of n−3 LC-PUFAs in transgenic plants. Metab Eng 17:30–41
117. Petrie JR et al (2012) Metabolic engineering plant seeds with fish oil-like levels of DHA. PLoS One 7(11):e49165
118. Sayanova O et al (2012) The role of Delta6-desaturase acyl-carrier specificity in the efficient synthesis of long-chain polyunsaturated fatty acids in transgenic plants. Plant Biotechnol J 10(2):195–206
119. Petrie JR et al (2014) Metabolic engineering Camelina sativa with fish oil-like levels of DHA. PLoS One 9(1):e85061
120. Ruiz-Lopez, N., et al. (2013) Successful high-level accumulation of fish oil omega-3 long-chain polyunsaturated fatty acids in a transgenic oilseed crop. The Plant Journal
121. Usher S et al (2017) Tailoring seed oil composition in the real world: optimising omega-3 long chain polyunsaturated fatty acid accumulation in transgenic Camelina sativa. Sci Rep 7(1):6570
122. Metz JG et al (2001) Production of polyunsaturated fatty acids by polyketide synthases in both prokaryotes and eukaryotes. Science 293 (5528):290–293
123. Barclay W, Weaver C, Metz J (2005) Development of a docosahexaenoic acid production technology using schizochytrium. In Single cell oils. AOCS Publishing
124. Hauvermale A et al (2006) Fatty acid production in Schizochytrium sp.: involvement of a polyunsaturated fatty acid synthase and a type I fatty acid synthase. Lipids 41(8):739–747
125. Walsh TA, et al. (2016) Canola engineered with a microalgal polyketide synthase-like system produces oil enriched in docosahexaenoic acid. Nat Biotech. Advance online publication

126. Ruiz-Lopez N et al (2013) Successful high-level accumulation of fish oil omega-3 long chain polyunsaturated fatty acids in a transgenic oilseed crop. Plant J 77(2):198–208
127. Broun P, Somerville C (1997) Accumulation of ricinoleic, lesquerolic, and densipolic acids in seeds of transgenic Arabidopsis plants that express a fatty acyl hydroxylase cDNA from castor bean. Plant Physiol 113(3):933–942
128. Smith MA et al (2003) Heterologous expression of a fatty acid hydroxylase gene in developing seeds of Arabidopsis thaliana. Planta 217(3):507–516
129. Burgal J et al (2008) Metabolic engineering of hydroxy fatty acid production in plants: RcDGAT2 drives dramatic increases in ricinoleate levels in seed oil. Plant Biotechnol J 6(8):819–831
130. van Erp H et al (2015) Reducing isozyme competition increases target fatty acid accumulation in seed triacylglycerols of transgenic arabidopsis. Plant Physiol 168(1):36–46
131. Bayon S et al (2015) A small phospholipase A2-alpha from castor catalyzes the removal of hydroxy fatty acids from phosphatidylcholine in transgenic Arabidopsis seeds. Plant Physiol 167(4):1259–1270
132. Wayne LL, Browse J (2013) Homologous electron transport components fail to increase fatty acid hydroxylation in transgenic Arabidopsis thaliana. F1000Res 2:203
133. Bates PD, Browse J (2011) The pathway of triacylglycerol synthesis through phosphatidylcholine in Arabidopsis produces a bottleneck for the accumulation of unusual fatty acids in transgenic seeds. Plant J 68 (3):387–399
134. Bates PD et al (2012) Acyl editing and headgroup exchange are the major mechanisms that direct polyunsaturated fatty acid flux into triacylglycerols. Plant Physiol 160 (3):1530–1539
135. Bates PD et al (2014) Fatty acid synthesis is inhibited by inefficient utilization of unusual fatty acids for glycerolipid assembly. Proc Natl Acad Sci U S A 111(3):1204–1209
136. Broadwater JA, Whittle E, Shanklin J (2002) Desaturation and hydroxylation. Residues 148 and 324 of Arabidopsis FAD2, in addition to substrate chain length, exert a major influence in partitioning of catalytic specificity. J Biol Chem 277(18):15613–15620
137. Dyer JM et al (2002) Molecular analysis of a bifunctional fatty acid conjugase/desaturase from tung. Implications for the evolution of plant fatty acid diversity. Plant Physiol 130 (4):2027–2038
138. Lee M et al (1998) Identification of non-heme diiron proteins that catalyze triple bond and epoxy group formation. Science 280(5365):915–918
139. Cahoon EB et al (2006) Conjugated fatty acids accumulate to high levels in phospholipids of metabolically engineered soybean and Arabidopsis seeds. Phytochemistry 67 (12):1166–1176
140. Mietkiewska E et al (2014) Combined transgenic expression of Punica granatum conjugase (FADX) and FAD2 desaturase in high linoleic acid Arabidopsis thaliana mutant leads to increased accumulation of punicic acid. Planta 240(3):575–583
141. Singh S et al (2000) Inhibition of polyunsaturated fatty acid accumulation in plants expressing a fatty acid epoxygenase. Biochem Soc Trans 28(6):940–942
142. Singh S et al (2001) Transgenic expression of a delta 12-epoxygenase gene in Arabidopsis seeds inhibits accumulation of linoleic acid. Planta 212(5–6):872–879
143. Rezzonico E et al (2004) Level of accumulation of epoxy fatty acid in Arabidopsis thaliana expressing a linoleic acid delta12-epoxygenase is influenced by the availability of the substrate linoleic acid. Theor Appl Genet 109(5):1077–1082
144. Li R et al (2010) Vernonia DGATs increase accumulation of epoxy fatty acids in oil. Plant Biotechnol J 8(2):184–195
145. Li R et al (2012) Vernonia DGATs can complement the disrupted oil and protein metabolism in epoxygenase-expressing soybean seeds. Metab Eng 14(1):29–38
146. Yu XH et al (2018) Identification of bottlenecks in the accumulation of cyclic fatty acids in camelina seed oil. Plant Biotechnol J 16:926–938
147. Zhu LH et al (2016) Dedicated industrial oilseed crops as metabolic engineering platforms for sustainable industrial feedstock production. Sci Rep 6:22181
148. Ruiz-Lopez N et al (2017) Tailoring the composition of novel wax esters in the seeds of transgenic Camelina sativa through systematic metabolic engineering. Plant Biotechnol J 15 (7):837–849
149. Roesler K et al (2016) An improved variant of soybean type 1 diacylglycerol acyltransferase increases the oil content and decreases the soluble carbohydrate content of soybeans. Plant Physiol 171(2):878–893
150. Ichihara Ki, Takahashi T (1988) Fujii, *Diacylglycerol acyltransferase in maturing safflower seeds: its influences on the fatty acid composition*

of triacylglycerol and on the rate of triacylglycerol synthesis. Biochim Biophys Acta 958 (1):125–129

151. Hobbs DH, Lu C, Hills MJ (1999) Cloning of a cDNA encoding diacylglycerol acyltransferase from Arabidopsis thaliana and its functional expression. FEBS Lett 452 (3):145–149
152. Routaboul JM et al (1999) The TAG1 locus of Arabidopsis encodes for a diacylglycerol acyltransferase. Plant Physiol Biochem 37 (11):831–840
153. Zou JT et al (1999) The Arabidopsis thaliana TAG1 mutant has a mutation in a diacylglycerol acyltransferase gene. Plant J 19 (6):645–653
154. Lardizabal KD et al (2001) DGAT2 is a new diacylglycerol acyltransferase gene family: purification, cloning, and expression in insect cells of two polypeptides from Mortierella ramanniana with diacylglycerol acyltransferase activity. J Biol Chem 276(42):38862–38869
155. Shockey JM et al (2006) Tung tree DGAT1 and DGAT2 have nonredundant functions in triacylglycerol biosynthesis and are localized to different subdomains of the endoplasmic reticulum. Plant Cell 18(9):2294–2313
156. Jako C et al (2001) Seed-specific over-expression of an Arabidopsis cDNA encoding a diacylglycerol acyltransferase enhances seed oil content and seed weight. Plant Physiol 126 (2):861–874
157. Taylor DC et al (2009) Molecular modification of triacylglycerol accumulation by over-expression of DGAT1 to produce canola with increased seed oil content under field conditions. Botany-Botanique 87(6):533–543
158. Weselake RJ et al (2008) Metabolic control analysis is helpful for informed genetic manipulation of oilseed rape (Brassica napus) to increase seed oil content. J Exp Bot 59 (13):3543–3549
159. Chen G et al (2017) High-performance variants of plant diacylglycerol acyltransferase 1 generated by directed evolution provide insights into structure function. Plant J 92 (2):167–177
160. Lardizabal K et al (2008) Expression of <em>Umbelopsis ramanniana DGAT2A</em> in seed increases oil in soybean. Plant Physiol 148(1):89–96
161. Liu J et al (2010) Increasing seed mass and oil content in transgenic Arabidopsis by the over-expression of wri1-like gene from Brassica napus. Plant Physiol Biochem 48(1):9–15
162. Liu FS et al (2016) Modification of starch metabolism in transgenic Arabidopsis thaliana increases plant biomass and triples oilseed production. Plant Biotechnol J 14 (3):976–985
163. Vanhercke T et al (2014) Metabolic engineering of biomass for high energy density: oilseed-like triacylglycerol yields from plant leaves. Plant Biotechnol J 12(2):231–239
164. Vanhercke T et al (2017) Step changes in leaf oil accumulation via iterative metabolic engineering. Metab Eng 39:237–246
165. El Tahchy A et al (2017) Thioesterase over-expression in Nicotiana benthamiana leaf increases the fatty acid flux into triacylgycerol. FEBS Lett 591(2):448–456
166. Yurchenko O et al (2017) Engineering the production of conjugated fatty acids in Arabidopsis thaliana leaves. Plant Biotechnol J 15 (8):1010–1023
167. Fouillen L, Colsch B, Lessire R(2013) The lipid world concept of plant lipidomics. In Metabolomics coming of age with its technological diversity, Rolin D (Ed). p. 331–76
168. Horn PJ, Chapman KD (2014) Lipidomics in situ: insights into plant lipid metabolism from high resolution spatial maps of metabolites. Prog Lipid Res 54:32–52
169. Alonso AP, Val DL, Shachar-Hill Y (2011) Understanding fatty acid synthesis in developing maize embryos using metabolic flux analysis. Metab Eng 13(4):454–454
170. Tan G-Y, Liu T (2017) Rational synthetic pathway refactoring of natural products biosynthesis in actinobacteria. Metab Eng 39 (Supplement C):228–236
171. CropLife International (2011) Cost of bringing a biotech crop to market. https://croplife.org/plant-biotechnology/regulatory-2/cost-of-bringing-a-biotech-crop-to-market/. Accessed 11 Jun 2017

Part IV

Transgenic Event Characterization

Chapter 24

Molecular Analysis for Characterizing Transgenic Events

Wei Chen and PoHao Wang

Abstract

To develop a commercial trait product, a large number of transgenic events are often produced to obtain the event with desired level of expression. It is crucial to develop efficient and sensitive molecular characterization methods to advance events with stable transgene expression, free of vector backbone sequences and without major changes to the native genome caused by transgene insertion. Here, we discuss a variety of analytical tools, including quantitative PCR (qPCR), Southern blot analysis, and various sequencing technologies, which have been widely used to determine the insert copy number, presence/absence of vector backbone sequences, integrity of the T-DNA, and genomic location of the T-DNA insertion. Moreover, since the discovery of RNA interference in 1998 (Fire et al., Nature 391:806-811, 1998), RNAi has emerged as another powerful tool in in the development of a new transgenic trait for insect control. RNAi creates a double-stranded RNA duplex as the active molecule which forms a strong secondary structure, resulting in challenges for detection. In addition to molecular analysis at the DNA level, this chapter describes detection methods of the active molecules (i.e., double-stranded RNA) for RNAi-based traits.

Key words Transgenic events, Molecular characterization, qPCR, NGS, RNAi

1 Introduction

Transgenic trait product development is a lengthy and expensive process. As illustrated in Mumm's article, it includes several phases, from trait discovery, proof of concept (phase I), early development (phase II), advanced development (phase III), and pre-launch (phase IV) to commercial launch [2]. In the trait discovery phase, tens of thousands of candidates are screened to identify potential genes for further development. In phase I, selected genes are engineered into plasmids with combinations of regulatory elements. The plasmids are used to transform target crops for testing anticipated expression and efficacy. In phase II, large numbers of transgenic events are produced, most notably via particle bombardment [3] or *Agrobacterium*-mediated gene transfer [4] to identify

Sandeep Kumar et al. (eds.), *Transgenic Plants: Methods and Protocols*, Methods in Molecular Biology, vol. 1864, https://doi.org/10.1007/978-1-4939-8778-8_24, © Springer Science+Business Media, LLC, part of Springer Nature 2019

events delivering adequate levels of expression for the candidate genes that leads to the desired phenotype. Phase III is characterized by introgression of a trait into high-performing germplasm for field testing, preparing hybrid seed production and generating data for regulatory submission. Seed increase is the activity most prevalent in phase IV. Finally, market launch releases the product for commercialization. This chapter summarizes the analytical methods used for extensive event characterization at phase II. The major endpoints of the event characterization are (1) transgene insert copy number, (2) absence/presence of vector backbone sequences, (3) integrity of T-DNA at the structural and nucleotide levels, (4) identification and characterization of the T-DNA insertion site, and (5) expression of the target genes. Here we describe the technological options to achieve the above-mentioned endpoints at the DNA level and expression analysis for RNAi-based traits. Protein analysis is covered in another chapter.

2 DNA-Based Molecular Analysis

2.1 qPCR

Quantitative real-time PCR (qRT-PCR), often referred to as qPCR, has been widely used in event characterization to determine the copy number of the transgene and the presence/absence of undesirable vector backbone sequences. qPCR relies on the ability to progressively monitor the fluorescence emitted from either double-stranded DNA-binding dyes or fluorophore-based probes that hybridize to a target sequence, and can be cleaved during the exponential phase of the PCR reaction to enable the quantification. qPCR was adopted as a cost-effective technology to estimate gene copy number in the late 1990s for various crops [5–9]. TaqMan probes rely on the 5′–3′ exonuclease activity of Taq polymerase to cleave the dual-labeled probe, have much higher specificity compared to binding dyes, and are amenable for multiplexing and automation. Therefore, they have become an invaluable tool to achieve the throughput required to screen the large number of transgenic events generated during the product development process. With TaqMan assays, it is fairly simple to use the $2^{-\Delta\Delta C_T}$ methodology to estimate the transgene copy number [10–12]. For $2^{-\Delta\Delta C_T}$ method to work, two gene-specific assays need to be designed: one for the gene of interest (GOI) and the other for a low-copy endogenous gene, preferably a single-copy gene in the genome. A standard curve for copy number control also needs to be established. The control can be made by using the genomic DNA of a well-characterized transgenic event containing the target gene, or by spiking the plasmid DNA with the target gene into wild-type genomic DNA, proportionally based on the size of construct and the host genome [7, 13, 14]. Once the PCR amplification reaches the plateau phase due to the limited amount

Table 1
Example for calculation of $2^{-\Delta\Delta CT}$

	Average C_T for gene of interest	Average C_T for housekeeping gene	ΔC_T	$\Delta\Delta C_T$	$2^{-\Delta\Delta CT}$
Known copy number control	24.57	23.66	0.91		
Unknown sample	23.38	23.45	−0.07	−0.98	1.97

of reagents in the reaction, such as the polymerase, primers, and dNTPs, the C_T scores, the point at which the fluorescence signal crosses the background threshold, can be calculated using the fit point algorithm. All commercially available qPCR systems provide the software for automatic calls with the C_T value. Once the C_T scores for both genes (GOI and endogenous gene) are identified, the relative copy number can then be predicted. Table 1 is a simple example of how to calculate $2^{-\Delta\Delta C_T}$: first, take the average of C_T values for GOI and endogenous gene for copy number controls as well as experimental samples; next, calculate the ΔC_T by subtracting the average C_T value of endogenous gene from GOI; then, calculate the $\Delta\Delta C_T$ by subtracting ΔC_T of the known copy number control from unknown sample; finally, calculate the $2^{-\Delta\Delta C_T}$ to get the copy number for experimental samples by comparing to the known copy number control. In this particular example, the unknown sample contains twice as many copies of the GOI as the known copy number control.

2.2 Southern Blot Analysis

Southern blot [15] is a hybridization-based technique that can effectively identify transgenic insert/copy number, structural integrity of T-DNA, and presence/absence of undesirable vector backbone sequences. However, Southern blot analysis is a very resource- and labor-intensive procedure as it requires a large amount of high-molecular-weight genomic DNA, and takes a week or two from DNA extraction to data interpretation. The lengthy process includes (1) extraction of genomic DNA, (2) digestion of genomic DNA with restriction enzymes, (3) electrophoresis to separate DNA fragments, (4) transferring of DNA from agarose gel onto a membrane, and (5) hybridization to probes incorporated with either radioactivity or chemiluminescence for detection. Compared to qPCR which can be completed in a matter of hours, Southern blot is very inefficient. Southern blot analysis is not easily amenable to automation, and can become a bottleneck for event characterization that requires high-throughput platforms to analyze large numbers of events. It can also become very challenging to design the experiments and interpret the results, for complex events with rearrangements within the T-DNA [16].

2.3 Amplicon Sanger Sequencing

Neither qPCR nor Southern blot analysis is sensitive enough to detect single-nucleotide polymorphisms and small DNA rearrangements such as insertions, inversions, and deletions. Sanger sequencing is a technology that has been successfully used to determine the integrity of T-DNA at a nucleotide level, and is capable of addressing the above-mentioned issues unless DNA rearrangements within T-DNA cause the formation of an adverse secondary structure. The other two important endpoints that can be achieved by Sanger sequencing are characterization of the transgene integration site in the plant genome, and identification of the flanking border sequences (junction region of the transgene with the native plant genome). Knowledge of the flanking border sequences, as well as mutations which may occur at the junction region, is essential for the development of event-specific detection methods to monitor genetically modified (GM) crops [17, 18]. Historically, this was accomplished by "genome walking," which is simply Sanger sequencing of PCR-generated amplicons covering flanking borders and junction regions. Various genome walking approaches have been developed [19, 20]. The most recent method described in a patent issued to Dow AgroSciences for high-throughput analysis of transgene borders [21] includes the following steps: (1) digestion of genomic DNA with restriction enzymes; (2) ligation of a double-stranded adapter to the isolated and digested genomic DNA; (3) a primer extension reaction of the adapter-ligated genomic DNA; (4) the isolation of the primer extension reaction product via a streptavidin-biotin interaction; and (5) further characterization via subsequent PCR amplification reactions and DNA sequencing. The method improves the accuracy, sensitivity, and reproducibility for determining unknown DNA sequences flanking a known DNA sequence. However, genome walking using Sanger sequencing requires intricate experimental design and is time consuming, especially when rearrangement occurs in the T-DNA.

2.4 Next-Generation Sequencing

Next-generation sequencing (NGS) has evolved as another technological option to conduct high-throughput event characterization. It has been demonstrated that NGS, particularly whole-genome sequencing and target capture sequencing applications, could be successfully used in a cost-effective manner to achieve several endpoints of molecular characterization of transgenic events, including insert/copy number, integrity of T-DNA, presence/absence of backbone sequences, and identification and characterization of flanking border sequences and parental loci [22–24]. Additionally, it was demonstrated that due to its paired-end chemistry, NGS technologies could resolve complex rearrangements within T-DNA and junction regions [22]. For high-throughput event characterization, target capture sequencing appeared to be more feasible. Capture-based sequencing reduces complexity by the use of the specific probes for the enrichment of target sequences, and decreases the amount of sequences generated for subsequent bioinformatics analysis [25]. Both microarray- and liquid-based sequence capture have been successfully demonstrated for targeted sequencing [22, 24, 26, 27]. The ability to multiplex samples using target capture sequencing also makes this technology very appealing for high-throughput event characterization [22, 24]. However, the current state of NGS-based technologies (specifically the short length of reads) could limit the use of this technology for events transformed with endogenous regulatory elements or long repetitive sequences within T-DNA and borders. Long-range/read sequencing or alternative library preparation methods to generate large DNA fragments are being developed as reliable solutions to overcome this shortfall of the NGS technology [28, 29]. Meanwhile, third-generation sequencing also holds high promise to revolutionize the application of sequencing technology in the future. However, third-generation sequencing is not yet routinely used due to its inconsistency in quality and relatively high cost [30, 31].

2.5 Example of Event Sorting

The journey to launching commercial germplasm expressing efficacious and stable transgenes is a resource- and time-intensive process, which could take more than 10 years and over 100 million dollars [32]. An example of the rate of attrition in a typical event-sorting process using random insertion transformation technology can be cited from Monsanto's commercial event Mon87419, which was selected from over 13,000 events originating from only two plasmid vector sequences [33]. A series of event characterization methodologies were deployed at various stages of the product development process. For R0, the first-generation plants after a successful transformation, qPCR copy number analysis was conducted and almost 80% of the events were discarded due to the high copy number of transgenic T-DNA inserted in the host genome. More than half of the remaining ~3000 events did not pass the subsequent herbicide spray, indicating absence of a functional selectable marker gene. From the 1440 surviving events, only 184 met the event selection requirements after more detailed characterization of T-DNA sequences using NGS. Southern blot analysis further removed 72 events; therefore, only 112 events went into next-generation (R1) field trials. The R1 segregating population was tested with qPCR for zygosity. Forty-two events demonstrated normal Mendelian segregation and were selected for advanced field trials. After two more years of field testing and in-depth molecular analysis, one event (Mon87419) became the leading event for commercial product development.

3 RNA Quantitation for Active Molecules Derived from GM RNAi Traits

Current commercially available GM crops contain protein-coding traits (e.g., *Bacillus thuringiensis* (Bt) toxins and herbicide tolerance proteins). For expression analysis, immunoassays or LC-MSMS are used as the main analytical tool to quantify the proteins [34–36]. In recent years, in addition to protein traits, GM-insect-resistant traits using RNAi approach have been developed as a powerful tool for insect pest control (e.g., Coleoptera, reviews in [37, 38]). Since RNA is the active molecule of an RNAi trait, capability for RNA quantitation becomes increasingly important during the trait development process, and will be crucial for quality assurance of downstream GM crops for foods and feed. Therefore, development of a reliable analytical tool for RNAi trait quantitation is critical.

An RNAi trait in GM crops can be designed to produce hairpin RNA (hpRNA) with a loop sequence linked to the double-stranded RNA (dsRNA) duplex region. Previous studies in Coleoptera have demonstrated that only the RNAi trait-derived dsRNAs over 60 base pairs (bps) are efficacious for insect control [39, 40]. Such dsRNA is subject to form a strong secondary structure, thus resulting in challenges for quantitation. Therefore, analytical methods for successful quantitation of dsRNA must have two key attributes: (1) quantify the dsRNA with strong secondary structure and (2) differentiate dsRNA from single-stranded RNA. Here, we focus on the analytical methods that have been validated and discuss the platforms that have potential to enable reliable quantitation of RNAi traits for commercial trait product development.

3.1 Reverse-Transcription Quantitative PCR (qRT-PCR)

qPCR is an industry standard assay and has been adapted for numerous GM crop applications including DNA analysis to determine transgene copy numbers, as described above. In addition, gene expression analysis by reverse-transcription quantitative PCR (RT-qPCR) has been a prevalent technology to measure the levels of mRNAs, miRNAs, and other RNA species because of its accuracy and precision [41]. To quantify the dsRNA derived from an RNAi trait of a GM crop, some modifications of the protocol are required. Wang et al. (2018) [42] reported a novel method, named RNase I_f-treated quantitative real-time PCR (RNase I_f-qPCR, Fig. 1). This modified qPCR method was derived to overcome the obstacles for dsRNA detection including (1) strong secondary structure, and (2) undesired RNA intermediates (e.g., single-stranded RNA (ssRNA)) mixed with dsRNA. To achieve this, two critical modified steps were included in the qPCR procedures. First, prior to the cDNA conversion step, RNase I_f, an endonuclease which has preferential activity to digest ssRNA over dsRNA, is utilized to separate

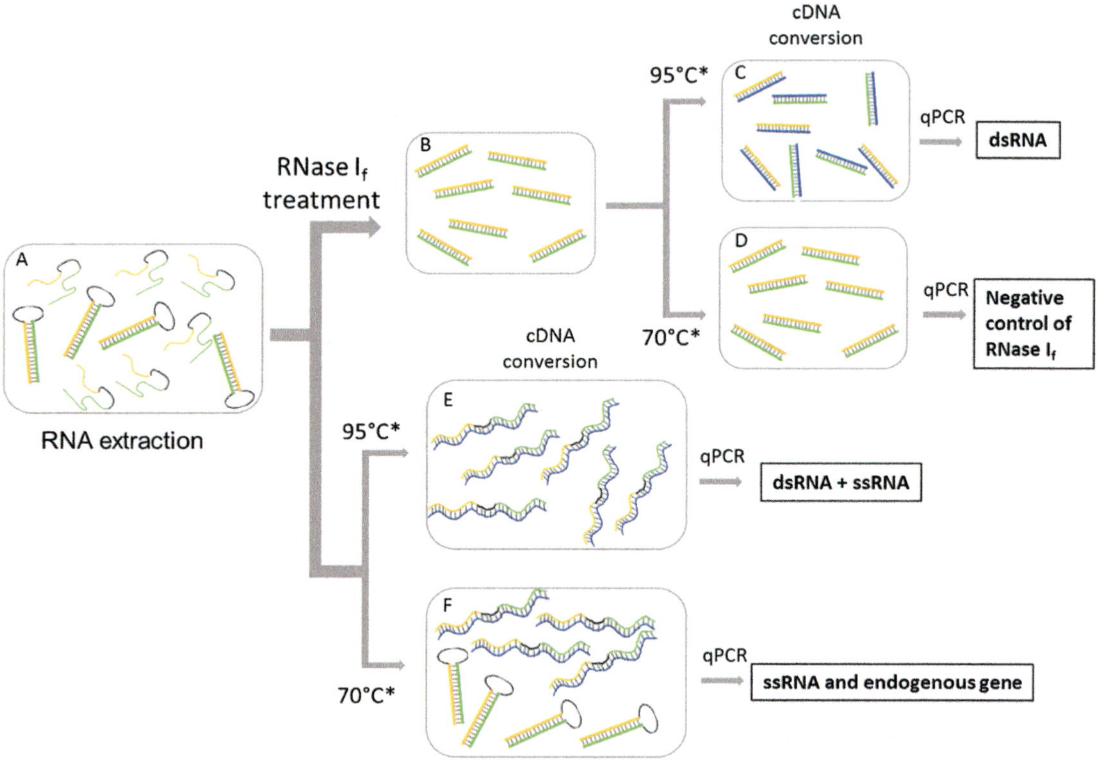

Fig. 1 RNase I_f-qPCR assay overview. The total RNA is isolated first (**a**) using the protocol described in materials and methods. Then the isolated total RNAs are treated with RNase I_f and followed by RNA cleanup to obtain the purified dsRNA (loop sequence is digested) (**b**). The RNase I_f-treated RNAs are proceeded for cDNA conversion. The RNAs are preincubated with random hexamers at 95 °C (**c**) and 70 °C (**d**), respectively, and followed by cDNA conversion. Blue strands represent the reversely transcribed cDNA for the RNA incubated at 95 °C. In contrast, no cDNA is produced from the RNA incubated in 70 °C if all ssRNA is digested by RNase I_f. In addition, non-RNase I_f-treated RNA was used for cDNA conversion at 95 °C (**e**) and 70 °C (**f**). For RNA preincubated in 95 °C, cDNAs (dark blue strand) are transcribed from both dsRNA and ssRNA. On the other hand, cDNAs are only converted from ssRNAs (including RNAi and endogenous gene). The qRT-PCR is performed after the cDNA conversion using a custom-designed TaqMan assay

dsRNA from ssRNA. Secondly, during the cDNA conversion, high (95 °C) and low (70 °C) temperatures are applied in the incubation of RNA and oligos prior to addition of reverse transcriptase in which dsRNA may be only reverse transcribed at 95 °C instead of 70 °C. RNaseI_f-qPCR allows the user to precisely distinguish dsRNA from ssRNA and effectively assay the dsRNA in a high-throughput manner.

3.2 Digital PCR (dPCR)

An alternative to qRT-PCR platform is digital PCR (dPCR) [43]. The dPCR enables the absolute quantification of targets through the use of limited dilutions, endpoint PCR, and Poisson statistics [44]. The dPCR works by dividing PCR mixtures into a large number of partitions containing zero, one, or more copies of the target nucleic acid, followed by conducting endpoint PCR to detect the target nucleic acid in each partition. Then, the signal is analyzed using Poisson statistics from the ratio of positive signal in total partitions and back-calculated for the absolute copy number of the original samples [45]. There are currently two dPCR platforms: chamber digital PCR (cdPCR) and droplet digital PCR (ddPCR). The two methods differ in the manner by which the partitions are separated and quantified. Recently, both dPCR platforms have been demonstrated to detect the copy numbers in GM crops [46, 47]. These examples show the applicability of dPCR for copy number detection in GM crops and imply the opportunity for using dPCR in RNA quantitation. The technology was further confirmed for absolute RNA quantitation (i.e., reverse transcription (RT)—dPCR) with low-copy RNA targets [48]. Since dPCR for RNA quantitation is used to quantify cDNA from RT conversion, the obstacles for accurately quantifying RNA using dPCR may be resolved from the quantitation sensitivity and sample partition pre/post-cDNA conversion. Although the application of RT-dPCR for quantitation of dsRNA for an RNAi trait is yet to be established, there is a high potential to develop this capability based on the nature of this technology.

3.3 Hybridization-Based Assay: Affymetrix QuantiGene Plex and SinglePlex

Different from the conventional PCR-based assay, hybridization-based technology is an alternative for transcript analysis. QuantiGene Plex (QGP) is one of the well-known technologies, as a Luminex/xMAP bead-based multiplex nucleic acid assay using branched-chain DNA(bDNA) technology for quantifying transcript expression. The QGP platform uses combinations of fluorescent-coated microsphere magnetic beads with unique spectral characteristics. Each type of bead may be assigned to bind to a specific target, thereby enabling a multiplex assay of up to 80 distinct target sequences of interest in one reaction.

Unlike the conventional qPCR, the QGP assay allows the detection directly from the tissue homogenates without RNA isolation, cDNA conversion, and PCR amplification. Therefore, it may reduce the bias that could potentially result from the qRT-PCR. A previous study also shows that QGP is a better platform to work on formalin-fixed paraffin-embedded (FFPE) samples due to its direct measurement to RNA without the RNA isolation step as compared with qRT-PCR [40]. The QGP procedure usually takes 2 days: it begins with an overnight incubation in which probe sets hybridize directly to both the RNA of interest and the fluorescent-coated microsphere magnetic bead. The resulting hybridization complex is then amplified using branched capture extenders and preamplifier probes to generate a branched DNA structure. The structure is labeled with phycoerythrin-conjugated streptavidin, and the signal is detected via a Luminex flow cytometer. The flow cytometry identifies a fluorescent signal that is proportional to hybridized RNA quantity, and the bead identifies the targets.

QGP is widely used in multiple applications such as diagnosis [40], drug discovery screening [49], and plant research [50]. Recently, the applicability of QGP or dsRNA quantitation of GM crop RNAi traits has been reported [51]. In this report, QGP displays reproducibility, precision, and a wide linear range for dsRNA quantitation. Studies have demonstrated good correlation ($R > 0.8$) between the QGP and RNase I_f-qPCR methodologies [42]. This implies that both platforms are adequate for dsRNA quantitation, although the QGP requires specific flow cytometry instrumentation that may not be accessible for every laboratory.

Similarly, in addition to the multiplexed QGP, QuantiGene SinglePlex (QGS) is a singleplex platform using the same branched DNA technology. Instead of using the microsphere magnetic beads for multiplexing like with QGP, QGS uses the capture probes coated on the surface of microplates and builds the hybridization complex in the microplates. The resulting luminescence signal is detected via microplate reader. The QGS may also be used for dsRNA quantitation.

3.4 NanoString Technologies

NanoString Technologies' nCounter is a novel platform that may be suitable for RNA expression analysis using hybridization-based technology [52]. The NanoString platform enables quantitation of multiplexed target molecules via color-coded molecular barcodes and a single-molecule imaging system. The advantages of this platform are (1) the capability for multiplexing up to 800 targets, (2) high sensitivity, and (3) non-amplification assay to avoid bias and increase the feasibility for FFPE samples [53–56]. Although the application of NanoString on dsRNA has not yet been reported, the application of NanoString for RNA expression has been reported in multiple studies such as FFPE samples [57], cancer research [58], and forensic science [59], implying its capability and flexibility for RNA analysis. The potential of NanoString for dsRNA quantitation requires further investigation.

In summary, the methods described above display the potential of distinct platforms for dsRNA quantitation of GM-crop RNAi traits. However, each platform has its advantages and drawbacks. For their use in GM crops trait development in the agriculture industry, several key attributes are desired: (1) reproducibility and precision, (2) cost-effectiveness, (3) robustness, (4) feasibility, and (5) high throughput. The comparison of platforms is listed in Table 2.

Table 2
Comparison for different technologies

	PCR based		Hybridization based	
	qPCR	dPCR	QuantiGene Plex	NanoString
Overview	Measures PCR amplification as it occurs	Measures the fraction of negative replicates to determine absolute copies	Measures amplifying reporter fluorescence of hybridized probes	Count-based transcript detection technology
PCR amplification	Real time	Endpoint	No PCR amplification	No PCR amplification
Sample preparation	RNA purification and RT conversion	RNA purification and RT conversion	Tissue lysate/ purified RNA, no RT conversion	Tissue lysate/ purified RNA, no RT conversion
Quantification	Capable of detection down to a twofold change	Linear response to the number of copies present to allow for small fold change differences to be detected	Distinguishes percentage differences—linear assay	Copy counts
Multiplex	Low multiplex	Low multiplex	Multiplexed up to 80 targets	Multiplexed up to 800 targets

4 Summary

Molecular analysis, as the key analytical tool complementing field trials for event characterization at early stages of commercial transgenic trait product development, has quickly shifted from traditional Southern blot analysis to high-throughput qPCR and next-generation sequencing for DNA-based detection. In addition, with the emerging RNAi trait development in Ag biotechnology, a reliable analytical method with precision and accuracy is required for quantitation of active molecules (i.e., dsRNA). The aforementioned methods provide pros and cons for dsRNA quantitation and may serve distinct purposes in different stages of trait development. This chapter summarizes the state-of-the-art molecular analyses used for the current GM trait development in agriculture biotechnology industry in both DNA and RNA levels. With rapid advancement of new technologies and in-depth understanding of GM crops, future methodologies for analyzing the transgenes will become even more powerful to ensure that only the best event goes into commercial production, hence reducing the resources for expansive late-stage studies. Furthermore, in light of the rapid advancement of genome editing technology (e.g., CRISPR, zinc finger) and its application in trait development [60], the demand for high-resolution molecular analysis to characterize the genome modified trait, and investigate off-target effects, remains a key focus for the agriculture industry to promote the next-generation trait development.

References

1. Fire A, Xu S, Montgomery MK, Kostas SA (1998) Potent and specific genetic interference by double-stranded RNA in Caenorhabditis elegans. Nature 391(6669):806–811
2. Mumm RH (2013) A look at product development with genetically modified crops: examples from maize. J Agric Food Chem 61 (35):8254–8259
3. Klein TM, Kornstein L, Sanford JC, Fromm ME (1989) Genetic transformation of maize cells by particle bombardment. Plant Physiol 91(1):440–444
4. Gould J, Devey M, Hasegawa O, Ulian EC, Peterson G, Smith RH (1991) Transformation of Zea mays L. using agrobacterium tumefaciens and the shoot apex. Plant Physiol 95 (2):426–434
5. Chiang P-W, Song W-J, Wu K-Y, Korenberg JR, Fogel EJ, Van Keuren ML, Lashkari D, Kurnit DM (1996) Use of a fluorescent-PCR reaction to detect genomic sequence copy number and transcriptional abundance. Genome Res 6(10):1013–1026
6. Ingham DJ, Beer S, Money S, Hansen G (2001) Quantitative real-time PCR assay for determining transgene copy number in transformed plants. BioTechniques 31(1):132–141
7. Song P, Cai C, Skokut M, Kosegi B, Petolino J (2002) Quantitative real-time PCR as a screening tool for estimating transgene copy number in WHISKERS?-derived transgenic maize. Plant Cell Rep 20(10):948–954
8. Schmidt M, Parrott W (2001) Quantitative detection of transgenes in soybean [Glycine max (L.) Merrill] and peanut (Arachis hypogaea L.) by real-time polymerase chain reaction. Plant Cell Rep 20(5):422–428
9. Vaïtilingom M, Pijnenburg H, Gendre F, Brignon P (1999) Real-time quantitative PCR detection of genetically modified maximizer maize and roundup ready soybean in some representative foods. J Agric Food Chem 47 (12):5261–5266
10. Livak KJ, Schmittgen TD (2001) Analysis of relative gene expression data using real-time quantitative PCR and the 2(−Delta Delta C (T)) method. Methods 25(4):402–408. https://doi.org/10.1006/meth.2001.1262
11. Bubner B, Baldwin IT (2004) Use of real-time PCR for determining copy number and zygosity in transgenic plants. Plant Cell Rep 23 (5):263–271. https://doi.org/10.1007/s00299-004-0859-y
12. Schmittgen TD, Livak KJ (2008) Analyzing real-time PCR data by the comparative CT method. Nat Protoc 3(6):1101–1108
13. Winer J, Jung CKS, Shackel I, Williams PM (1999) Development and validation of real-time quantitative reverse transcriptase–polymerase chain reaction for monitoring gene expression in cardiac myocytesin vitro. Anal Bioanal Chem 270(1):41–49
14. Bubner B, Gase K, Baldwin IT (2004) Two-fold differences are the detection limit for determining transgene copy numbers in plants by real-time PCR. BMC Biotechnol 4 (1):14. https://doi.org/10.1186/1472-6750-4-14
15. Southern EM (1975) Detection of specific sequences among DNA fragments separated by gel electrophoresis. J Mol Biol 98 (3):503–517
16. Cantsilieris S, Baird PN, White SJ (2013) Molecular methods for genotyping complex copy number polymorphisms. Genomics 101 (2):86–93. https://doi.org/10.1016/j.ygeno.2012.10.004
17. Miraglia M, Berdal K, Brera C, Corbisier P, Holst-Jensen A, Kok E, Marvin H, Schimmel H, Rentsch J, Van Rie J (2004) Detection and traceability of genetically modified organisms in the food production chain. Food Chem Toxicol 42(7):1157–1180
18. Sparrow P (2010) GM risk assessment. Mol Biotechnol 44(3):267–275

19. Leoni C, Volpicella M, De Leo F, Gallerani R, Ceci LR (2011) Genome walking in eukaryotes. FEBS J 278(21):3953–3977
20. Harkey MA, Kaul R, Jacobs MA, Kurre P, Bovee D, Levy R, Blau CA (2007) Multiarm high-throughput integration site detection: limitations of LAM-PCR technology and optimization for clonal analysis. Stem Cells Dev 16 (3):381–392
21. Cao Z, Novak S, Zhou N (2014) High through-put analysis of transgene borders. US Patent 8:911–943
22. Guttikonda SK, Marri P, Mammadov J, Ye L, Soe K, Richey K, Cruse J, Zhuang M, Gao Z, Evans C (2016) Molecular characterization of transgenic events using next generation sequencing approach. PLoS One 11(2): e0149515. https://doi.org/10.1371/journal.pone.0149515
23. Kovalic D, Garnaat C, Guo L, Yan Y, Groat J, Silvanovich A, Ralston L, Huang M, Tian Q, Christian A, Cheikh N, Hjelle J, Padgette S, Bannon G (2012) The use of next generation sequencing and junction sequence analysis bioinformatics to achieve molecular characterization of crops improved through modern biotechnology. Plant Genome 5(3):149–163
24. Zastrow-Hayes GM, Lin H, Sigmund AL, Hoffman JL, Alarcon CM, Hayes KR, Richmond TA, Jeddeloh JA, May GD, Beatty MK (2015) Southern-by-sequencing: a robust screening approach for molecular characterization of genetically modified crops. Plant Genome 8(1):1–15. https://doi.org/10.3835/plantgenome2014.08.0037
25. Fu Y, Springer NM, Gerhardt DJ, Ying K, Yeh CT, Wu W, Swanson-Wagner R, D'Ascenzo M, Millard T, Freeberg L (2010) Repeat subtraction-mediated sequence capture from a complex genome. Plant J 62(5):898–909
26. Warr A, Robert C, Hume D, Archibald A, Deeb N, Watson M (2015) Exome sequencing: current and future perspectives. G3 5 (8):1543–1550
27. García-García G, Baux D, Faugère V, Moclyn M, Koenig M, Claustres M, Roux A-F (2016) Assessment of the latest NGS enrichment capture methods in clinical context. Sci Rep 6:20948. https://doi.org/10.1038/srep20948
28. Putnam NH, O'Connell BL, Stites JC, Rice BJ, Blanchette M, Calef R, Troll CJ, Fields A, Hartley PD, Sugnet CW (2016) Chromosome-scale shotgun assembly using an in vitro method for long-range linkage. Genome Res 26(3):342–350
29. Eisenstein M (2015) Startups use short-read data to expand long-read sequencing market. Nat Biotechnol 33(5):433–435
30. Liu L, Li Y, Li S, Hu N, He Y, Pong R, Lin D, Lu L, Law M (2012) Comparison of next-generation sequencing systems. Biomed Res Int 2012:11. https://doi.org/10.1155/2012/251364
31. Branton D, Deamer DW, Marziali A, Bayley H, Benner SA, Butler T, Di Ventra M, Garaj S, Hibbs A, Huang X (2008) The potential and challenges of nanopore sequencing. Nat Biotechnol 26(10):1146–1153
32. McDougall P (2011) The cost and time involved in the discovery, development and authorisation of a new plant biotechnology derived trait. Crop Life Int
33. Goley ME, Burns WC, Huang J, MCCANN MC, Shao A, SPARKS OC, Stoecker MA, Wei L (2015) Transgenic maize event mon 87419 and methods of use thereof. EP3119186 A1
34. Fantozzi A, Ermolli M, Marini M, Scotti D, Balla B, Querci M, Langrell SR, Van den Eede G (2007) First application of a microsphere-based immunoassay to the detection of genetically modified organisms (GMOs): quantification of Cry1Ab protein in genetically modified maize. J Agric Food Chem 55(4):1071–1076
35. Roda A, Mirasoli M, Guardigli M, Michelini E, Simoni P, Magliulo M (2006) Development and validation of a sensitive and fast chemiluminescent enzyme immunoassay for the detection of genetically modified maize. Anal Bioanal Chem 384(6):1269–1275. https://doi.org/10.1007/s00216-006-0308-6
36. García-Cañas V, Simó C, León C, Ibáñez E, Cifuentes A (2011) MS-based analytical methodologies to characterize genetically modified crops. Mass Spectrom Rev 30(3):396–416
37. Terenius O, Papanicolaou A, Garbutt JS, Eleftherianos I, Huvenne H, Kanginakudru S, Albrechtsen M, An C, Aymeric JL, Barthel A, Bebas P, Bitra K, Bravo A, Chevalier F, Collinge DP, Crava CM, de Maagd RA, Duvic B, Erlandson M, Faye I, Felfoldi G, Fujiwara H, Futahashi R, Gandhe AS, Gatehouse HS, Gatehouse LN, Giebultowicz JM, Gomez I, Grimmelikhuijzen CJ, Groot AT, Hauser F, Heckel DG, Hegedus DD, Hrycaj S, Huang L, Hull JJ, Iatrou K, Iga M, Kanost MR, Kotwica J, Li C, Li J, Liu J, Lundmark M, Matsumoto S, Meyering-Vos M, Millichap PJ, Monteiro A, Mrinal N, Niimi T, Nowara D, Ohnishi A, Oostra V, Ozaki K, Papakonstantinou M, Popadic A, Rajam MV, Saenko S, Simpson RM, Soberon M, Strand MR, Tomita S,

Toprak U, Wang P, Wee CW, Whyard S, Zhang W, Nagaraju J, Ffrench-Constant RH, Herrero S, Gordon K, Swevers L, Smagghe G (2011) RNA interference in Lepidoptera: an overview of successful and unsuccessful studies and implications for experimental design. J Insect Physiol 57(2):231–245

38. Price DRG, Gatehouse JA (2008) RNAi-mediated crop protection against insects. Trends Biotechnol 26(7):393–400. https://doi.org/10.1016/j.tibtech.2008.04.004
39. Bolognesi R, Ramaseshadri P, Anderson J, Bachman P, Clinton W, Flannagan R, Ilagan O, Lawrence C, Levine S, Moar W, Mueller G, Tan J, Uffman J, Wiggins E, Heck G, Segers G (2012) Characterizing the mechanism of action of double-stranded RNA activity against western corn rootworm (Diabrotica virgifera virgifera LeConte). PLoS One 7(10):e47534. https://doi.org/10.1371/journal.pone.0047534
40. Hall JS, Usher S, Byers RJ, Higgins RC, Memon D, Radford JA, Linton KM (2015) QuantiGene Plex represents a promising diagnostic tool for cell-of-origin subtyping of diffuse large B-cell lymphoma. J Mol Diagn 17(4):402–411
41. Wong ML, Medrano JF (2005) Real-time PCR for mRNA quantitation. BioTechniques 39(1):75–88
42. Wang P-H, Schulenberg G, Whitlock S, Worden A, Zhou N, Novak S, Chen W (2018) RNase if -treated quantitative PCR for dsRNA quantitation of RNAi trait in genetically modified crops. BMC Biotechnol 18(1):3. https://doi.org/10.1186/s12896-018-0413-6
43. Vogelstein B, Kinzler KW (1999) Digital PCR. Proc Natl Acad Sci U S A 96(16):9236–9241. https://doi.org/10.1073/pnas.96.16.9236
44. Hindson BJ, Ness KD, Masquelier DA, Belgrader P, Heredia NJ, Makarewicz AJ, Bright IJ, Lucero MY, Hiddessen AL, Legler TC (2011) High-throughput droplet digital PCR system for absolute quantitation of DNA copy number. Anal Chem 83(22):8604–8610
45. Pinheiro LB, Coleman VA, Hindson CM, Herrmann J, Hindson BJ, Bhat S, Emslie KR (2011) Evaluation of a droplet digital polymerase chain reaction format for DNA copy number quantification. Anal Chem 84(2):1003–1011
46. Morisset D, Štebih D, Milavec M, Gruden K, Žel J (2013) Quantitative analysis of food and feed samples with droplet digital PCR. PLoS One 8(5):e62583. https://doi.org/10.1371/journal.pone.0062583
47. Corbisier P, Bhat S, Partis L, Xie VRD, Emslie KR (2010) Absolute quantification of genetically modified MON810 maize (Zea mays L.) by digital polymerase chain reaction. Anal Bioanal Chem 396(6):2143–2150
48. Sanders R, Mason DJ, Foy CA, Huggett JF (2013) Evaluation of digital PCR for absolute RNA quantification. PLoS One 8(9):e75296. https://doi.org/10.1371/journal.pone.0075296
49. Ogasawara A, Torimoto N, Tsuda N, Aohara F, Ohashi R, Yamada Y, Taniguchi H (2016) New screening criteria setting on evaluation of cytochrome P450 induction using HepaRG cells with multiplex branched DNA technologies in early drug discovery. Drug Metab Lett 10(3):152–160
50. Preuss SB, Meister R, Xu Q, Urwin CP, Tripodi FA, Screen SE, Anil VS, Zhu S, Morrell JA, Liu G (2012) Expression of the Arabidopsis thaliana BBX32 gene in soybean increases grain yield. PLoS One 7(2):e30717. https://doi.org/10.1371/journal.pone.0030717
51. Armstrong TA, Chen H, Ziegler TE, Iyadurai KR, Gao A-G, Wang Y, Song Z, Tian Q, Zhang Q, Ward JM, Segers GC, Heck GR, Staub JM (2013) Quantification of transgene-derived double-stranded RNA in plants using the quantigene nucleic acid detection platform. J Agric Food Chem 61(51):12557–12564
52. Kulkarni MM (2011) Digital multiplexed gene expression analysis using the nanostring nCounter system. Curr Protoc Mol Biol Chapter 25B:Unit25B.10. https://doi.org/10.1002/0471142727.mb25b10s94
53. Cesano A (2015) nCounter® PanCancer immune profiling panel (NanoString technologies, Inc., Seattle, WA). J Immunother Cancer 3(1):42. https://doi.org/10.1186/s40425-015-0088-7
54. Geiss GK, Bumgarner RE, Birditt B, Dahl T, Dowidar N, Dunaway DL (2008) Direct multiplexed measurement of gene expression with color-coded probe pairs. Nat Biotechnol 26:317. https://doi.org/10.1038/nbt1385
55. Nielsen T, Wallden B, Schaper C, Ferree S, Liu S, Gao D (2014) Analytical validation of the PAM50-based prosigna breast cancer prognostic gene signature assay and ncounter analysis system using formalin-fixed paraffin-embedded breast tumor specimens. BMC Cancer 14:177. https://doi.org/10.1186/1471-2407-14-177
56. Liu L, Mayes PA, Eastman S, Shi H, Yadavilli S, Zhang T (2015) The BRAF and MEK inhibitors dabrafenib and trametinib: effects on immune function and in combination with

immunomodulatory antibodies targeting PD1, PD-L1 and CTLA-4. Clin Cancer Res 21:1639–1651. https://doi.org/10.1158/1078-0432.ccr-14-2339

57. Reis PP, Waldron L, Goswami RS, Xu W, Xuan Y, Perez-Ordonez B, Gullane P, Irish J, Jurisica I, Kamel-Reid S (2011) mRNA transcript quantification in archival samples using multiplexed, color-coded probes. BMC Biotechnol 11(1):46. https://doi.org/10.1186/1472-6750-11-46
58. Beard RE, Abate-Daga D, Rosati SF, Zheng Z, Wunderlich JR, Rosenberg SA, Morgan RA (2013) Gene expression profiling using nanostring digital RNA counting to identify potential target antigens for melanoma immunotherapy. Clin Cancer Res 19(18):4941–4950
59. Moreno LI, Tate CM, Knott EL, McDaniel JE, Rogers SS, Koons BW, Kavlick MF, Craig RL, Robertson JM (2012) Determination of an effective housekeeping gene for the quantification of mRNA for forensic applications. J Forensic Sci 57(4):1051–1058
60. Zhang K, Raboanatahiry N, Zhu B, Li M (2017) Progress in genome editing technology and its application in plants. Front Plant Sci 8:177. https://doi.org/10.3389/fpls.2017.00177

Chapter 25

Detection of Transgenic Proteins by Immunoassays

Satyalinga Srinivas Gampala, Bryant Wulfkuhle, and Kimberly A. Richey

Abstract

Rapid development and global cultivation of genetically modified (GM) crops necessitated the use of analytical tools that detect GM crops throughout the product life cycle. Antibody-based immunoassays such as enzyme-linked immunosorbent assays (ELISA) are instrumental in detecting protein expression in transgenic plants. These analytical tools are used throughout development, regulatory registration, commercialization, and stewardship of biotech products. Here we describe the Cry1F ELISA analytical method protocol, data generation and review, and troubleshooting of technical challenges.

Key words Monoclonal and polyclonal antibodies, Immunoassays, Sandwich ELISA, Transgene expression, Transgenic proteins, Genetically modified (GM) crops, Cry1F

1 Introduction

Commercialization of genetically modified crops will continue at a fast pace using biotechnology-derived traits, and there is a significant need for developing and implementing appropriate detection tools [1–4]. Immunoassays, and in particular enzyme-linked immunosorbent assays (ELISAs), have served as primary protein detection tools for scientists in agricultural biotechnology [5–8]. Sandwich ELISA involves binding of antigen (transgenic protein expressed in plants) to target-specific capture antibodies followed by creation of an immuno-sandwich complex using target-specific enzyme-labeled detection antibodies. The immunocomplex is subsequently detected using spectrophotometric methods [5–8]. Each ELISA is different, and an example of performing the Cry1F assay is described in this chapter.

2 Materials

2.1 Materials/Supplies

1. Bead, 1/8″ chrome steel.
2. Cap, for 2.0 mL conical tube.

Sandeep Kumar et al. (eds.), *Transgenic Plants: Methods and Protocols*, Methods in Molecular Biology, vol. 1864, https://doi.org/10.1007/978-1-4939-8778-8_25,

3. Multichannel pipette, 12-channel, 10–300 μL.
4. Pipette tips, various sizes.
5. Plate covers or equivalent.
6. Reagent reservoirs, non-sterile.
7. Single-channel pipettes of various sizes, 10 μL–1.0 mL.
8. Tubes, polypropylene, 5 mL.
9. Tube, 15 mL polypropylene centrifuge with cap.
10. Tube, 2.0 mL conical microcentrifuge.
11. U-bottom plates, nonbinding 96-well.

2.2 Equipment (See Note 1)

1. Balance, analytical, Model AB54-S, Mettler Instrument Corporation or equivalent.
2. Centrifuge, capable of holding 2 mL Eppendorf tubes, Eppendorf–5417C or equivalent.
3. Freezer, capable of maintaining −20 °C, Model 75F, U-Line Corporation, Milwaukee, WI, or equivalent.
4. Plate reader, capable of reading 450 nm, Molecular Devices, catalog no. 0200-2018, or equivalent.
5. Refrigerator, capable of maintaining temperature at 2–8 °C.
6. Vortex, Genie-2 Model, catalog number 12-812, Fisher Scientific, or equivalent.
7. Shaker/Grinder, Model Geno-grinder, catalog number 2000-115, Certiprep, Metuchen, New Jersey, or equivalent.
8. Washer, 96-well microplate, Model Elx 405, Bio-Tek Instruments, Inc., or equivalent.

2.3 Reagents (See Notes 2–4)

1. Microtiter Plate Cry1F ELISA Test Kit (commercially purchased from kit vendors): Store test kits at 2–8 °C. Typically kits include antibody-coated 96-well microtiter plates, Cry1F Antibody HRP-Conjugate (lyophilized), conjugate diluent, color reagent, stop solution, extraction buffer PBST (PBS Tween, pH 7.4, Sigma P-3563), and Cry1F Microtiter Plate ELISA Assay User's Guide.
2. Protein standard stock solution: It may need to be aliquoted and stored in freezer (such as −20 °C). Prepare a 1000 ng/mL stock solution based on the liquid standard concentration in PBST. Keep it in ice to be used within 2 h. Discard if any visible contamination is observed. Discard the remainder protein standard stock; do not refreeze it for reuse.

3 Methods

3.1 Handling of ELISA Kit

1. Bring ELISA kit reagents to 20–25 °C by removing them from the refrigerator at least 30 min prior to performing the assay. Ensure that the kit has a certificate of analysis and met the provider's quality control criteria.
2. ELISA kits typically have all the necessary reagents in solution. If not, follow the kit insert instructions to prepare solutions. For example, reconstitute the lyophilized protein antibody conjugate by dissolving the lyophilized powder in the vial with suggested volume of the conjugate diluent.

3.2 Preparation of Protein Standards (See Note 5)

1. Prepare protein standards in 5 mL polystyrene tubes as described in Table 1.
2. Vortex each dilution of protein standard for a few seconds to mix well before transferring to the next dilution.
3. Store protein standards on ice; prepare new standards for each assay immediately before adding samples to the ELISA plate.

3.3 Preparation of Test Samples (See Notes 6 and 7)

1. Extract the samples (2 mL tube containing 15 mg of lyophilized tissue powder in a 1.5 mL of extraction buffer) using the Geno/Grinder automatic shaker/grinder at a dial setting of 500 and the toggle switch at the 1× setting (approximately 1500 strokes per minute) for 3 min as one cycle. An alternative equivalent grinding or extraction method may be used.
2. Centrifuge the samples at 2000 × *g* (or greater) rpm for 5 min or until separated (no visible particles in the supernatant). The supernatant can be transferred to a separate tube or aliquoted for analysis. Keep the solution on ice and assay it within 2 h.

3.4 Addition of Protein Samples to ELISA Plate(s)

1. For Cry1F ELISA, use the simultaneous ELISA assay format which involves simultaneous addition of samples and antibody HRP-conjugate to pre-coated binding plates.
2. Transfer the ELISA standard dilutions to columns 1–3 on a non-binding 96-well u-bottom microtiter plate (approximately 130 μL/well). For each plate tested, run standard solutions in triplicate to ensure that the plate passes the acceptance criteria (*see* **Notes 8** and **9**).
3. Prepare sample dilutions as needed and transfer diluted samples to the non-binding 96-well microtiter plate (130 μL/well) containing the standard calibration solutions and record the location on the 96-well assay template sheet.
4. Pipet 100 μL of the protein antibody conjugate to each well of the antibody-coated 96-well microtiter plate.

Table 1
Example preparation of the protein standard curve

Concentration of stock solution (ng/mL)	Aliquot of stock solution (μL)	Starting buffer volume (μL)	Final solution volume (μL)	Final standard concentration (ng/mL)	Remaining volume after aliquot[a] (μL)
1000	15	1485	1500	10	540
10	960	240	1200	8	450
8	750	250	1000	6	450
6	550	330	880	3.75	480
3.75	400	350	750	2	430
2	320	320	640	1	440
1	200	200	400	0.5	400
0	0	500	500	0	500

[a]The final solution volume is the remaining volume in the container after it has served as the stock solution for the next standard concentration and the relevant amount of solution is transferred

5. Transfer 100 μL of the ELISA standard solutions and diluted samples from the u-bottom microtiter plate to a pre-coated plate, keeping the same orientation as the samples are transferred to the pre-coated plate. Change pipette tips with each row.
6. Cover the pre-coated plate, and gently swirl on the benchtop or a plate shaker for approximately 5 s to mix. Allow to incubate at ambient temperature for 1 h (± 5 min) (*see* **Note 10**).

3.5 Washing and Stopping of ELISA Reaction

1. Wash the pre-coated plate 3–5 times by filling each well with PBST. Tap out excess liquid on a paper towel. It may be washed by plate washer.
2. Add 100 μL of the TMB substrate to each well of the ELISA plate. Cover and gently mix. Allow to incubate at ambient temperature in the dark for 15 ± 2 min.
3. Add 100 μL of stop solution to each well to stop the reaction. Mix the plate gently and read the absorbance at 450 nm using the plate reader.
4. Save the raw data file and perform the data analysis.

3.6 Data Analysis and Calculations

3.6.1 Calibration Curve

Use absorbance values from the reference standards to develop a calibration curve (*see* **Note 11**). For data analysis and calculation, use either SOFTmax PRO software or Microsoft Excel. The calibration curve for the protein ELISA is constructed fitting a quadratic curve of the expected concentrations of the standards and their subsequent absorbance (optical density) using regression (*see* **Note 12**).

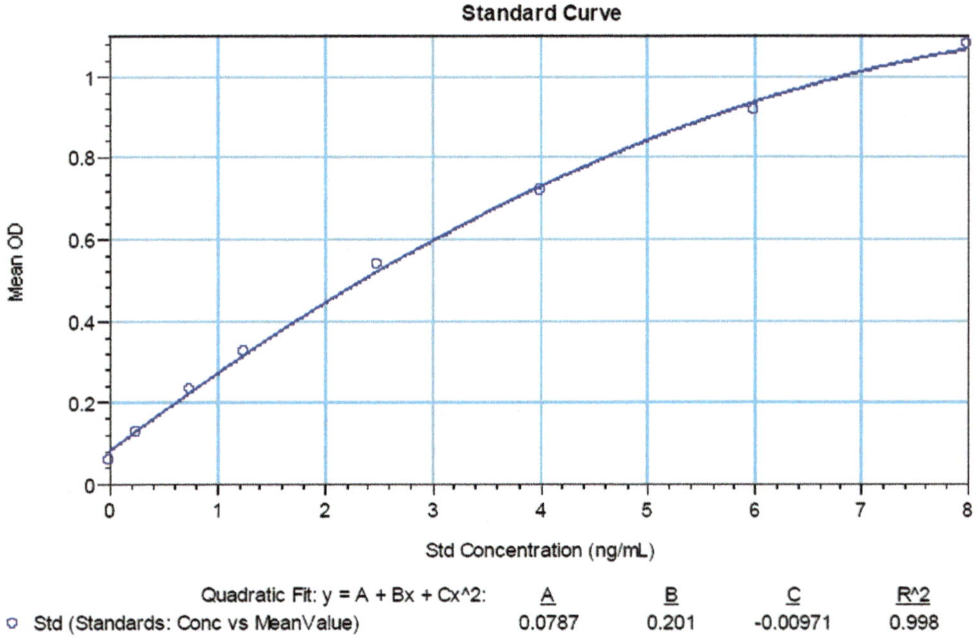

Fig. 1 An example for utility of quadratic equation for determining the protein concentration

Example of such methodologies for data analysis is included in the following graph (Fig. 1). Data analysis can also be adapted for throughput and specific applications (*see* **Notes 13** and **14**).

3.6.2 The Equation Fits the Best Parabola to the Standard Curve Based on the Quadratic Equation

$$y = A + Bx + Cx^2$$

where y = mean absorbance value (OD), and x = reference standard concentration.

3.6.3 Calculation of Protein in Test Samples

Use SOFTmax PRO or Microsoft Excel to calculate the protein concentration of each test sample. The predicted concentration is determined using the coefficients of the curve and optical density (OD) readings in the quadratic equation. The regression equation is applied as follows:

$$\text{Predicted concentration} = \frac{-B + \sqrt{B^2 - 4C^*(A - \text{OD})}}{2C}$$

Obtain the estimated protein concentration of a test sample (i.e., the individual replicates of a single dilution) by multiplying the predicted concentration by the dilution factor used.

Table 2
Criteria for acceptance of an analytical batch

0 ng/mL standard	Absorbance 450 nm < 0.100
10 ng/mL Cry1F standard	Absorbance 450 nm ≥ 1.00
Calibration curve	r^2 (correlation of determination) >0.990
All positive reference standards, OD	CV (OD) of triplicates ≤15%
Unknown or QC samples, solution	CV (OD) of replicates ≤20%
Quality control samples, solution (if applicable)	Measured value ≤ ± 20% expected value

3.6.4 Criteria for Acceptance of an Analytical Batch

Each analytical batch shall meet the acceptance criteria in the procedure to be valid (some representative criteria are listed in Table 2). If the data fail to meet these performance criteria, the analyst should evaluate the results, determine the potential source of the variation, and repeat the analysis if necessary.

4 Notes

1. Ensure that periodic maintenance is done for all instruments including plate washers, plate readers, Geno-grinders, etc., and check to make sure that instruments passed the acceptance criteria prior to running ELISA. Pipettes (single channel and multichannel) should be calibrated prior to use.
2. Safety precautions must be taken during handling of the reagents and all institutional policies must be followed to ensure safe disposal of the reagent waste.
3. ELISAs are temperature sensitive and using hot or cold reagents can influence the final results. Hence, bring ELISA reagents (kits) to room temperature prior to use.
4. ELISA kits may come with a lyophilized conjugate to increase shelf life and reduce kit size. Proper reconstitution of the conjugate will help with assay success.
5. Each ELISA kit has an established quantitative range and it can vary widely from protein to protein. Protein standard is diluted to produce a standard curve appropriate to the given kit. Proper mixing of each step of the standard curve will allow for accurate dilutions. Proteins are temperature sensitive so these dilutions should be done on ice.
6. Sample to extraction buffer volume can be optimized by the user. In general for a lyophilized sample size range of 5–30 mg, a buffer volume of 500–3000 μL along with 2–4 metal beads and 1–3 Geno-grinder cycles would work. Make sure to use

appropriate tube/vial to accommodate the needed buffer volume.

7. Plant samples offer a variety of challenges such as difficulty with grinding and separation of extract supernatant. Check to see that samples are well ground, and the supernatant (without debris) is transferred to a 96-well plate for running the assay.
8. If multiple plates are being run, include standard curve on each plate to account for plate-to-plate variability. Watch for analyst-to-analyst variation, plate edge effect, and hook effect during analysis of high-expressing plant samples.
9. Analysis of each sample in duplicates is recommended to minimize variability.
10. Based on geographical location and ambient lab temperature, substrate incubation time can be adjusted (higher temperature requires lower incubation time and lower temperature requires higher incubation time).
11. Ensure that the absorbance of the highest reference standard does not fall outside of the linear range of the spectrophotometer (plate reader). Matrix effects are observed with plant tissues, and appropriate dilutions are recommended to achieve accurate measurement of transgenic proteins.
12. For data analysis, either quadratic or four-parameter curve-fit methodologies can be applied.
13. ELISA workflow can be optimized for medium- and high-throughout applications based on proximity of instrumentation and/or utility of liquid handlers as part of automation.
14. If needed, quantitative ELISA kits can be used for qualitative (yes/no answer) analysis of samples.

References

1. Kamle S, Ali S (2013) Genetically modified crops: detection strategies and biosafety issues. Gene 522:123–132
2. Parisi C et al (2016) The global pipeline of GM crops out to 2020. Nat Biotechnol 34:31–36
3. James C (2016) Global Status of Commercialized Biotech/GM Crops: 2016. ISAAA Brief No. 52, International Service for the Acquisition of Agri-biotech Applications, Ithaca, NY
4. Kamle et al (2017) Current perspectives on genetically modified crops and detection methods. 3 Biotech 7:219
5. Grothaus et al (2006) Immunoassay as an analytical tool in agricultural biotechnology. J AOAC Int 89:914–928
6. Shan G (ed) (2011) Immunoassays in agricultural biotechnology. Wiley, New York
7. Lipton et al (2000) Guidelines for the validation and use of immunoassays for determination of introduced proteins in biotechnology enhanced crops and derived food ingredients. Food Agric Immunol 12:153–164
8. Keith et al (1983) Principles of environmental analysis. Anal Chem 55:2210–2218

Chapter 26

Systematic Evaluation of Field Crop Performance Using Modern Phenotyping Tools and Techniques

Christopher R. Boomsma and Vladimir A. da Costa

Abstract

The genetic improvement of field crops through plant breeding and genetic modification is highly dependent on understanding, measuring, selecting, and manipulating phenotypes. Most phenotypes result from the complex interaction of a crop's genetics with the environment and management practices in which that crop is grown. Linking gene to phenotype in field environments to create superior crop varieties can therefore be challenging, particularly for genetically complex traits that are difficult to measure. This chapter is designed to help readers overcome these difficulties by describing tools and techniques used in successful crop improvement programs. It provides methodologies that can be broadly applied across numerous situations irrespective of field crop, environment, modest financial resources, or other factors. The chapter's focus is primarily on small- and large-scale, replicated, research plot-based screening trials since these trials are crucial, ubiquitous, and costly for both public- and private-sector crop improvement programs. To ease the understanding of the protocols discussed, this chapter's materials and methods section is composed of ten subsections, with each subsection covering a critical portion of the field crop phenotyping process: regulatory, environmental, and safety considerations; trait identification and prioritization; environment characterization; field site selection; experimental design; field design, preparation, and management; crop and soil measurements; environmental monitoring; in-field data recording; and data management and analysis.

Key words Field crop, Environment characterization, Field selection, Experimental design, Research plot, Phenotyping, Secondary trait, Remote sensing, Unmanned aerial vehicle

1 Introduction

The improvement of major field crops such as maize (*Zea mays* L.), wheat (*Triticum aestivum* L.), rice (*Oryza sativa* L.), and soybean [*Glycine max* (L.) Merr.] through genetic methods (e.g., conventional and marker-assisted breeding, transgenesis, cisgenesis, genome modification) is often contingent upon the evaluation of these crops in a field setting. Such work is sometimes part of a larger performance evaluation pipeline composed of growth chamber, greenhouse, managed stress environment (MSE), replicated research plot and strip plot trials, and target population of

Sandeep Kumar et al. (eds.), *Transgenic Plants: Methods and Protocols*, Methods in Molecular Biology, vol. 1864,
https://doi.org/10.1007/978-1-4939-8778-8_26,

environment (TPE) screening [1]. The tools and techniques described in this chapter focus primarily on small- and large-scale, replicated, research plot-based screening trials since these trials are crucial, ubiquitous, and costly for both public- and private-sector crop improvement programs. Methodology improvements in these field screening efforts can therefore have considerable financial and scientific impact on crop improvement programs, thereby hastening the rate of genetic advancement of the world's major field crops.

Discussion of specific methodologies for effective and efficient replicated strip trials and multi-region TPE screening is beyond the scope of this chapter, though many of the methodologies discussed herein can be applied to these field research programs. For more information on proper tools and techniques for large-scale replicated strip trials and multi-region TPE screening, readers should consult articles by [1–4] and others. While substantial benefits can come from employing high-quality controlled environment screening programs, methodologies for these crop evaluation systems will not be reviewed in this chapter since such tools and techniques are less frequently employed in crop improvement programs and are already discussed at length by others [5].

Phenotyping technologies and techniques are integral to most small- and large-scale, replicated, research plot-based screening trials. Key phenotyping methodologies and example technologies are therefore discussed in various parts of this chapter. Readers should note that the phenotyping tools and techniques mentioned are broadly applicable examples commonly found in the existing phenotyping literature and routinely employed by public- and private-sector research groups [1, 6]. Any actual or implied reference to an organization or technology is neither an endorsement nor criticism of the stated party or technology.

Proper evaluation of crop performance in a field research setting can be performed using a wide variety of tools and techniques. Often specific methodologies vary by crop, location, research group (e.g., company), financial resources, research personnel experience and training, and numerous other factors. It is therefore impossible to prescribe highly detailed materials and methods without providing an onerously long chapter replete with examples and caveats covering numerous, complex screening situations. This chapter resultantly provides methodologies that can be broadly applied across numerous situations regardless of crop, environment, modest financial resources, or other factors. To ease the understanding of the protocols discussed, this chapter's materials and methods section is composed of ten subsections that highlight key items a researcher should consider when systematically planning and implementing a field crop performance evaluation trial involving the use of modern phenotyping tools and techniques: regulatory, environmental, and safety considerations; trait identification and prioritization; environment characterization; field site selection; experimental design; field design, preparation, and management; crop and soil measurements; environmental monitoring; in-field data recording; and data management and analysis. References are provided for readers that desire greater conceptual and/or technical detail.

2 Materials and Methods

2.1 Regulatory, Environmental, and Safety Considerations

1. It is critical to examine all applicable government regulations related to the design and execution of field crop performance evaluation trials for the crop of interest. Such an examination should be performed for all countries and other government administration divisions (e.g., states, provinces) in which the research trial is to occur (*see* **Note 1**). If working with regulated research trial inputs (e.g., experimental transgenic seed), carefully explore laws and regulations related to seed movement and packaging, weed control, crop destruction, equipment cleanup, and volunteer crop monitoring. Strictly follow all regulations related to the storage, movement, and use of pesticides. Lastly, consider and follow any regulations imposed by the organization performing the research (e.g., private-sector seed company), as some organizations impose regulations that exceed the standards required by government agencies.
2. Give careful consideration to the environmental impacts of the field crop performance evaluation trial. Consider impacts on human health, animal life, insects, native plant species, above- and below-ground water resources, and soils. Take necessary measures to protect all environmental resources in and around the research trial in accordance with crop management best practices and government rules and regulations. This should include following product (e.g., herbicide, insecticide, fungicide) label directions, minimizing pesticide drift and off-target damage, and limiting fertilizer and irrigation overapplication. The erection of physical barriers around research trials (e.g., electrified fences) can be used to protect the research trial from animal foraging and limit the destruction of wild or domesticated animals that may threaten the trial. Properly placing a research trial within a landscape can help minimize negative impacts on and interactions with humans, animals, beneficial insects, native plants, soil, and water resources.
3. Strong emphasis should be placed on the health and safety of all individuals involved in the execution of the field crop performance evaluation trial. Workers should be provided with all necessary personal protective equipment (PPE). The equipment should be task specific and properly fitted. Standard

operating procedures (SOPs) should be drafted, reviewed, disseminated, employed, and enforced for all research trial field activities. All government and organizational regulations related to occupational hazards and worker safety should be strictly followed.

2.2 Trait Identification and Prioritization

When planning a field crop performance evaluation trial, substantial effort must be placed on identifying what phenotypic traits need to be measured to meet the objectives of the trial and the larger crop improvement program. Once this is complete, all other activities described in the following subsections can commence.

1. Both primary and secondary traits should be identified during this phase. Measurement of the primary trait of interest (often grain yield) should be performed when possible and appropriate (*see* **Note 2**). When this is not advisable, one or more secondary traits should be measured.
2. Considerable effort should be expended on prioritizing secondary trait measurements since these measurements are often time consuming, laborious, and costly (*see* **Note 3**). Resources can be easily wasted if secondary trait measurements are not carefully planned. Selection and prioritization of secondary traits should be based on a variety of factors including the reason for the measurement, the advantages and disadvantages of the associated measurement tool(s), the cost of the tool (s) and accompanying labor, and the time for field and laboratory data collection and processing. A variety of tables modeled after the examples provided by [7] should be constructed to help prioritize, schedule, and efficiently resource secondary trait measurements (*see* **Note 4**). These tables should be constructed for each trial and should account for measurement activities throughout the entire growing season. If multiple trials are being simultaneously performed within a larger crop improvement program, it is prudent to compare tables across trials, assess the availability of time and resources, adjust the objectives and scope of each trial, and resultantly prioritize and alter time and resource allocations among the trials so that the entire research program is run efficiently and effectively. Many secondary trait measurements are likely to occur during those developmental stages that highly impact the primary trait of interest (*see* **Note 5**). Considerable attention should be given to these periods when identifying and prioritizing secondary trait measurements.

2.3 Environment Characterization

Before establishing and performing a small- or large-scale, replicated, research plot-based screening trial, the environment (*see* **Note 6**) in which a trial may be located must be characterized and well understood.

1. Climate and soil characteristics must be researched and documented. Such characteristics include, but are not limited to, average annual, seasonal, and monthly rainfall; average monthly low and high temperatures; heat accumulation trends; typical spring and fall frost and freeze dates; and predominant soil series and associated soil properties. Depending on the country in which the trial is to be located, these data may be freely available. In other instances, the data may need to be gathered over one or more years using weather stations, remote sensing technology, soil sampling, and data sharing agreements. Whatever the method utilized for capturing environmental data, it is critical that the information used is from a reputable source and of high quality (*see* **Note** 7).
2. Once gathered and organized, it must be determined if the environmental characteristics permit proper growth and development of the crop; effective use of proper field management equipment and techniques; timely, precise, and repeatable employment of the trial's treatment types [e.g., nitrogen (N) fertilization rate, water stress level]; and effective measurement and characterization of the trait or traits of interest. If these criteria are not met, the environment is not suitable for the field trial and a different environment must be utilized.

2.4 Field Site Selection

Once an environment has been characterized and deemed acceptable, it is critical to choose field sites within that environment that are suitable for field crop performance evaluation trials (*see* **Note 8**). Selection of these sites is a multistep process.

1. An extensive list of site selection factors should be drafted. These factors should include agronomic, environmental, soil, and social attributes. Examples include, but are not limited to, previous crop rotation, prior tillage and fertilization activities, latitude, soil type, soil organic matter and pH, soil texture, multi-property soil variability, topography, irrigation water availability, water table depth, site accessibility and privacy, landowner and/or tenant equipment availability and quality, and landowner and/or tenant attitude. It is important to choose factors that are pertinent to the environment in which the field site is located and supportive of the overall crop improvement program's objectives.
2. A scoring system should be developed using a simple numerical scale that ranges from completely unacceptable to ideal. The same numerical scale should be used for each factor. When developing this scoring system, it is imperative to define what constitutes unacceptable and ideal values for each factor. Intermediate values should also be defined (*see* **Note 9**).

3. Each factor should be scored based on available information. If information for a particular factor is not available, pertinent measurements should be taken, an alternative but related factor should be used, or the factor should simply be discarded. A variety of tools and techniques can be used to assess site suitability when existing data are not sufficient. These include, but are not limited to, historical soil maps, zone- and grid-based soil sampling techniques, soil electroconductivity measurements, aerial and satellite imagery, grain yield maps, farmer records and questionnaires, and visual observations (*see* **Note 10**).
4. The scores should be tallied and an objective rating for each field site should be generated.

2.5 Experimental Design

The next logical step in planning and implementing a field crop performance evaluation trial is choosing and implementing an experimental design. There are many design options for a trial including, but not limited to, a complete randomized design, randomized complete-block design, incomplete-block design, alpha lattice design, row-column design, or augmented design [8, 9]. Choosing the proper experimental design for a trial is both crucial and difficult. Proper selection requires the consideration of a number of factors including field space and dimensions, project goals and resources, soil variability patterns, management patterns, and phenotyping plans. It is important to collaborate with a statistician on this step before beginning a trial. A statistician can help maximize the efficient use of field space and financial resources. Furthermore, he/she can help design the trial so that it answers key experimental questions while minimizing unwanted impacts.

2.5.1 Design Selection

The design selection process involves some critical steps that should be followed [9].

1. Define what is being investigated and measured.
2. Identify treatment factors and their respective levels.
3. Assign treatment factors and their respective levels to the experimental units (e.g., genotype).
4. Identify and adjust for other sources of variation besides treatment factors or experimental error.
5. Plan how the data should be analyzed.

2.5.2 Key Design Principles

A number of key principles should guide the design process.

1. Replication, randomization, and blocking should be frequently and properly used (*see* **Note 11**).
2. The experimental design must consider whether a trial is an early- or late-stage field screening effort (*see* **Note 12**). If it is an

early-stage trial, it may be advisable to use fewer replications and locations and more genotypes. If it is a late-stage trial, it may be prudent to use more replications and locations and fewer genotypes.

3. Plot-to-plot competition (i.e., phenotypic interference or interaction between neighboring research plots) and alley effects should be minimized. This is done by insuring that genotypes that are phenotypically similar are grown adjacent to each other and by removing plants along plot alleys during mid-season growth or making the width of trial alleys as small as possible.
4. If a research trial includes complex treatment factors such as water or N stress, it is important to make every possible effort to keep experimental error as low as possible. Under these stress conditions, genotype differences are frequently small, heritability is often low, genotype by environment interactions are commonly pronounced, and field variability is magnified.

2.5.3 Treatments

A key aspect of experimental design is properly imposing the desired treatment factors and their respective levels. Items to consider when designing and administering treatments vary by treatment, but a few key principles should be noted and applied when imposing nongenetic treatments (e.g., N stress, water stress).

1. Treatment factors and levels should be chosen that have good potential to properly test the trial's hypotheses.
2. Treatments should be uniformly applied at the proper time, location, and rate with the correct product.
3. Application equipment should be in good working order and well calibrated.
4. Buffer rows should be placed between research plots with different treatment levels if "bleeding over" of the applied material could potentially occur.

2.5.4 Plot Design and Identification

An often-overlooked component of experimental design is research plot design and identification. When designing a plot, it is crucial to consider the plot row number, plot inter-row spacing, plot row length, and alley number and width. These metrics are often dictated by research trial stage and objectives, land availability, treatment buffer needs, people and equipment movement demands, crop damage and soil compaction potential, sensor and other measurement equipment requirements, and measurement numbers. Each of these items should be carefully assessed when designing research plots. During the growing season, it is critical to clearly and accurately identify all research plots. This can be done using handwritten tags or stakes, barcoded tags or stakes, radio-frequency identification (RFID) tags, and other methods. Plot identification should be done early in the growing season and tags, stakes, and other plot identification materials should be maintained throughout the trial.

2.6 Field Design, Preparation, and Management

Field design is key for a successful small- or large-scale, replicated, research plot-based screening trial. When designing a field for hosting a trial, it is important to consider a number of factors.

2.6.1 Field Design

1. Note the number of blocks and research plots and their respective sizes. Make sure that the field site can fit the research trial when using only high-quality land. If this is not possible, the trial may need to be split and either spatially separated at a site or placed at multiple sites. If splitting a trial is required, it is recommended to not split apart individual blocks.
2. Take note of and adjust for any natural and/or artificial impediments of research and general field activities. Consider the arrangement of blocks and research plots so they are not positioned on or near obstacles such as rocky areas, field roads, and drainage ditches. Pay particular attention to overhead and buried electrical lines since they can have an adverse impact on equipment operation [e.g., unmanned aerial vehicle (UAV) flights, soil tillage].
3. Insure adequate accessibility for and proper movement of people and equipment. Design the field so that paths are available for the movement of workers, measurement equipment, and field management tools. This is crucial for limiting damage to crops and compaction of soil in research plots. However, keep paths to a minimum to limit wasted field space and compacted soil.
4. Make certain that the field design provides necessary infrastructure and safety equipment for the well-being of workers. This often includes shaded areas, water and lavatory accessibility, and safety information posting sites.
5. Make certain that the field design provides the security and isolation necessary for the trial. This is particularly important when screening regulated research trial inputs (*see* Subheading 2.1). Adhering to government regulations is particularly important in this situation.
6. Adjust the field design to account for issues that could arise from the presence and features of neighboring properties. Extreme care should be taken to prevent herbicide drift onto adjacent properties. Trees are an important aspect to consider since their root systems may negatively interfere with nearby trials and their foliage can cause temporary to permanent shading of research plots.
7. Determine wildlife awareness of and accessibility to the area. If wildlife is likely to damage the crop, take appropriate measures (*see* Subheading 2.1).

8. Arrange the field so it is possible to impose the desired treatments. Potential items to consider include the placement of irrigation lines, the coverage pattern of overhead irrigation, and the possible pathways of fertilizer or sprayer equipment. If treatments are likely to be placed in the same location over multiple years, insure precise and consistent placement using staking and flagging equipment or global positioning system (GPS), automatic guidance, and other mapping and precision agriculture technologies.

Once all of these items are considered and addressed, the trial and its associated experiments, blocks, treatments, research plots, and other features can be mapped out with the aid of computer software (*see* **Note 13**) and modern precision agriculture technologies (e.g., GPS, satellite imagery).

2.6.2 Field Preparation

Prior to the start of a research trial, the land on which a trial is to be placed must be properly prepared. Many techniques are used to prepare the land and it is not possible in this chapter to discuss all options. However, a few items are worth mentioning.

1. A proper crop rotation must be established before the start of a trial. In some instances, a diverse crop rotation should be employed to break down any present or potential weed and pest infestations. In other instances, the use of nonleguminous continuous cropping accompanied with full biomass removal is required over multiple years. This, for example, can enable the drawdown of soil N in preparation for a N stress screening trial. The proper crop rotation therefore varies based on the objectives and hypotheses of a trial.
2. Weed problems must be controlled prior to the establishment of a trial. Weed seed banks must be drawn down over one or more years using herbicides and other methods of weed control.
3. Land leveling may be necessary for high-value, long-term research locations that involve drip irrigation. The ideal slope should ideally be less than 1% after leveling activities.

2.6.3 Field Management

When managing a research trial, it is critical to:

1. Manage the trial for maximum spatial uniformity and minimal error. This includes uniform tillage, pre- and in-season irrigation, fertilization, and harvest residue distribution.
2. Control all weeds and pests in a timely and appropriate manner.
3. Insure good stand establishment through the use of proper planter equipment and settings, appropriate tillage activities, and, if needed, uniformly applied pre-irrigation.

4. Make all nutrients non-limiting except when nutrient stress is a treatment factor.
5. Minimize field traffic and use controlled traffic methods.
6. Keep accurate, multi-year records (print or digital) of all crop management activities performed before, during, and after the growing season.

Clearly management of a trial before, during, and after the growing season is extremely important and requires a deep knowledge of agronomy, entomology, plant pathology, soil science, weed science, and various other allied disciplines. Having personnel that have the requisite skills in crop management to successfully manage a research trial is key to a trial's success. Thus readers should consult the vast body of crop management literature or hire a consultant before beginning a research trial.

2.7 Crop and Soil Measurements

Crop and soil measurements are a key focus for most crop improvement programs. Crop measurements in the field are often dictated by the primary and secondary traits of interest (see Subheading 2.2). Soil measurements are often used to better understand observed crop phenotypes, assess treatment impacts (e.g., N fertilization) and genotype differences, examine genotype by environment interactions, and guide crop management decisions.

2.7.1 Key Measurement Principles

While a comprehensive examination of potential crop and soil measurements and their associated protocols is beyond the scope of this chapter, it is important to mention some key plant sampling guidelines along with a number of basic concepts central to the proper measurement of crop phenotypes [10, 11].

1. All crop measurements should be taken as precisely, accurately, and consistently as possible using SOPs. While measurement errors and worker fatigue can be expected, substantial effort should be taken to minimize these problems. SOPs should be developed for all measurements and all workers (i.e., observers, operators, and recorders) should be trained on these SOPs.
2. Workers should have good familiarity with all tools and techniques before taking any measurements and should know how to carefully navigate research plots with sensitive or bulky equipment.
3. All workers should understand what the expected range of values is for a given plant measurement so that erroneous measurements can be quickly detected and issues can be remedied.
4. Workers should be diligent in checking that each measurement is taken in the proper plot and at the correct crop developmental stage and time of day (*see* **Note 14**).

5. Workers should not change measurement duties when taking measurements within a block or replication.
6. In nearly all circumstances, workers should take more than one measurement per research plot. The sample size for any given measurement should be chosen based on available labor and financial resources, equipment availability and throughput, research objectives, and statistical guidance.
7. Depending on the trait of interest and trial objectives, measurements can be taken from a designated location within a research plot or in a random fashion within each plot.
8. Adequate buffers should be left between in-season sampling areas. Measurements should be avoided in buffer plots or rows, in unrepresentative areas (e.g., areas impacted by flooding or abnormally high disease incidence), or on alley-adjacent plants.

2.7.2 Key Equipment Principles

Given the plethora of technologies available for crop phenotyping, proper operation of measurement equipment (e.g., digital caliper, leaf area index meter) is key for phenotyping plants in a field trial and thus requires a key set of guiding principles [12, 13].

1. All equipment should be calibrated and in good working order before entering the field trial and equipment should be cleaned and serviced upon exiting the field trial.
2. All operators should be fully trained and proficient with the equipment before taking measurements and should insure that data are recording properly while measurements are being taken.
3. When possible workers should not change equipment or alter equipment settings within a block or replication.
4. Equipment should never be operated outside of its environmental specifications and sensitive equipment should never be left in direct sunlight, rain, or other environmental hazards.

2.7.3 Proximal and Remote Sensing

The use of proximal and remote sensing techniques is now common in research plot-based screening trials. This popularity arises from the many benefits of these methods including the potential for high-throughput phenotyping of various diverse traits (e.g., green biomass, photosynthetic canopy size, chlorophyll and carotenoid concentrations, leaf water content, grain yield), the availability of equipment for nearly any size budget, the rapid pace of technological advances in sensor and platform capabilities, and the ability for swift, remote scouting of crop health and development and treatment efficacy. While a review of all instrumentation and techniques used for proximal and remote sensing is not possible in this chapter, some key principles apply to many situations in which such instrumentation is used for crop phenotyping [14–16].

1. It is important to use the proper spectroradiometer, fiber-optic cable (if necessary), lens, and accessory equipment for the trial based on prior pertinent research and manufacturer recommendations.
2. Workers should fully understand the capabilities and limitations of all equipment. Equipment manufacturers often provide this information, so consulting operator manuals is strongly encouraged.
3. Workers should employ the proper techniques for the equipment, experimental design, crop, and trait of interest (including calibration activities). Information on these techniques can easily be found in the large body of scientific literature dedicated to the subject.
4. It is prudent to use spectral reference panels with a fixed, geo-referenced, horizontal position in the field when taking remote sensor measurements.
5. Workers should note the environmental conditions in which they intend to operate the equipment and consider their impacts on equipment, basic physics, and crop and trait of interest. If conditions are unsuitable for use of the equipment, no measurement should be taken.
6. During measurement collection, remember the role of time of day and the need to revisit reference panels installed in the field.
7. Remember that canopy reflectance can be influenced by crop canopy structure, morphology, and cover; crop water status; incident radiation geometry; shading; clouds; and nearby objects.
8. Consider the optimal number of measurements per research plot.
9. Work with a remote sensing expert, consult the literature, and/or use standard tools and techniques if there are any doubts about the ability to properly perform the work.

2.7.4 UAVs

For a variety of reasons, the use of UAVs is becoming very common in small- and large-scale, replicated, research plot-based screening trials (*see* **Note 15**). However, there is limited information on proper UAV selection, operation, and maintenance compared to other phenotyping technologies. Thus special consideration is given in this chapter to UAV-based phenotyping. Proper UAV selection, UAV program setup, and system operation and maintenance are crucial for any crop improvement program attempting to use this technology.

UAV Selection

When choosing a UAV (*see* **Note 16**), it is imperative to consider the intended use of the technology (e.g., simple observational phenotyping and plot scouting versus plot- and plant-level canopy measurements), the crop phenotypes likely to be measured, the spatial resolution required for phenotyping key traits, the probable sensor payload for most flight operations, the number of research plots or land area to be covered, the required and existing infrastructure for flight and maintenance operations, the skill level of the likely operator, and the typical environment for flight operations (e.g., weather, topography, nearby structures). Consideration of these factors largely insures that the UAV selected meets the payload, range, flight time, airspeed, agility, operational ease and space, and safety requirements of the crop improvement program.

Key UAV Principles

When constructing a UAV-based phenotyping effort, it is important to abide by the following set of key principles largely adapted from manned aviation programs (*see* **Note 17**). Adherence to these principles creates a safe, effective, and efficient working environment.

1. Terms of UAV use (e.g., allowable altitude and operating space) should be drafted and understood by all personnel involved in the program. Checklists, protocols, operator training, and regulatory compliance should be mandatory. Assessment of the safety of nearby personnel and property should be performed prior to each flight.
2. A UAV operator must be experienced and closely supervised. All personnel, and particularly the operator, should plan for failure scenarios.
3. Program changes should be incorporated in a controlled, logical manner. UAV improvements should be evolutionary, not revolutionary. A configuration control program for UAV software and hardware changes must be created and enforced.
4. Electronic systems should be proven and closely monitored. Tests for electromagnetic interference should be performed on a periodic basis. Restrictions on the use of mobile phones and other electronic devices during flight operations must be drafted and enforced.
5. Airframe attributes should be conservative, stable, and rugged.
6. Documentation of UAV activities should be performed for each flight and mishaps should be reported when they occur. Rules for communication with air traffic control authorities should be developed and enforced.

UAV Operational Guidelines

Any UAV-based crop phenotyping effort should include a set of operational principles that guide preflight, in-flight, and postflight activities. While this list should be modified for individual crop improvement programs, it should at least include the following key tenets.

1. Preflight principles:
   - (a) Go through all checklists and preflight procedures. Such checklists and procedures should be based on material and insights from manned aviation and remote sensing resources.
   - (b) Prepare the flight plan. This should be done with input from other key personnel involved in the phenotyping effort. The plan should balance data needs and UAV and sensor capabilities.
   - (c) Check weather, field, and crop conditions. Adjust the flight plan in accordance with these conditions.
   - (d) Communicate with key personnel and authorities regarding the flight plan.
   - (e) Carefully prepare, package, and transport all equipment (e.g., UAV, anemometer, laptop computer, binoculars) to the research location.
2. In-flight principles:
   - (a) Monitor the UAV. This includes continually observing the UAV, watching any computer displays, and heeding system or equipment warnings.
   - (b) Continually hold the UAV controller while the UAV operates. This should be done by the operator even when the GPS autopilot system is engaged on a UAV equipped with such a system.
   - (c) Maintain conversation discipline. This includes using standard terminology, making timely remarks, and avoiding unnecessary discussion.
   - (d) Monitor the changing environment. This involves continuously scanning the area being measured, scanning primary and secondary landing areas, and avoiding other air traffic.
   - (e) Avoid any hazards including people, structures, vehicles, and crops.
3. Postflight principles:
   - (a) Go through all checklists and postflight procedures. Such checklists and procedures should include material from manned aviation and remote sensing resources.
   - (b) Download and immediately back up sensor data.
   - (c) Log appropriate information including weather, field, and crop conditions along with any pertinent in-flight information.

(d) Carefully package all equipment (e.g., UAV, anemometer, laptop computer, binoculars) and transport it to a secure location.

(e) Examine UAV components for wear, perform equipment repairs, and implement UAV improvements.

(f) Improve checklists and procedures based on pre- and in-flight experiences.

2.8 Environmental Monitoring

As with crop and soil measurements, environmental monitoring is a key component of any crop improvement program. Such measurements are often used to plan and schedule phenotyping efforts, better understand observed crop phenotypes, assess treatment impacts and genotype differences, examine genotype by environment interactions, and guide crop management decisions. Environmental monitoring within the context of a small- or large-scale, replicated, research plot-based screening trial typically includes the automated measurement of atmospheric and, often to a lesser extent, soil phenomena (e.g., precipitation, solar radiation, wind speed and direction, relative humidity, air and soil temperature, soil moisture). This is often done with a diverse set of instrumentation. Procedures for the installation, operation, removal, and maintenance of environmental monitoring equipment are often specific to an instrument, making a thorough review of methodologies on this topic beyond the scope of this chapter. Readers should therefore consult an instrument's operator manual and the general scientific literature for methodology specifics. Nevertheless, there are some key principles that should be universally considered when monitoring environmental characteristics within a screening trial.

1. Each instrument should be calibrated, maintained, installed, and removed in accordance with its respective operator manual.
2. All instruments should be calibrated and in good working order before placing them at a field trial. Equipment should be cleaned and serviced upon removal from a field trial.
3. All personnel working with the equipment should be fully trained on and proficient with the instrumentation prior to setup. Personnel should understand how to install, operate, remove, and maintain the equipment.
4. When data logging is done automatically by the instrument, personnel should periodically insure that data are recording properly. When the instrument experiences an error, the problem should be remedied as quickly as possible.
5. Equipment should never be operated outside of its environmental specifications.
6. All personnel should be made aware of the appearance and location of all environmental monitoring instrumentation.

Flags or other visual aids should mark the location of any instrumentation.

7. Instrumentation should be placed in locations that provide measured data representative of the environmental phenomena occurring in a trial's research plots. Enough instruments should be employed to insure that representative data can be captured, interpolated, estimated, and/or modeled for all research plots.

2.9 In-Field Data Recording

The accurate recording of all measurements taken on phenotypic traits is a key but often overlooked component of a small- or large-scale, replicated, research plot-based screening trial. Erroneous data recording (e.g., duplicate, skipped, and/or inaccurate data entry) is a relatively common but easily remedied problem, particularly with the introduction of key technologies in recent years. Proper data recording requires both appropriate equipment and proper methodologies. Traditional equipment for recording data includes a pen or pencil, paper, and clipboard. Modern equipment typically involves a consumer-grade or ruggedized tablet computer or smartphone with stylet and/or mobile app capabilities and custom or off-the-shelf software. Many measurement tools (e.g., leaf chlorophyll meter) have integrated data recording capabilities. Workers tasked with taking measurements and recording data should strongly consider using tablet computers, smartphones, and measurement instruments with embedded data recording for all data logging activities. The use of software specifically designed for data recording in small- or large-scale, replicated, research plot-based screening trials is available free of charge or from commercial vendors [17]. The use of these software tools is also strongly encouraged. If traditional tools are likely to be used, personnel should develop and employ data-logging sheets that are clear and logical.

It is critical that personnel adhere to a key set of data recording guidelines when measuring crop phenotypes.

1. Data recording equipment should be well maintained and tested prior to in-field use.
2. Workers should have good familiarity with all hardware and software before recording data and should know how to carefully navigate research plots.
3. SOPs developed for phenotype measurements should be closely followed since they should contain instructions for proper data recording. All workers (i.e., observers, operators, and recorders) should be trained on these SOPs.
4. All workers should understand what the expected range of values is for a given plant measurement so that erroneous measurements can be quickly detected and issues can be

remedied. Data that fall outside the anticipated value range should not be recorded.

5. Workers should be diligent in checking that data are recorded for the proper plot.
6. Workers should not change data recording duties within a block or replication.

2.10 Data Management and Analysis

Modern field crop phenotyping programs generate considerable amounts of data. Proper management of these data is critical for the success of the overall crop improvement program. A full review of data management within the context of field crop performance evaluation is well beyond the scope of this review. Readers should therefore consult articles [17, 18] and others for helpful guidance and information on this topic including example data storage and maintenance systems. Still, a few data management guidelines are warranted in this chapter.

1. Prior to beginning a trial, personnel should develop a detailed data management plan that outlines the roles and responsibilities of all involved personnel, describes the types of data likely to be collected, and provides any legal or regulatory disclaimers about data type, storage, use, and sharing. The document should also outline all data management protocols that should be used.
2. The data management plan must be followed and enforced on a daily basis. All personnel should follow the plan without exception.
3. User access and read/write permissions should be set and maintained to prevent inadvertent deletion, alteration, or sharing of data. Only key personnel should have access to any critical data storage location(s). Sensitive data (e.g., intellectual property, trade secrets) should be carefully managed and have very restricted access.
4. Adequate storage space should be maintained. Unnecessary data should be purged by authorized personnel.
5. Data management hardware and software should be capable of handling diverse data types from a multitude of instruments and sensors.
6. All critical data should be backed up to multiple storage locations on a routine basis. The data storage locations should be spatially separated if possible to protect against physical hazards.
7. Data should be transferred from instruments, sensors, and other data recording devices to secure storage resources while phenotyping is occurring or shortly after it is completed.

8. Key data management processes (e.g., data transfer and backup) should be automated whenever possible.
9. A robust file naming system should be created and employed by all personnel. File names should be alphanumeric with only lowercase letters and no spaces. The file name could specify the data collection instrument (e.g., sensor, platform); the data collection date, time, and location; experiment information (e.g., lead researcher, experiment, sub-experiment); document version; and file extension.
10. Protocol, hardware, and software should help insure easy file sharing with strong data security.

Data analysis is key to any field crop performance evaluation trial and thus many resources are available for understanding and implementing an effective data analysis effort. A collaborative investigation of the data with a statistician is encouraged and should likely involve an examination of data quality using tools such as histograms and descriptive statistics (e.g., skewness, kurtosis, standard deviation, mean, median, range, outliers) [19]. It should also involve an exploration of treatment factor impacts and treatment-level differences using techniques such as analysis of variance (ANOVA) and means analysis [20]. Correlation matrices, principal component analysis, and cluster analysis may be used to look at relationships between phenotypes or group a trial's genotypes by phenotype [19, 21, 22], though many other applications exist for these tools. Proximal and remote sensing data can be analyzed using regression, inverse modeling, image analysis, multivariate analysis, and many other techniques to derive or calculate novel phenotypes from measured phenotypes or sensor output [23]. Given the multitude of options for analyzing data and large number of image analysis tools [18, 24], it is critical to work with trained statisticians and image analysis experts.

3 Notes

1. The number and type of regulations governing a research trial can be broad, deep, and complex. Substantial time must be spent investigating these regulations, particularly if working with regulated seed. Internal and/or external organization guidance should be sought if necessary.
2. A primary trait is often the principal trait of interest in a crop improvement program. In a majority of instances, the primary trait is grain yield. Unfortunately, it can be difficult, costly, and inefficient to measure a primary trait due to low heritability values under field conditions and poor correlations between

high-stress and low-stress field environments. Thus secondary traits are often measured to evaluate crop performance.

3. For a secondary trait to be measured in a research trial, it should (1) be genetically correlated with the primary trait, (2) be less impacted by environment than the primary trait (and thus have a higher heritability than the primary trait), (3) exhibit sufficient genetic variability in the appropriate germplasm, (4) have no negative impact on crop performance, (5) be measured more quickly and reliably than the primary trait, (6) be amenable to nondestructive measurement at the individual plant or canopy level of organization, (7) be cost effective to measure, and (8) have a set of tools available for measurement [6]. If a secondary trait does not meet these criteria, it should not be measured.
4. Reference [7] provides excellent examples of the types of tables that should be constructed when prioritizing secondary trait measurements. Specifically refer to Tables 1, 2, and 3 in reference [7] and create tables using these formats.
5. For example, a maize performance evaluation trial for drought tolerance would potentially involve multiple, intensive measurements during the flowering period including anthesis-silking interval (ASI), canopy temperature, leaf chlorophyll content, and aboveground vegetative and reproductive biomass. Measuring all of these traits and potentially others takes considerable time, labor, and financial resources. The use of a large, skilled workforce and/or high-throughput phenotyping technologies can enable the measurement of all of these traits, but such resources are often unavailable and thus prioritization is necessary.
6. In this instance, the word "environment" is used to describe the climate and soil characteristics of a farm-, multi-farm, or county-scale geographical area. A characterization of larger areas (e.g., ecoregions, multistate areas, individual states) is encouraged for multi-region TPE screening trials, but this topic lies beyond the scope of this chapter. Refer to [2, 4] for more information on environment characterization in a multi-region TPE context.
7. Data that are used to characterize an environment are likely to have varying spatial granularity. For example, soil series and property information may be available at the sub-field level while frost and freeze information may only be available at course spatial scales. It is critical to have information at the proper spatial scales in order to characterize an environment. Determine the needed spatial granularity for each data type.
8. It is prudent to choose and employ more than one field site for a field crop performance evaluation trial. This enables the study

of genotype by environment interactions and the use of backup locations in the event of research errors or destructive weather events.

9. For example, a maize performance evaluation trial for low N tolerance may require a low soil N environment. A key factor for evaluating this property would be soil nitrate content prior to planting. An unacceptable value would be associated with a very high soil nitrate situation while an ideal value would be associated with a very low soil nitrate situation. Intermediate values would be rated relative to their proximity to the unacceptable and ideal values.
10. Examination of many of these sources of information requires an understanding of geographic information system (GIS) software and hardware. Collaboration with GIS experts is recommended, though simple, preliminary analyses can be performed by novices using many software programs.
11. Replication and randomization are very effective against unwanted, naturally present field variability. Replication and blocking reduce the standard error of the estimates and parameters and enable control of the error variance. Replication also improves the significance of the results. Blocking should be used against a smooth field variability pattern. It should also be used to minimize variation among research plots of the same block and maximize variation between plots in different blocks. Randomization helps reduce the chance of systematic bias in an experimental design.
12. In this instance, the term "stage" is used in the context of a research pipeline. In a commercial crop improvement program, "early stage" and "late stage" typically refer to genotypes which are farther from and closer to launch in the marketplace, respectively.
13. The software used for this process does not need to be costly or complex. Simple, effective maps can be generated using basic spreadsheets if necessary.
14. Circadian cycle fluctuations, remote sensing principles, and environmental conditions must be considered when determining the time of day when a measurement should be taken. The ideal time varies by trait of interest and measurement tool and technique. The existing body of literature should be consulted when determining the time at which to take a measurement.
15. The ease of operation (e.g., if configured with a GPS autopilot) and maintenance of UAVs make them appealing for field phenotyping activities, as does their relatively low purchase price relative to other aerial remote sensing tools (e.g., manned aircraft). UAVs can provide detailed images of the crop when flown at low altitudes and outfitted with high-resolution

sensors. Their ability to be quickly deployed at an operator's will make them a very flexible, time-sensitive phenotyping tool.

16. Selection of a UAV typically begins with choosing between a fixed- and rotary-wing platform. After this selection has been made, a specific model is often chosen by the researcher, operator, and other involved parties.
17. These principles should be adapted for each crop improvement program and must abide by all applicable government regulations.

Acknowledgment

The information, data, or work presented herein was funded in part by the Advanced Research Projects Agency-Energy (ARPA-E), United States Department of Energy, under Award Number DE-AR0000593. The views and opinions of the authors expressed herein do not necessarily state or reflect those of the US Government or any agency thereof.

References

1. Bruce W, Edmeades G, Barker T (2002) Molecular and physiological approaches to maize improvement for drought tolerance. J Exp Bot 53:13–25
2. Chenu K, Cooper M, Hammer G et al (2011) Environment characterization as an aid to wheat improvement: interpreting genotype-environment interactions by modelling water-deficit patterns in North-Eastern Australia. J Exp Bot 62:1743–1755
3. Cooper M, Gho C, Leafgren R et al (2014) Breeding drought-tolerant maize hybrids for the US corn-belt: discovery to product. J Exp Bot 65:6191–6204
4. Löffler C, Wei J, Fast T et al (2005) Classification of maize environments using crop simulation and geographic information systems. Crop Sci 45:1708–1716
5. Fahlgren N, Gehan M, Baxter I (2015) Lights, camera, action: high-throughput phenotyping is ready for a close-up. Curr Opin Plant Biol 24:93–99
6. Araus J, Serret M, Edmeades G (2012) Phenotyping maize for adaptation to drought. Front Physiol 3:305
7. Reynolds M, Pask A, Pietragalla J (2012) Introduction. In: Pask A, Pietragalla J, Mullan D et al (eds) Physiological breeding II: a field guide to wheat phenotyping. CIMMYT, El Batán
8. Altman N, Krzywinski M (2015) Points of significance: split plot design. Nat Meth 12:165–166
9. Crossa J (2012) Field experimental designs in agriculture. In: Reynolds M, Pask A, Mullan D (eds) Physiological breeding I: interdisciplinary approaches to improve crop adaptation. CIMMYT, El Batán
10. Pask A, Pietragalla J (2012) General recommendations for good field practice. In: Pask A, Pietragalla J, Mullan D et al (eds) Physiological breeding II: a field guide to wheat phenotyping. CIMMYT, El Batán
11. Tuberosa R (2012) Phenotyping for drought tolerance of crops in the genomics era. Front Physiol 3:347
12. Pietragalla J, Pask A (2012) General recommendations for the use of instruments. In: Pask A, Pietragalla J, Mullan D et al (eds) Physiological breeding II: a field guide to wheat phenotyping. CIMMYT, El Batán
13. Da Costa V, Cothren J (2011) Drought effects on gas exchange, chlorophyll, and plant growth of 1-methylcyclopropene treated cotton. Agron J 103:1230–1241
14. Mullan D (2012) Spectral radiometry. In: Reynolds M, Pask A, Mullan D (eds) Physiological breeding I: interdisciplinary approaches to improve crop adaptation. CIMMYT, El Batán

15. Pinter P Jr, Hatfield J, Schepers J et al (2003) Remote sensing for crop management. Photogramm Eng Rem S 69:647–664
16. Lillesand T, Kiefer R, Chipman J (2014) Remote sensing and image interpretation. John Wiley & Sons, Hoboken
17. Shrestha R, Matteis L, Skofic M et al (2012) Bridging the phenotypic and genetic data useful for integrated breeding through a data annotation using the crop ontology developed by the crop communities of practice. Front Physiol 3:326
18. Cobb J, DeClerck G, Greenberg A et al (2013) Next-generation phenotyping: requirements and strategies for enhancing our understanding of genotype-phenotype relationships and its relevance to crop improvement. Theor Appl Genet 126:867–887
19. Burton A, Brown K, Lynch J (2013) Phenotypic diversity of root anatomical and architectural traits in *Zea* species. Crop Sci 53:1042–1055
20. Boomsma C, Santini J, Tollenaar M et al (2009) Maize morphophysiological responses to intense crowding and low nitrogen availability: an analysis and review. Agron J 101:1426–1452
21. Trachsel S, Kaeppler S, Brown K et al (2010) Shovelomics: high throughput phenotyping of maize (Zea mays L.) root architecture in the field. Plant Soil 341:75–87
22. Ciampitti I, Vyn T (2012) Grain nitrogen source changes over time in maize: a review. Crop Sci 53:366–377
23. White J, Andrade-Sanchez P, Gore M et al (2012) Field-based phenomics for plant genetics research. Field Crop Res 133:101–112
24. Rahaman M, Chen D, Gillani Z et al (2015) Advanced phenotyping and phenotype data analysis for the study of plant growth and development. Front Plant Sci 6:619

Index

Sandeep Kumar et al. (eds.), *Transgenic Plants: Methods and Protocols*, Methods in Molecular Biology, vol. 1864, https://doi.org/10.1007/978-1-4939-8778-8, © Springer Science+Business Media, LLC, part of Springer Nature 2019

D

E

F

G

H

I

K

L

M

N

O

P

Q

R

S

Printed by Printforce, the Netherlands